BEACHES OF THE QUEENSLAND COAST:
COOKTOWN TO COOLANGATTA

A guide to their nature, characteristics, surf and safety

ANDREW D SHORT

Coastal Studies Unit
School of Geosciences F09
University of Sydney
Sydney NSW 206

Research Officer and cartographer:

Katherine McLeod
Surf Life Saving Australia

SYDNEY UNIVERSITY PRESS

Coastal Studies Unit   and   **Surf Life Saving Australia Ltd**
School of Geosciences F09   1 Notts Ave
University of Sydney   Locked bay 2
Sydney NSW 2006   Bondi Beach NSW 2026

Short, Andrew D
    Beaches of the Queensland Coast: Cooktown to Coolangatta   0-9586504-1-1
    A guide to their nature, characteristics, surf and safety   Published 2000
                                                                            Reprinted 2005

Other books in this series by A D Short:
- *Beaches of the New South Wales Coast, 1993*   0-646-15055-3
- *Beaches of the Victorian Coast and Port Phillip Bay, 1996*   0-9586504-0-3
- *Beaches of the South Australian Coast and Kangaroo Island, 2001*   0-9586504-2-X
- *Beaches of the Western Australian Coast: Eucla to Roebuck Bay, 2005*   0-9586504-3-8

Forthcoming books:
    Beaches of the Tasmania Coast and Islands (publication. 2006)   1-920898-12-3
    Beaches of the Northern Australian Coast: The Kimberley, Northern Territory and Cape York
                                                                            1-920898-16-6
    Beaches of the New South Wales Coast (2nd edition)   1-920898-15-8

Published by:
Sydney University Press
University of Sydney
www.sydney.edu.au/sup

Copies of all books in this series may be purchased online from Sydney University Press at:

http://www.purl.library.usyd.edu.au/sup/marine

Queensland beach database:
Inquiries about the Queensland beach database should be directed to Surf Life Saving Australia
at sls.com.au

Cover photograph   Gold Coast (A D Short)
Book Layout   Werner Hennecke

Table of Contents

To Brett Williamson

CEO Surf Life Saving Queensland, for all his help and support in making this book a reality.

Foreword

"Queensland Beaches"

Queensland is blessed with some of the most beautiful beaches in the world. The white sandy beaches, crystal clear waters and great surf, have made going to the beach a national pastime and a major attraction for all Australians and increasing numbers of overseas visitors. With more than 18 million visitations to Queensland beaches every year, our beaches are a significant recreational and commercial entity for the state.

Volunteer surf lifesavers make a significant contribution to providing a safe environment on the beach. The Queensland Government is proud to be a strong supporter of this voluntary movement, and help lifesavers keep the beaches safer each year.

The Australian Beach Safety and Management Program is the first of its kind in the world. This program is a comprehensive guide to beach conditions across Queensland covering geology, coastal evolution, coastal processes, climate and atmosphere, and beach, wave and surf conditions.

The Australian Beach Safety and Management Program identifies all popular beaches from Cooktown in the north to Rainbow Bay in the south including the hazard and safety concerns of different areas.

The Queensland Government is indebted to the work of all contributors to this book. It is the culmination of many years of research and analysis which will enable surf lifesavers, local government, coastal planners, developers, in fact anyone working or enjoying recreational activities along the coast, to use our beaches more safely.

On behalf of the Government and the people of Queensland, I extend congratulations to Surf Life Saving Australia, Surf Life Saving Queensland, and the University of Sydney for this commendable publication on beach safety.

The Hon. Merri Rose, MLA
Minister for Emergency Services
Queensland Government

Preface

This book is the third in a series of book produced by the Australian Beach Safety and Management Program (ABSMP). One of the aims of the program is to develop a better understanding of the location, type, characteristics, nature, hazards and public risk along all of Australia's 10 685 mainland beaches. To this end the program is developing a database on the beaches of each state and the Northern Territory, with the databases for New South Wales, Victorian, Queensland and South Australia now complete. It is also publishing books summarising the main characteristics of each beach on a state by state basis. This book is the third in the series with *Beaches of the New South Wales Coast* published in 1993, and *Beaches of the Victorian Coast and Port Phillip Bay* in 1996, while the *Beaches of the South Australia Coast and Kangaroo Island*, the fourth in the series is well underway.

The ABSMP was initiated in New South Wales in 1986 as the New South Wales Beach Safety Program in collaboration with Andrew May of Surf Life Saving New South Wales (SLSNSW), with Chris Hogan as Research Officer. In 1990 with support from the Australian Research Council and in cooperation with Surf Life Saving Australia Ltd, the project was expanded to encompass all states and the Northern Territory.

The Coastal Studies Unit commenced scientific investigations on the Queensland coast in 1986, followed by fieldwork, aimed at understanding Queensland beach systems, in the Mackay and Capricorn coast regions in 1987, and subsequently more intensive investigations in 1992 and 1993. Field investigations specifically for this project commenced in June 1990, with a land and boat based field trip from Burnett Heads to Cooktown, followed by a boat based second trip from Agnes Waters to Cooktown in July 1994. The southeast Queensland beaches were investigated in 1990, 1992, 1994 and 1999, and the beaches of the south east sand islands were visited by helicopter in 1994. Finally the entire coast was photographed from a small plane in 1994.

In compiling a book of this magnitude there will be errors and omissions, particularly with regard to the names of beaches, many of which have no official name, and many local factors. If you notice any errors or wish to comments on any aspects of the book please communicate them to the author at the Coastal Studies Unit F09, University of Sydney, Sydney NSW 2006, phone (02) 9351 3625, fax (02) 9351 3644, email: A.Short@geosci.usyd.edu.au or via Surf Life Saving Queensland (07) 3846 8000 or Surf Life Saving Australia (02) 9130 7370. In this way we an update the beach database and ensure that future editions are more up to date and correct.

Andrew D Short
Narrabeen Beach, December 1999

Acknowledgements

This book while researched in Queensland was written in Sydney and many people from both states have contributed to the final publication.

The three major field trips to visit and study most of the beaches between Cooktown and Coolangatta were all conducted with tremendous support from Surf Life Saving Queensland (SLSQ), and particularly the CEO Brett Williamson. Brett has maintained a strong, and at times patient interest in the project since its inception in 1990.He always ensured we had the full support of both the staff at SLSQ and the many surf clubs along the coast, and it is to Brett that this book is dedicated. Other personnel at SLSQ who rendered valuable assistance were Peter Wadcoat who steered the open boat along much of the coast from Burnett Heads to Cardwell during the 1990 expedition, accompanying it all the way to Cooktown. Gaps left after this trip were filled during a second boat trip, this time the boat run by John Wadley, a lifeguard from Sunshine Beach who managed to keep the boat running and afloat all the way from Agnes Waters to Cooktown. On both these trips Chris Hogan (SLSNSW) towed the boat trailer and usually managed to find us at the end of each day's leg, as well as looking after all the shore-based logistics. During our beach investigation along the Gold and Sunshine Coasts in 1990 and 1992 and on visits to the Beach Protection Board we were also able to use the SLSQ headquarters as a base. In 1999 Bob Starkey showed me round the Moreton Bay beaches and finally George Cook proofread the entire text. Many thanks to all.

The Beach Protection Board and Coastal Management Branch, permitted access to their colour aerial photograph collection of the Queensland coast, which provided an invaluable photographic record of each conditions along the entire coast. I particularly acknowledge the assistance of David Robinson. I am also indebted to my former colleague, Professor J L Davies, who generously provided a complete set of black and white aerial photographs of the entire coast, as he did for the two previous books.

In November 1994 the Australian Aerial Patrol Cessna piloted by Harry Mitchell and Steve Conock, flew the author along the entire Queensland coast while he photographed every beach. This was a memorable flight, including a most memorable stop over on Thursday Island, many thanks Harry. A helicopter flight on the SLSQ Rescue Helicopter of the Sunshine and Gold Coasts was undertaken in June 1990, and a flight to each of the south-east sand islands (Fraser to South Stradbroke) in December 1994. This flight was piloted by Royce Lanham and funded by Kevin Weldon, at the time President of International Life Saving.

At Surf Life Saving Australia (SLSA) where the project is based it received full support from the Scott Derwin (CEO of SLSA, 1990-1995), during which time the project was supervised by Darren Peters and Stephen Leahy. The project also received an Australian Research Grant from 1990-1992, which enabled it to expand nationally. Since 1996 CEO Greg Nance has supervised the project at SLSA and has provided tremendous support leading to an expansion of the SLSA-University of Sydney collaboration with additional project funding from a joint Australian Research Council Collaborative Research Grant (1996-1998) and a Strategic Partnership with Industry- Research and Training (SPIRT) Grant (1999-2002).

Also at SLSA the Project Research Officer Katherine McLeod has been an essential component of ABSMP since 1996. She has worked full-time on the project ensuring all the data collected is recorded and managed in the ABSMP database. Katherine also drafted most of the figures in the book and edited the entire text. Werner Hennecke (University of Sydney) did an excellent and speedy job at laying out the entire text and 174 figures to produce the following book.

Finally, as the entire beach database was complied and the book was written at my home office, I thank my wife and children for putting up with its intrusion into our home life.

Abstract

This book is about all the Queensland beaches south of Cooktown. It begins with three chapters that provide a background to the physical nature and evolution of the entire Queensland coast and its 1600 mainland beach systems. Chapter 1 covers the geological evolution of the coast and the role climate, wave, tides and wind in shaping the present coast and beaches. Chapter 2 presents in more detail the twelve types of beach systems that occur along the Queensland coast, while chapter 3 discusses they types of beach hazards along the coast and the role of Surf Lifesaving Queensland in mitigating these hazards. Finally the long chapter 4 presents a description of every one of the 604 mainland beaches between Cooktown and Coolangatta, as well as 110 beaches on eighteen major islands, in all 714 beaches. The description of each beach covers its name, location, physical characteristics, access and facilities, with specific comments on its surf zone character and physical hazards, as well as its suitability for swimming, surfing and fishing. Based on the physical characteristics each beach is rated in terms of the level of beach hazards from the least hazardous rated 1 (safest) to the most hazardous 10 (least safe). The book contains 174 figures which include 137 photographs, which illustrate all beach types ,as well as beach maps and photographs of all beaches patrolled by surf lifesavers and many other popular beaches.

Keywords: beaches, surf zone, rip currents, beach hazards, beach safety, Queensland

---

Australian Beach Safety and Management Program (ABSMP)

Awards

NSW Department of Sport, Recreation and Racing
Water Safety Award – Research 1989
Water Safety Award – Research 1991

Surf Life Saving Australia
Innovation Award 1993

International Life Saving
Commemorative Medal 1994

New Zealand Coastal Survey
In 1997 Surf Life Saving New Zealand adopted and modified the ABSMP in order to compile a similar database on New Zealand beaches.

Great Britain Beach Hazard Assessment
In 2002 the Royal National Lifeboat Institute adopted and modified the ABSMP in order to compile a similar database on the beaches of Great Britain.

Hawaiian Ocean Safety
In 2003 the Hawaiian Lifeguard Association adopted ABSMP as the basis for their Ocean Safety survey and hazard assessment of all Hawaiian beaches.

Handbook on Drowning 2005
This handbook was product of the World Congress on Drowning held in Amsterdam in 2002. The handbook endorses the ABSMP approach to assessing beach hazards as the international standard.

AUSTRALIAN BEACH BOOKS

Published by the Sydney University Press for the
Australian Beach Safety and Management Program
a joint project of
Coastal Studies Unit, University of Sydney and Surf Life Saving Australia

by

Andrew D Short
Coastal Studies Unit, University of Sydney

BEACHES OF THE NEW SOUTH WALES COAST
Publication: 1993   **ISBN:** 0 646 15055 3
358 pages, 167 original figures, including 18 photographs; glossary, general index, beach index, surf index.

BEACHES OF THE VICTORIAN COAST & PORT PHILLIP BAY
Publication: 1996   **ISBN:** 0 9586504 0 3
298 pages, 132 original figures, including 41 photographs; glossary, general index, beach index, surf index.

BEACHES OF THE QUEENSLAND COAST: COOKTOWN TO COOLANGATTA
Publication: 2000   **ISBN**   0 9586504 1 1
369 pages, 174 original figures, including 137 photographs, glossary, general index, beach index, surf index.

BEACHES OF THE SOUTH AUSTRALIAN COAST & KANGAROO ISLAND
Publication: 2001   **ISBN**   0 9586504 2-X
346 pages, 286 original figures, including 238 photographs, glossary, general index, beach index, surf index.

BEACHES OF THE WESTERN AUSTRALIAN COAST : EUCLA TO ROEBUCK BAY
Publication: 2005   **ISBN**   0-9586504-3-8
433 pages, 517 original figures, including 409 photographs, glossary, general index, beach index, surf index.

Order online from **Sydney University Press** at

http://www.sup.usyd.edu.au/marine

forthcoming titles:

BEACHES OF THE TASMANIAN COAST AND ISLANDS (publication 2006)

BEACHES OF THE NORTHERN AUSTRALIA: THE KIMBERLEY, NORTHERN TERRITORY & CAPE YORK (publication 2007)

BEACHES OF THE NEW SOUTH WALES COAST (revised, expanded and updated, publication 2007)

Beaches of the Queensland Coast

from
Cooktown

to
Coolangatta

1.   THE QUEENSLAND COAST

1.1   Introduction

Queensland's mainland coast is 6089 km in length, extending from the Northern Territory border through 1479 km of Gulf Coast to Cape York, then for 4610 km of shoreline south to the New South Wales border at Point Danger. Sixty percent (3200 km) of the coast consists of sandy beaches. The beaches range in size from a few tens of metres to 90 km long on Fraser Island, with an average length of 2 km. They are usually bordered by either rocky headlands and/or tidal inlets and in the northeast may be fringed by coral reefs.

This book is concerned with the more developed and accessible 3589 km section of the coast between Cooktown and Coolangatta. This section contains 59% of the coast and 961 beaches (60% of all beaches). In addition there are more than 350 high islands off this coast, that also contain hundreds of beaches. This book is about the mainland beaches between Cooktown and Coolangatta, together with descriptions of beaches on 18 of the high islands along this section of coast; in all 961 mainland and 110 island beaches.

This book is a product of the Australian Beach Safety and Management Program, a cooperative project of the University of Sydney's Coastal Studies Unit, Surf Life Saving Australia and Surf Life Saving Queensland. It is part of the most comprehensive study ever undertaken of beaches on any part of the world's coast. In Queensland it has investigated every beach on the mainland coast (Cooktown to Coolangatta), including the large bays such as Broad Sound, Shoalwater Bay and Moreton Bay, and beaches on 18 islands.

This book begins by examining the nature of the Queensland coast, including its geological evolution, its climate and ocean processes (Chapter 1). Chapter 2 details the types of beaches that occur around the coast, while Chapter 3 looks at beach usage and hazards, both physical and biological. The bulk of the book (Chapter 4) is devoted to a description of every beach located on the mainland coast between Cooktown and Coolangatta, together with selected island beaches (Table 1.1). Information is provided on each beach's name, location, access, facilities, physical characteristics and surf conditions. Specific comments are made regarding each beach's suitability for swimming, surfing and fishing, together with a beach hazard rating from 1 (least hazardous) to 10 (most hazardous).

1.1.1   What is a beach?

A *beach* is a wave deposited accumulation of sediment, usually sand, but possibly cobbles and boulders. They extend from the upper limit of wave swash, approximately 3 m above sea level, out across the surf zone and seaward to the depth to which average waves can move sediment shoreward. On the higher wave energy, southeast Queensland coast, this means they usually extend seaward to depths between 15 and 25 m and as much as 1 to 3 km offshore (Fig. 1.1a). However, in lee of Fraser Island and inside the Great Barrier Reef, wave height and energy decrease substantially and the beaches range from the more exposed, with waves averaging less than 1 m, to essentially waveless, sandy tidal flats. The higher energy protected beach systems may extend at most a few hundred metres offshore, while the low energy beaches and sand flats usually terminate at low tide (Fig. 1.1b).

To most of us however, the beach is that part of the dry sand we sit on, or cross to reach the shoreline and the adjacent surf zone. It is an area that has a wide variety of uses and users (Table 1.2). This book will focus on the dry or subaerial beach plus the surf zone or area of wave breaking, typical of the beaches illustrated in Figure 1.1.

Most Queenslanders live within an hour of a beach and even those who live inland often travel to the coast for their holidays. Many are frequent beachgoers and have their favourite beach. The beaches most of us have been to, or go to, are close by our home or holiday area. They are usually at the end of a sealed road, with a car park and other facilities. Often they are patrolled by lifesavers or professional lifeguards. These popular, developed beaches, however, represent only a minority of the state's beaches. In Queensland, 57 surf lifesaving clubs are located on 53 mainland and four island beaches. In addition, 70 beaches are also patrolled by professional lifeguards (Table 1.3). In total however, only 75 beaches (4%) are patrolled. Furthermore while 253 (26%) of the beaches have sealed road access and 75 lie at the end of a gravel road, 236 (25%) can only be reached by offroad vehicle and 335 (34%) have no vehicle access and are only accessible on foot or by boat. What this means is that there are still many beaches in a totally natural state.

Table 1.1 Queensland coastal characteristics

	Mainland Coast	NT border to Cooktown	Cooktown to Coolangatta	18 selected Islands	Total
Total length (km)	6089	2599	3589		6089
Sandy (beach) coast (km)	3199	1523	1676	357	3556
Other (rocky or tidal flat) coast (km)	2890	977	1913		2890
Number of beaches	1601	640	961	110	1711

Australian Beach Safety Management Program

Figure 1.1   *(a) Braydon Beach on Moreton Island is one of Queensland's more exposed and high energy beaches with waves averaging 1.4 m, while (b) Airlie Beach in the Whitsundays, shown here at low tide, is one of the lowest energy popular beaches, with an average wave height of less than 0.1 m.*

Table 1.2   Types of Queensland beach users and their activities

Type	User	Location
Passive	sightseer, tourist	road, car park, lookout
Passive-active	sunbakers, picnickers beach sports	dry beach
Active	beachcombers, joggers	swash zone
Active	fishers, bathers	swash, inner surf zone
Active	surfers, water sports	breakers & surf zone
Active	skis, kayaks, windsurfers	breakers & beyond
Active	IRB, boats	beyond breakers

Table 1.3 Queensland beaches: Cooktown to Coolangatta - lifesaving facilities and access

	Mainland coast	Islands	Total
Surf Lifesaving Clubs	53	4	57
Lifeguards (SLSQ)	67	3	70
Sealed road access	253	9	262
Gravel road access	75	4	79
Dirt road access (4WD)	236	49	285
Foot access only	196	18	214
No vehicle access	335	26	361
	1095	106	1201
Total beaches	961	110	1071

1.2   Evolution and Geology of the Coast

Beaches are a part of the coastal environment. They always occur at the shoreline, with part extending landward as the dry beach and part extending seaward as the surf and nearshore zone. For a beach to form however, a number of parameters and processes are required. These include contributions from all the world's major spheres; the lithosphere or geology, which supplies sediment and boundaries; the atmosphere, which contributes the climate; the hydrosphere or ocean, source of waves and tides; and the biosphere, the source of the marine and dune biota (Fig. 1.2). The remainder of this chapter outlines the nature of the Queensland coastal environment and the contribution of each of these spheres to the evolution and nature of Queensland's beaches.

The Beach Environment, Boundaries and Processes

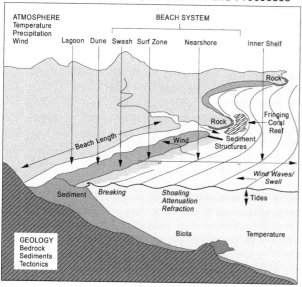

Figure 1.2   *The coastal environment is the most dynamic part of the earth's surface. It contains elements of all four spheres that make up the earth, namely the atmosphere, hydrosphere or ocean, the lithosphere or geology and the biosphere. As the four spheres interact at the coast, they produce a wide spectrum of coastal environments, ranging from rocky coast to muddy tidal flats to coral reefs to sandy beaches.*

Figure 1.3   *Part of the Don River delta at Bowen, showing multiple recurved spits migrating north of the delta and actively supplying river sand to the downdrift beaches.*

1.2.1   Queensland coastal geology

The beaches and geology of a region are inseparable, as without an underlying geological environment, there would be no beaches. The geology of a coast, and often its hinterland, is essential for two reasons. Firstly, beaches are of a finite size; they have a certain width, height and depth below. This means they must rest on some other surface, and this surface is usually the bedrock geology. On the Queensland coast, where beaches average about 2 km in length, many are bounded by prominent rock headlands, while others occupy drowned coastal river valleys. The bedrock or geology therefore provides what is called the coastal boundary within which, or on which, the beach rests. In addition, approximately 2890 km of the coast is rock, much of which is exposed as headlands, bluffs and cliffs, while 350 'high' rocky islands lie off the coast.

Secondly, most beaches on the mainland coast are composed predominantly of quartz (silica) sand grains, together with other minerals and rock fragments. These sediments originated in the coastal hinterland and have been delivered to the east coast by 32 rivers and over 200 smaller creeks and streams, in addition to minor local sources from eroding headlands. The river sediments have two routes to the modern shoreline. Some have been delivered to the coast and continental shelf during periods of lower sea level, and subsequently moved by waves on or along-shore to be deposited as beaches and dunes during the periods of rising sea level. This is particularly the case in southeast Queensland, where the great sand islands have been built up over the past two million years by successive layers of sand added during each rise in sea level. In addition along most of the coast, many rivers and streams are also supplying sediment directly to the coast, particularly during floods, continuing to build out the numerous river and stream mouths and supply the adjoining beaches with sand (Fig. 1.3). The regional and hinterland geology is therefore the major source of sand for all Queensland's mainland and most high island beaches, while the inner continental shelf is also dominated by river-derived sediments.

The outer continental shelf and the low, coral-algal islands however, have a very different source of sediments. The low islands are a product of coral and algal growth over the past 6000 years, and these organisms not only build up the islands, but also supply carbonate sediment to the island beaches and adjacent continental shelf. Therefore, unlike the predominately quartz-rich mainland beaches, the low islands are dominated by locally produced coral, algal and shell detritus and often have sands that are 100% carbonate. Finally, wherever fringing reefs occur along the coast (Fig. 1.4), they also supply carbonate material to the adjoining mainland beaches leading to a mixture of quartz and carbonate sands.

Figure 1.4   *Fringing reefs at Donovan Point lower wave height at the backing beaches, resulting in ridged sand flats with scattered mangroves, as well as supplying carbonate sediment to the beaches.*

Queensland's beaches are therefore closely intertwined with the longer term geological history and setting of the coast and continental shelf and, in the case of the Great Barrier Reef, with the biological production on the shelf. On the mainland and high islands, the geology provides the basic boundaries, shape and often the bulk of the beach sand, while on the low islands, the geology provided the shelf foundations for the reefs, while the reefs themselves provide the structure and sediments that make up the massive reef system.

1.2.2   Geological evolution

The Queensland coast has a long and interesting geological history. The bedrock that forms the headlands and cliffs ranges from 1500 million to 100 million years in age, while the coastline began taking on its present shape between 75 million and 50 million years ago. The evolution of the coast can be divided into three broad geological periods, which are discussed in section 1.2.3.

Four hundred million years ago, the Queensland region consisted of ancient low-lying shield rocks in the western cape and gulf regions, while the entire eastern third of Australia from the cape through to Tasmania was deep ocean. Sedimentation was active in this region until about 200 million years ago.

During and following this marine phase were two periods of tectonic activity. The first, between 300 and 200 million years ago, consists of a volcanic chain from Bowen south to Newcastle, with marine sedimentary basins to either side. This was followed by a period of continental stability, until 100 million years ago. During this period the massive Gondwanaland complex, centred on Antarctica, of which Australia was part, began to break up, commencing about 150 million years ago. Australia separated from Antarctica about 120 million years ago and began moving northward at a rate of 5-6 cm/yr (Fig. 1.5). This initiated the opening of the Southern Ocean, followed by rifting that commenced in the Bass Strait region about 75 million years ago and spread north, opening up the Tasman and southern Coral Seas by 55 million years ago and forming the eastern seaboard of southeast Queensland and NSW (Fig. 1.6). Finally, about 50 million years ago, rifting in the northeast region opened up the remainder of the Coral Sea (Fig. 1.7).

The two periods of rifting also resulted in uplift of the Eastern Highlands and associated volcanic activity along the Great Dividing Range. The present geology east of the range, including the coast and continental shelf, consists of a series of uplifted basins, separated by higher block and arch regions. The major geological elements, their age, location and rock types are listed in Table 1.4.

1.2.3   Coastal geology

1.2.3.1   Precambrian shield

The oldest rocks in Queensland are generally not exposed at the coast. They lie in three areas of ancient Precambrian shield occurring in the northern part of the state, namely the Mt Isa-Cloncurry area (1700 million years old), the Georgetown region and the Coen Inlier, a narrow part of Cape York Peninsula (1500 million years old). The Coen Inlier is only exposed at the coast between Temple Bay and Cape Sidmouth (Fig. 1.8) where it is composed of clastic and chemical sediments and volcanic rocks.

1.2.3.2   Great Artesian Basin

The ancient shield rocks are bordered by younger sediments (180-100 million years old) of the Great Artesian Basin, which dominate the entire gulf and western and northern cape region. The basin sediments are both continental and marine sediments, together with some Tertiary and Pliocene laterites, most notably at Weipa in the east and Groote Eylandt and the Gove Peninsula in the Northern Territory.

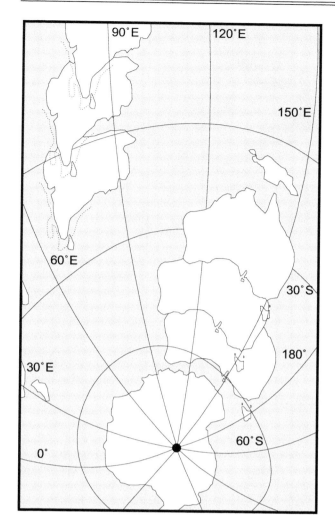

Figure 1.5   *During the past 120 million years the Australian plate, containing Australia, New Guinea and half of New Zealand, has been moving northward at a rate of a few centimetres per year. This figure shows the movement of Australia and India as they have both moved northward away from Antarctica, the core of the once supercontinent - Gondwanaland.*

Figure 1.6   *The coast of southeast Australia began forming when seafloor spreading occurred in what is now the southern Coral and Tasman Seas. The spreading centre shown in the figure was active for 20 million years and caused the Lord Howe Rise, including New Zealand, to move up to 1500 km east of the present eastern Queensland coast. At the same time, the movement of the oceanic plate under the east coast resulted in the uplift of Australia's eastern highlands, forming the Great Dividing Range.*

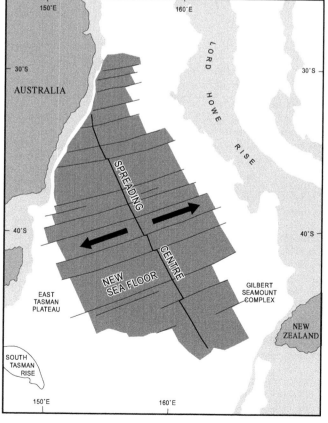

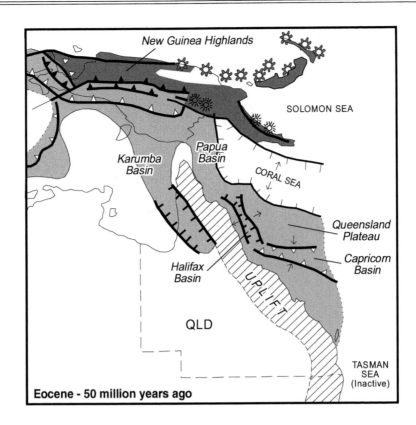

Figure 1.7   *Seafloor rifting during the Eocene led to the opening of the Coral Sea, together with deepening of the Halifax, Capricorn and Karumba basins, and uplift of the Eastern Highlands. The general shape of the Queensland coast came into being about this time.*

Table 1.4 Queensland coastal geology (major source: Day et al., 1983)

Coastal geological element (west to east)	Age (M yr = million years)	Coastal Location	Rock types
Great Artesian Basin			
Carpentaria Sub-Basin	180-100 M yr	NT border - Gulf	continental & marine sediments
Precambrian Shield			
Cape York Inlier	300 M yr	Torres Strait (high islands)	granite & metamorphic
Coen Inlier	1500 M yr	Temple Bay - Cape Sidmouth	clastic & chemical sediments, volcanics
Laura Basin	180-100 M yr	Princess Charlotte Bay	continental & marine sediments
TASMAN FOLD BELT			
North Coast Structural High	410-350 M yr	Cape Melville-Hinchinbrook Is	volcaniclastic & carbonate sediments, volcanics
Lolworth-Ravenswood Block	500 M yr	north Halifax Bay	volcanics & clastic sediments
Burdekin Basin	370-320 M yr	mid Halifax Bay	clastic & carbonate sediments
Bowen Basin	270-200 M yr	Townsville-Upstart Bay	clastic sediments, limestone
Connor Arch	350 M yr	Cape Upstart to Bowen	volcanics
Strathmuir Syncline	270 M yr	Edgcumbe Bay	volcaniclastic
Campwyn Block	360-250 M yr	Gloucester Is -Broad Sound	volcaniclastic sediments, limestone
Whitsunday Block	130 M yr	Whitsunday Islands & coast	volcanics
Styx Basin	130 M yr	St Lawrence (Broad Sound)	clastic sediments
Stanage Block	410-370 M yr	Long Island (Broad Sound)	volcanics, limestone
Coastal Block	400-300 M yr	Stanage Point-Rodds Bay	arenite, chert, basalt, conglomerate, limestone
Gympie Block	270-200 M yr	Rodds Bay-Wreck Pt & Caloundra	clastic sediments, volcanics, limestone
Maryborough Basin	200-100 M yr	Wreck Pt-Coolum	clastic sediments & volcanics
Nambour Basin	200-160 M yr	Moreton Bay	clastic sediments
Beenleigh Block	400-300 M yr	Gold Coast	arenite, chert, basalt, conglomerate, limestone

1.2.3.3 Tasman Fold Belt

The east coast south of the Coen Inlier initially consists of a sequence of uplifted basins, the Laura, Hodgkinson, Burdekin, and Bowen Basins, then a long section of folded blocks and arches that dominate much of the coast between Cape Upstart and the NSW border. In amongst the blocks are the small Styx Basin, around St Lawrence, and the Marlborough and Nambour Basins in the southeast.

The structural trend of the rifting and the rocks of the Tasman Fold Belt is NNW to SSE. This trend is evident in the orientation of the entire east coast, as well as the trend of many structural and drainage systems along the coast. The structural trends extend offshore, such that the geology of the shelf is also related to the onshore geology.

1.2.3.4 The Gulf of Carpentaria

Fifteen hundred kilometres of the Queensland coast borders the Gulf of Carpentaria. The gulf is an epicontinental sea approximately 300 000 km² in area, with a 500 km wide mouth in the north between eastern Arnhem Land and the cape, where it borders the Arafura Sea (Fig. 1.9). Its maximum north-south reach is 750 km between Cape York and Karumba. As part of the Australian continent it is however, relatively shallow, reaching a maximum depth of 60 m in the eastern-central gulf, while much is less than 40 m in depth. In addition, there is a 50 m high ridge

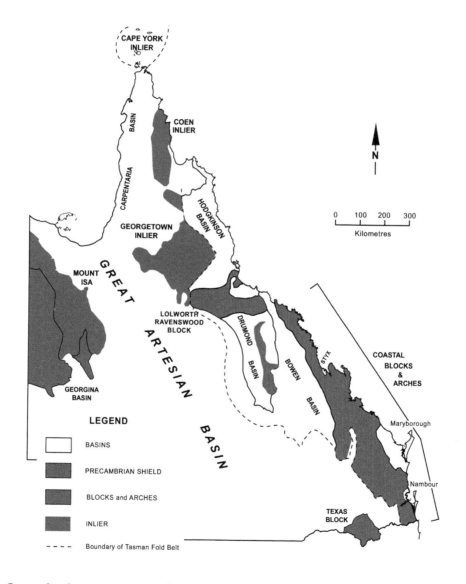

Figure 1.8   *The Queensland coast consists of eight geological provinces ranging in age from 550 million to two million years. The nature of the rocks and strata in each province, in association with the climate and erosion processes over millions of years, has largely shaped the bedrock relief of the coast. The map indicates the location and extent of each province, while Table 1.4 indicates the name, age and geology of each province.*

across the north, which at low sea levels dams a 125 000 km^2 lake, up to 15 m deep, known as Lake Carpentaria, while the rest of the gulf and continental shelf is exposed.

The entire Queensland gulf shore and hinterland consists of low gradient Pliocene through to Quaternary fluvial sediments, together with some Tertiary laterites in the northeast, while the entire coast is fringed by Quaternary marine deposits (<2 million years old) that increase in width in the south.

Because of the landlocked nature of much of the gulf, it receives no deep ocean swell and all coastal processes are related to the generally low wind waves generated within the gulf by the southeast Trades and summer northwest winds. The Trades blow offshore along the Queensland gulf coast, while the summer northwest monsoonal winds are low in velocity resulting in low short waves. Along parts of the gulf coast north and south of Weipa refracted waves generated in the lower gulf by strong trade winds do reach the coast as low swell. This swell, while a short 4-5 s, periodically reaches 1 to 2 m in height, and has considerable impact on the more exposed beaches. At the more protected Weipa waves average only 0.26 m, with a short two second period, indicating low wind chop. Only occasional tropical cyclones produce waves of a significant height along the eastern gulf shore.

1.2.3.5   The Great Barrier Reef

Corals began growing on plateaus in the Coral Sea as it formed during the periods of rifting that opened up the sea (Fig. 1.7), with the earliest reef deposits dating back 60 million years. Extensive coral growth began in the sea about 30 million years ago and by 18 million years ago were well established along the subsiding Queensland continental shelf. Shelf subsidence continued until about one million years ago, by which time coral reefs occupied essentially the entire Great Barrier Reef area.

Today the Great Barrier Reef extends from Torres Strait at 9°15'S to Lady Elliot Island at 24°07'S, over a distance of about 2300 km and cover an area of 230 000 km^2. The outer reef ranges between 23 km to 260 m from the coast (Fig. 1.10), while the reef consists of over 2500 individual reefs ranging in size from small reefs to larger reefs up to 25 km long and 125 km^2 in area.

While this book is not directly concerned with the reef, the reef does have a tremendous impact on the coast and beaches of Queensland between Torres Strait and Bundaberg. The reef, combined with Fraser Island, essentially prevents most to all ocean swell from reaching the mainland coast, resulting in a far lower energy coast and beaches than if the reef was not present.

1.2.3.6   The southeast sand islands and lowlands

The southeast coast of Queensland between Baffle Creek and Southport consists of an extensive accumulation of both river and marine sediments, the latter also composing the great Fraser, Cooloola, Bribie, Moreton and Stradbroke sand islands (Fig. 1.11). The lowlands extend along 280 km of coastline, while the islands paralleling the coast occupy 322 500 ha and reach a maximum height of 280 m (Table 1.6). These deposits have been deposited by the Kolan, Burnett, Burrum, Mary and Brisbane Rivers, which drain into the Nambour and Beenleigh Basins. Here they form fluvial, lacustrine and estuarine deposits, which at the coast have been reworked since the late Pliocene into the massive marine deposits contained in the five sand islands, and adjacent Moreton and Harvey Bay's tidal deposits. These deposits have largely accumulated during periods of rising sea level that have triggered wave mobilisation of shelf and coastal sands. The waves deposited the sands as island beach

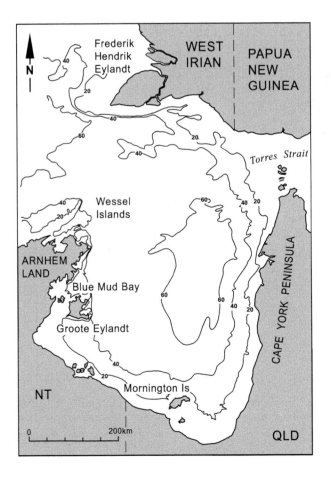

Figure 1.9   *The Gulf of Carpentaria reaches a maximum depth of 70 m, 100 km off Weipa. In the north a 40-50 m deep sill encloses the deeper gulf floor, the deeper parts of which form a freshwater lake during periods of low sea level.*

systems, with the southeast winds then moving it onshore as massive transgressive dunes on the higher islands. In more protected areas such as Bribie and southern Fraser Island the sand accumulated as areas of foredune ridges as the shoreline built seaward. Between the islands the waves and tides also produced the extensive subtidal deposits in Breaksea Spit, and Hervey and Moreton Bays.

1.2.3.7   Quaternary (past two million years)

Sea level transgressions and deposits

While ancient rock and more recent lavas form the hard core of the Queensland coast including all the rocky headlands and high islands, the coastline itself is dominated by Quaternary terrestrial and marine deposits. These occur in the form of extensive estuarine and deltaic deposits, with the shoreline dominated by tidal flats, beaches and backing coastal dunes, together with areas of fringing reefs and the massive Great Barrier Reef system offshore.

The Quaternary deposits have been deposited during the past two million years, particularly during periods of high sea level, as at the present time. Figure 1.12 illustrates the major oscillations in sea level that have occurred over the past 250 000 years, together with the rapid rise in sea level to near its present position about 6500 years ago. During the periods of rising sea level waves and tidal currents moved sediment landward, building beaches and, in places, dune systems. Once sea level stabilised, the waves continued to move more marine sand on and along-shore to build out the coast, while the numerous streams and major rivers deposited terrestrial sediments at the present river mouths, some of which are reworked alongshore by waves to also build out the coast. At the same time, the predominantly southeasterly winds have continued to build and modify the coastal dunes along the coast.

In all, over 50% of the coast is of Quaternary origin (less than two million years old). Most of this comprises Holocene marine deposits (tidal flats, beaches and dunes), associated with the most recent sea level rise, and is less than 7000 years old. However the largest single deposits, the southeast sand islands, have predominantly older sand cores, dating back two million years, and represent the accumulation of layers of windblown sand over many episodes of rising and high sea level.

1.2.4   Coastal processes

Coastal processes are the marine, atmospheric and biological activities that interact over time with the geology and sediments to produce a particular coastal system or environment. At the shoreline, the four great spheres; the atmosphere, the hydrosphere or ocean, the lithosphere or earth's surface, and the biosphere; all coexist. Consequently it is often the most dynamic part of the earth's surface and these dynamics are most evident on sandy beaches, particularly those exposed to high waves.

Sandy beaches consequently consist of lithospheric elements, namely sand, lying ultimately on bedrock geology; they are acted on by the waves, tides and currents of the hydrosphere; by the wind, rain and temperature of the atmosphere; and play host to the fauna of the biosphere. The following sections examine the climate and marine processes that provide the energy to build and maintain Queensland's beach systems.

Figure 1.10   *The Great Barrier Reef consists of over 2500 individual reefs that line the outer edge of the continental shelf from Torres Strait to Gladstone, as well as parts of the mid-shelf (source: Hopley, 1982).*

Table 1.5   Age of Queensland coastal features

- rocks from   1500 million years old
- shape of coast   70 million to 50 million years ago
- beaches from   120 000 years old
- most beaches less than   6500 years old
- coral reefs   6000 to present

Table 1.6   South East Queensland sand islands

	Area (ha)	Shoreline (km)	Mean ht (m)	Max ht (m)
Fraser	184 000	206[1]	100	244
Cooloola	60 000	56[2]	60	237
Bribie	18 000	36[1]	5	10
Moreton	15 000	87[1]	100	280
Stradbroke	26 000	50[1]	100	218
South Stradbroke	19 500	21[2]	10	20
Total	322 500	456		

[1] entire island

[2] ocean shore only

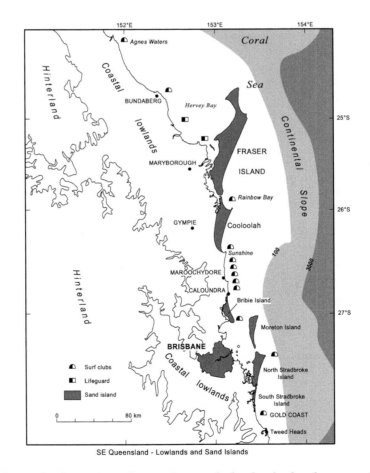

SE Queensland - Lowlands and Sand Islands

Figure 1.11   *Southeast Queensland consists of extensive sandy lowlands that have received sediments from the Burnett, Mary and Brisbane Rivers. They are fronted by the six major sand islands - Fraser, Cooloola, Bribie, Moreton, North Stradbroke and South Stradbroke, together with the Sunshine and Gold Coast mainland beaches, and Hervey and Moreton Bays. The whole region represents a massive accumulation of Quaternary river, marine and aeolian (wind blown) sediments. The symbols represent the general location of surf clubs and lifeguards.*

1.3 Climate and Atmospheric Processes

Climate's contribution to beaches is in two main areas. Climate interacting with the geology and biology provides the geo- and bio-chemistry to weather the land surface, which, together with the physical forces of rain, runoff, rivers and gravity, erode and transport sediments to the coast. At the coast it is also the climate, particularly winds that interact with the ocean to generate waves and currents that are essential to move and build this sediment into beaches and dunes.

On a global scale, beaches can be classified by their climate. The *polar beaches* of the Arctic Ocean and those surrounding Antarctica are all dark in colour and composed of coarse sand, cobbles and even boulders. They receive only low waves and have little or no surf. The beaches have steep swash zones and, because of the coarse beach, there are no dunes. All these characteristics are a product of the cold polar climate.

Temperate middle latitude beaches typical of southern Australia have sediments composed predominantly of well-weathered quartz sand grains, with variable amounts of shell fragments. In addition, wind and wave energy associated with the westerly wind stream is relatively high.

The waves produce energetic surf zones, while the winds can build massive coastal dune systems.

Tropical beaches reside in areas of lower winds of the equatorial low pressure area (the doldrums) and the great Trade winds of moderate velocity. Consequently wave energy is low to moderate at best, unless the shore is exposed to higher waves generated in the mid-latitudes. The areas of lower waves and winds tend to have steep high tide beaches with little or no surf and few dunes. Their sediments, however, are often white, being composed of well-weathered quartz sands derived from plentiful tropical rivers and bleached coral and algal fragments derived from coral and algal reefs. Areas exposed to the Trades can have more energetic beaches and, in places, extensive coastal dunes.

The Queensland coast extends from 9° to 28°S latitude, placing it within the tropics and into the sub-tropics; the realm of tropical beaches, with temperate influences only in the southeast. The climate is controlled seasonally by three major pressure systems: the equatorial low and associated monsoons and tropical cyclones; the sub-tropical high; and in winter by the occasional passage of fronts associated with the sub-polar lows (Fig. 1.13)

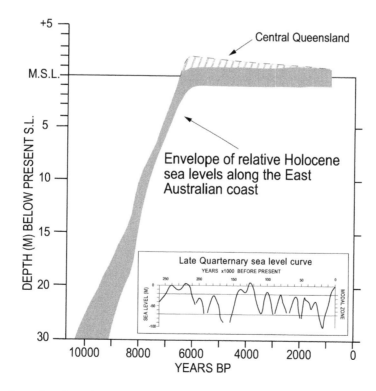

Figure 1.12   *Two plots showing the recent rise in sea level, between 10 000 and 6000 years ago, obtained from evidence along the southeast Australian coast; together with (insert) the regular oscillations in global sea level that have taken place over the past 250 000 years. Along parts of the central-northern Queensland coast, sea level stood up to 1-3 m higher about 4000 years ago. The longer-term oscillations in sea level are a result of fluctuating earth surface temperature and its impact on the growth and decay of continental ice sheets, or ice ages.*

1.3.1   Equatorial low

The equatorial low lies beneath the overhead sun and shifts from the northern to southern hemisphere with the seasonal shift in the sun. During the southern summer it extends across northern Australia in a broad low pressure trough generated by the high surface temperatures of northern-inland Western Australia, Northern Territory and Queensland. The heat low separates the moist air to its north, associated with the summer monsoon, and the dry air of the arid interior to the south associated with the subtropical high.

1.3.2   Summer Monsoons

The Wet of northern Australia is the manifestation of the summer monsoons. The wet-humid conditions are produced by southward movement of moist air from the seas north of Australia toward the heat low (Fig. 1.14). The air flows from the north-west and is characterised by low wind velocities (Fig. 1.15), with precipitation often occurring as scattered thunderstorms, which may locally produce short periods of strong winds, lightning and heavy rain. The monsoons reach the coast during November to February and penetrate inland as far as the low.

1.3.3   High pressure systems

In summer the highs are concentrated around 36°S, which permits the equatorial low to centre itself over west-central Queensland. The counterclockwise flow of air around the highs brings a southeasterly flow of humid Coral Sea air onto the east coast, also attracted by the inland low pressure. The moist air brings humid conditions and summer rain to the east coast. In winter the high moves north to 30°S, still centered below the Queensland border. For this reason Queensland is largely impacted by the northern, and in particular northeastern, section of the high pressure systems, which is manifest by a broad strong flow of southeast to easterly air, originating in the Tasman and Coral Seas. The southeast Trades are the most dominant feature of Queensland weather systems and the source of most of the wave and wind energy along the east and gulf coasts, particularly in the winter months (Fig.1.15). While the highs produce the most dominant winds on the coast, their low to moderate velocity results in the generation of only 10% of waves in excess of 2.5 m (Table 1.7). The Trades do however produce most waves inside the Great Barrier Reef Lagoon and the gulf, and have built the great sand dune systems from Stradbroke Island through to Cape York.

Table 1.7 Frequency of storm waves (H>2.5 m) in southeast Queensland produced by cyclonic and anticyclonic systems (Source PWD, 1985)

Cyclonic:	
Tropical cyclones	27%
East coast cyclones	47%
Mid-latitude cyclones	14%
Anticyclonic:	
Highs (Trade winds)	10%

1.3.4   Sea breeze systems

Associated with the highs and the warmer weather they bring are the coastal sea and land breezes. The sea breeze

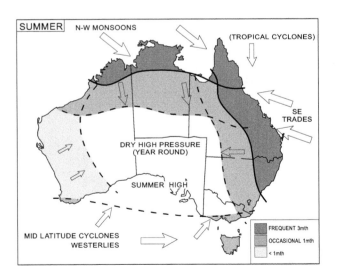
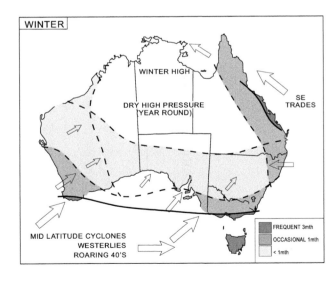

Figure 1.13   *The Australian climate is dominated year round by the dry high pressure anticyclones that sit over much of Australia. In summer, the equatorial low forms over northern Australia, bringing the north-west monsoons (The Wet), while the dry subtropical high pressure systems dominate over southern Australia. In winter, the dry high moves north to dominate central and northern Australia, and permits mid latitude cyclones and their cold fronts to bring cool weather and rain across southern Australia and occasionally as far north as southeast Queensland. This figure shows the source, season and extent of influence of the major air masses around the Australian coast.*

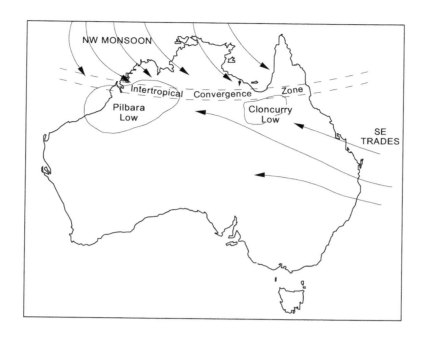

Figure 1.14   *A generalised map showing the summer wind circulation across northern Australia. The Intertropical Convergence Zone (ITCZ) lies over the great heat lows in the Pilbara and west-central Queensland. The lows draw in a warm-humid northwest flow from the north, while the moist southeast Trades flow onshore to the south.*

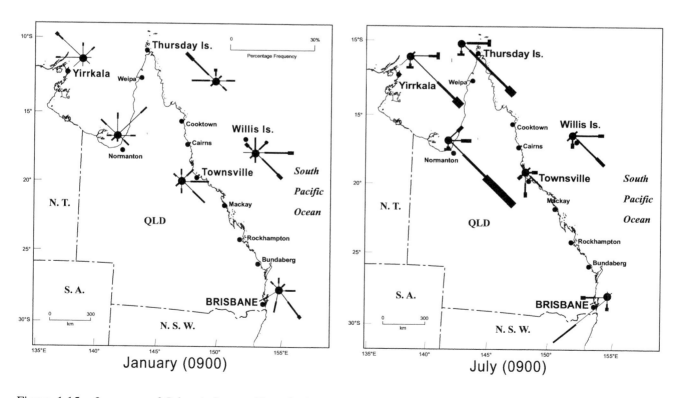

Figure 1.15   *January and July wind roses. Note the January dominance of the northwest monsoonal winds at Yirrkala and Thursday Island. Elsewhere the southeast Trades dominate, apart from the cooler winter southwesterlies at Brisbane.*

results from the daily warming of the coastal land surface. The air above the land warms up and begins to rise. As the land gets hotter and more air rises, it is replaced by air moving in from above the adjacent, but relatively cooler, coastal ocean waters. Consequently a local circulation cell is initiated, resulting in the cooler sea breezes replacing the hotter land air. At night and in the early morning hours, the reverse can occur as the now cooler land surface and relatively warmer ocean causes the air over the ocean to rise, which is in turn replaced by the cooler land breeze. This produces a light offshore breeze, more common on clear, still mornings.

Around the Queensland coast, the direction of the sea breeze is predominantly southeasterly along the east coast, while they are westerly on the eastern gulf shore (Fig. 1.16). The sea breezes are more frequent and intense in summer and consequently make an important contribution to the Queensland coastal environment, particularly through the generation of low wind waves, which reinforce the Trade winds and waves.

1.3.5 Cyclonic systems

In addition to the high pressure systems, three types of low pressure cyclones can impact on the Queensland coast. They are seasonal tropical cyclones which form in the northern Coral Sea and gulf and directly impact the coast, and the east coast cyclones and mid latitude

cyclones which form in the Tasman Sea and Southern Ocean respectively and only have indirect wave impact on the more exposed sections of the eastern and southern coast.

1.3.5.1 Tropical Cyclones

Tropical cyclones are the best known of Queensland's weather systems. They are intense cyclonic systems that form in the Arafura Gulf and northern Coral Sea, generally between 5° and 10°S. They can occur between November and May, with most occurring in February and March. Once formed, they rapidly intensify and tend to move initially westerly under the influence of the southeast Trades, then may move toward the south and can make landfall along much of the gulf and northern to central Queensland coast (Fig. 1.17). At the coast they are accompanied by very strong winds, storm surges up to 3 m and waves up to several metres, as well as heavy rains that bring coastal flooding. They tend to weaken into tropical depressions when over land, however if they remain over the Coral Sea they can maintain their integrity as far south as the Tasman Sea, bringing big seas to southeast Queensland and northern NSW coast. In southeast Queensland they are responsible for 27% of waves greater than 2.5 m in height.

1.3.5.2 East coast cyclones

East coast cyclones are relatively poorly understood phenomena, yet they are the cyclones that wreak most havoc on the southeast Queensland and NSW coast. They can occur at any time of the year, but are more prevalent in early to mid-winter. They usually last for four to five days, and are highly variable in frequency, with some years having only one east coast cyclone, while others may have several such cyclones. East coast cyclones generally form over the central coastal region of NSW and rapidly intensify, possibly reaching the strength of a tropical cyclone. They then meander in a southwest path across the Tasman Sea (Fig. 1.18). When near the coast they produce very strong winds, heavy rainfall and big seas and swell. They are responsible for 47% of high (>2.5 m) waves along the southeast Queensland coast and outer reef, nearly twice the frequency of the better known tropical cyclones.

1.3.5.3 Mid-latitude cyclones

Mid-latitude cyclones are part of the subpolar low pressure system that extends around the entire southern hemisphere in a belt centred on 40° to 50°S latitude, the so called 'Roaring Forties'. This system forms the southern boundary of the subtropical high pressure system. Like the high it shifts with the seasons, moving closer to the southern Australian continent in winter and further south in summer. The lows or cyclones that are embedded in this

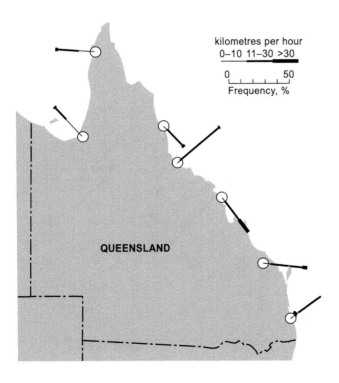

Figure 1.16 Sea breeze directions around northeast Australia. Note the predominance of onshore southeast winds along the east coast and westerly winds in the eastern gulf.

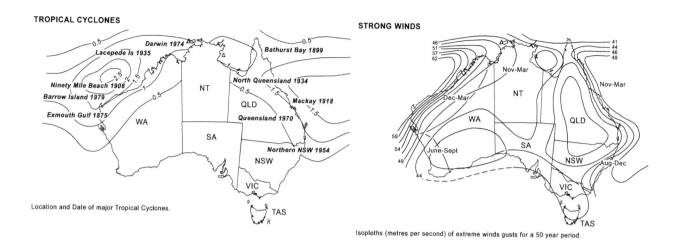

Figure 1.17   (a) Annual frequency of tropical cyclones in the Australian region including location of some major cyclones; and (b) Maximum wind velocities in the Australian region.

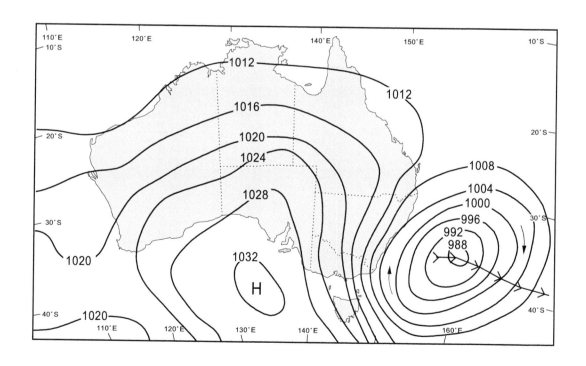

Figure 1.18   An example of an East Coast Cyclone off the eastern Queensland coast. These cyclones can occur at any time of the year, usually originating over and just off the central NSW coast. They produce the strongest winds, heaviest rain and coastal flooding, and biggest waves and swell along the NSW coast and into southeast Queensland. As they slowly meander off towards the southwest the weather clears, however big swell continues to batter the coast.

belt are continually moving from west to east. On average, one passes south of Australia every three to four days. These cyclones are responsible for the majority of southerly waves arriving along the southeast Australian coast. The waves decrease in height and in southeast Queensland produce only 14% of high waves.

In summer, when they are kept south by the high, they have very little impact on Queensland's weather. However, during winter when they move further north and when there are gaps between the highs, they can penetrate the continent with their effect being felt as far north as central Queensland. Their arrival is usually heralded by a cold front, accompanied by strong west through south winds and at times some rain. As the lows pass, the following highs take over, as the winds lighten and tend more easterly. These lows are, however, responsible for the low to moderate southeast swell that arrives along the southeast coast (Fig. 1.19).

1.3.6   Queensland's coastal climate

The climate of the Queensland coast is a composite of all the above weather systems averaged over a period of decades. The coastal summer climate is characterised by hot, humid conditions and periods of heavy rain. The winters are warm in the north with extended dry periods, particularly in the gulf and cape, while conditions are milder in the southeast, with some rain occurring along the eastern coastal fringe.

The *rainfall* along the coast varies from a low of 600 mm in the southern gulf, with most of this rain falling in summer, to a maximum of over 1600 mm in the Tully region. Figure 1.20 shows the annual rainfall for Queensland, with its distinctive coastal maximum. In addition the rainfall maximum for the entire state occurs in summer in association with the monsoon in the gulf, the Trades along the east coast and occasional tropical cyclones.

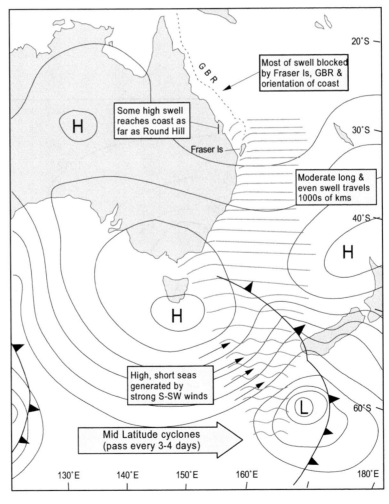

Figure 1.19   *An example of a mid-latitude cyclone and cold front passing south of Australia. As the cyclone traverses the Southern Ocean, its strong west to southwesterly winds, blowing over a long stretch of ocean, produce high seas. As the sea waves travel north of the cyclone, they become more regular swell. In summer when the lows are located further south of the continent, the swell can take two to three days to reach the southeast Queensland coast. In winter the lows are closer to the coast and may even cross the coast, producing bigger seas and swell.*

Winds are very important for much of the Queensland coast, as southeast Trade winds, sea breezes and the summer northwest monsoons are responsible for much of the wave energy along the gulf and the coast north of Fraser Island. These winds generate the short wind waves that provide most to all of the waves and surf around the coast. Only south of Round Hill, and particularly south of Sandy Cape on Fraser Island, does ocean swell, generated in the southern Coral and Tasman Seas and Southern Ocean, provide most of the wave energy. In summer the Trade winds dominate south of Cooktown, while the northwest monsoons dominate in the gulf and northern cape (Fig. 1.13). In winter the Trade winds increase in velocity and dominate the entire state north of Mackay, while in the south, southwesterly winds, associated with the southern half of the high pressure system, dominate.

1.4   Ocean Processes

Oceans occupy 71% of the world's surface. They therefore influence much of what happens on the remaining land surfaces. Nowhere is this more the case than at the coast, and nowhere are coasts more dynamic than on sandy beach systems. The oceans are the immediate source of most of the energy that drives coastal systems. Approximately half the energy arriving at the world's coastlines is derived from waves; much of the rest arrives as tides, with the remainder contributed by other forms of ocean and regional currents. In addition to supplying physical energy to build and reshape the coast, the ocean also influences beaches through its temperature, salinity and the rich biosphere that it hosts (Fig. 1.2).

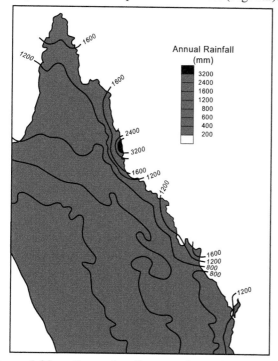

Figure 1.20   Queensland's annual rainfall, showing the distinct coastal maximum. Most rain in the gulf and cape arrives in summer.

The Queensland coast is bordered by the Gulf of Carpentaria and adjoining Arafura Sea to the north and west, and the Coral Sea and associated Great Barrier Reef Lagoon to the east. Both areas are warm tropical seas, releasing vast amounts of moisture for coastal precipitation, and capable of producing summer tropical cyclones.

There are seven types of ocean processes that impact the Queensland coast, namely: wind waves and storm surges, tides, ocean currents, local wind driven currents, upwelling and downwelling, sea surface temperature and ocean biota (Table 1.8). Upwelling refers to the movement of deep, cool ocean waters to the coast and their arrival, or upwelling, at the shore, while downwelling refers to the movement of warmer, surface ocean waters toward the coast and their turning down, or downwelling, at the shore. Ocean biota refers to all marine organisms at the shore and in the coastal ocean.

1.4.1   Ocean waves

There are many forms of waves in the ocean ranging from small ripples to wind waves, swell, tidal waves, tsunamis and long waves including standing and edge waves; the latter lesser known but very important for beaches. In this book, the term 'waves' refers to the wind waves and swell, while other forms of waves are referred to by their full name. The major waves and their impact on beaches are discussed in the following sections.

1.4.1.1   Wind waves

Wind waves, or sea, are generated by wind blowing over the ocean. They are the waves that occur in what is called the area of wave generation; as such they are called '*sea*'. Five factors determine the size of wind waves:

Wind velocity - wave height will increase exponentially as velocity increases;
Wind duration - the longer the wind blows with a constant velocity and direction, the larger the waves will become until a fully arisen sea is reached; that is, the maximum size sea for a given velocity and duration;
Wind direction will determine, together with the Coriolis force, the direction the waves will head;
Fetch - the sea or ocean surface is also important; the longer the stretch of water the wind can blow over, called the fetch, the larger the waves;
Water depth is important as shallow seas will cause wave friction and possibly breaking. This is not a problem in the deep ocean, which averages 4.2 km in depth, but is very relevant in the Gulf of Carpentaria (Fig. 1.9) and Great Barrier Reef Lagoon (Fig. 1.10), both of which average much less than 100 m in depth.

The biggest seas occur in those parts of the world where strong winds of a constant direction and long duration blow over a long stretch of ocean. The part of the globe where these factors occur most frequently is in the southern oceans between 40° and 50°S, where the Roaring Forties and their westerly gales prevail. Satellite sensing of all the world's oceans found that the world's biggest waves, averaging 6 m and reaching up to 20 m, occurred most frequently in the Southern Ocean, south and west of Australia. The waves generated by the same cyclones as they cross the southern Tasman Sea reach the southeast Queensland coast, as low to moderate southeast swell. This swell can arrive throughout the year, with a slight winter maximum. Likewise, the southeast Trade winds blowing over the Coral Sea have a long fetch, but only low to moderate wind velocity; consequently waves rarely exceed 2 m, and then most break on the Great Barrier Reef and do not reach the coast.

Inside the Gulf and Great Barrier Reef lagoon all waves are wind waves, that is, generated by either the light northwest monsoons and/or fresher southeast Trades and sea breezes, blowing over the confined gulf and lagoon waters. For this reason these waves rarely exceed 1 m.

1.4.1.2   Swell

Wind waves become swell when they leave the area of wave generation, by either travelling out of the area where the wind is blowing or when the wind stops blowing. Wind waves and swell are also called free waves or progressive waves (Fig. 1.21). This means that once formed, they are free of their generating mechanism, the wind, and they can travel without it. They are also progressive, as they can move or progress unaided over great distances.

Once swell leaves the area of wave generation, the waves undergo a transformation that permits them to travel great distances with minimum loss of energy. Whereas in a sea the waves are highly variable in height and length, in swell the waves decrease in height, increase in length and become more uniform. As the speed of a wave is proportional to its length, they also increase in speed.

A quick and simple way to accurately calculate the speed of waves in deep water is to measure their period, that is, the time between two successive wave crests. The speed

is equal to the wave period multiplied by 1.56 m. Therefore, a 10 second wave travels at 10 x 1.56 metres per second, which equals 15.6 m/s or 56 km per hour. In contrast, a 5 second wave travels at 28 km per hour and a 12 second wave at 67 km per hour. What this means is that as sea and swell propagates across the ocean, the longest and fastest waves arrive first.

Ocean Wave Generation, Transformation and Breaking

Wave type	Breaking	Shoaling	Swell	Sea
Environment	Shallow water - surf zone	Inner continental shelf	Deep water >> 100m	Deep water >>100m Long fetch = sea/ocean surface Wind velocity waves Wind duration waves Wind direction = wave direction
Distance travelled	100 m	1 to 100 km	100's to 1000's km	100's to 1000's km
Time required	seconds	minutes	hours to days	hours to days
Wave profile				
Water depth	1.5 x wave height	< 100m	>> 100m	>> 100m
Wave character	wave breaks wave bore swash	higher shorter steeper same speed	regular lower longer flatter faster	variable height high short steep slow
Example; height (m) period (sec) length (m) speed (km/hr) distance travelled (km/day)	2.5 to 3 12 0 to 50 0 to 15	2 to 2.5 12 50 to 220 15 to 60	2 to 3 12 220 66 1600	3 to 5 6 to 8 50 to 100 33 to 45 800 to 1100

Figure 1.21 *Waves begin life as 'sea waves' produced by winds blowing over the ocean or sea surface. If they leave the area of wave generation they transform into lower, longer, faster and more regular 'swell', which can travel for hundreds to thousands of kilometres. As all waves reach shallow water, they undergo a process called 'wave shoaling' which causes them to slow, shorten, steepen and finally break. This figure provides information on the characteristics of each type of wave. Most of the Queensland coast receives low, short sea waves, with swell only reaching the southeast coast.*

Table 1.8 Major ocean processes impacting coast

Process	Area of coastal impact	Type of impact
waves - sea & swell	shallow coast & beach	wave currents, breaking waves, wave bores
Tides	shoreline & inlets	sea level, currents
ocean currents	continental shelf	currents
local wind currents	nearshore & shelf	currents
Upwelling & downwelling	nearshore & shelf	currents & temperature
ocean temperature	entire coast	temperature
Biota	entire coast	varies with environment

Swell also travels in what surfers call 'sets' or more correctly *wave groups,* that is, groups of higher followed by lower waves. These wave groups are a source of long, low waves (the length of the groups) that become very important in the surf zone, as discussed later.

Swell and seas will move across the ocean surface through a process called orbital motion. This means the wave particles move in an orbital path as the wave crest and trough pass overhead. This is the reason the wave form moves while the water, or a person or boat floating outside the breakers, simply goes up and down, or more correctly around and around. However when waves enter water where the depth is less than 25% of their wave length (wave length equals wave period squared, multiplied by 1.56; for example, a 10 sec wave will be 10 x 10 sec x 1.56 = 156 m in length, and a 5 sec wave only 39 m long) they begin to transform into shallow water waves, a process that may ultimately end in wave breaking. The variation in length is significant on the Queensland coast, where the longer waves in the southeast impact the seabed to depths of 20 m and more, while the shorter waves inside the Great Barrier Reef lagoon are restricted in impact to usually just a few metres depth.

As waves move into shallower water and begin to interact with the seabed or feel the bottom, four processes take place, affecting the wave speed, length, height, energy and ultimately the type of wave breaking (Fig. 1.22).

Wave speed decreases with decreasing water depth.

Variable wave depth produces variable wave speed, causing the wave crest to travel at different speeds over variable seabed topography. At the coast this leads to *wave refraction.* This is a process that bends the wave crests, as that part of the wave moving faster in deeper water overtakes that part moving slower in shallower water.

At the same time that the waves are refracting and slowing, they are interacting with the seabed, a process called *wave attenuation.* At the seafloor, some potential wave energy is transformed into kinetic energy as it interacts with the seabed, doing work such as moving sand. The loss of energy causes a decrease or attenuation in the overall wave energy and therefore a lower height of the wave.

Finally, as the water becomes increasingly shallow, the waves shoal, which causes them to slow further, decrease in length and increase in height. The speed and distance over which this takes place determines the type of *wave breaking.*

1.4.1.3   Wave breaking

Waves break basically because the wave trough reaches shallower water (such as the sand bar) ahead of the following crest. The trough therefore slows down, while the crest in deeper water is still travelling a little faster. Depending on the slope of the bar and the speed and distance over which this occurs, the crest will attempt to 'overtake' the trough by spilling or even plunging forward, and thereby breaking.

There are three basic types of breaking wave:

Plunging or dumping waves, which surfers know as a tubing wave, occur when shoaling takes place rapidly, such as when the waves hit a reef or a steep bar and/or are travelling fast . As the trough slows, the following crest continues racing ahead and as it runs into the stalling trough, its forward momentum causes it to both move upward, increasing in height, and throw forward, producing a curl or tube.

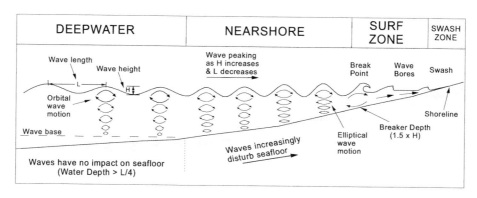

Figure 1.22   *A beach cross-section illustrating the transformation of deepwater waves as they cross the nearshore, surf and swash zones. On the surface they can be seen to slow, shorten, steepen and perhaps increase in height. At the break point they break and move across the surf zone as wave bores (broken white water) and finally up the beach face as swash. Below the surface, the orbital wave currents are also interacting with the seabed, doing work by moving sand and ultimately building and forever changing the beach systems.*

Wave types: sea and swell
Waves are generated by wind blowing over water surfaces.
Large waves require very strong winds, blowing for many hours to days, over long stretches of deep ocean.
Sea waves occur in the area of wave generation. Most waves north of Fraser Island are sea waves.
Swell are sea waves that have travelled out of the area of wave generation. Swell dominates the southeast coast.

Spilling breakers on the other hand, occur when the sea-bed shoals gently and/or waves are moving slowly, resulting in the wave breaking over a wide zone. As the wave slows and steepens, only the top of the crest breaks and spills down the face of the wave. Whereas a plunging wave may rise and break in a distance of a few metres, spilling waves may break over tens or even hundreds of metres.

The third type is *surging breakers, which* occur when waves run up a steep slope without appearing to break. They transform from an unbroken wave to beach swash in the process of breaking. Such waves can be commonly observed on steeper beaches when waves are low, or after larger waves have broken offshore and reformed in the surf zone. They then may reach the shore as a lower wave, which finally surges up the beach face as swash.

1.4.1.4   Broken waves

As waves break they are transformed from a progressive wave to a mass of turbulent white water and foam called a *wave bore*. It is also called a 'wave of translation', as unlike the unbroken progressive wave, the water actually moves or translates shoreward. Boardriders, assisted by gravity, surf on the steep part of the breaking wave. Once the wave is broken, boards, bodies and whatever can be propelled shoreward with the leading edge of the wave bore, while the turbulence in the wave bore is well known to anyone who has dived into or under the white water.

1.4.2   Surf zone processes

Ocean waves can originate thousands of kilometres from a beach. They can travel as swell for days to reach their destination. However on reaching the coast, they can undergo wave shoaling and breaking in a matter of seconds.

Once broken and heading for shore the wave has been transformed from a progressive wave containing potential energy, to a wave bore or wave of translation with kinetic energy, which can do work in the surf zone.

There are three major forms of wave energy in the surf zone - broken waves, surf zone currents and long waves (Table 1.9).

Broken waves consist of wave bores and perhaps reformed or unbroken parts of waves. These move shoreward to finally run up and down the beach as swash and backwash. Some of the backwash may reflect out to sea as a reflected wave, albeit a much smaller version of the original source.

Surf zone currents are generated by broken and unbroken waves, wave bores and swash. They include orbital wave motions under unbroken or reformed waves; shoreward moving wave bores; the up and down of the swash across the beach face; the concentrated movement of the water along the beach as a longshore current; and where two converging longshore or rip feeder currents meet, the seaward moving rip currents.

The third mechanism is a little more complex and relates to *long waves* produced by wave groups and, at times, other mechanisms. Long waves accompany sets of higher and lower waves. However, the long waves that accompany them are low (a few centimetres high), long (perhaps a few hundred metres) and invisible to the naked eye. As sets of higher and lower waves break, the accompanying long waves also go through a transformation. Like ocean waves they also become much shorter as they pile up in the surf zone, but unlike ocean waves, they do not break, but instead increase in height toward the shore. Their increase in height is due to what is called 'red shifting'; a shift in wave energy to the red or lower frequency

Table 1.9   Waves in the surf zone

Wave form	Motion	Impact
Unbroken wave	orbital	stirs sea bed
Breaking wave	crest moves rapidly shoreward	wave collapses
Wave bore	all bore moves shoreward	shoreward moving turbulence
Surf zone currents	water flows shoreward, longshore and seaward (rips)	moves water and sediment in surf
Long waves	slow on-off shore	location of bars & rips

part of the wave energy field. These waves become very important in the surf zone, as their dimensions ultimately determine the number and spacing of bars and rips.

As waves arrive and break every few seconds, the energy they release at the break point diminishes shoreward, as the wave bores decrease in height toward the beach. The energy released from these bores goes into driving the surf zone currents and into building the long wave. The long wave crest attains its maximum height at the shoreline. Here it is visible to the naked eye in what is called *wave set-up* and *set-down*. These are low frequency, long wave motions, with periods in the order of several times the breaking wave period, that are manifest as a periodic rise (set-up) and fall (set-down) in the water level at the shoreline. If you sit and watch the swash, particularly during high waves, you will notice that every minute or two the water level and maximum swash level rises then rapidly falls.

The height of wave set-up is a function of wave height, and also increases with larger waves and lower gradient beaches, to reach as much as one third to one half the height of the breaking waves. This means that if you have 1, 2, 5 and 10 m waves, the set-up could be as much as 0.3, 0.6, 1.5 and 3.0 m high, respectively. For this reason, wave set-up is a major hazard at the shoreline during high waves, particularly where the beach has a low gradient or slope. Because the waves set up and set down in one place, the crest does not progress. They are therefore also referred to as a *standing wave*, one that stands in place with the crest simply moving up and down. These standing long waves are extremely important in the surf zone as they help determine the number and spacing of bars and rips. This interaction is discussed in Section 2.3 on beach dynamics.

1.4.3 Wave measurements

While it is easy to see waves and to make an estimate of their height, period, length and direction, accurate measures of these statistics are more difficult. Yet we need to know just what type of waves are arriving at the coast, if we are to properly design for and manage the coast. Traditionally, wave measurements have been made by observers on ships and at lighthouses visually estimating wave height, length and direction. All of Queensland's lighthouses used to make such measurements, until they were progressively automated. Since the 1950's however, increasingly sophisticated electronic wave measuring devices have been developed and installed at some coastal locations. Regrettably, there are fewer electronic wave stations than there used to be manned lighthouses, so our record of Australian wave conditions has in fact diminished as lighthouses have been automated.

The present state-of-the-art on-site wave recording device is the Datawell *Waverider buoy, which* was devel-

oped by the British Oceanographic Institute in the 1960's. It operates using an accelerometer housed in a watertight buoy, about 50 cm in diameter. The buoy is chained to the sea floor, usually in about 80 m water depth. As the waves cause the buoy to rise and fall, the vertical displacement of the buoy is recorded by the accelerometer. This information is transmitted to a shore station and then by phone line to a central computer, where it is recorded.

The first Australian Waverider buoy was installed off the Gold Coast in 1968. Today, Queensland and New South Wales have a network of Waverider buoys stretching from Weipa to Eden; the most extensive in the world. Figure 1.23 locates the site of Queensland's Waverider buoys, together with COPE sites (volunteer manned Coastal Observation Program supervised by the Beach Protection Authority).

While wave rider buoys provide the best real-time measure of actual wave height, period and direction, since 1978, satellites, using laser beams, can sense the height and direction of ocean waves on a global scale. This information is rapidly revolutionising our knowledge of ocean wave generation, dimensions and travel.

1.4.4 Queensland wave climate

Wave climate refers to the seasonal variation in the source, size, and direction of waves arriving at a location. Waves on the Queensland coast originate from six possible sources, and are both seasonal and regional in impact. Three are cyclonic - tropical, mid-latitude and east coast cyclones; and two are associated with the high pressure systems - the southeast Trades and the local sea breeze; while in the gulf the north-west monsoons generate low waves from the north-west during the wet.

1.4.4.1 Wave sources: north to south

Gulf of Carpentaria

There are three sources of waves in the gulf – the summer northwest monsoons, the dominant southeast Trades and occasional tropical cyclones. The northwest monsoons arrive along the gulf and eastern cape coast between November and April (Fig. 1.14). They flow as low to occasionally moderate winds from the north-west that produce low, short waves along the southern to eastern gulf shore. The Trades dominate most of the year, peaking in velocity during the winter months. While they blow offshore along the western cape coast, strong trades winds generate winds waves in the southern gulf which refract to arrive a short, low to occasionally moderate swell along the central to northern cape shore. These waves can reach 2 m and have considerable impact on the more exposed beaches either side of Weipa. Over then entire year the Weipa Waverider buoy records an

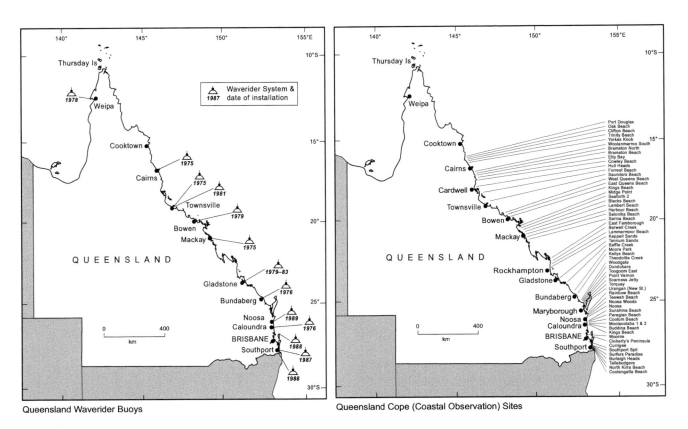

Figure 1.23 *Queensland Waverider (a) and COPE sites (b). The former automatically record offshore wave height and period, the latter are volunteer-based sites that record a number of wave and beach conditions (source: Beach Protection Authority).*

How to estimate wave height from shore

Wave height is the vertical distance from the trough to the crest of a wave. It is easier to make a visual measurement when waves are relatively low, and if there are surfers in the water to give you a reference scale. If a surfer is standing up and the wave is waist height, then it's about 1 m; if as high as the surfer, about 1.5 m; if a little higher, then it's about 2 m. For big waves, just estimate how many surfers you could 'stack' on top of each other to get a general estimate, i.e. two surfers, about 4 m, three, about 6 m and so on. Many surfers prefer to underestimate wave height by as much as 60% and a substantial number still estimate height in the old imperial measure of 'feet', as suggested below:

Actual wave height	Surfer's (under) estimate	% underestimate
0.5 m	1 ft (0.3 m)	60%
1.0 m	2 ft (0.6 m)	60%
1.5 m	3 ft (0.9 m)	60%
2.0 m	5 ft (1.5 m)	80%
2.5 m	6 ft (1.8 m)	70%
3.0 m	8 ft (2.4 m)	80%

average wave of 0.25 m, with a H_{max} of 0.7 m, implying waves greater than 1 m occur less than 10% of the time.

Tropical cyclones

Tropical cyclones occur during November to May, with the likelihood of one impacting some part of the Queensland coast each year (Fig. 1.17). The greatest impact is in the region of landfall where storm surges, accompanied by high waves, can cause severe flooding and shoreline erosion. However because of their low frequency of occurrence at any one location, they do not have a significant impact on the long term wave climate. As Table 1.10 indicates, the maximum wave height for most sites north of Gladstone is only between 0.8 and 1.4 m.

South East Trades

The southeast Trades are the major source of wave energy for the east coast north of Bundaberg, particularly in lee of the Great Barrier Reef, which essentially filters out all ocean swell. Figure 1.24 and Table 1.11 indicate the typical 'deepwater' Trade wind waves. They average about 0.7 m at Gladstone and Mackay, dropping to 0.5 m at Bowen and Cairns. Wave periods average 4 to 5 sec, indicating the short sea waves generated. They arrive almost exclusively from the southeast. In addition, southeast to easterly sea breeze along the east coast acts to reinforce the Trade winds and their waves.

At the shoreline the waves are generally lower still, as they refract, attenuate and shoal along headlands and islands, over reefs and tidal shoals. Table 1.11 lists the observed breaker wave heights at all Queensland COPE sites. In Table 1.12 the same sites are grouped according to their location and level of exposure. Beaches more fully exposed to the Trades (Table 1.12a) have waves averaging 0.2 m with a maximum of 0.4 m. Those in lee of headlands and islands and/or facing north average 0.06 m, with a maximum of 0.4 m (Table 1.12b). Table 1.12c shows the impact of Fraser Island as the average wave height decreases from 0.8 m at Baffle Creek to 0.4 m at Bargara and Woodgate, to 0.15 m at Urangan. These are all significantly lower that the most exposed southeast ocean beaches which average 1.2 m in height (Table 1.12d).

Ocean Swell

Ocean swell arrives along much of the Queensland coast, however north of Fraser Island most of it breaks on the outer Great Barrier Reef. Large swell does occasionally move north of Fraser Island to reach the mainland as far north as Round Hill, and some large swell does move through gaps in the reef, permitting swell to occasionally reach parts of the southeast facing shoreline south of Mackay. Only south of Sandy Cape does the full force of the predominately southeast swell arrive unimpeded

at the coast. This swell averages 1.4 m, with a maximum of 2.4 m. At the coast it is lowered slightly, averaging between 1.1 and 1.2 m (Table 1.12d & e). Furthermore the swell averages 9 to 10 s in period. As a result the wave energy available to do work in the surf zone is 100 times more than received at Weipa, and between 5 to 20 times that received in lee of the Great Barrier Reef (Table 1.10). This swell has four main sources - the great high pressure systems, and tropical, east coast and mid-latitude cyclones.

The high pressure system that dominates the eastern Queensland weather systems provides a flow of south to southeast air across the Tasman Sea, resulting in low to moderate sea and swell along the southeast and outer reef coast. It arrives, on average, 240 days a year (Table 1.13). Tropical cyclones are Queensland's best known weather system, but fortunately have a low frequency of occurrence and only generate swell on the southeast coast on average 12 days a year. Mid-latitude cyclones move continuously across the Southern Ocean and generate waves that travel up to 2000 km and take two days to reach the southeast coast (Figure 1.17). Each month between five and nine cyclones cross the southern Tasman Sea, with the waves they generate arriving on average about 75 days each year (Table 1.13). East coast cyclones can occur in all months, with a preference for the early winter months of May, June and July. On average, three to four form each year and last for four to five days. The waves generated by these cyclones arrive on average about 35 days each year (Table 1.13) and produce amongst the biggest waves on the southeast Queensland coast and some of the biggest waves in the world. In January 1978, a wave of 17 m was recorded on the Newcastle Waverider buoy.

Queensland, therefore has three basic deepwater or off-shore wave climates:
1. The **eastern gulf** with low, short waves (H = 0.2 m, T = 2 sec, dir = W) generated by the summer north-west monsoons, with calms or low Trade wind swell the rest of the year. Note the western coast of the Northern Territory has substantially higher waves as the Trades blow across the gulf and onshore.
2. **Cape York to Hervey Bay**, where ocean swell breaks on the Great Barrier Reef and Fraser Island, with low to moderate Trade wind waves generated in their lee (H = 0.5 to 0.8 m, T = 5-6 sec, dir = SE).
3. The **southeast islands - Sunshine and Gold coasts**, which receive predominantly east to southerly ocean sea and swell averaging 1-1.2 m, T = 9-10 sec.

As indicated in Table 1.12, many sections of the coast receive considerably lower waves at the shore owing to protection from offshore reefs, islands, orientation and wave attenuation and refraction. The end result is a relatively low energy wave climate for much of the coast north of Fraser Island.

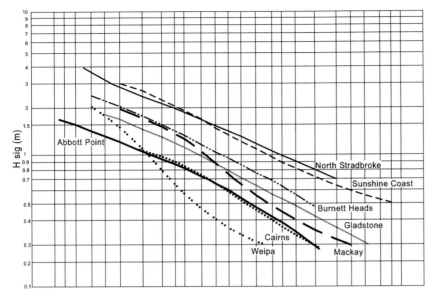

(a)

Figure 1.24 *Wave exceedence curves for (a) Queensland offshore wave rider sites, and (b) Queensland COPE sites that record breaker wave height at the beach. Each curve indicates the percentage of time a particular wave height will occur at each site. The higher the curve on the table the overall high level of wave height. Note the decrease in wave height northward, owing primarily to increasing protection from the Great Barrier Reef. (Source: Beach Protection Authority, Wave Data Recording Program.)*

A Percentage of Time Exceedance

(b)

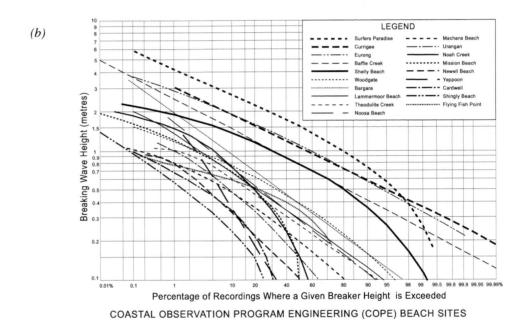

Percentage of Recordings Where a Given Breaker Height is Exceeded

COASTAL OBSERVATION PROGRAM ENGINEERING (COPE) BEACH SITES

Table 1.10   Typical deepwater wave characteristics on the Queensland coast (Source: Beach Protection Authority, Wave Data Recording Program.)

Location	Latitude (S)	average 50% wave height (m)	max 10% wave height (m)	average wave period (sec)	relative wave energy*	dominant wave direction
Gulf of Carpentaria						
Weipa	$12^0 37'$	0.25	0.7	2	1.0	W
Great Barrier Reef Lagoon						
Cairns	$16^0 55'$	0.50	0.9	4	6.8	SE
Bowen	$20^0 04'$	0.50	0.9	4	7.1	SE
Mackay	$21^0 09'$	0.65	1.4	5	15.1	SE
Gladstone	$23^0 51'$	0.70	1.3	6	21.0	SE
Burnett Heads	$24^0 46'$	0.85	1.5	5	25.7	SE
Coral Sea						
Sunshine Coast	$26^0 00'$	1.30	2.4	10	120.7	SE
Stradbroke Is	$27^0 30'$	1.40	2.2	9	125.7	SE
Mean		0.77	1.4	5.6	40.4	SE

* all wave energy relative to Weipa ($E = H^2 T$ arbitrary units)

Table 1.11   Breaker wave characteristics at Queensland COPE sites, listed north to south (Source: Beach
Protection Authority, Coastal Observation Program - Engineering).

	Latitude (S)	Hb (m) 50%	Hb 10% (m)	T (s)	Dir*	Comment
Noah Ck	16^{0}10'	0.15	0.75	6.5	3	in lee of GBR
Newell	16^{0}26'	0.11	0.42	5	3	in lee of GBR
Machans	16^{0}51'	0.2	0.48	4	3	in lee of GBR
Flying Fish Pt	17^{0}30'	0.35	0.6	6	3	in lee of GBR
Mission	17^{0}52'	0.1	0.7	4	3	in lee of GBR & Dunk Island
Cardwell	18^{0}16'	0.05	0.35	3	4	in lee of GBR & Hinchinbrook Island
Shingly	20^{0}16'	0.05	0.25	3	3	in lee of GBR, faces north
Yeppoon	23^{0}08'	0.05	0.37	6	3	in lee of GBR & Great Keppel Island
Lammermoor	23^{0}10'	0.3	0.7	7	3	in lee of GBR
Baffle Creek	24^{0}31'	0.8	1.5	8	3	some protection from Fraser Island
Bargara	24^{0}49'	0.39	0.92	5	3	moderate protection from Fraser Island
Woodgate	25^{0}07'	0.4	0.8	5	3	in lee of Fraser Island
Urangan	25^{0}20'	0.15	0.5	6	2	in lee of Fraser Island
Eurong (Fraser Is)	25^{0}31'	1.2	1.8	10	3	fully exposed
Noosa	26^{0}23'	0.3	0.77	8	3	in lee of Noosa Head
Shelly (Caloundra)	26^{0}48'	0.8	1.4	9	3	in lee of Caloundra Head
South Stradbroke	27^{0}55'	1.1	1.9	10	3	fully exposed
Surfers Paradise	28^{0}00'	1.2	1.85	10	3	fully exposed
Mean		0.42	0.87	6.5	E	

*   direction: 1=NE, 2=E, 3=SE

Table 1.12   Breaker wave characteristics at Queensland COPE sites sorted by exposure/location (Source:
Beach Protection Authority, Coastal Observation Program - Engineering).

a. In lee of Great Barrier Reef - exposed

	Latitude (S)	Hb 50% (m)	Hb 10% (m)	T (s)	Dir*	Comment
Noah Ck	16^{0}10'	0.15	0.75	6.5	3	in lee of GBR
Newell	16^{0}26'	0.11	0.42	5	3	in lee of GBR
Machans	16^{0}51'	0.2	0.48	4	3	in lee of GBR
Flying Fish Pt	17^{0}30'	0.35	0.6	6	3	in lee of GBR
Lammermoor	23^{0}10'	0.3	0.7	7	3	in lee of GBR
Mean		0.2	0.4	5.7	3	

b. In lee of Great Barrier Reef, plus protection from island and/or headland

Mission	17^{0}52'	0.1	0.7	4	3	in lee of GBR & Dunk Island
Cardwell	18^{0}16'	0.05	0.35	3	4	in lee of GBR & Hinchinbrook Island
Shingly	20^{0}16'	0.05	0.25	3	3	in lee of GBR, faces north
Yeppoon	23^{0}08'	0.05	0.37	6	3	in lee of GBR & Great Keppel Island
Mean		0.06	0.41	4	3	

c. Increasing protection from Fraser Island

Baffle Creek	24^{0}31'	0.8	1.5	8	3	some protection from Fraser Island
Bargara	24^{0}49'	0.39	0.92	5	3	moderate protection from Fraser Island
Woodgate	25^{0}07'	0.4	0.8	5	3	in lee of Fraser Island
Urangan	25^{0}20'	0.15	0.5	6	2	in lee of Fraser Island
Mean		0.43	0.93	6	3	

d. Protected southeast coast, in lee of headland

Noosa	26^{0}23'	0.3	0.77	8	3	in lee of Noosa Head
Shelly (Caloundra)	26^{0}48'	0.8	1.4	9	3	in lee of Caloundra Head
Mean		0.5	1.0	8.5	3	

e. Fully exposed southeast coast

Eurong (Fraser Is)	25^{0}31'	1.2	1.8	10	3	fully exposed
South Stradbroke	27^{0}55'	1.1	1.9	10	3	fully exposed
Surfers Paradise	28^{0}00'	1.2	1.85	10	3	fully exposed
Mean		1.16	1.85	10	3	

*   direction: 1=NE, 2=E, 3=SE

Australian Beach Safety Management Program

Table 1.13   Wave and wind climate of the Gold Coast (based on Phinn, 1992)

Wave source	Location	days/year	Accompanying winds
High pressure systems	Tasman Sea	240	moderate southeast through southwest
Mid-latitude cyclones	southern Tasman Sea	75	moderate to strong southwest
East coast cyclones	South Coral-northern Tasman Seas	35	moderate to strong southeast
Tropical cyclones	Coral Sea	12	moderate to strong east to southeast

---

Freak waves, king waves, rogue waves, and tidal waves

Freak waves do not exist.

All waves travel in wave groups or sets. A so-called 'freak', 'king' or 'rogue' wave is simply the largest wave or waves in a wave group.

However, unusually high waves are more likely in a sea than a swell. For this reason they are more likely to be encountered by yachtsmen than surfers or rock fishermen.

Tidal waves arrive on the Queensland coast twice each day. These are related to the predictable movement of the tides and not the damaging *tsunamis* with which they are commonly confused. Tidal waves are discussed in Section 1.4.5.

1.4.5   Tides

Tides are the periodic rise and fall in the ocean surface, due to the gravitational force of the moon and the sun acting on a rotating earth. The amount of force is a function of the size of each and their distance from the earth. While the sun is much larger than the moon, the moon exerts 2.16 times the force of the sun because it is much closer to earth. Therefore, approximately 2/3 of the tidal forces are due to the moon and are called the *lunar* tides. The other 1/3 are due to the sun and these are called *solar* tides.

Because the rotation and orbit of the earth and the orbit of the moon and sun are all rigidly fixed, the *lunar* tidal period, or time between successive high or low tides, is an exact 12.4 hours; while the *solar* period is 24.07 hours. Because these periods are out of phase, they progressively go in and out of phase. When they are in phase, their combined forces act together to produce higher than average tides, called *spring tides*. Fourteen days later they are 90^0 out of phase they counteract each other to produce lower than average tides, called *neap tides*. The whole cycle takes 28 days and is called the lunar cycle, over a lunar month.

The actual tide is in fact a wave, correctly called a *tidal wave,* not to be confused with tsunamis. They consist of a crest and trough, but are hundreds of kilometres in length. When the crest arrives it is called *high tide*, and the trough *low tide*. Ideally, the tidal waves would like to travel around the globe. However the varying size, shape and depth of the oceans, plus the presence of islands, continents, continental shelves and small seas complicate matters. The result is that the tide breaks down into a series of smaller tidal waves that rotate around an area of zero tide called an *amphodromic point*. In the southern hemisphere, the Coriolis force causes the tidal waves to rotate in a clockwise direction, and anticlockwise in the northern hemisphere.

Tides in the deep ocean are zero at the amphodromic point and average less than 20 cm over much of the ocean. However three processes cause them to be amplified in shallow water and at the shore. First is due to shoaling of the tidal waves across the relatively shallow (< 150 m deep). Like breaking wave they are amplified due to wave shoaling processes and increase in height (tide range) up to 1 to 3 m, and in some locations even break as a tidal bore. Secondly, when two tidal waves arriving from different directions converge, as occurs in Broad Sound and Torres Strait, they will also be amplified. Finally in certain large embayments the tidal wave can be amplify by a process of wave resonance, which causes the tide to reach heights of several metres, as occurs in parts of northwest Australia.

Tides are classified as being micro-tidal when their range is less than a 2 m, meso-tidal when between 2 and 4 m, and macro-tidal when greater than 4 m.

1.4.5.1   Queensland tides

The Queensland coast has a micro- to macro-tide range, with a spring range that varies from 1.1 m on the Gold Coast to over 6 m in Broad Sound (Table 1.14).

Tide range

There is a substantial variation in both the height of the tide around the Queensland coast and its time of arrival (Fig. 1.25). This is due to a number of factors:

- the tidal wave approaches from the southeast, reaching southeast Queensland first and the gulf last;
- it slows down moving through the shallow Great Barrier Reef and into Broad Sound and Shoalwater bays;
- it slows considerably moving through the narrow, shallow Torres Strait;

- it is amplified by the shallow shelf waters of the Great Barrier Reef Lagoon;
- it is further amplified to a maximum in Moreton Bay (3 m), Broad Sound (8 m) and Torres Strait (3 m)

Whereas all southern Australian tides are less than 3 m in range, and most are less than 1-2 m, across Queensland they generally exceed 2-3 m and reach heights of 8 m in Broad Sound. Australia's highest tides are in the northwest, reaching 10-11 m, while the world's highest tide reaches 17 m in Canada's Bay of Fundy.

Time of tide

The Coral Sea tidal wave arrives from the southeast, first reaching the border at Point Danger, and one hour later at Brisbane. The wave then arrives about 1 hour later along much of the east coast south of Cooktown, apart from Broad Sound-Shoalwater Bay, where it is slowed by up to 4 hours as well as being amplified. It then slows as it moves north of Cooktown, taking up to 6 hours to move from the Coral Sea through Torres Strait to the gulf. Within the gulf, the Coral Sea tidal wave also interacts with a Gulf of Carpentaria tidal wave, with a point of rotation located just south of New Guinea (Fig. 1.25).

Tidal currents

Tidal currents are a prominent feature of much of the Queensland coast and continental shelf. The combination of higher tide ranges, a relatively shallow shelf, with flows constrained by the islands and reefs of the Great Barrier Reef, together with over 200 inlets along the coast, all result in both substantial tidal flows, often constrained and thereby strengthened by reef, island and inlet constrictions.

In general, tidal currents along the east coast flow to the north with the flooding or rising tide, and reverse to the south with the ebbing tide. However this flow is modified when the currents are flowing through reef channels, into large bays and into all inlets and creek mouths. In all these cases the flow tends to parallel the direction of the major channel, passage or inlets, and is often more perpendicular to the coast. In coastal inlets, the net result is for the tidal currents to commonly dominate over the wave processes, producing extensive tidal sand deltas, particularly on the seaward side of the inlets (Fig. 1.26). These constricted tidal flows can average 1 to 2 m/sec (3-7 km/hr) and can, in places like Torres Strait, can reach 4-5 m/sec.

1.4.6   Ocean Currents

Ocean currents refer to the wind driven movement of the upper 100 to 200 m of the ocean. The major wind systems blowing over the ocean surface drive currents that move in large ocean gyres, spanning millions of square kilometres.

Ocean currents generated in both the South Pacific and the Arafura Sea impact the Queensland coast. In the South Pacific, the ocean gyre moves in a giant anticlockwise circulation. Around Antarctica it is called the West Wind Drift, driven by the westerlies of the Roaring 40's. In the Southeast Pacific, the drift is deflected equatorward by

Table 1.14   Tidal characteristics of Queensland ports

Location	Spring tide range (m)	Max spring tide range (m)	Relative time of arrival 0 hr = Point Danger + = after, add hours
Karumba	1.1	2.6	+11
Weipa	2.2	2.9	+10.5
Thursday Is	1.9	2.8	+5
Cooktown	1.8	2.2	+2
Cairns	1.1	2.5	+2
Lucinda	2.3	2.7	+2
Townsville	2.5	2.9	+2
Bowen	2.1	2.7	+2
Mackay	4.9	5.5	+2.5
Broad Sound	6.3	8.0	+4
Port Clinton	3.9	4.1	+2
Gladstone	3.2	3.9	+1.5
Bundaberg	1.7	2.5	+1.5
Brisbane Bar	1.8	2.1	+1
Pt Danger	1.1	1.5	0

South America and becomes the northward moving Peru Current. It carries cool, polar water along the west coast of South America, before being deflected westward in the sub-tropical latitudes (10-20 S) by the southeast Trade winds. The Trade winds power the westerly moving Equatorial Drift, that travels for more than 10 000 km across the Pacific, warming as it goes. The drift pools in the Coral Sea as warm (25-30°C) tropical water, and is deflected southward (toward Australia) by the land masses of New Guinea and northern Australia. As this warm water moves south, it is joined by equally warm water from the Great Barrier Reef lagoon, to form the warm *East Australian Current*. The current flows south outside the Great Barrier Reef at speeds of 2 to 5 km per hour. It only directly impacts the coast from Sandy Cape on Fraser Island south, where the continental shelf narrows.

Inside the Great Barrier Reef and the Gulf of Carpentaria, the ocean currents have limited direct impact. The depths are generally less than 100 m and 50 m respectively, and entrance to the GBR lagoon is blocked by the reef itself, while Torres Strait prevents the current from flowing into the gulf. Consequently currents in these two regions, and consequently for most of the Queensland coast, are

dependent on local to regional winds, especially the southeast Trades, and tidal currents, particularly close to the coast and in all areas of restricted flows. As a result the Trades tend to generate a northerly current along the coast, while the tides produce a net north to onshore current with the flooding tide, and net offshore flow with the ebbing tide. For these reasons, currents in coastal Queensland waters are both highly variable and weather and tide dependent.

1.4.7 Other currents

There are several other forms of ocean currents driven by winds, density, tides, shelf waves and ocean waves. It is not uncommon to have several operating simultaneously. Each will have a measurable impact on the overall current structure and must be taken into account if one needs to know the finer detail of the coastal currents, their direction, velocity and temperature.

1.4.8 Sea surface temperature

The sea temperature along the Queensland coast is a product of three main processes. Firstly, the latitudinal

---

Note: Spring tides are also called 'king' tides and are highest around New Year and Christmas.
Spring or *king tides* are not responsible for beach erosion, unless they happen to coincide with large waves.
In areas of higher tide range, tides can influence wave height, as high tide and deeper nearshore waters result in less wave shoaling and higher waves at the shore, while shallower water at low tide induces greater wave shoaling and lower waves at the shore.

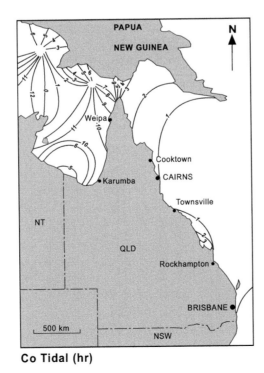

Co Tidal (hr)

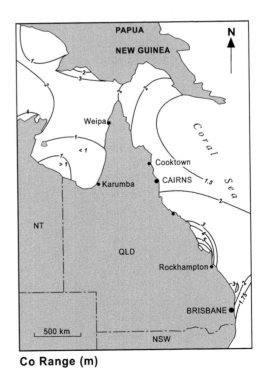

Co Range (m)

Figure 1.25   *Co-tidal (time of tide in hours) and co-range (height of tide in meters) lines for the Queensland coast*

location of the coast between 9-28°S determines the overhead position of the sun and the amount of solar radiation available to warm the ocean water; secondly, the seasonal response to the movement of the sun and its warming influence on the East Australian Current; and thirdly, the impact of El Niño-Southern Oscillation Index (SOI), which produces pulses of relatively warmer (positive SOI/La Nina) and cooler water (negative SOI/ El Nino) across the western Pacific every few years.

1.4.9 Salinity

All oceans and seas contain dissolved salts derived from the erosion of land surfaces over hundreds of millions of years. Chlorine and sodium dominate and, together with several other minerals, account for the dissolved 'salt'. The salts are well mixed and globally average 35 parts per thousand, increasing slightly into the dry sub-tropics and decreasing slightly in the wetter Roaring 40's.

1.5 Biological Processes

At first glance, sandy beaches resemble an arid, desert type landscape. While fish live in the sea and birds fly overhead, one might wonder what actually lives in and on the beach. Over the past decades our whole concept of beach ecology has changed dramatically, as scientists have looked closely at just what inhabits our beaches. As a result, we now know that beaches can be highly productive ecosystems, producing and exporting nutrients to the adjacent ocean, and contributing shell and algal fragments to beach sediments.

1.5.1 Beach ecology

The basis of the beach ecosystem are the microscopic diatoms that live in the water column (called phytoplankton) and microscopic meiofauna that live on and between sand grains; including bacteria, fungi, algae, protozoans and metazoans. Feeding on these are larger organisms that live in the sand, including meiofauna such as small worms and shrimp; and filter feeding benthos such as molluscs and worms. In the water column they are also preyed upon by zooplankton such as amphipods, isopods, mysids, prawns and crabs. At the top of the food chain are the fish and sea birds, and the occasional mammal such as dolphins, dugongs and even whales.

A number of hard bodied organisms also contribute their skeletal material to the beach in the form of sediment. When all organisms die, their internal and/or external skeletons can be broken up, abraded and washed onto beaches by waves. In addition along parts of the Queensland coast, coral-algal reefs form close to shore, in places as fringing reefs, and contribute their often substantial structure, as well as generally coarse carbonate detritus that is washed onto adjacent beaches. Offshore on coral atolls and reefs the beaches and cays usually consist of 100% carbonate, all derived from the reefs.

Marine organisms are a major factor in Queensland beach usage owing to the prevalence of a number of dangerous stingers, as well as sharks and poisonous shell fish and corals. These organisms are discussed in section 3.4.

Table 1.15 Queensland sea temperatures

	Summer max. (°C)	Winter min. (°C)
Cairns	30	23
Rockhampton	28	21
Gold Coast	25	19

Figure 1.26 *Eastern Queensland has over 200 tidal inlets/creeks and associated ebb tide deltas. (a) shows a 1 km wide tidal delta off the mouth of Carmilla Creek in Broad Sound, while (b) shows the well developed tidal channel and tidal shoals at Round Hill.*

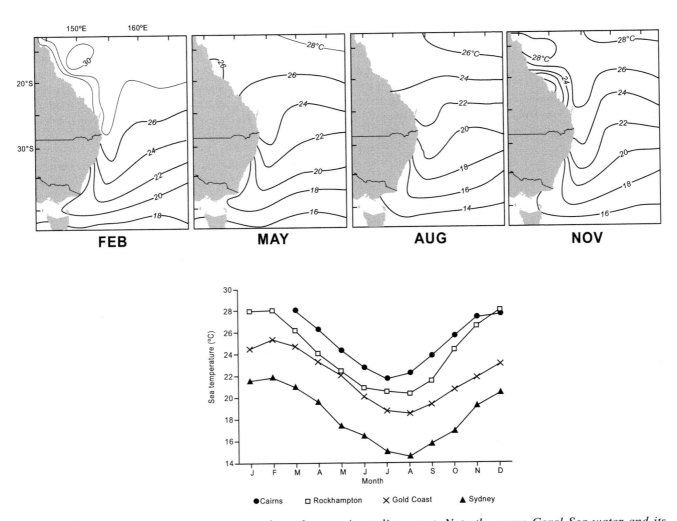

Figure 1.27　*(a) Sea surface temperatures along the east Australian coast. Note the warm Coral Sea water and its movement to the south as the East Australian Current. (b) Mean monthly sea surface temperature at Cairns, Rockhampton, Gold Coast and Sydney.*

2. BEACH SYSTEMS

Beaches throughout the world consist of wave deposited sediment and lie between the base of wave activity and the limit of wave run up or swash. They therefore include the dry (subaerial) beach, the swash or intertidal zone, the surf zone, and beyond the breakers the nearshore zone. Usually only the dry beach and swash zone is clearly visible, while bars and channels are often present in the surf zone but are obscured below waves and surf, while the nearshore is always submerged. The shape of any surface is called its morphology; hence beach morphology refers to the shape of the beach, surf and nearshore zone (Fig. 2.1).

2.1 Beach Morphology

As all beaches are composed of sediment deposited by waves, beach morphology reflects the interaction between waves of a certain height, length and direction and the available sediment; whether it be sand, cobbles or boulders; together with any other structures such as headlands, reefs and inlets.

Queensland beaches can be very generally divided into three groups: wave dominated, tide modified and tide dominated. The *wave dominated* beaches occur along the open south east coast south of Sandy Cape. They are exposed to persistent ocean swell and waves and lower tides (<2 m). The *tide-modified* beaches occupy the remainder of the coast north of Fraser Island, in lee of the Great Barrier Reef and in the Gulf of Carpentaria. These beaches are exposed to lower, shorter waves, and generally higher tide ranges (>2 m). The *tide-dominated* beaches are those beaches in the area of tide modification, which receive such low waves, owing to some form of protection, that they become increasingly dominated by the tides and are

a mix of beach and tidal flats. The remainder of this chapter is devoted to a description of first the higher energy wave dominated beaches (section 2.2), followed by the tide-modified (section 2.5), and then the tide-dominated (section 2.6).

The simplest way to describe a beach is in two dimensions, as shown in Figure 2.2. The beach consists of three zones: the subaerial beach, the surf zone and the nearshore zone.

2.1.2 Subaerial Beach

The *subaerial beach* is that part of the beach above sea level, which is shaped by wave run-up or swash. It starts at the shoreline and extends up the relatively steep swash zone or beach face. This may be backed by a flatter berm or cusps, which in turn may be backed by a runnel, where the swash reaches at high tide. Behind the upper limit of spring tide and/or storm swash usually lies the beginning of the vegetated dunes. The subaerial, or dry, beach is that part which most people go to and consider 'the beach'. However, the real beach is far more extensive, with the subaerial beach being literally the tip of the iceberg.

2.1.3 Swash or intertidal zone

On wave dominated beaches the swash zone connects the dry beach with the surf, it is the steeper part of the shoreline across which the broken waves run up and down across the beach face. As wave height decreases and tide range increases, this zone tends to become flatter and considerably wider, and is termed the intertidal zone on tide-modified to tide dominated beaches.

Figure 2.1   *The nature and shape of beaches along the Queensland coast are products of both the inherited bedrock geology, which usually forms the boundary headlands and rock and coral reefs, and the rivers, creeks, waves and tides that have brought the sand to the shore to build the dynamic beaches and surf zones. (a) In this view, Mick Ready Beach is bordered by headlands and influenced by both waves (note the sand ridges), tides and fringing coral reefs. (b) Waddy Point on Fraser Island forms a barrier to longshore sand transport. Some sand crosses the point via wind blown dunes, while waves move submerged sand around the headland to form the elongate spits and sand waves that move into the northern Marloo Bay.*

2.1.4 Surf Zone

The *surf zone* extends seaward of the shoreline and out to the area of wave breaking. This is one of the most dynamic places on earth. It is the zone where waves are continuously expending their energy and reshaping the seabed. It can be divided into the area of wave breaking, often underlain by a bar; and immediately shoreward, the area of wave translation where the wave bore (white water) moves toward the shoreline, transforming along the way into surf zone currents and, at the shoreline, into swash. Surf zones are widest is the south east where they often contain two bars, decreasing in size to be transient (with the tide) on tide-modified beaches, and almost non-existent on tide-dominated beaches.

2.1.5 Nearshore Zone

The *nearshore zone* is the third and, on wave dominated beaches, the most extensive part of the beach. It extends seaward from the outer breakers to the maximum depth at which average waves mobilise beach sediment and moves it shoreward. The point where this begins is called the wave base, referring to the base of wave activity. On the high energy south-east Queensland coast it usually lies at a depth between 15 and 25 m and may extend 1, 2

or even 3 km out to sea, while along the protected central, north and Gulf coasts it is less than 10 m deep on the more exposed beaches, decreasing to 0 m or low tide on tide dominated beaches. As wave height and period decrease the depth to wave base also decreases, causing the nearshore zone to diminish in depth and extent. On tide-modified beaches it is usually only a few metres deep, while on tide-dominated beaches it may terminate at low tide.

2.1.6 Three dimensional beach morphology

In three dimensions, wave dominated beaches become more complex. This is because most beaches are not uniform alongshore, but vary in a predictable manner.

Beaches vary longshore on two scales. On the Queensland coast, most beaches more than a few hundred metres long (and particularly those with prominent headlands) tend to curve toward their southern end. This is because the shape or plan of the beach reflects the interaction of the waves with the seabed. On the Queensland coast, the predominantly southerly waves reach the southern headlands first, causing the waves to begin to 'feel' the shallower seabed and slow down, while the part of the wave crest that is still in deeper moves more rapidly.

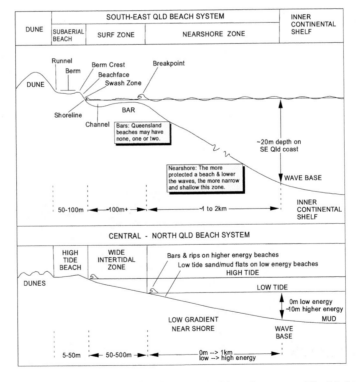

Figure 2.2   *This cross section illustrates two typical Queensland beach system. The higher energy south east beaches consists of the dry subaerial beach above the shoreline; the surf zone containing bars, troughs and breaking waves; and the nearshore zone which extends seaward of the breaker zone out to modal wave base. Wave base is the depth to which ocean waves can move beach sands. Seaward lies the inner continental shelf. The approximate width and depth of each zone are indicated. Along the more protected central-north coast the lower waves and higher tide result in a lower gradient and shallower beach, intertidal and nearshore zone, with wave base as shallow as low tide.*

As a result, the waves bend or refract toward the shore and decrease in height around southern headlands or reefs, while they are less impeded and arrive higher and straighter toward the northern end of beaches. The overall effect of the bending wave crests is to cause a spiral in the shape of the beach, so that the curvature increases toward the more protected southern end. This bending of the wave crests is called *wave refraction*, while the loss of wave energy and height is called *wave attenuation*. The decrease in wave height down the beach results in lower breakers, a narrower and shallower surf zone, a narrower and often steeper swash zone and a lower and narrower subaerial beach.

The second scale of longshore beach variation relates to any and all undulations on the beach and in the surf zone, usually with scales in the order of tens of metres to as much as 500 m. Variable longshore forms produced at this scale include regular beach cusps located in the high tide swash zone and spaced between 15 - 40 m, and all variation in rips, bars, troughs and any undulations along the beach, usually with spacing of between 50 - 500 m. These features are associated with rip circulation and are known as rip channels, crescentic and transverse bars, and megacusp horns and embayments. Each beach type is described in the section 2.4.

2.2   Beach Dynamics

Beach dynamics refers to the dynamic interaction between the breaking waves and currents and the sediments that compose the beach. This interaction not only builds all beaches, but, as waves change, causes continual changes in beach response and shape. Over shorter periods of time (days to months) it causes all the changes in the shape of the beach and surf zone you usually see when you go to the beach. Over a long period of time (10's-1000's of years) it can lead to a building out of the beach, called progradation, or erosion of the beach, and perhaps ultimately complete removal of a beach.

Five factors determine the character of a beach. These are the size of the sediment, the height and length of the waves, the characteristics of any long waves present in the surf zone, and the tide range. The impact of each is briefly discussed below.

2.2.1   Beach sediment

The size of beach sediment determines its contribution to beach dynamics. Unlike in air where all objects fall at the same speed, sediment falls through water at a speed which is proportional to its size. Very fine sediment, like clay, will simply not sink but stay in suspension for days or weeks causing turbid, muddy water. Silt; sediment coarser than clay but finer than sand; takes up to two hours to settle in a laboratory cylinder (Table 2.1). As a consequence fine sediments usually stay in

suspension through the energetic surf zone and tidal inlets and are carried out to sea to settle in deep water on the continental shelf.

Sand which composes most beaches takes between a few seconds for coarse sand, to five minutes for very fine sand, to settle through 1 m of water. For this reason it is fine enough to be episodically carried by rivers and creeks to the sea, and then alongshore and onshore by waves, to build beaches, yet coarse enough to settle quickly to the seabed as soon as waves stop breaking.

Most Queensland beaches consist of sand. However, a few are composed of cobbles or even boulders. Cobbles and boulders require enormous energy to be lifted or moved, and then settle immediately. They are therefore only rarely moved, such as during extreme storms; and then only very slowly and over short distances. Consequently, beaches containing such coarse sediment always have a nearby source, usually an eroding cliff or bluff.

In the energetic breaking wave environment, anything as fine or finer than very fine sand stays in continual suspension and is flushed out of the beach system into deeper, quieter water. This is why ocean beaches never consist of silts or moods. Most beaches consist of sand because it can be transported in large quantities to the coast and settle fast enough to remain in the surf zone. A few Queensland beaches that consist of cobbles and boulders all derived their material from an immediate source, usually an adjacent cliff.

Depending on the nature of its sediment, each beach will inherit a number of characteristics. Firstly, the sediment will determine the mineralogy or composition of the beach; usually quartz sand or silica on mainland Queensland beaches, with significant quantities of carbonate (coral, algal, shell detritus etc.) in lee of fringing reefs and on all coral reef beaches. Secondly, the size of the sediment will, along with waves, determine beach shape and dynamics. Fine sand produces a low gradient (1 to 3°) swash zone, wide surf zone and potentially highly mobile sand. Medium to coarse sand beaches have a steeper gradient (5 to 10°), a narrower surf zone and less mobile sand. Cobble and boulder beaches are not only very steep, but have no surf zone and are almost immobile. Therefore, identical waves arriving at adjacent fine, medium and coarse sand beaches will interact to produce three distinctly different beaches.

Table 2.1   Sediment size and settling rates

Material	Size - diameter	Time to settle 1 m
clay	0.001-0.008 mm	hours to days
silt	0.008-0.063 mm	5 min to 2 hours
sand	0.063-2 mm	5 sec to 5 min
cobble	2 mm-6.4 cm	1 to 5 sec
boulder	> 6.4 cm	< 1 sec

Likewise, three beaches having identical sand size, but exposed to low, medium and high waves, will have three very different beach systems. Therefore, it is not just the sand or the waves, but the interaction of both, along with long waves and tides that determine the nature of our beaches.

2.2.2　Wave energy - long term

Waves are the major source of energy that build and change beaches. Seaward of the breaker zone, waves interact with the sandy seabed to stir sand into suspension and, under normal conditions, slowly move it shoreward. The wave-by-wave stirring of sand across the nearshore zone and its shoreward transport has been responsible for the delivery of all the sand that presently composes the beaches and coastal sand dunes of Queensland.

Waves are therefore responsible for supplying the sand to build beaches. The higher the waves, the deeper the depth from which they can transport sand and the faster they can transport it. Consequently, the largest volumes of sand that build the biggest beaches and dunes are all in areas of very high waves. Along the south-east sand islands, massive amounts of up to 100 000 - 200 000 m³ of sand have been transported onshore for every metre of beach, amounting to 1 000 000 m³ for every few metres of beach. However, these same large waves can just as rapidly erode the very dynamic beaches they initially built.

Lower waves can only transport sand from shallow depths and at slower rates. Consequently, they build smaller barrier systems, usually delivering less than 10 000 m³ for every metre of beach, as is typical of much of the central and north coast. However, their beaches are less dynamic, more stable and are less likely to be eroded.

2.2.3　Wave energy - short term

Waves are not only responsible for the long term evolution of beaches, but also the continual changes and adjustments that take place as wave conditions vary from day to day. As noted above, a wave's first impact on a beach is felt as soon as the water is shallow enough for wave shoaling to commence, usually in less than 20 m water depth. As waves shoal and approach the break point, they undergo a rapid transformation which results in the waves becoming slower and shorter, but higher; and ultimately breaking, as the wave crest literally overtakes the trough.

As waves break, they release kinetic energy; energy that may have been derived from the wind some hundreds or even thousands of kilometres away. This energy is released as turbulence, sound (the roar of the surf) and even heat. The turbulence stirs sand into suspension and

carries it shoreward with the wave bore. The wave bore decreases in height shoreward, eventually turning into swash as it reaches the dry beach.

Breaking waves, wave bores and swash, together with unbroken and reformed waves, all contribute to a shoreward momentum in the surf zone. As these waves and currents move shoreward, much of their energy is transferred into other forms of surf zone currents, namely longshore, rip feeder and rip currents, and long waves and associated currents. The rip currents are responsible for returning the water seaward while the long wave currents play a major role in shaping the surf zone.

It is therefore the variation in waves and sediment that produces the seemingly wide range of beaches present along the coast, ranging from the steep, narrow, protected beaches, to the broad, low gradient beaches with wide surf zones, large rips and massive breakers. Yet every beach follows a predictable pattern of response, largely governed by its sediment size and prevailing wave height and length. The types of beaches that can be produced by waves and sand is discussed in the section 2.4.

2.3　Beach types

Beach type refers to the prevailing nature of a beach, including the waves and currents, the extent of the nearshore zone, the width and shape of the surf zone including its bars and troughs, and the dry or subaerial beach. *Beach change* refers to the changing nature of a beach or beaches along a coast as wave, tide and sediment conditions change.

The first comprehensive classification of wave dominated beaches was developed by the Coastal Studies Unit (CSU), at the University of Sydney in the late 1970's. This classification is now used internationally, wherever tide range is less than 2 m. In Australia this classification applies to most of the southern coast, from Fraser Island in the east around to Exmouth Peninsula in the west. Across northern Australia waves are lower to non-existent and tides are generally higher, producing a different range of beaches. In the early 1990's the CSU undertook research along the central Queensland coast which resulted in the identification of a tide-modified and tide-dominated range of beach types. The full range of Queensland beach types are summarised in Table 2.2. In the following sections each of the twelve beach types is described, together with examples and photographs of the ten beach types which occur along the Queensland coast.

2.3.1　Wave dominated beach types

Wave dominated beaches consist of three types: reflective, intermediate or dissipative, with the intermediate having four states as indicated in Figures 2.3 and 2.4. In Queensland, only the moderate to lower energy intermediate and reflective beach types occur, with the higher energy dissipative and intermediate longshore bar and trough types usually absent (Table 2.2).

2.3.2 Dissipative beaches (D)

There are no dissipative beaches in Queensland, as waves are not sufficiently high to form and maintain such beaches. They can occur, however, for short periods of time during and immediately following periods of high cyclonic waves on exposed south east beaches. They more commonly occur on a few exposed fine sand beaches along parts of the high energy western Victorian, west Tasmanian, South Australian and south Western Australian coast.

When *dissipative* beaches occur, the combination of high waves and fine sand ensures that they have wide surf zones and usually two to occasionally three shore parallel bars, separated by subdued troughs. The beach face is composed of fine sand and is always wide, low and firm, firm enough to support a 2WD drive car. Much of Fraser Island and Cooloola has firm beaches composed of fine sand.

Wave breaking begins as high spilling breakers on the outer bar, which reform to break again and perhaps again on the inner bar or bars. In this way they dissipate their energy across the surf zone, which may be up to 300-500 m wide. This is the origin of the name 'dissipative'.

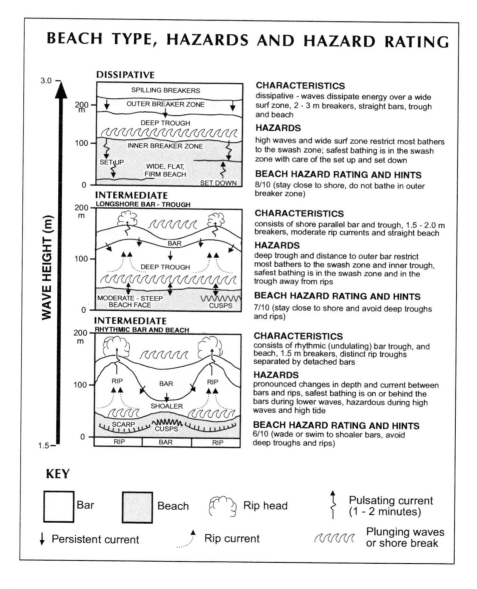

Figure 2.3 *A plan view of the rhythmic bar and beach, longshore bar and trough and dissipative beach types. Note how as wave height increases between the rhythmic and dissipative beaches, the surf zone increases in width, rips initially increase and then are replaced by other currents, and the shoreline becomes straighter. The physical characteristics and beach and surf hazards associated with each type are indicated.*

Table 2.2   Queensland beach types (Cooktown to Coolangatta) by number and length

Beach Type	Code	No. of beaches		% of total beaches		Mean length (km)		Total length (km)		Proportion of length of all beaches (%)		
		SE*	GBR	SE	GBR	SE	GBR	SE	GBR	SE	GBR	Total
Wave dominated												
1. Dissipative	6	0	0	0	0	-	-	-	-	0	0	0
2. Longshore bar trough	5	0	0	0	0	-	-	-	-	0	0	0
3. Rhythmic bar beach	4	7	0	6.6	0	15.95	-	111.8	-	18.8	0	5.7
4. Transverse bar rip	3	34	0	32.1	0	7.08	-	240.6	-	40.5	0	12.2
5. Low tide terrace	2	13	0	12.3	0	6.46	-	96.9	-	16.3	0	4.9
6. Reflective	1	19	10	17.9	1.1	3.35	0.25	63.6	2.5	10.7	0.2	3.3
Tide modified												
7. Reflective + LTT	7	9	372	8.5	41.9	3.56	1.15	32.0	428.4	5.4	31.0	23.3
8. Reflective + TBR	8	0	71	0	8.0	-	3.62	-	257.4	0	18.6	13.0
9. Ultradissipative	9	0	81	0	9.1	-	2.28	-	185.0	0	13.4	9.4
Tide dominated												
10. HT + sand ridges	10	1	70	0.9	7.9	0.35	2.45	0.4	171.5	0.1	12.4	8.7
11. HT + sand flats	11	22	249	20.8	28.1	3.64	1.16	80.0	288.0	13.5	20.9	18.6
12. Tidal sand flats	12	1	34	0.9	3.8	8.50	1.40	8.5	47.7	1.4	3.5	2.8
TOTAL		106	887	100	100	5.61	1.56	594.5	1380.5	100	100	100

*   SE   refers to all south-east beaches between Sandy Cape (Fraser Island) and Coolangatta
**   GBR   refers to all beaches in lee of the Great Barrier Reef, between Cooktown and Hervey Bay

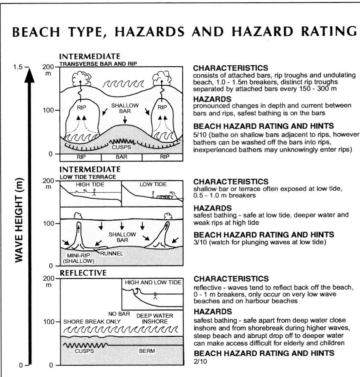

Figure 2.4   A plan view of the reflective, low tide terrace and transverse bar and rip beach types. Note how as wave height increases between the reflective and transverse beaches, the surf zone and bar increase in width, rips form and increase in size, and the shoreline becomes crenulate. The physical characteristics and beach and surf hazards associated with each type are indicated.

In the process of continual breaking and re-breaking across the wide surf zone, the incident or regular waves decrease in height and may be indiscernible at the shoreline. The energy and water that commenced breaking in the original wave is gradually transferred in crossing the surf zone to a lower frequency movement of water, called a standing wave. This is known as red shifting, where energy shifts to the lower frequency, or red end, of the energy spectrum.

At the shoreline, the standing wave is manifest as a periodic (every 60 to 120 seconds) rise in the water level (set-up), followed by a more rapid fall in the water level (set-down). As a rule of thumb, the height of the set-up is 0.3 to 0.5 times the height of the breaking waves (i.e. 1 to 1.5 m for a 3 m wave). Because the wave is standing, the water moves with the wave in a seaward direction during set-down, with velocities between 1 to 2 m/sec closer to the seabed. As the water continues to set down, the next wave is building up in the inner surf zone, often to a substantial wave bore, 1 m+ high. The bore then flows across the low beach face and continues to rise, as more water moves shoreward and sets up. This process continuously repeats itself every one to two minutes.

Because of the fine sand and the large, low frequency standing wave, the beach is planed down to a wide, low gradient; with the high tide swash reaching to the back of the beach, often leaving no dry sand to sit on at high tide.

2.3.3 Intermediate beaches

Intermediate beaches refer to those beach types that are intermediate between the lower energy reflective beaches and the highest energy dissipative beaches. The most obvious characteristic of intermediate beaches is the presence of a surf zone with bars and rips. On the open south east coast 54 of the 73 wave dominated beaches are intermediate and they dominate the exposed sand

islands, as well as the Sunshine and Gold Coasts. They are usually several kilometres long and occupy 450 km of the south-east sandy coast. This dominance is a result of the presence of the long exposed beaches, which are composed of fine to medium sand and receive waves between 0.5 and 1.5 m high, all of which combine to generate the rip dominated beach systems. However, between those beaches produced by 0.5 m waves and those by 1.5 m waves, there is quite a range in the shape and character of the beach. For this reason, intermediate beaches are classified into four beach states. The lowest energy one is called 'low tide terrace'; then as waves increase, the 'transverse bar and rip'; then the 'rhythmic bar and beach' and finally the 'longshore bar and trough'. Each of these beaches is described below.

2.3.4 Longshore bar and trough (LBT)

The *longshore bar and trough* beach type is rare on the Queensland coast, only occurring as the outer bars on the most exposed section of the south-east sand islands (Fig. 2.5). They very rarely occur as the inner bar, where people swim, and then only following high cyclone seas.

Longshore bar and trough beaches are characterised by waves averaging 1.5 m or more, which break over a near continuous longshore bar located between 100 and 150 m seaward of the beach, with a 50 to 100 m wide, 2 to 3 m deep longshore trough separating it from the beach. The beach face is straight alongshore and usually has a low gradient. While the bar, trough and beach may look straight and devoid of rips, the bar is usually crossed by rips every 250 to 500 m. The deep trough and the presence of less obvious rips make this a particularly hazardous swimming beach.

Higher waves tend to break continuously along the bar, with lower waves not breaking in the vicinity of the rip gaps. The wave bores cross the bar and enter the deeper

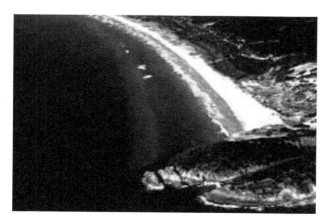

Figure 2.5   *Longshore bar and trough beaches are characterised by an offshore shore parallel bar, fronted by a deep trough and usually relatively straight beach. (a) Point Lookout on North Stradbroke Island with waves barely breaking on the outer longshore bar, and (b) Cooloola Beach with a well developed longshore (outer) bar and trough.*

longshore trough, where they quickly reform and continue shoreward as a lower wave to break or surge up the beach face. The water moving shoreward into the trough returns seaward using two mechanisms. Firstly, the water piles up along the beach face as wave set-up. As the water sets down, it moves both seaward as a standing wave, and longshore to feed the rip current. Secondly, as the converging feeder currents approach the rip, they accelerate, causing additional set-up in lee of the rip. As this set-up sets down, the rip pulses seaward.

2.3.5   Rhythmic bar and beach (RBB)

The *rhythmic bar and beach* type is the highest energy beach type that commonly occurs on the south east ocean coast. In all, seven long beaches, averaging 16 km in length, are of this type. They only occur on the longer, more exposed island and Sunshine Coast beaches and consequently occupy 112 km (19%) of the total south-east beach length.

These energetic beaches require two primary ingredients for their formation: relatively fine sand and south-east exposure to the highest deepwater waves. On the south-east coast, they occur where waves average at least 1.5 m and sand is more fine than medium (Fig. 2.6).

Rhythmic beaches consist of a rhythmic longshore bar that narrows and deepens where the rips cross the breakers, and in between broadens, shoals and approaches the shore. It does not, however, reach the shore, with a continuous rip feeder channel feeding the rips to either side of the bar. The shoreline is usually rhythmic with protruding megacusp horns in lee of the detached bars and commonly scarped megacusp embayments behind the rips. The surf zone may be up to 100 to 250 m wide and

---

Rhythmic bar and beach hazards

- Bar - just to reach the bar requires crossing the rip feeder channel. This may be an easy wade at low tide or a difficult swim at high tide. Be very careful once water depth exceeds waist depth, particularly if a current is flowing. Also, as you reach the bar, water pouring off the bar may wash you back into the channel.
- The centre of the bar is relatively safe at low tide, but at high tide you run the risk of being washed into the rip feeder or rip channel.
- Rip feeder channel - depth varies with position and tide; both depth and velocity increase toward the rip.
- Rip - the rip channel is usually 2 to 3 m deep, with a continuous, but pulsating, rip current.
- High tide - deeper bar and channels, but weaker currents and rip.
- Low tide - waves break more heavily and may plunge dangerously; shallower bar and channels, but stronger currents and rip.
- Oblique waves - skew bar and rips alongshore.
- Higher waves - intensify wave breaking and strength of all currents.

- Summary: Caution is required by the young and inexperienced on rhythmic beaches, as the bar is separated from the beach by often deep channels and strong currents. Do not venture out to the bar unless you are between the red and yellow flags and you are a strong and experienced swimmer.

 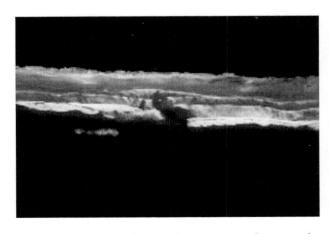

Figure 2.6   *Rhythmic bar and beach systems have a sinuous or rhythmic outer bar, with rips exiting between the protruding bars, and a more rhythmic shoreline, with protruding megacusp horns in lee of the bars. (a) a rhythmic outer bar along Sunshine-Peregian Beach, and (b) detail of a protruding bar section and adjacent rip (Cooloola Beach).*

the bar and rips are spaced every 250 to 1000 m along-shore.

The shallower rhythmic bars causes waves to break more heavily, with the white water flowing shoreward as a wave bore. The wave bore flows across the bar and into the backing rip feeder channel. The water from both the wave bore and the swash piles up in the rip feeder channel and starts moving sideways toward the adjacent rip embayment, which may be several to more than 100 m alongshore. The feeder currents are weakest where they diverge behind the centre of the bar, but pick up in speed and intensity toward the rip. In addition, the rip feeder channels deepen toward the rip.

In the adjacent rip channels, waves break less or often not at all. They may move unbroken across the rip to finally break or surge up the steeper rip embayment swash zone. The strong swash often causes slight erosion of the beach face and cuts an erosion scarp.

In the rip embayment, the backwash returning down the beach face combines with flow from the adjacent rip feeder channels. This water builds up close to shore (called wave set-up), then pulses seaward as a strong, narrow rip current. The currents pulse every 30 to 90 seconds, depending on wave conditions. The rip current accelerates with each pulse and persists with lower velocities between pulses. Rip velocities are usually less than 1 m per second (3.5 km/h), but will increase up to 2 m per second under higher waves.

To identify this beach type, looks for the pronounced longshore beach rhythms, i.e. the shoreline is very sinuous. The shallowest widest bars and heaviest surf lies off the protruding parts of the shore (the megacusps). Water flows off the bars, into the feeder channel, along the beach to the deeper rip embayment then seaward in the rip current.

Hazards

This is the most hazardous beach type commonly occurring along the south east coast. Most people are put off entering the surf by the deep longshore trough containing rips and their feeder currents. If you are swimming or surfing on a rhythmic beach, the following highlights some common hazards.

2.3.6   Transverse bar and rip (TBR)

The *transverse bar and rip* beach type is the most common beach type on the south east coast, with 34 beaches (32%) occupying 241 km (40%) of the south-east beach length. They occur on the south east coast where sand is fine to medium and waves average 1 to 1.5 m. As most south east beaches have such sand, and as waves average 1.4 m overall, they occur on all beaches well exposed to ocean waves, including most of the Sunshine and Gold Coast beaches (Fig. 2.7).

Transverse beaches receive their name from the fact that as you walk along the beach, you will see bars transverse or perpendicular to, and attached to, the beach, separated by deeper rip channels and currents. The bars and rips are usually regularly spaced every 150 to 250 m, but can reach spacings of 500 m. Their surf zones range from 50 to 100 m wide.

Transverse beaches and bars are discontinuous alongshore, being cut by prominent rips. The alternation of shallow bars and deeper rip channels causes a longshore variation in the way waves break across the surf zone. On the shallower bars waves break heavily, losing much of their energy. In the deeper rip channels they will break less, and possibly not at all, leaving more energy to be expended as a shorebreak at the beach face. Consequently, across the inner surf zone and at the beach face, there is an alternation of lower energy swash in lee of the bars,

Figure 2.7   *Transverse bar and rip beaches are characterised by alternating attaching to attached bars with well developed rip channels in between, as well as a highly rhythmic shoreline in lee of the bars and rips. (a) Cooloola Beach, with highly rhythmic attaching bars; (b) rhythmic shoreline and adjacent rips on North Stradbroke Island.*

Rip currents

Rip currents are a relatively narrow, seaward moving stream of water. They represent a mechanism for returning water back out to sea that has been brought onshore by breaking waves. They originate close to shore as broken waves (wave bores) flow into longshore rip feeder troughs. This water moves along the beach as rip feeder currents. On normal beaches, two currents arriving from opposite directions usually converge in the rip embayment, turn and flow seaward through the surf zone. The currents usually maintain a deeper rip feeder trough close to shore, and a deeper rip channel through the surf zone. As the confined rip current flows seaward of the outer breakers, it expands and may meander as a larger rip head. Its speed decreases and it will usually dissipate within a distance of two to three times the width of the surf zone. Rip currents will exist in some form on ALL beaches where there is a surf zone. Their spacing is usually regular and ranges from as close as 50 m on central and north coast beaches, up to 500 m during high waves on south east beaches. Headlands and reefs in the surf will induce additional rips, called topographically controlled rips, and megarips can form during big seas.

Rip current spacing

- spacing approximately = surf zone width x 4
- on south east ocean coast, rip spacing is commonly 150 to 250 m
- on the GBR coast, where rips occur, they are usually spaced between 50-100 m
- also a function of beach slope; the lower the slope (hence wider the surf zone), the wider the rip spacing

---

and higher energy swash/shorebreak in lee of the rips.

This longshore variation in wave breaking and swash causes the beach to be reworked, such that slight erosion usually occurs in lee of the rips, and slight deposition in lee of the bars. This results in a rhythmic shoreline, building a few metres seaward behind the attached bars as deposition occurs, and being scoured out and often scarped in lee of the rips. The rhythmic undulations are called megacusp horns (behind the bars) and embayments (behind the rips). Whenever you see such rhythmic features, which have a spacing identical to the bar and rips (150 to 500 m+), you know rips are present.

The transverse surf zone has a cellular circulation pattern. Waves tend to break more on the bars and move shoreward as wave bores. This water flows both directly into the adjacent rip channel and, closer to the beach, to the rip feeder channels located at the base of the beach.

The water in the rip feeder and rip channel then returns seaward in two stages. Firstly, water collects in the rip feeder channels and the inner part of the rip channel, building up an hydraulic head against the lower beach face. Once high enough, it pulses seaward as a relatively narrow accelerated flow, the rip. The water usually moves through the rip channel, out through the breakers and seaward for a distance usually less than the width of the surf zone, that is, a few tens of metres.

The velocity of the rip currents varies tremendously. However, on a typical beach with waves less than 1.5 m, they peak at about 1 m per second, or 3.5 km per hour, about walking pace. However, under high waves they may double that speed. What this means is that under average conditions, a rip may carry someone out from the shore

to beyond the breakers in 20 to 30 seconds. Even an Olympic swimmer would only be able to maintain their position, at best, when swimming against a strong rip.

Two other problems with rips and rip channels are their depth and their rippled seabed. They are usually 0.5 to 1 m deeper than the adjacent bar, reaching maximum depths of 3 m. Furthermore, the faster seaward flowing water forms megaripples on the floor of the rip channel. These are sand ripples 1 to 2 m in length and 0.1 to 0.3 m high, that slowly migrate seaward. The effect of the depth and ripples on bathers is to provide both variable water depth in the rip channels and a soft sand bottom, compared to the more compact bar. As a result, it is more difficult to maintain your footing in the rip channel for three reasons: the water is deeper, the current is stronger and the channel floor is less compact. Also, someone standing on a megaripple crest that is suddenly washed or walks into the deeper trough, may think the bottom has 'collapsed'. This may be one source of the 'collapsing sand bar' myth; an event that can not, and does not, occur.

Hazards

Transverse beaches are one of the main reasons south east Queensland beaches have such good surf, as well as why there are over 1000 rescues a year. The shallow bars tempt people into the surf, while lying to either side are the deeper, more treacherous rip channels and currents.

2.3.7   Low tide terrace (LLT)

Low tide terrace beaches are the lowest energy intermediate beach type. They occur on the south east coast where sand is fine to medium and wave height averages between

---

Transverse bar and rip beach hazards

- Bars - the centres of the attached bars are the safest place to swim. They are shallow, furthest from the rip channels, and the wave bores move toward the shore.
- Rips - the rips are the cause of most rescues on the south east coast, so they are best avoided unless you are a very experienced surfer.
- Rip feeder channels - any channel close to shore has been carved by currents and is part of the surf zone circulation. It will be carrying water that is ultimately heading out to sea. Rip feeder channels usually run along behind and to the sides of the bar, adjacent to the base of the beach.
- In the rip embayment, the feeder currents converge and head out to sea. If you are not experienced, stay away from any channels, particularly if water is greater than waist depth and is moving.
- Children on floats must be very wary of feeder channels as they can drift from a seemingly calm, shallow, inner feeder channel located right next to the beach rapidly out into a strong rip current.
- Breakers - waves will break more heavily on the bar at low tide, often as dangerous plunging waves or dumpers. In the rip embayment, the shorebreak will be stronger at high tide.
- Higher waves - when waves exceed 1 to 1.5 m, both wave breaking and rip currents will intensify.
- Oblique waves - these will skew both the bars and rips alongshore and may make the rips more difficult to spot.
- Low tide - rip currents intensify at low tide, but are more confined to the rip channel.
- High tide - rip currents are weaker and may be partially replaced by a longshore current, even across the bar.
- Summary: This beach type is one of the main reasons why south east Queensland has surf lifesaving clubs and so many rescues. It is relatively safe on the bars during low to moderate waves, but beware, as many hazards, particularly rips, lurk for the young and inexperienced. Stay on the bar/s and well away from the rips and their side feeder currents.

0.5 and 1 m. There are 13 modally low tide terrace beaches on the south-east ocean coast, representing 12% of south east beaches and occupying 97 km of coast. They tend to occur toward the lower energy, more protected end of long beaches (Fig. 2.8) such as southern Fraser Island, Rainbow Bay and Noosa and in moderately protected areas such as all of Bribie Island and in lee of headlands, as at Greenmount.

Low tide terrace beaches are characterised by a moderately steep beach face, which is joined at the low tide level to an attached bar or terrace, hence the name - low tide terrace. The bar usually extends between 20 and 50 m seaward and continues alongshore, attached to the beach. It may be flat and featureless; have a slight central crest, called a ridge; and may be cut every several tens of metres by small shallow rip channels, called *mini rips.*
At high tide when waves are less than 1 m, they may pass right over the bar and not break until the beach face, behaving much like a reflective beach. However at spring low tide, the entire bar is usually exposed as a ridge or terrace running parallel to the beach. At this time, waves break by plunging heavily on the outer edge of the bar. At mid tide, waves usually break right across the shallow bar.

Under typical mid-tide conditions, with waves breaking across the bar, a low 'friendly' surf zone is produced. Waves are less than 1 m and most water appears to head toward the shore. In fact it is also returned seaward, both by reflection off the beach face and via the mini rips,

even if no rip channels are present. The rips however, are usually weak, ephemeral and shallow.

Hazards

Low tide terrace beaches are the safest of the intermediate beaches, because of their characteristically low waves and shallow terrace. However, changing wave and tide conditions do produce a number of hazards to swimmers and surfers.

2.3.8   Reflective beaches (R)

Reflective sandy beaches lie at the lower energy end of the wave dominated beach spectrum. They are characterised by steep, narrow beaches usually composed of coarse sand and low waves. On the Queensland coast, sandy beaches require waves to be less than 0.5 m to be reflective. For this reason they are usually found inside the entrance to harbours and estuaries and at the lower energy end of some ocean beaches, and in lee of many of the reefs and rock platforms that front many central Queensland beaches. As a result, reflective beaches occur both along the more protected section of the south-east coast, as well as in lower wave environments such as Harvey and Moreton Bays. Figure 2.9 illustrates two reflective beaches.

In Queensland there are 29 reflective beaches. Nineteen occur along the south-east coast, and another ten in Harvey and Moreton Bays. They tend to be shorter, aver-

Figure 2.8   Low tide terrace beaches are characterised by a continuous attached bar which will be exposed on spring low tides. (a) Woorim Beach on Bribie Island usually consists of a low tide terrace, seen here at low tide; (b) the southern end of Fraser Island grades in this view from a rhythmic bar and beach, to transverse bar and rip, to an irregular low tide terrace (foreground).

Low tide terrace beach hazards

- High tide - deep water close to shore; behaves like a reflective beach.
- Low tide - waves may plunge heavily on the outer edge of the bar, with deep water beyond. Take extreme care if body surfing or body boarding in plunging waves, as spinal injuries can result.
- Mid tide - more gently breaking waves and waist deep water, however weak mini rips return some water seaward.
- Higher waves - mini rips increase in strength and frequency and may be variable in location.
- Oblique waves - rips and currents are skewed and may shift along the beach, causing a longshore and seaward drag.
- Most hazardous at mid to high tide when waves exceed 1 m and are oblique to shore, such as during a strong summer north-east sea breeze.

- Summary: One of the safer beach types when waves are below 1 m high, at mid to high tide. Higher waves, however, generate dumping waves, strong currents and ephemeral rips; called 'side drag', 'side sweep' and 'flash' rips by lifesavers.

Reflective beach hazards

- Steep beach face - may be a problem for toddlers, the elderly and disabled people.
- Relatively strong swash and backwash - may knock people off their feet.
- Step - causes a sudden drop off from shallow into deeper water.
- Deep water - absence of bar means deeper water close into shore, which can be a problem for non-swimmers and children.
- Surging waves and shorebreak - when waves exceed 0.5 m, they break increasingly heavily over the step and lower beach face. They can knock unsuspecting swimmers over. If swimming seaward of the break, swimmers may experience problems returning to shore through a high shorebreak.
- Most hazardous when waves exceed 1 m and shorebreak becomes increasingly powerful.
- Where fronted by a rock platform or reef, additional hazards are associated with the presence of the rock/reef.

- Summary: Relatively safe under low wave conditions, so long as you can swim. Watch children as deep water is close to shore. Hazardous shorebreak and strong surging swash under high waves (> 1 m).

---

aging 3.3 km in length in the south-east and only 250 m elsewhere. Therefore, in terms of total beach length, they represent only 3.3% (66 km) of the east Queensland sandy coast.

Reflective beaches are a product of both coarser sand and lower waves. Consequently, all beaches composed of gravel, cobble and boulders are always reflective, no matter what the wave height.

Reflective beaches always have a steep, narrow beach and swash zone. Beach cusps are commonly present in the upper high tide swash zone. They also have no bar or surf zone as waves move unbroken to the shore, where they collapse or surge up the beach face.

Their beach morphology is a product of four factors. Firstly, low waves will not break until they reach relatively shallow water (< 1 m); secondly, the coarse sand results in a steep gradient beach and relatively deep nearshore zone (> 1 m); thirdly, because of the low waves and deep water, the waves do not break until they reach the base of the beach face; finally, because the waves break at the beach face, they must expend all their remaining energy over a very short distance. Much of the energy goes into the wave swash and backwash; the rest is reflected back out to sea as a reflected wave, hence the name reflective.

The strong swash, in conjunction with the usually coarse sediment, builds a high, steep beach face. The *cusps* which often reside on the upper part of the beach face are a product of sub-harmonic edge waves, meaning the waves have a period twice that of the incoming wave. The edge wave period and the beach slope determine the edge wave length, which in turn determines the cusp spacing, which on the Queensland coast can range from 20 to 40 m.

Another interesting phenomenon of most reflective beaches is that all those containing a range of sand sizes

have what is called a *beach step*. The step is always located at the base of the beach face, around the low water mark. It consists of a band containing the coarsest material available, including rocks, cobbles, even boulders and often numerous shells. Because it is so coarse, its slope is very steep; hence the step-like shape. They are usually a few decimetres in height, reaching a maximum of perhaps a metre. Immediately seaward of the step, the sediments usually fine markedly and assume a lower slope.

The reason for the step is twofold. The unbroken waves sweep the coarsest sediment continuously toward the beach and the step. The same waves break by surging over the step and up the beach face. However, the swash deposits the coarsest, heaviest material first, only carrying finer sand up onto the beach, then the backwash rolls any coarse material back down the beach. The coarsest material is therefore trapped at the base of the beach face by both the incoming wave and the swash and backwash.

Hazards

The low waves and protected locations that characterise reflective beaches usually lead to relatively safe swimming locations. However, as with any water body, particularly one with waves and currents, there are hazards present that can produce problems for swimmers and surfers.

2.3.9   Determining wave-dominated beach type

The type of wave-dominated beach that occurs along the south-east coast is a function of the modal wave height, which has a maximum of 1.4 m, the wave period, which

Figure 2.9   *Reflective beaches have no bar and waves break at the base of the beach and surge up the usually steep beach face. (a) waves lapping against the beach immediately south of Round Hill; (b) strong wave surging and swash on reflective and cusped Shelly Beach (Caloundra), together with waves breaking over adjoining rock reefs.*

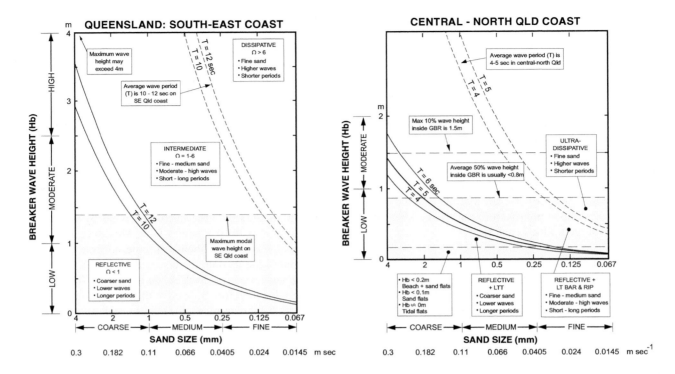

Figure 2.10 *A plot of breaker wave height versus sediment size, together with wave period, that can be used to determine approximate Ω and beach type. Ω = Hb/T Ws where Hb = breaker height (m), T = wave period (sec), Ws = sand fall velocity (m/sec). To use the chart, determine the breaker wave height, period and grain size/fall velocity (mm or cm/sec). Read off the wave height and grain size, then use the period to determine where the boundary of reflective/intermediate, or intermediate/dissipative beaches lies. Ω = 1 along solid T lines, and 6 along dashed T lines. Below the solid lines Ω < 1 and the beach is reflective; above the dashed lines Ω > 6 and the beach is dissipative; between the solid and dashed lines Ω is between 1 and 6 and the beach is intermediate.*

averages 10 seconds, and finally the sand size. While the wave period is essentially constant for the south-east coast, the wave height is at a maximum on exposed locations, decreasing into more protected environments, and sand size can range from fine to coarse sand.

Figure 2.10 provides a method for determining the predicted beach type based on these three parameters. The highest energy beaches, with waves averaging 1.4 m, cannot produce longshore bar-trough or dissipative beaches in Queensland, or New South Wales for that matter. They do however form rhythmic bar and beach systems on beaches composed of fine sand, and transverse bar and rip beaches where sand is a little more medium. Low tide terrace beaches occur where waves have been reduced to around 1 m, and reflective beaches when waves are lowered still further to average about 0.5 m.

2.4   Tide-modified beaches

The majority of Queensland beaches north of Fraser Island are influenced by a combination of waves and tides, becoming tide-modified to in places tide dominated. This is because wave energy is lower. They receive little or no ocean waves, and are exposed to only lower wind waves

generated by the Trades, which average 0.5 to 0.8 m in height and only occasionally reach 1.5 m, as well as being shorter in period and length (Table 1.11). At the same time, tide range increases up the coast, in places reaching to several metres in height (Fig. 1.25). As a consequence, the tide range is commonly several times greater than the wave height. By definition, tide-modified beaches occur when the tide range is between 3 and 15 times the wave height. These conditions produce three beach types along the coast - ultradissipative, reflective plus bar and rips, and reflective plus low tide terrace.

The impact of the lower and shorter waves is to produce a smaller version of the three low energy wave dominated beaches; that is the TBR, LTT and REF. Owing to

the shorter wave period; average 5 sec as opposed to 10 sec on the open coast; when rips do form, their spacing is usually less than 100 m, down to as close as 50 m. At the same time the higher tide range acts to essentially dislocate the surf zone from the swash zone. What this means is that all tide-modified beaches have a relatively steep high tide beach, a wide lower to very low gradient intertidal zone, and only at low tide a surf zone that can

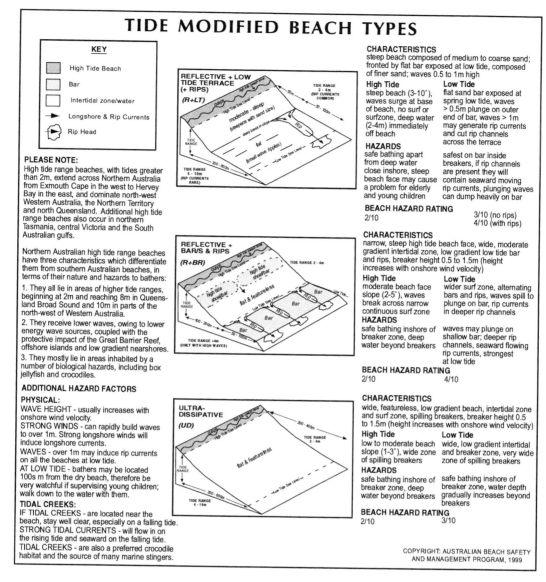

Figure 2.11 *A three dimensional sketch of the three tide-modified Queensland beaches: beach and low tide terrace; beach with low tide bars and rips; and ultradissipative.*

Figure 2.12 *Ultradissipative beaches are exposed to the Trade winds and their waves and are composed of finer sand, resulting is wide, low gradient intertidal beaches, as seen here at Cherry Tree Beach (Cooktown) and Mulambin Beach near Yeppoon; the latter also showing the narrow, steep high tide beach.*

resemble the bars and rips of the open coast. The wide intertidal zone therefore separates the upper 'beach' from the lower 'surf zone' by as much as a few hundred metres. Basically three types of tide-modified beaches occur along the central and northern coasts, generally in areas moderately to well exposed to the prevailing south-east Trades and their associated waves. These are illustrated in Figure 2.11 and described below.

2.4.1 Ultradissipative

Ultradissipative beaches occur in locations exposed to the full force of the south-east wind waves where the beaches are composed of fine to very fine sand. They are characterised by a very wide (200-400 m) intertidal zone, with a low gradient high tide beach, and a very low gradient to almost horizontal low tide beach. Because of the low gradient right across the beach, waves break across a relatively wide, shallow surf zone as a series of spilling breakers (Fig. 2.12). This wide spilling surf zone dissipates the waves to the extent that they are known as 'ultradissipative' beaches. During period of higher waves (>1 m), the surf zone can be well over 100 m wide, though still relatively shallow. There are 81 ultradissipative Queensland beaches, all in lee of the GBR. They average 2.3 km in length and occupy 13% of the GBR beach-coastline.

Basically the fine sand induces the low gradient, while the tide range moves the higher waves backwards and forwards across the wide intertidal zone every six hours. The two act to plane down the beach, while the lack of stationarity or stability of the shoreline precludes the formation of bars and rips.

Hazards

The major hazards associated with ultradissipative beaches are their usually higher waves, the relatively deep water off the high tide beach, the long distance from the shore to the low tide surf, and the often considerable distance from the shoreline out to beyond the breakers. Currents run along the beach when waves arrive at angles, however strong rip currents are generally absent. Seaward of the breakers however, shore parallel tide currents also increase in strength.

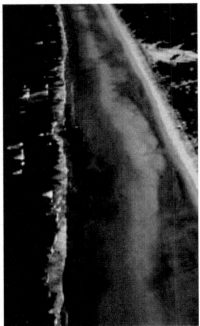

Figure 2.13   *The more exposed Trade wind beaches with fine to medium sand usually have a well developed high tide beach, a moderately wide, low gradient intertidal beach, and transverse to rhythmic bars and rips in the low tide surf zone, as seen here at (a) Nine Mile Beach (Byfields National Park); (b) close-up of Nine Mile Beach with rip highlighted by dye; (c) closely spaced rips along the low tide at Ramsey Bay Beach (Hinchinbrook Island).*

2.4.2   Reflective plus bar & rips

Reflective beaches fronted by low tide bars and rips occur in similar environments to the ultradissipative, only on beaches with fine to medium sand. The slightly coarser sand results in more moderate beach gradients. These beaches are consequently characterised by a relatively steep high tide beach, a moderately wide, lower gradient intertidal beach, which may be featureless or with low sand shoals, and a low gradient low tide surf zone containing either transverse bars and rips, or occasionally rhythmic bars and rips (Fig. 2.13). These beaches occur in 71 locations along the northern-central coast, where they form 8% of the beaches. They average 3.6 km in length and occupy 19% of the sandy shoreline in lee of the reef and 13% of the total east Queensland sandy shoreline.

The morphology of these beaches is in part due to the steeper gradients, which produce the steep high tide beach. At spring high tide, waves usually surge against the beach face with no surf. During mid tide, the waves spills across the lower gradient intertidal zone, while at low tide the cellular circular flow associated with the rips dominates the surf. The surf and rips are able to form on this beach type owing to the period of time (2-4 hr) that the surf zone is roughly located over the low tide bar. This provides the waves with sufficient time to generate a surf zone similar to the intermediate beaches of the wave-dominated coast.

Hazards

As with all tide-modified beaches, the nature of the beach, surf zone and associated hazards changes with the state of the tide as well as the height of the waves. Under normal moderate wave conditions (Hb=1 m) the high tide beach is relatively safe, apart from the surging waves and deeper water off the beach; the mid tide beach has a lower gradient and wider, more dissipative surf zone; while the low tide beach is the most hazardous with more heavily breaking waves, rip feeder and rip currents, as well as being up to 300 m seaward of the high tide shoreline. The rip currents are, however, only active at low tide, while at mid to high tide the rip morphology is simply submerged below the zone of wave shoaling, seaward of the breaker zone.

2.4.3   Reflective plus low tide terrace

The lowest energy of the tide-modified beaches is the reflective high tide beach fronted by a low tide terrace. This beach type occurs in areas afforded slight to moderate protection from the south-east wind waves, and/or when sand is medium to even coarse. These beaches are characterised by the steepest high tide beach, at the base of which is an abrupt change in slope and a wider, low gradient low tide terrace, which may occasionally contain rips (Fig. 2.14).

This is the most common of the tide-dominated beaches, with a total of 372 occurring in lee of the GBR (42%). They average 1.2 km in length and occupy 31% of the GBR beach shoreline, and 23% of the total east coast sandy shoreline, making this the most common beach type south of Cooktown.

At high tide, waves surge up the steep beach face. This continues as the tide falls, until the shallower water of

Figure 2.14   Many of the less exposed and coarser sand beaches consist of a relatively steep high tide beach, with a sharp break in slope and a low tide terrace, as seen here at (a) Trinity Beach at high tide, (b) Bramston and (c) Grasstree beaches at low tide.

the terrace induces wave breaking increasingly across the terrace. At low tide, waves spill over the outer edge of the terrace, with the inner portion remaining dry during spring low tide. If rips are present, they will cut a channel across the terrace. Like the bar and rip beaches, the rips are only active at low tide.

Hazards

This beach undergoes a marked change in morphology between high and low tide. At high tide hazards are associated with the waves surging against the steep high tide beach, together with the deeper water off the beach, while at low tide, waves spill across the broad, shallow terrace, with hazards associated with the deeper water off the terrace, and the rips, if present.

2.5   Tide-dominated beaches

When the tide range begins to exceed the wave height by between 10 and 15 times, then the tide becomes increasingly important in the beach dynamics, at the expense of the waves. Basically there are three beach types - two with a high tide beach and sand flats, while the third is essentially a transition from sand flat to a tidal flat.

These 'beaches' represent a transition from the high tide wave dominated beach to the mid to low tide, more tide dominated tidal sand flats (Fig. 2.15). They still receive sufficient wave energy to both build the high tide beach, remove any fine silts and mud to maintain sand, rather than mud flats, and on some beaches to build multiple sand ridges.

2.5.1   Beach plus sand ridges

The high tide beach fronted by multiple sand ridges occurs in 70 locations along the north-central coast. They occupy 172 km of the coast and account for 8% of the tidally influenced beaches. These systems usually have a moderate to steep high tide beach, with shore parallel, sinuous, low amplitude, evenly spaced sand ridges extending out across the inter- to sub-tidal sand flats (Figs.2.15 & 2.16). These are not to be confused with the higher relief (sand) bars of the higher energy beaches. The beach is only active at high tide when either very low waves or calms prevail. The intertidal zone and ridges are usually inactive. The exact mode of formation is unknown, though it is suspected that the ridges are active and formed during infrequent periods of higher waves acting across the intertidal ridged zone.

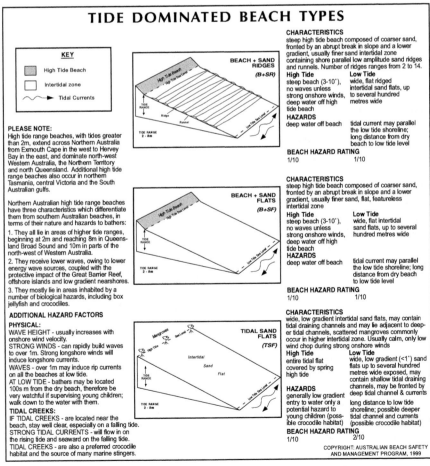

Figure 2.15   *A three dimensional sketch of the three tide-dominated Queensland beaches: beach+ridged sand flats, beach+sand flats and tidal sand flats.*

Figure 2.16   *In protected areas exposed to occasional wave attack, high tide beaches are fronted by wide sand flats containing two to sixteen very low amplitude, parallel to sinuous sand ridges, shown here at (a) Queens Beach, site of Bowen surf club and (b) just north of Clairview in Broad Sound.*

The number of ridges averages 7, but can range up to 19 (Table 2.3). The ridges are very low amplitude, no more than a few centimetres to a decimetre from trough to crest, and average between about 50 m in spacing. They tend to parallel the coast, but can at times lie obliquely to the shore, while at other places they merge into more complex patterns.

Hazards

The major hazard with these low energy beaches is the relatively deep water off the high tide beach and, in places, associated tidal currents. Low tide is dominated by the wide, shallow to exposed ridges and sand flats.

2.5.2   Beach plus sand flats

The most common tide-modified beach is the high tide beach fronted by very low gradient, flat, featureless sand flats, which average 700 m in width (Figs. 2.15 & 2.17). There are 249 such beaches in lee of the GBR, averaging 1.2 km in length, and occupying 21% of the GBR coast, in addition to 22 beaches in the south-east, all located in Hervey and Moreton Bays. They are the second most dominant beach type, after the reflective and low tide terrace. The main difference between the two is that the beach and sand flats are exposed to significantly lower

Table 2.3   Beach and number of sand ridges

Location	Beaches	Ridges	Mean	Range
GBR	98	682	7	19
South-east	1	9	9	0
Islands	2	3	3	1
Total	101	685	7	19

*standard deviation

waves, with calms common, and they are fronted by the wide sand flats, which are usually several hundred metres wider than the LTT.

The only hazards associated with these beaches are the deeper water off the high tide beach and the slight chance of tidal currents off the beach. Low tide reveals a wide, shallow to exposed tide flat.

2.5.3   Tidal sand flats

Sand flats are not by definition 'beaches', however they are included here for two reasons. First, these sand flats represent a gradation from the true high tide beaches of the above two beach types, to the mangrove vegetated tidal flats that dominate the tide dominated sections of the coast from the cape to Hervey Bay. Secondly, a number of sand flats are labelled 'beaches' and known locally as such. The main difference between the beach-sand flats, and tidal-sand flat, is that on tidal-sand flats there is no high tide beach; rather the sand flat maintains its low gradient all the way to the high tide shoreline. In addition, tidal-sand flats always have some mangroves and many of the 35 'beaches' classified as sand flats do in fact have a scattering of mangroves (Figs. 2.15 & 2.18). These 'beaches' occupy over 50% of the coast, though true tidal flats are far more extensive.

Hazards

The only physical hazards associated with tidal sand flats are the deeper water over the flats at high tide, the increased likelihood of tidal currents moving across or parallel to the flats, and their wide distance at low tide, which increases the chance of being caught in the intertidal zone by the rising tide.

Figure 2.17 *The most protected beach consists of a usually steep high tide beach fronted by flat, relatively feature-less intertidal sand flats, with mangroves commonly found at the more protected ends, as shown here at (a) Sarina Inlet, and (b) Dingo Beach, showing the common sharp break in slope and sediment size.*
Hazards

Figure 2.18 *Very low energy tidal sand flats, by defi-nition, have no beach, just a wide expanse of inter-tidal sand flats, seen here in Edgecumbe Bay.*

Figure 2.19 *Beachrock in the intertidal swash zone of Dingo Beach, Abbott Point.*

2.6 Beachrock

A feature of many tropical coasts, including sections of the Queensland coast, is the formation of beachrock. This term is applied to beaches that have undergone ce-mentation of their intertidal sands. Such cementation occurs commonly in the tropics where the prime ingre-dients exist; that is, warm sea water and sediments rich in carbonate (shelly or coral) material. The cementation is a result of carbonate precipitating out in the intertidal zone and forming a cementing agent between the sand grains (Fig. 2.19). Over a period of decades to centuries, the entire intertidal zone can be cemented, replacing the once sandy beach with a solid rock. It occurs on most coral cay beaches and a number of mainland beaches.

2.7 Larger scale beach systems

The beach systems described above are all part of larger scale beach and barrier systems; the barriers including

backing beach and foredune ridges and, in places, sand dunes, as well as adjoining tidal creeks and inlets, and headlands and reefs. Figures 2.20 provides a schematic overview of the typical arrangement of some of these beach and associated barrier systems along the central and northern Queensland coast, in lee of the Great Bar-rier Reef. Because of the predominance of south-east waves and winds, together with the many prominent head-lands and bays, there are many sections of coast with a pronounced gradient in wave-tide energy from south to north. In this figure, beginning in the south, the headland leads to substantial wave refraction and attenuation, such that essentially zero wave energy reaches the southern corner, and tides dominate. The tides produce wide, low gradient tidal mud flats (Fig. 2.20a) in conjunction with a tidal creek channel and ebb tide delta. Over time, sedi-ments delivered by the creek build out the mud flats, which is also impacted by occasional tropical cyclone storm waves and surges that emplace occasional sandy, shelly cheniers on top of the mud flats (b).

Moving up the embayment, wave height begins to increase slightly, resulting in the formation of a high tide sand beach fronted by wide sand flats, which are backed by beach ridges associated with coastal progradation over the sandflats (c). As wave energy continues to increase, the central tidal delta is restricted in its seaward extent by the waves, while the beach to the north becomes a tide-modified beach and low tide terrace. Also, the higher onshore wind energy builds more substantial foredunes in lee of the beach, resulting, as the shoreline progrades, in a series of wind deposited foredune ridges (d).

Toward the northern, south-east facing end of the embayment, where wave and wind energy are at a maximum, the waves produce a beach with low tide bar and rips, or ultradissipative, depending on tide range and sand size, while the strong winds mobilise the sand into massive parabolic dune systems, which in places extend several kilometres inland (e). The northern creek is restricted to draining across the beach, with no tidal influence. Rips might not only dominate the low tide surf, but also form a permanent rip against the northern headland.

Such an idealised sequence is rarely found in entirety along the coast, however parts or variations on this sequence do occur repetitively within all embayments exposed to variable wave conditions. Consequently beach and barrier types can vary considerably, and predictably, within an embayment and along a single beach-tidal flat system.

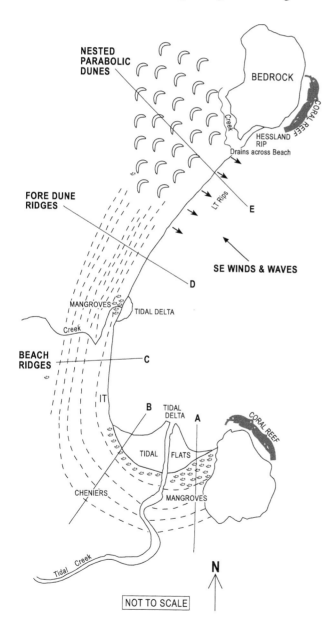

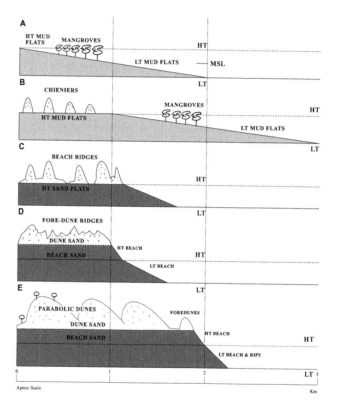

Figure 2.20 *Schematic sketch of typical central-north Queensland embayment, which (left) grades from a very protected southern end dominate initially by tidal flats. As exposure and wave height increase up the shoreline cheniers, then beach ridges back the beach, and as the shoreline swings to face the south east Trades, higher energy beaches baked by transgressive dune systems represent the highest energy of the tide-modified beach systems. Cross-sections of transects A-E shown above.*

3. QUEENSLAND BEACHES AND HAZARDS

3.1 Beach usage and surf life saving

The Australian coast remained untouched until the arrival of the first aborigines some tens of thousands of years ago. The coast and beaches they found, however, no longer exist. Probably crossing to Australia during one of the glacial low sea level periods, they not only reached a far larger, cooler, drier and windier continent, but one where the shoreline was some tens of metres below present sea level. Hence the coast they walked and fished lies out below sea level on the inner continental shelf.

The present Australian coast only obtained its position some 6500 years ago, when sea level rose to approximately its present position. As it was rising, at about 1 m every 100 years, the aborigines no doubt followed its progress, by slowly moving inland and to higher ground. Therefore, we can assume that usage of the present Queensland beaches began as soon as they began forming some 6000 years ago and continued in the traditional way until the 1800's.

As the long term aboriginal inhabitants were removed and dispossessed, the European new-comers found the beaches to be of little use; except for running aground during storms, or loading goods and launching small boats in more sheltered locations. Timber-cutters and fishermen, however, did find (and often occupy) every anchorage suitable for small craft along the coast, within a few decades of settlement. However, the new settlers generally thought the beaches to be economically of little value, the waves a nuisance and hazard to all coastal sailors, and the dunes behind the beaches a waterless, infertile wasteland.

As European settlement spread along the Queensland coast, the beaches were in places used as a more attractive route for foot and horse travel along the coast, particularly compared to hinterland which varied from dense impenetrable rain forests to equally impenetrable mangroves. The tradition of using the beaches for transport continued well into the 19th century, when many of the early coastal 'highways' followed beaches for parts of their routes.

The first profitable use of beaches by European settlers was therefore for transport between suitable settlements and for loading of cargo in protected locations. However, as ports grew up along the coast and coastal shipping and roads improved, even these uses were overtaken by progress and once again beaches were left to the fishermen.

Australia celebrated its bicentenary in 1988 and the Surf Life Saving Association of Australia (now known as Surf Life Saving Australia) celebrated 92 years in 1999; however, for the first 114 years of European settlement, the beaches were unwanted, unpopulated and for a time it was even illegal to swim in daylight hours.

Only in the early 1900's did beach and surf swimming become a popular recreational pastime. The popular beaches of the time were all either within walking distance of residential areas, or serviced by public transport. The public tended to surf at the nearest suitable beach, regardless of whether it was hazardous, as along much of the Sunshine and Gold Coast, or relatively safe, as with most of the central and northern Queensland beaches. As people got into difficulties and drowning occurred the establishment of surf life saving clubs soon followed.

3.1.1 The Surf Life Saving Movement

To understand the formation of the surf life saving clubs and the broader Surf Life Saving Associatin is to realise two things. Firstly, in their rush to the surf, most beachgoers could not swim and had little or no knowledge of the surf and its danger. Secondly, the coast of southern Australia, including south-east Queensland, is as dangerous as it is inviting to the unprepared.

The first club in Queensland was established at Tweed Heads in 1911, though the origins go back to 1909 at Coolangatta Beach. By 1924, 15 clubs were established as far north as Cairns. Approximately half of Queensland surf clubs were established before World War II, the remainder after the war. Table 3.1 lists the date of establishment of all Queensland surf life saving clubs, and Table 3.2 the establishment of the six state branches. Today 57 clubs patrol beaches between Port Douglas in the north and Rainbow Bay in the south, including clubs on Magnetic, Bribie and North Stradbroke Island. In addition, lifeguards patrol most of the beaches with surf clubs, together with 13 additional beaches totalling 70 patrolled beaches. All beaches patrolled by surf life savers and lifeguards, together with the location of north Queensland stinger nets, are listed in Table 3.3.

While people flocked to the beaches and the clubs were forming, all was still not happy with the law. It decreed that not only should both sexes wear neck-to-knee swimming costumes, but they should also cover their lower parts with a skirt. The neck-to-knees law was soon removed. However, four decades were to pass before men, after World War II, were legally allowed to swim topless; and another four decades before women were allowed the same rights in the 1980's.

The first surf life saving clubs and their Association had embarked upon the establishment and growth of an organisation that has become an integral part of Australian beach usage and culture, and through which it is so readily identified internationally.

Table 3.1   Formation of Queensland Surf Lifesaving Clubs

Year established	Surf Lifesaving Club	Branch
1911	Tweed Heads & Coolangatta	Point Danger
1912	Southport	South Coast
1915	Maroochydore	Sunshine Coast
1916	Kirra	Point Danger
1919	Coolum	Sunshine Coast
1919	Burleigh Heads/Mowbray Park	South Coast
1919	Currumbin	Point Danger
1922	Bilinga	Point Danger
1923	Mooloolaba	Sunshine Coast
1924	Cairns	North Queensland
1924	Bundaberg	South Barrier
1924	Alexandra Headland	Sunshine Coast
1924	Metropolitan Caloundra	Sunshine Coast
1924	Surfers Paradise	South Coast
1924	Tugun	Point Danger
1925	Emu Park	South Barrier
1926	Ayr	North Barrier
1926	Yeppoon	South Barrier
1926	Bribie Island	Sunshine Coast
1927	Noosa	Sunshine Coast
1927	Picnic Bay (Magnetic Island)	North Barrier
1928	Forrest Beach	North Barrier
1928	Arcadian (Magnetic Island)	North Barrier
1928	Mackay	North Barrier
1930	Palm Beach	Point Danger
1935	Etty Beach	North Queensland
1935	Mission Beach	North Queensland
1935	Broadbeach	South Coast
1936	Tannum Sands	South Barrier
1945	Mermaid Beach	South Coast
1946	Miami Beach	South Coast
1946	Tallebudgera	Point Danger
1947	Point Lookout (North Stradbroke)	South Coast
1947	Northcliffe	South Coast
1947	Pacific North Palm Beach	Point Danger
1949	North Burleigh	South Coast
1949	North Kirra	Point Danger
1950	North Caloundra	Sunshine Coast
1952	Ellis Beach	North Queensland
1954	Moore Park	South Barrier
1954	Nobbys Beach	South Coast
1955	Hervey Bay District	South Barrier
1958	Kurrawa	South Coast
1958	Coolangatta	Point Danger
1962	Peregian Beach	Sunshine Coast
1963	Rainbow Bay	Point Danger
1965	Elliott Heads	South Barrier
1965	Rainbow Beach	Sunshine Coast
1968	Sarina	North Barrier
1969	Marcoola	Sunshine Coast
1980	Kawana Waters	Sunshine Coast
1982	Sunshine Beach	Sunshine Coast
1984	Port Douglas	North Queensland
1989	Agnes Water	South Barrier
1991	Redcliffe Peninsula	Sunshine Coast
1995	Bowen	North Barrier
1996	Mudjimba	Sunshine Coast

Table 3.2   Queensland Surf Lifesaving Branches

Year established	Branch	Number of Clubs	Patrol season
1924	Point Danger	11	Sept - May
1931	South Barrier	8	Sept - May
1931	Sunshine Coast	15	Sept - May
1936	North Barrier	7	Sept - May
1945	South Coast	11	Sept - May
1950	North Queensland	5	July - May
		57	

Table 3.3   Queensland SLSC & Lifeguard patrolled beaches and stinger nets seasons

Branch/beach	SLSC Patrol Season	Lifeguard Patrol Season	Stinger Net season
North Queensland Branch			
Port Douglas (Four Mile)	July-May	Full-time	Nov-May
Ellis Beach	July-May	Nov-May	Nov-May
Cairns (Palm Cove)	July-May	Full-time	Nov-May
Clifton		Nov-May	Nov-May
Kewarra		Nov-May	Nov-May
Trinity	July-May	Full-time	Nov-May
Yorkeys Knob		Nov-May	Nov-May
Holloways		Sept-May	Nov-May
Brampston		Peak only	Nov-May
Etty Bay	July-May	Christmas,Easter	Nov-May
Ku*rrimine Beach*			Nov-May
Mission Beach	July-May	Christmas,Easter	Nov-May
North Barrier Branch			
Forrest Beach	Sept-May	Christmas,Easter	Nov-May
Arcadian (Alma Bay)	Sept-May	Christmas,Easter	
Picnic Bay	Sept-May	Christmas Easter	Nov-May
The Strand	Sept-May	All Year	Nov-May
Ayr (Alva Beach)	Sept-May	Christmas	
Bowen (Queens Beach)	Sept-May		Nov-May
Lamberts		Christmas, Easter	
Mackay (Harbour Beach)	Sept-May	All Year	Nov-May
Sarina	Sept-May	Christmas,Easter	Nov-May
Blacks		Christmas,Easter	Nov-May
Eimeo		Christmas,Easter	Nov-May
South Barrier Branch			
Yeppoon	Sept-May	Christmas,Easter	
Emu Park	Sept-May	Christmas,Easter	
Tannum Sands	Sept-May	School Hols	
Agnes Water	Sept-May	Christmas,Easter	
Moore Park	Sept-May	School Hols	
Oaks		School Hols	
Bundaberg (Bargara/Neilsons Park)	Sept-May	School Hols	
Kellys		School Hols	
Elliott Heads	Sept-May	School Hols	
Hervey Bay	Sept-May	School Hols	
Sunshine Coast Branch			
Surfair Resort		School Hols	
North Shore Camp		On Request	
Rainbow Beach	Sept-May	School Hols	
Noosa west		Full-Time	
Noosa Heads	Sept-May	Full-Time	
Sunshine Beach	Sept-May	Full-Time	
Sunrise		Wkend,Hols	
Peregian Beach	Sept-May	Full-time	
Hyatt Coolum		Full-time	
Coolum Beach	Sept-May	Full-time	
Marcoola	Sept-May	School Hols	

Australian Beach Safety Management Program

Maroochydore	Sept-May	Full-time
Alexandra Headland	Sept-May	Full-time
Mooloolaba	Sept-May	Full-time
Kawana Waters	Sept-May	Season
Mudjimba SLSC		Season
Twin Waters		Full-time
North Caloundra	Sept-May	Season
Caloundra (Kings Beach)	Sept-May	Full-time
Bulcock Beach		Holidays
Bribie Island (Woorim)	Sept-May	Nov-Feb

South Coast Branch

Point Lookout	Sept-May	School Holidays
Redcliffe Peninsula	Sept-May	
Southport	Sept-May	
Surfers Paradise	Sept-May	
Northcliffe	Sept-May	
Broadbeach	Sept-May	
Kurrawa	Sept-May	
Mermaid Beach	Sept-May	
Nobby Beach	Sept-May	
Miami Beach	Sept-May	
North Burleigh	Sept-May	
Burleigh Hd-Mowbray Pk	Sept-May	

Point Danger Branch

Tallebudgera	Sept-May
Pacific	Sept-May
Palm Beach	Sept-May
Currumbin	Sept-May
Tugun	Sept-May
Bilinga	Sept-May
North Kirra	Sept-May
Kirra	Sept-May
Coolangatta	Sept-May
Tweed Heads	Sept-May
Rainbow Bay	Sept-May

Gold Coast Lifeguards	
Couran Cove	365 days
45 Seaway	Sept-April+hols
42 Seaworld	Sept-April
41 Shearton	365 days
40 Main Beach	365 days
39 Breaker Street	Sept-April
38 N Narrowneck	Sept-April
37 S Narrowneck	365 days
36 Staghorn	365 days
35 Elkhorn	365 days
34 Surfers Paradise	365 days
33 Clifford Street	Sept-April
32 Northcliffe	365 days
31 Wharf St	365 days
30 Broadbeach	Sept-April+hols
28 Kurrawa	365 days
27 Margaret Street	Sept-April
26 Mermaid Beach	365 days
25 Hilda Street	Sept-April
24 Seashell Avenue	Sept-April
23 Nobby's Beach	365 days
21 Miami	365 days
20 North Burleigh	365 days
19 Fourth Avenue	365 days
18 Burleigh Heads	365 days
17 Tallebudgera Ck	Sept-April+hols
16 Tallebudgera	Sept-April
15 Pacific	Sept-April+hols
14 Palm Beach	Sept-April+hols
13 Currumbin Alley	Sept-April+hols
10 Currumbin	365 days
9 Tugun	Sept-April
7 Bilinga	Sept-April+hols
5 North Kirra	Sept-April
4 Kirra	Sept-April+hols
3 Coolangatta	365 days
2 LifeguardCentre	365 days
1 Rainbow Bay	365 days

Now, nearly a century after the initial rush to the beaches and the foundation of the early surf life saving clubs, both beach usage and the 269 surf life saving clubs around the coast are accepted as part of Australian beaches and beach life. However, as we near the 21st century, beach usage is undergoing yet another surge as the Australian population and visiting tourists increasingly concentrate on the coast. This is resulting in more beaches being used, more of the time by more people, many of whom are unfamiliar with beaches and their dangers. In addition, most of the newly used beaches have no surf life saving clubs or patrols.

There is now a greater need than ever to maintain public safety on these beaches, a service that is provided on patrolled beaches by volunteer life savers and professional lifeguards. This book is the result of a joint Surf Life Saving Australia, Surf Life Saving Queensland and the Coastal Studies Unit project that is addressing this problem. The book is designed to provide information on each and every ocean beach in Queensland between Cooktown and Coolangatta, including beaches on 18 islands; in all, 1061 beaches. It contains information on each beach's general characteristics and suitability for swimming, surfing and fishing. In this way, swimmers may be forewarned

---

Oldest and Youngest Clubs

The oldest surf lifesaving club in Queensland is Tweed Heads & Coolangatta, formed in 1911, while the youngest is Mudjimba, formed 85 years later in 1996.

---

of potential hazards before they get to the beach and consequently swim more safely.

If you are interested in joining a surf life saving club or learning more about surf life saving, contact Surf Life Saving Australia, your state centre (listed below) or nearest surf life saving club.

3.1.3 Physical beach hazards

Beach hazards are elements of the beach-surf environment that expose the public to danger or harm. *Beach safety* is the mitigation of such hazards and requires a combination of common sense, swimming ability and beach-surf knowledge and experience. The following section highlights the major physical hazards encountered in the surf, with hints on how to spot, avoid or escape from such hazards, followed by the biological hazards.

There are seven major physical hazards on Queensland beaches:

1. water depth (deep and shallow)
2. breaking waves
3. surf zone currents (particularly rip currents)
4. tidal currents
5. strong winds
6. rocks, reefs and headlands
7. water temperature

In the surf zone, three or four hazards, particularly water depth, breaking waves and currents, usually occur together (Fig. 3.1). In order to swim safely, it is simply a matter of avoiding or being able to handle the above when they constitute a hazard to you, your friends or children.

3.1.3.1 Water depth

Any depth of water is potentially a hazard. Water greater than chest depth can drown the non-swimmer or inexperienced, while water too shallow can cause injury to the unwary.

Knee depth water can be a problem to a toddler or young child. Chest depth is hazardous to non-swimmers, as well as to panicking swimmers. In the presence of a current, it is only possible to wade against the current when water is below chest depth. Be very careful when water begins to exceed waist depth, particularly if younger or smaller people are present, and if in the vicinity of rip or tidal currents.

Shallow water is also a hazard when people are diving in the surf or catching waves. Both can result in spinal injury if people hit the sand head first.

Surf Life Saving Australia 128 The Grand Parade Brighton-Le-Sands NSW 2216 Phone (02) 9597 5588 Fax (02) 9599 4809 Email: info@slsa.asn.au	**Surf Life Saving Queensland** PO Box 3747 South Brisbane QLD 4101 Phone (07) 3846 8000 Fax (07) 3846 8008 Email: slsq@lifesaving.com.au
Surf Life Saving New South Wales PO Box 430 Narrabeen NSW 2101 Phone: (02) 9984 7188 Fax: (02) 9984 7199 Email: experts@surflifesaving.com.au	**Surf Life Saving Victoria** A W Walker House Beaconsfield Parade St Kilda VIC 3182 Phone (03) 9534 8201 Fax (03) 9534 0311 Email: slsv@slsv.asn.au
Surf Life Saving South Australia PO Box 82 Henley Beach SA 5022 Phone (08) 8356 5544 Fax (08) 8235 0910 Email: surflifesaving@surfrescue.com.au	**Surf Life Saving Tasmania** GPO Box 1745 Hobart TAS 7001 Phone: (03) 6231 5380 Fax: (03) 6231 5451 Email: slstedo@trump.net.au
Surf Life Saving Western Australia PO Box 1048 Osborne Park Business Centre WA 6017 Phone (09) 9244 1222 Fax (09) 9244 1225 Email: slswa@slswa.asn.au	**Surf Life Saving Northern Territory** PO Box 43096 Casuarina NT 0811 Phone (08) 8941 3501 Fax (08) 8981 3890 Email: slsnt@topend.com.au

Australian Beach Safety Management Program

Water Depth
- Safest: knee deep - can walk against a strong rip current
- Moderately safe: waist deep - can maintain footing in rip current
- Unsafe: chest deep - unable to maintain footing in rip current

Remember: what is shallow and safe for an adult can be deep and distressing for a child.

Shallow water hazards

Spinal injuries are usually caused by people surfing dumping waves in shallow water, or by people running and diving into shallow water.

To avoid these:
- Always check the water depth prior to entering the surf.
- If unsure, WALK, do not run and dive into the surf.
- Only dive under waves when water is at least waist deep.
- Always dive with your hands outstretched.

Also
- Do not surf dumping waves.
- Do not surf in shallow water.

a)   b)   c)   d)

Figure 3.1   *Queensland's physical beach hazards include deep water, breaking waves, rips and tidal currents and rocks and reefs. In this series of photographs, (a) the beach at Red Rock has low waves but deep water right off its reflective beach face; (b) oblique waves at North Spit, Fraser Island produce a strong drag down the beach; (c) breaking waves and a deep tide channel at Caloundra; (d) rocks and reefs make even a low energy beach potentially dangerous, Red Cliff Point (Cairns).*

3.1.3.2 Breaking waves

As waves break, they generate turbulence and currents which can knock people over, drag and hold them under water, and dump them on the sand bar or shore. If you do not know how to handle breaking waves (as most people don't), stay away from them; stay close to shore and on the inner part of the bar.

If you are knocked over by a wave, remember two points – the wave usually holds you under for only two to three seconds (though it may seem like much longer), therefore do not fight the wave; you will only waste energy. Rather, let the wave turbulence subside, then return to the surface. The best place to be when a big wave is breaking on you is as close to the seafloor as possible. Experienced surfers will actually 'hold on' by digging their hands into the sand as the wave passes overhead, then kick off the seabed to speed their return to the surface.

If a wave does happen to gather you up in its wave bore (white water) it will only take you towards the shore and will quickly weaken, allowing you to reach the surface after two to three seconds, and usually leave you in a safer location than where you started.

3.1.3.3 Surf zone currents and rip currents

Surf zone currents and particularly *rip currents* are the biggest hazards to most swimmers. They are the hardest for the inexperienced to spot and when swimmers are caught by them, they can generate panic. See the following section on rips, on how to identify and escape from rip currents.

The problem with currents, particularly rip currents, is that they can move you unwillingly around the surf zone and ultimately seaward (Fig. 3.2). In moving seaward they will also take you into deeper water and possibly toward and beyond the breakers. As mentioned earlier, currents are manageable when the water is below waist level, but as water depth reaches chest height they will sweep you off your feet.

3.1.3.4 Tidal currents

Much of the Queensland coast is dominated by high tide ranges (>2 m) and consequently substantial tidal currents. Tidal currents tend to parallel the coast offshore, flowing north with the flooding (rising) tides, and south with the ebbing (falling) tides. However their direction is also

---

Breaking waves and wave energy
Surging waves - safe when low
· break by surging up beach face; usually less than 50 cm high
· can be a problem for children and elderly, who are more easily knocked over
· become increasingly strong and dangerous when over 50 cm high

Spilling waves - relatively safe
· break slowly over a wide surf zone
· are good body surfing waves

Plunging (dumping) waves - the most dangerous wave
· break by plunging heavily onto sand bar
· strong wave bore (white water) can knock swimmers over
· very dangerous at low tide or where water is shallow
· waves can dump surfers onto sandbar, causing injury
· most spinal injuries are caused by people body surfing or body boarding on dumping waves
· to avoid spinal injury, do not surf dumping waves or in shallow water; if caught by a wave
   do not let it dump you head first; turn sideways and cover your head with your arms
· only very experienced surfers should attempt to catch plunging waves

---

Wave energy ≈ square of the wave height
wave energy represents the amount of power in a wave of a particular height

0.3 m wave	=	1 unit wave energy/power
1.0 m wave	=	11 units
1.5 m wave	=	25 units
2.0 m wave	=	44 units
2.5 m wave	=	70 units
3.0 m wave	=	100 units

Therefore, a 3 m wave is 10 times more powerful than a 1 m wave and 100 times more powerful than a 0.3 m wave.

---

modified by obstacles such as islands, reefs and head-lands, and by all inlets, and river and creek mouths. Tides must flow into and out of every coastal entrance twice a day, and in doing so generate strong constricted currents flowing in and out of every inlet and creek on the coast, which also maintain deeper tidal channels (Fig. 3.1c). As most settlements are located on or near rivers and inlets, and many beaches are located on or adjacent to inlets and river mouths, these strong currents and their deep channels are a very real hazard on all beaches located close

to inlets. They are particularly hazardous on a falling tide as the currents flow seaward.

When swimming or even boating in tidal creeks, always check the state and direction of the tide and be prepared for strong currents. Swimmers should not venture beyond waist deep water. In addition, as mentioned later, croco-diles do inhabit most Queensland estuaries north of Yeppoon and are an additional biological hazard in these areas.

Rip and surf zone current velocity

Breaking waves   travel at 3-4 m/sec   (10-15 km/hr)
Wave bores (white water) travel at 1-2 m/sec   (3-7 km/hr)
Rip feeder and longshore currents travel at 0.5 - 1.5 m/sec (2-5 km/hr)
Rip currents under average wave conditions (< 1.5 m high) attain maximum velocities of 1.5 m/sec = 5.4 km/hr
(Note: Olympic swimmers can swim at 7 km/hr)
An average rip in a surf zone 50 m wide can carry you outside the breakers in as little as 30 seconds.

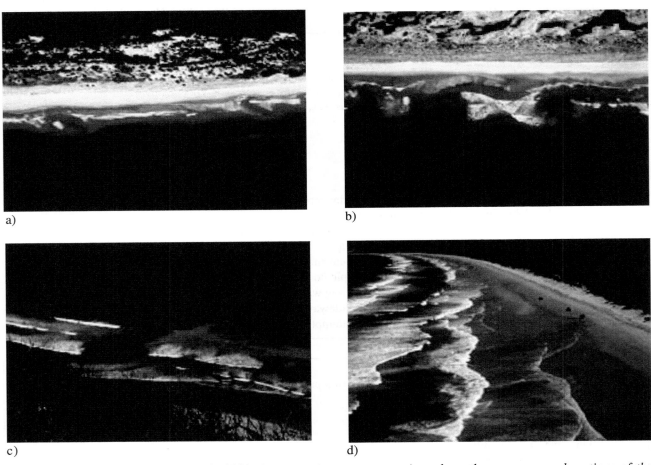

a)   b)

c)   d)

Figure 3.2   *There are approximately 3000 rips operating at any one time along the more exposed sections of the Queensland coast. These photographs illustrate a number of typical rips. Well developed rip channels and currents usually lie between shoaler bars as shown at: (a) inner bar and rips, with outer parallel bar (North Stradbroke Island); (b) highly rhythmic bar and beach with two rips (Moreton Island); (c) skewed rip flowing between two attached bars (Cooloola); (d) a narrow skewed rip flowing across a near continuously attached bar (Rainbow Beach).*

3.1.3.5 Strong winds

Strong winds can be a major hazard on exposed beaches. *Onshore winds* will help pile more water onto the beach and increase the water level at the shore. They also produce more irregular surf, which makes it more difficult to spot rips and currents. During tropical cyclone conditions, very strong winds can also generate a storm surge, raising the level of the sea by up to a few metres.

Longshore winds, particularly strong south-east winds, will cause wind waves to run along the beach, with accompanying longshore and rip currents also running along the beach. The waves and currents can very quickly sweep a person along the beach and into deeper rip channels and stronger currents.

Offshore winds tend to clean up the surf. They are generally restricted to winter westerly conditions in the south-east. However, if you are floating on a surfboard, bodyboard, ski or wind surfer, it also means it will blow the board offshore. In very strong offshore winds, it may be difficult or impossible for some people to paddle against this wind.

3.1.3.6 Wind generated waves and currents

In lee of Fraser Island and the Great Barrier Reef, the beaches are largely protected from ocean swell and dependent on wind blowing between the island or reef and the mainland to generate waves. As a rule of thumb, calms result in no waves; light winds build low waves up to 0.3 m high; moderate winds build waves up to 0.5 m; strong Trade winds build waves up to 1 m and more; and only gale force winds associated with tropical cyclones build waves up to 3 m and more. Winds along most of the central and northern coast are related to the generally light to moderate Trade winds and sea breeze, both of which tend to blow from the south-east, through to a lesser percentage from the north east (Figs. 1.15 & 1.16). As a result, most of the waves arrive from the south-east through east and on exposed beaches average between 0.5 and 1 m high. The net result is a low to at best moderate energy surf and a predominantly northerly drift along most of Queensland's beaches, a phenomenon very evident on beaches from the Tweed River to Cape York (Figs. 1.3 & 2.1b).

3.1.3.7 Rocks, reefs and headlands

Many Queensland beaches have some rocks, rock or coral reefs and headlands. These pose problems on higher energy beaches because they cause additional wave breaking; generate more (and stronger) rips; and have hard and often dangerous surfaces. When they occur in shallow water and/or close to shore they are also a danger to people walking, swimming or diving because of the hard seabed, and the fact they may not be visible from the surface.

Rocks, reefs and headlands

- if there is surf against rocks or a headland, there will usually be a rip channel and current next to the rocks
- rocks and reefs can be hidden by waves and tides, so be wary
- all coral reefs will be submerged at high tide
- do not dive or surf near rocks, as they generate greater wave turbulence and stronger currents
- rocks often have sharp, shelled organisms growing on their surface which can inflict additional injury
- if walking or fishing from rocks, be wary of being washed off by sets of larger waves

---

Remember these points:
- **Do** swim on patrolled beaches.
- **Do** swim between the red and yellow flags.
- **Do** swim in the net enclosure (where present).
- **Do** observe and obey the instructions of the lifesavers or lifeguards.
- **Do** swim close to shore, on the shallow inshore and/or on sand bars.
- **Always** have at least one experienced surf swimmer in your group.
- **Never** swim alone.
- **Do** swim under supervision if uncertain of conditions.
- **Do Not** enter the surf if you cannot swim or are a poor swimmer.
- **Do Not** swim or surf in rips, troughs, channels or near rocks.
- **Do Not** enter the surf if you are at all unsure where to swim or where the rips are.
- **Be Aware** of hypothermia caused by exposure to cold air and water, particularly on bare skin, and with small children. **Wind** will add to the chill factor.

3.1.3.8   Safe swimming

The safest place to swim is on a patrolled beach, between the red and yellow flags. If there are no flags then stay in the shallow inshore or toward the centre of attached bars, or close to shore, if water is deep. However, remember that rip feeder currents are strongest close to shore, and rip currents depart from the shore. The most hazardous parts of a beach are in or near rips and/or rocks, outside the flags or on unpatrolled beaches, and alone.

PATROLLED BEACHES
- swim between the red and yellow flags
- obey the signs and instructions of the lifesavers or lifeguards
- still keep a check on all the above, as over 1200 people are rescued from patrolled beaches in Queensland each year

UNPATROLLED BEACHES
- always look first and check out the surf, bars and rips

- select the safest place to swim, do not just go to the point in front of your car or the beach access track
- try to identify any rips that may be present
- select a spot behind a bar and away from rips and rocks
- on entering the surf, check for any side currents (these are rip feeder currents) or seaward moving currents (rip currents)
- if they are present, look for a safer spot
- it's generally safer to swim at low tide, if you avoid the rips

Children
- NEVER let them out of your sight
- ADVISE them on where to swim and why
- ALWAYS accompany young children or inexperienced children and teenagers into the surf
- REMEMBER they are small, inexperienced and usually poor swimmers and can get into difficulty at a much faster rate than an adult

---

MODAL WAVE HEIGHT AND MODAL HAZARD RATING FOR QUEENLAND BEACHES

WAVE DOMINATED	< 0.5	0.5	1.0	1.5	2.0	2.5	3.0 m	← Wave Height
DD — does not occur in QLD						9	10	
LBT — only occurs as outer bars				7	7			
RBB				6				
TBR			5	6	**MAXIMUM AVERAGE WAVE HEIGHT IN SOUTH - EAST = 1.5m**			**SOUTH EAST COAST**
LTT		3	4					
R	2	3						
TIDE-MODIFIED								
UD		2	4	**MAXIMUM AVERAGE WAVE HEIGHT IN lee of GBR = 1m**				
R+BR		2	3					
R+LT		1	2					**LEE of GBR**
TIDE - DOMINATED								
B+SR	1	1						
B+SF	1		**MAXIMUM AVERAGE WAVE HEIGHT PROTECTED LOCATIONS = < 0.5m**					
TSF	1							

Figure 3.3   *A plot listing the ten Queensland beach types, their modal or average wave conditions and associated modal beach hazard rating. See pages 34-49 for full description of each beach type.*

3.1.4   Beach Hazard Rating

The *beach hazard rating* refers to the scaling of a beach according to the physical hazards associated with its beach type together with any local physical hazards. It ranges from the low, least hazardous rating of 1 to a high, most hazardous rating of 10. It does not include biological hazards. These are discussed later in the chapter. The beach characteristics and hazard rating for wave dominated, tide modified and tide dominated beaches are shown in Figures 2.3, 2.4, 2.11 and 2.15.

Each beach, depending on its beach type and typical wave conditions, will have a *modal beach hazard rating*, that typifies that beach. Figure 3.3 plots both the wave height required to generate the ten Queensland beach types, as well as the hazard rating associated with those beach type-wave conditions. The hazard rating ranges from a high of 6 on rip dominated RBB and TBR beaches exposed to waves averaging 1.5 m, to a low of 1 on tide-dominated sand flats.

Figure 3.4 shows several beaches between Bilinga and Duranbah which, while composed of similar sediment, have varying orientations and exposure to the dominant southerly waves. As wave height varies between the beaches, so too does the nature of the beach and surf zone, and consequently the beach type and beach hazard rating, which ranges from a low of 4 to a high of 7.

As the modal hazard rating refers to average wave conditions, any change in wave height, length or direction will change the surf conditions and accompanying hazards. This will in turn change the beach's hazard rating. Figure 3.5 provides a matrix for calculating the actual or *prevailing beach hazard rating* for wave dominated, tide-modified and tide-dominated Queensland beaches. It assumes you know the beach type and can estimate the breaker wave height. With these two factors, the prevailing beach hazard rating can be read off the chart. This figure also describes the general hazards associated with each beach type as wave conditions rise and fall.

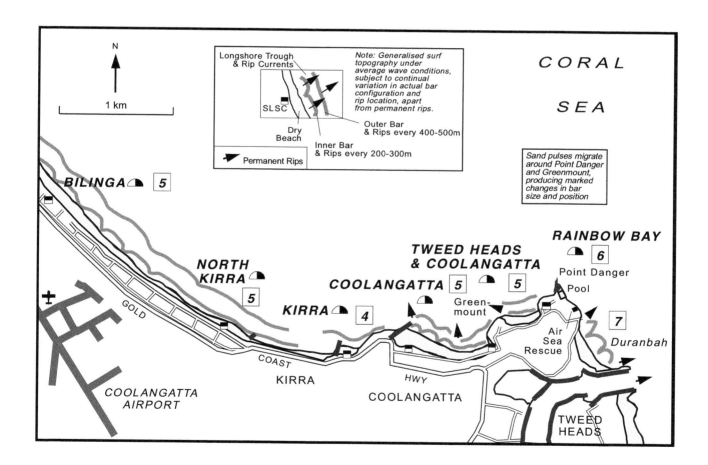

Figure 3.4   *The beaches between Tweed Heads and Bilinga range in orientation from east to north, and in exposure to the dominant southerly waves. Duranbah receives higher waves and has strong, persistent rips and a beach hazard rating of 7, whereas the beaches to the north have some protection from the waves, and consequently a lower wave height and rating between 4 and 5.*

BEACH HAZARD RATING GUIDE
Impact of changing breaker wave height on hazard rating for each beach type
Wave Dominated Beaches

BEACH TYPE \ WAVE HEIGHT	< 0.5 (m)	0.5 (m)	1.0 (m)	1.5 (m)	2.0 (m)	2.5 (m)	3.0 (m)	> 3.0 (m)
Dissipative	4	5	6	7	8	9	10	10
Long Shore Bar Trough	4	5	6	7	7	8	9	10
Rhythmic Bar Beach	4	5	6	6	7	8	9	10
Transverse Bar Rip	4	4	5	6	7	8	9	10
Low Tide Terrace	3	3	4	5	6	7	8	10
Reflective	2	3	4	5	6	7	8	10

Tide Modified Beaches
(at high tide - at low tide add 1)

	< 0.5 (m)	0.5 (m)	1.0 (m)	1.5 (m)	2.0 (m)	2.5 (m)	3.0 (m)
Ultradissipative	1	2	4	6	8	10	10
Reflective + Bar & Rips	1	2	3	5	7	9	10
Reflective + LTT	1	1	2	4	6	8	10

Tide Dominated Beaches
(at high tide - at low tide add 1)

	< 0.5 (m)	0.5 (m)	1.0 (m)		
Beach + Sand Ridges	1	1	2	Waves unlikely to exceed 0.5 - 1m	
Beach + Sand Flats	1	1		*Note: if adjacent to tidal channel, beware of deep water and strong tidal currents.*	
Tidal Sand Flats	1				

BEACH HAZARD RATING

Least hazardous: 1 - 3
Moderately hazardous: 4 - 6
Highly hazardous: 7 - 8
Extremely hazardous: 9 - 10

KEY TO HAZARDS

- Water depth and/or tidal currents
- Shorebreak
- Rips and surfzone currents
- Rips, currents and large breakers

NOTE: All hazard level ratings are based on a swimmer being in the surf zone and will increase with increasing wave height or with the presence of features such as inlet, headland or reef induced rips and currents. Rips also become stronger with falling tide.

BOLD gradings indicate the average wave height usually required to produce the beach type and its average hazard rating.

Figure 3.5   *Matrix for calculating the prevailing beach hazard rating for wave dominated, tide-modified and tide-dominated beaches, based on beach type and prevailing wave height and, on tide-modified beaches, state of tide.*

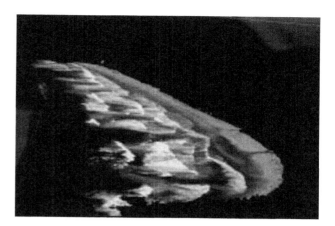

Figure 3.6   *The impact of changing wave conditions is illustrated here at Alexandria Bay during (a) lower waves and well developed transverse bars and rips (hazard rating 5), and (b) higher waves and larger but less well defined rips (hazard rating 6).*

Beach Hazard Ratings

> 1 - least hazardous beach

> 10 - most hazardous beach

Beach hazard rating is the scaling of a beach according to the physical hazards associated with its beach type and local beach and surf environment.

Modal beach hazard rating is based on the beach type prevailing under average or modal wave conditions, for a particular beach type (Figs 2.3, 2.4, 2.11, 2.15) or beach.

Prevailing beach hazard rating refers to the level of beach hazard associated with the prevailing wave and beach conditions on a particular day or time (Fig. 3.5).

Table 3.4 summarises the rating for all Queensland beaches.
Note: **Biological hazards** are not included in this rating system.

Table 3.4   Queensland beaches, by beach hazard rating at (a) high tide and (b) low tide (excluding wind hazard) for the south-east coast (SE), the mainland coast in lee of the Great Barrier Reef (GBR) and the islands of the Great Barrier Reef (GBR I).

a.

Beach Hazard Rating HT*	No. of beaches			% of total beaches			Mean length (km)			Total length (km)			Proportion of length of beaches (%)			
	SE	GBR	GBRI	SE	GBR	GBRI	SE	GBR	GBRI	SE	GBR	GBRI	SE	GBR	GBRI	Total
1	52	635	62	43	72	100	2.9	1.2	0.6	151	760	37	25	55	100	47
2	12	249	0	10	28	-	3.2	2.5	-	38	612	-	6	44	-	32
3	7	3	0	6	<1	-	2.7	3.0	-	19	9	-	3	1	-	1
4	7	0	0	6	-	-	7.2	-	-	50	-	-	8	-	-	3
5	11	0	0	9	-	-	3.2	-	-	35	-	-	6	-	-	2
6	30	0	0	25	-	-	10.7	-	-	320	-	-	52	-	-	16
7	2	0	0	2	-	-	0.1	-	-	0.2	-	-	<1	-	-	-
8	1	0	0	1	-	-	0.03	-	-	0.03	-	-	-	-	-	-
9	0	0	0	-	-	-	-	-	-	-	-	-	-	-	-	-
10	0	0	0	-	-	-	-	-	-	-	-	-	-	-	-	-
TOTAL	122	887	62	100	100	100	5.0	1.6	0.6	613	1381	37	100	100	100	100

b.

Beach Hazard Rating LT*	No. of beaches			% of total beaches			Mean length (km)			Total length (km)			Proportion of length of beaches (%)			
	SE	GBR	GBRI	SE	GBR	GBRI	SE	GBR	GBRI	SE	GBR	GBRI	SE	GBR	GBRI	Total
1	20	280	32	16	32	52	2.3	1.4	0.6	45	394	19	7	29	51	23
2	33	352	26	27	40	42	3.3	1.1	0.6	109	379	16	18	27	43	25
3	17	216	4	14	24	6	2.8	2.7	0.5	48	577	2	8	42	5	31
4	7	38	0	6	4	-	7.4	0.8	-	52	31	-	9	2	-	4
5	12	1	0	10	<1	-	2.3	0.2	-	25	0.2	-	4	-	-	1
6	23	0	0	19	-	-	9.3	-	-	223	-	-	36	-	-	11
7	9	0	0	7	-	-	12.4	-	-	112	-	-	18	-	-	6
8	1	0	0	1	-	-	0.03	-	-	0.03	-	-	-	-	-	-
9	0	0	0	-	-	-	-	-	-	-	-	-	-	-	-	-
10	0	0	0	-	-	-	-	-	-	-	-	-	-	-	-	-
TOTAL	122	887	62	100	100	100	5.0	1.6	0.6	613	1381	37	100	100	100	100

* Beach Hazard Rating: 1 = least hazardous; 10 = most hazardous

Local factors must also be considered in calculating the prevailing beach hazard, particularly hazards such as rocks, reefs, inlets and headlands, and the presence of strong winds. These will generally increase the rating by 1.

To illustrate how one beach can change both beach type and beach hazard rating as a result of changing wave height and/or wave direction, Figure 3.6 shows Alexandria Bay Beach under two contrasting wave conditions and associated rip characteristics and hazard ratings. All beaches will change their beach and surf zone morphology and dynamics in response to changing wave (and tide) conditions, resulting in a high degree of variation in the prevailing, as opposed to the average, beach hazard rating. For this reason Figures 2.3, 2.4, 2.11 and 2.15 should be used as a guide only, with Figure 3.5 providing a more accurate assessment of the prevailing conditions and hazards.

Queensland beaches predominantly have a low hazard rating, particularly the lower energy beaches on the north and central coast. However the popular south east coast is dominated by moderately to occasionally very hazardous beaches. The main reason for the higher south east rating is the combination of higher waves and rip currents. The following section provides additional information on rip currents and their hazards.

3.1.5 Rip identification - how to spot rips

To the experienced surfer rips are not only easy to spot, but they are the surfer's friend, providing a quick way (and at times the only way) to get out the back; as well as carving channels to produce better waves. To the inexperienced however, rips are not only unknown or 'invisible' to them, but if caught in one it can be a terrifying and even fatal experience. Most recreational swimmers and visitors do not have the time or desire to become experienced swimmers and surfers. In order to assist them, a check list of features that indicate a rip or rips are present on the beach, is noted below:

3.1.5.1 Rips - how to escape if caught in one

- If the water is less than chest deep, adults should be able to **walk out** of a rip. Conversely, avoid going into deeper water. So if you are in any surf current, become very careful once the water exceeds waist depth. Get out while the water is shallow.
- Most people rescued in rips are **children**. Never let them out of your sight and if they get into difficulties, go to them immediately while the water is still relatively shallow.
- As long as you can swim or **float**, the rip will not drown you. There is no such thing as an undertow associated with rips, or for that matter, with surf zone currents. Only breaking waves can drive you under water. Most swimmers caught in rips drown because they panic. So **stay calm**, tread water, float and conserve your energy.
- If there are people/life savers on the beach, raise one arm to **signal** for assistance.
- **Do not** try to swim or wade in deep water directly against the rip, as you are fighting the strongest current; there are easier ways out.
- Where possible, **wade rather than swim**, as your feet act as an anchor and help you fight the current.
- If it is a relatively weak and/or shallow rip, **swim or wade sidewards to the nearest bar**. Once on the bar, walk or let the waves or wave bores return you to shore.
- If it is a strong and/or deep rip, go with it through the breakers. Do not panic. When **beyond the breakers**, slowly and calmly swim alongshore in the direction of the nearest bar, indicated by heavier wave breaking. If you are not a surfer, simply wait for a lull in the waves, swim into the break and allow the waves to wash you to shore. Stay near the surface so the broken wave can wash you shoreward. Do not dive under the waves as they will wash over you.

Rip Current Spotting Check List

Note: any one of these features indicates a rip, but not all will necessarily be present.
* indicates always present
+ indicates may be present

* **A seaward movement of water** (Fig. 3.2) either at right angles to or diagonally across the surf. To check on currents, watch the movement of the water or throw a piece of driftwood or seaweed into the surf and follow its movement.

* Rips only occur when there are **breaking waves** seaward of the beach. If water is moving shoreward, it must return seaward somewhere.

* **Disturbed water surface** (ripples or chop) above the rip; caused by the rip current as it pushes against incoming waves and water. May be difficult to spot.

\+ Longshore **rip feeder channels** and/or currents running alongshore, hard against the base of the sloping beach face. Rips are usually supplied by two rip feeder channels, one on either side of the rip.

\+ **Rhythmic or undulating beach topography**, with the rips located in the indented rip embayments.

\+ **Areas where waves are not breaking**, or are breaking less across a surf zone, owing to the deeper rip channel.

\+ **A deep channel or trough**, usually located between two bars or against rocks. The channel may contain inviting clear, calmer water compared to the adjacent turbulent surf on the bars. However, do not be fooled. In the surf, calm water usually means it's deeper and contains currents.

\+ **A low point in the bar** where waves are not breaking, or break less. This is where the rip channel exits the surf zone.

\+ **Turbid, sandy water** moving seaward; either across the surf zone and/or out past the breakers.

\+ In the rip feeder and rip channel, the stronger currents produce a **rippled seabed**. These ripples are called megaripples and are sandy undulations up to 30 cm high and 1.5 m long. If you see or can feel large ripples on the seabed, then strong currents are present, so stay clear.

\+ If you see one rip on a **long beach**, there will be more if wave height remains the same along the beach. Rip spacing can vary from 150 to 500 m, depending on wave conditions.

\+ If there is surf and **rocks, reef or a headland**, a rip will always flow out close to or against the rocks. These rips are often permanent and have local names like the 'express', the 'accelerator', the 'garbage bowl', etc.

- To summarise: stay calm, swim sideways toward breakers or the bar and let the broken waves return you to shore. Raise an arm to signal for help if people are on the beach.
- If rescued, thank the rescuer.

3.2   Surf Safety

Surfing, as opposed to swimming, requires the surfer to go out to and beyond the breakers, so he or she can catch and ride the waves; in other words, go surfing. This can be done using just your body (body surfing) or a range of surfboards, bodyboards and skis.

Surfing safely requires a substantially greater knowledge of the surf, compared to swimming on the bar or close to shore. The following points should be observed before you begin to surf:

1. You must be a strong swimmer.
2. You must also be experienced at swimming in the surf.
3. You must be able to tell if and when a wave will break.
4. You must know the basics of how breaking waves and the surf zone operate. You should be able to **spot rip currents**.

5. Equally, you should know what **hazards** are associated with the surf, including breaking waves, rips, reefs, rocks and so on.
6. You must only use **equipment** that is suitable for you, i.e. the right size and level.
7. You must know how to use your equipment; whether it be flippers, bodyboard, surfboard or wave ski.
8. You should use **safety equipment** as appropriate, including a legrope or handrope, wetsuit, flippers and in some cases a helmet.
9. You should ensure your **equipment is in good condition**, with no broken fibreglass, frayed legrope, etc.

3.2.1.1   Some tips on safe surfing:

- Remember surfing is fun, but it is also hazardous.
- **Never surf alone**. If you get into trouble, who will help you?
- Before you enter the water, always **look at the surf for at least five minutes**. This will enable you to first, gauge the true size of the sets, which may come only every few minutes; and secondly, besides picking out the best spot to surf, you can also check out the breaker

pattern, channels, currents and rips; in other words, the circulation pattern in the surf. This is important as you can use this to your advantage.

- On patrolled beaches, **observe the flags**, surfboard signs and directions of the life savers. Do not surf between the red and yellow flags or near a group of swimmers.
- If you are surfing out the back, the safest, quickest and usually the easiest way to get out is via a rip. This is because the water is moving seaward, making it easier to paddle; the rip flows in a deeper channel, resulting in lower waves; and the rip will keep you away from the bar or rocks where waves break more heavily.
- Once out past the breakers, particularly if paddling out in a rip, move sidewards and position yourself behind the break.
- Buy and read the **Surf Survival Guide**, published by Surf Life Saving Australia and available at all newsagents.
- Obtain an **SLSA Surf Survival Certificate** from your school.

3.2.1.2   Some general tips:

Surfers conduct many rescues around the Australian coast, so be prepared to assist if required. Remember, if you are on a surfboard, bodyboard or wave ski and have a legrope and wetsuit, you are already kitted out to perform rescues. The board is a good flotation device that can be used to support someone in difficulty. The wetsuit will keep you warm and buoyant and thereby give you more energy and flotation to assist someone in distress; and the legrope or board can be used to tow someone in difficulty, while you paddle them toward safety.

The simplest way to get someone back to shore is to lay them on your board while you swim at the side or rear of the board, and let the waves wash the board, patient and you back to shore.

Some surfing hazards to watch out for when paddling out:

```
* Heavily breaking/plunging waves, particularly the lip
  of breaking waves.
* Rocks and reefs.
* Strong currents, particularly in big surf.
* Other surfers and their equipment.
```

...when you are surfing:

```
* Other surfers - the surfer on the waves has more
  control and is responsible for avoiding surfers
  paddling out, or in the way.
* Heavily breaking waves.
* Your own and other surfboards. They can and do hit
  you, and can knock you out.
* Shallow sand bars.
* Rocks and reefs.
* Close-out sets and big surf.
```

...when returning to shore:

```
* Heavy shorebreaks.
* Rocks and reefs.
* Strong longshore/rip feeder currents.
```

Remember: The greatest danger to surfers is to be knocked out and drown. Most surfers are knocked out by their own boards or by hitting shallow sand or rocks. This can be avoided by always covering your head with your arms when wiped out, by wearing a wetsuit for flotation, by wearing a helmet and by surfing with other surfers who can render assistance if required.

3.3   Rock fishing safety

The rocky sections are the most hazardous part of the Queensland coast, with most fatalities due to fishermen drowning after being washed off the rocks.

Rock fishing is hazardous because:
- Deep water lies immediately off the rocks, often containing submerged reefs and rocks, and heavily breaking waves.
- Occasional higher sets of waves can wash unwary fishermen off the rocks.
- Most fishermen are not prepared or dressed for swimming, as they are often wearing heavy waterproof clothing, shoes and tackle.
- Many fishermen are not experienced surf swimmers, and some cannot even swim.

To minimise your chances of joining this distressing statistic, two points must be heeded. Firstly, avoid being washed off and secondly, if you are washed off, make sure you know how to handle yourself in the waves until you can return to the rocks or await rescue.

The biggest problems usually occur when inexperienced fishermen are washed off rock platforms. To compound the problem, they either cannot swim or are not prepared for a swim. You only need to watch experienced board and body surfers surfing rocky point and reef breaks, to realise rocks are not a serious hazard to the experienced and the properly equipped.

So the rules are:-
1. Before you leave home:

- Check the weather forecast. Avoid rock fishing in strong winds and rain.
- Phone the boat or surf forecast and check the wave height. Avoid waves greater than 1 m.
- Check the tide state and time. Avoid high and spring tides.
- Are you suitably attired for rock fishing?
- Are you suitably attired in case of being washed off the rocks?

- A loose fitting wet suit is both comfortable and warm; and it will keep you afloat and protected if washed off the rock platform.

2. Before you start fishing:

- Check the waves for ten minutes, particularly watching for bigger sets.
- Choose a spot where you consider you will be safe.
- When choosing a spot to fish: if the rocks are wet, then waves are reaching that spot; if the rocks are dry, waves are not reaching them, but may if the tide is rising or wave height is increasing.
- Ensure you have somewhere to easily and quickly retreat to, if threatened by larger waves.
- Place your tackle box and equipment high and dry.

3. When you are fishing:

- Never turn your back on the sea, unless it is a safe location.
- Watch every wave.
- Be aware of the tide; if it is rising, the rocks will become increasingly awash.
- Watch the waves, to check for:
  - increasing wave height, leading to more hazardous conditions;
  - the general pattern of wave sets; it is the sets of higher waves that usually wash people off rocks.
- Remember, 'freak waves' exist only in media reports. No waves are freak; all that happens is that a set of larger waves arrives, as any experienced fisherman or surfer can tell you. These larger sets are likely to arrive every several minutes.
- Do not fish alone; two can watch and assist better than one.
- If you see a larger set of waves approaching - retreat. If you cannot retreat, lie flat and attach all your limbs to the rock. Forget your gear; you are more valuable. As soon as the wave has passed, get up and retreat.
- Wear sensible clothing. A wetsuit provides warmth, protection and safety; particularly if you are washed off or knocked over.

4. If you are washed off, here are some hints:

- If you have sensible clothing; that is, clothing that will keep you buoyant, such as a wetsuit or life jacket; then you should do the following:
- Head out to sea away from the rocks, as they are your greatest danger.
- Abandon your gear, it will not keep you afloat.
- Take off any shoes or boots and you will be able to swim better.
- Tread water and await rescue, assuming there is someone who can raise the alarm.
- If you are alone, or can only be saved by returning to the rocks, try the following:

- Move seaward of the rocks and watch the waves breaking over the rocks in the general area, then:

Choose a spot where there is either:

> a **channel** - this may offer a safer, more protected route;
>
> a **gradually sloping rock** - if waves are surging up the slope, you can ride one up the slope, feet and bottom first, then grab hold of the rocks as the swash returns;
>
> or a **steep vertical face** with a flat top reached by the waves - swim in close to the rocks, wait for a high wave that will surge up to the top of the rocks, float up with the wave, then grab the top of the rocks and crawl onto the rock as the wave peaks. As the wave drops, you can stand or crawl to a safer location.

> **Fishing Information:** The *Queensland Recreational Fishing Guide* is published by the Department of Conservation and Natural Resources and is available for free from all DCNR offices.

3.4 Coastal biological hazards

The major reference for biological hazards discussed in this book is:

Venomous and Poisonous Marine Animals - a Medical and Biological Handbook, 504 pp., edited by J A Williamson, P J Fenner, J W Burnett and J F Rifkin, published by University of New South Wales Press, Sydney, 1996, ISBN 0 86840 279 6. Available through UNSW Press or the Medical Journal of Australia.

This is an excellent and authoritative text, which provides the most extensive and up-to-date description and illustrations of these marine animals and the treatment for their envenomation.

Biological hazards pose a risk to beach-goers in northern Australia owing to the prevalence of a greater number and abundance of venomous and poisonous marine animals. Whereas poisonous animals may cause illness in victims, venomous animals are capable of causing fatalities. This section briefly summarises some of the types of poisonous marine organisms together with statistics on the actual number of incidents involving illness and fatalities. For more thorough information, see the above book.

3.4.1 Bluebottles

Bluebottles and other stinging jellyfish are the most common cause of first aid treatment on Queensland and all Australian beaches (Table 3.5). In summer, the warm tropical waters of the East Australian Current and onshore winds can deliver them in their thousands to the coast and beaches. The tentacles of the bluebottle contain hundreds of minute, poisonous, pressure sensitive harpoons,

Table 3.5 Number of jellyfish stings on Australian beaches 1987-1994

Jellyfish	number	percent	average # stings/yr
Physalia Bluebottle	9776	85	1400
Cyanea 'hair jelly'	821	7	120
Catostylus 'blubber'	739	6	110
Carukia 'Irukandji'	119	1	20
Chironex 'box jellyfish'	36	0.3	5
	11491	99.3	1650

Source: Williamson et al., 1996

Table 3.6 Risk period for Chironex 'box jellyfish' in northern Australia

Lat	Place	July	Aug	Sept	Oct	Nov	Dec	Jan	Feb	Mar	Apr	May	June
12^0	Darwin	x	x	x	x	x	x	x	x	x	x	x	x
16^0	Cairns					x	x	x	x	x	x	x	x
19^0	Townsville						x	x	x	x	x	x	
21^0	Mackay						x	x	x	x	x	x	
23^0	Rockhampton						x	x	x	x			

Source: Williamson et al., 1996

Table 3.7 Jellyfish fatalities

Location	number
New Guinea	2
Borneo	3
Philippines	9
Malaya	1
Total	*15*
Box jellyfish fatalities in Australia	
East Queensland	30
Gulf & Torres Strait	5
NT - Darwin	8
NT - Islands	9
NT - Arnhemland	10
Total Australia	*62*

Source: Williamson et al., 1996

Table 3.8 Cone shell fatalities

Location	number
Australia (north Qld)	1
India	1
New Caledonia	1
Fiji	1
Okinawa	4
New Hebrides	1
Total	*9*

Source: Williamson et al., 1996

Table 3.9 Summary of poisonous marine fatalities in Australia

Cause	number
Box jellyfish	61
Blue-ringed octopus	2
Stingray	2
Cone shell	1
Ciguatera (fish poisoning)	1
Puffer fish	14
Total	*81*

Source: Williamson et al., 1996

Table 3.10 Irukandji sting records

	Number of stings reported	Percentage of stings
Total	301	
morning	45	18%
afternoon	213	82%
Males	185	63%
Females	110	37%
Children under 10 years	42	14%
Fine weather	60	79%
Cloudy weather	16	21%
Site of stings (Total)	234	
Trunk	92	39%
Leg	60	26%
Arms	58	25%
Head/neck	24	10%
State		
Queensland	238	80%
Northern Territory	35	11%
Western Australia	28	9%

Source: Williamson et al., 1996

which upon contact are fired and injected into the skin. They immediately appear as small white beads, which soon swell into a red welt. The pain is felt instantly and usually lasts for about an hour, while the welts may remain for a few days. If stung by a bluebottle swim to shore, remove any tentacles (either pick off with the fingers or wash off with salt (sea) water), seek first aid from the life savers and stay calm for the next hour. Recommended first aid is to pack ice or cold packs on the stung area. Do not apply vinegar.

3.4.2 Box jellyfish

Most local beach users in northern Australia are familiar with the type and occurrence of marine stingers. Table 3.6 indicates their likely period of occurrence of the deadly box jellyfish across northern Australia. They occur year round along the north coast of the Territory (12⁰S), decreasing to between December to March by Rockhampton (23⁰S).

In addition to the more well known and feared box jellyfish, there are, however other marine animals that can cause fatalities. Table 3.7 lists the fatalities in Australia and south-east Asia due to jellyfish and box jellyfish, Table 3.8 lists fatalities due to cone fish, while Table 3.9 lists all other marine animals that have caused fatalities in Australia. While box jellyfish are the major risk, so too are puffer fish, and also lethal but less frequent are the blue ringed octopus, stingray, cone shell and ciguatera (fish poisoning).

While much attention is focused on the occasional fatalities due to marine animals, far more people are affected by illness stemming from contact with various marine organisms. Table 3.10 lists the relative frequency or likelihood of fatalities caused by marine animals, together with those that are likely to cause illness due to infection, wounds and poison. By far the most common are due to cuts and abrasions, together with stings from poisonous, but not venomous, animals.

Table 3.5 lists the cause of all jellyfish stings in Australia. Bluebottles which spread down into the more populous southern states are by far the major causes, averaging 1400 stings (85%) each year, likely a very conservative figure, as most stings are unreported. Fortunately the more deadly box jellyfish, an inhabitant of less populated areas, stings only five people on average each year. This low number is due in part however to the efforts of surf life saving clubs across northern Australia who maintain stinger enclosures, encourage the wearing of protective lycra swimming suits and educate the public in the risks associated with coastal swimming.

Table 3.10 provides an overview of the nature of Irukandji stings across northern Australia. Most stings occur on the most populated Queensland coast, where most people stung are male (82%), with most stings occurring in fine weather (when more people go to the beach), and the stings more likely on the upper body (head, neck, arms and trunk - 74%).

In order to provide additional protection for north Queensland swimmers, seasonal stinger enclosures (Fig. 3.7) are located at 14 beaches listed in Table 3.11.

3.4.3 Sharks, crocodiles and sting rays

Sharks are the most feared fish in the sea and in most years they attack one or two victims around the Australian coast. On the east coast, the frequency of attacks has been reduced since the introduction of meshing in 1937. In Queensland, meshed beaches are located at Cairns, Townsville, Mackay, Yeppoon, Tannum Sands, Bundaberg, Rainbow Beach, and along the Sunshine and Gold Coasts.

Saltwater crocodiles inhabit most rivers and estuaries north of Yeppoon, and extreme care must be taken when near crocodile habitats. While crocodiles predominantly live in estuarine and stream environments, they can and do regularly move along the coast, occasionally landing on beaches. Basically you should never swim in rivers and estuaries north of Yeppoon, and use extreme care in known crocodile areas. Most people still use the beaches north of Yeppoon, particularly in winter, however care must still be taken, as crocodiles will occasionally land on most beaches.

Stingrays are a common resident of the surf zone, where they usually lie hidden below a veneer of sand, feeding on molluscs and crabs. If disturbed, they speed off in a cloud of sand. However, if you are unfortunate enough to stand directly on one, it might spear your leg with its sharp, serrated barb located below its tail. In extreme circumstances, the barb can lodge in the leg and require surgery to remove it. Fortunately, this occurrence is uncommon.

Table 3.12 summarises the relative frequency of fatalities and illness due to all marine animals. As can be seen, more fatalities occur as a result of eating poisonous animals than due to stings. Similarly with illness; most are a result of cuts and abrasions, followed by bluebottle stings.

The general rules in avoiding dangerous marine animals in north-central Queensland are:
- Avoid known crocodile areas
- Swim at beaches meshed for sharks
- Only swim in stinger enclosures during the stinger season (see Table 3.11)
- Wear a lycra suit during the stinger season
- Treat all cuts and abrasions in the ocean as potentially infectious

- Look along the swash line for evidence of stranded stingers, bluebottles, jellyfish, etc. If present, stay on the beach.
- If stung, follow recommended treatment guidelines

Table 3.11   *Location of North Queensland stinger nets*

Beach	Surf Life Saving Club	Lifeguard
Port Douglas (Four Mile)	X	X
Ellis	X	X
Palm Cove	X	X
Kewarra		X
Clifton		X
Kewarra		X
Trinity	X	X
Yorkeys Knob		X
Holloways		X
Brampston	X	X
Etty Bay	X	X
Kurrimine		X
South Mission	X	X
Forrest	X	X
Picnic Bay	X	X
The Strand		X

Table 3.12   *Relative frequency of fatalities and illnesses from marine animals (most to least frequent)*

Fatalities	Illness
Pufferfish poisoning	Coral cuts, abrasions. stings
Shellfish poisoning	Jellyfish stings
Box jellyfish stings	(esp. bluebottle)
Ciguatera poisoning	Ciguatera poisoning
Turtle flesh poisoning	Spiny fish stings
Sea snakes	Other marine animal
Spine fish injuries	poisoning from ingestion
Bluebottle stings	

Source: Williamson et al., 1996

Figure 3.7   *Stinger enclosure have been a feature of northern and central Queensland beaches for decades. Some of the older timber enclosures remain, as at Halliday Bay (a), while most of the patrolled beaches have the newer floating, mobile nets as seen at Etty Bay (b) and off Picnic Beach, Magnetic Island (c).*

4. QUEENSLAND BEACHES

4.1 How to find a beach in this book

This remainder of this book contains a description of every beach on the Queensland coast arranged in sequential order from north (Cooktown) to south (Coolangatta). All beaches are located by number on 12 regional maps, while individual beach maps and photographs are provided of all beaches patrolled by surf clubs and/or lifeguards.

To find a beach, you can use any of four systems:
1. By using the alphabetical **BEACH INDEX** at the rear of the book.
2. By location on the coast; using either the **REGIONAL MAPS** of the particular section of coast, or by following the beaches up or down the coast until you find the beach.
3. By name of the **SURF LIFE SAVING CLUB**, if the beach has one. If it differs from the beach name, then both are listed in the **BEACH INDEX**.
4. If you are a surfer and only know the name of the surfing break, which may differ from the beach name, then use the **SURF INDEX**.

Queensland Regional Maps

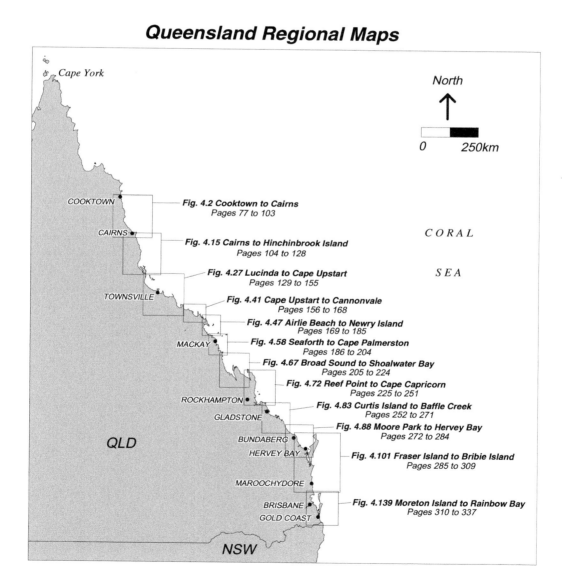

Figure 4.1   Map of Queensland showing the coverage of the 12 regional maps and their location in this book

What the beach description tells you

The description of each beach contains the following information:

- **Patrolled** or unpatrolled by lifesavers or lifeguards
- Patrol months on Patrolled beaches
- **No.** Beach number: 641 to 1601 on the coast, plus 110 beaches on 18 islands
- **Beach** (name) or names (or Beach number in no known name)
- **Surf Life Saving Club**: name of surf lifesaving club (if present)
- **Lifeguards**: lifeguard patrol hours (if present)
- **Rating** (beach hazard rating): 1 to 10,
  at HT (High tide) and LT (low tide)
- **Type**: see pages 34 to 50 for full explanations

    Wave dominated beach types (see pages 34 to 44 for explanation)
D	dissipative	
LBT	longshore bar & trough	(also as outer bar)
RBB	rhythmic bar & beach	(also as outer bar)
TBR	transverse bar & rip	(also as outer bar)
LTT	low tide terrace	
R	reflective	

    Tide-modified beach types (see pages 44 to 48 for explanation)
UD	ultradissipative
R+BR	reflective+bar & rips
R+LT	reflective+low tide bar

    Tide-dominated beach types (see pages 48 to 50 for explanation)
B+SR	(HT) beach+sand ridges
B+SF	(HT) beach+sand flats
TSF	tidal sand flats (no beach)

    other beach comments: TF - tidal flat (mud)
    sediment size (cobbles, boulders)
    tidal creek
    tidal shoals
    coral reef
    rocks/rock reef
    inlet

- **Length**   length of beach

- **Inner bar**: refers to inner sand bar, close to or attached to the beach
- **Outer bar**: refers to the outermost sand bar on double bar beaches, lying seaward of the inner bar

Comments on beach location, access, length, beach and surf conditions, including beach type and presence and location of rips.

Specific comments are made for most beaches on:
- **Swimming** - suitability and safety
- **Surfing** - good surfing spots
- **Fishing** - presence and location of gutters and holes
- **Summary** - general overview of beach

4.2 Surf lifesaving club patrol dates and hours

In Queensland, all surf lifesaving clubs voluntarily patrol the beach on weekends and public holidays over summer. The 57 surf lifesaving clubs are established on 31 of the 961 Queensland beaches.

Patrol dates

Patrols commence in July (North Queensland) or September and run each weekend until May (see Table 3.2).

Patrol days and hours

- Saturday   7 am to 5 pm
- Sunday   7 am to 5 pm
- Public holidays   7 am to 5 pm

Note: Times may vary in regional areas (see Table 3.2)

4.3 Lifeguards and Contract Lifesavers

Surf Life Saving Queensland and Gold Coast City lifeguards also patrol at 90 locations, including all beaches with surf clubs and many popular are resort beaches The patrol periods vary from the peak summer holiday period (Boxing Day to the end of the school holidays in late January), to 365 days a year on very popular beaches (see Table 3.2).

Most lifeguards patrol between the hours of 10 am and 6 pm.

COOKTOWN TO CAIRNS (Fig. 4.2)

Length of Coast:	254 km
Beaches:	98 (beaches 641-739)
Surf Life Saving Clubs:	Port Douglas, Ellis Beach, Cairns (Palm Cove)
Lifeguards:	Clifton Beach, Kewarra Beach, Trinity Beach, Yorkeys Knob, Holloway Beach
Major towns:	Cooktown, Mossman-Port Douglas, Cairns region

Regional Map 1: Cooktown to Cairns

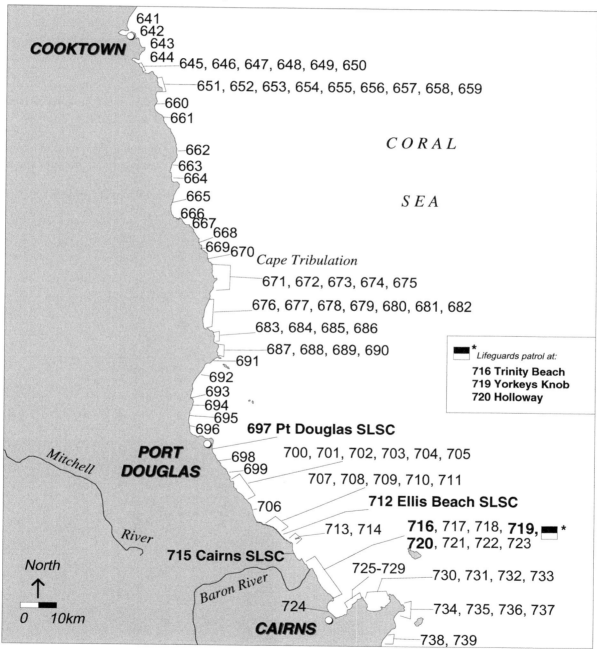

Figure 4.2   *Regional map 1 Cooktown to Cairns, beaches 641 to 739. Beaches with Surf Life Saving Clubs indicated in bold, beaches with lifeguard patrol only indicates by flag.*

Cooktown

Cooktown is a town of 1600 people located on the southern banks of the Endeavour River. It is famous as the location where Cook's ship the Endeavour was careened to repair a damaged hull between 17 June and 3 August 1770. The town itself followed more than 100 years later emerging as a port during the Palmer River gold rush, booming quickly to have a population of 35 000 by 1876. However the decline of the gold saw the population dwindle to 400 in the 1880's. Today Cooktown is again growing, owing to the increasing tourist trade and slowly improving road access. Cooktown is an historic, attractive and very interesting town with good facilities. It overlooks the river mouth and is backed by slopes rising 200 m to Grassy Hill, where Cook surveyed the surrounding waters. While it is not safe to swim in the river, there are two small beaches immediately east of the town which offer safe ocean swimming.

641 CHERRY TREE BAY

No.	Beach	Unpatrolled			
		Rating HT	LT	Type	Length
641	Cherry Tree	2	3	UD	200 m

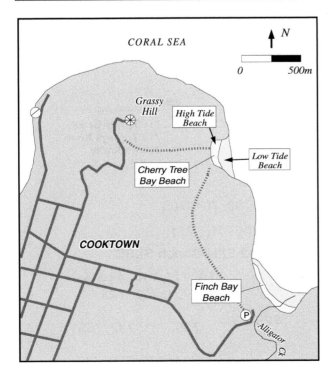

Figure 4.3   *Cooktown's two small ocean beaches are located at Cherry Tree Bay and Finch Bay. Two hundred metre long Cherry Tree can only be reached on foot, while a road runs out to Finch Bay. Both beaches have wide low tide sand bars.*

There are two swimming beaches at Cooktown; the secluded Cherry Tree Bay and more accessible and popular Finch Bay (Fig. 4.3). Cherry Tree Bay is an attractive small beach (641), 200 m long, lying at the foot of Captain Cook's lookout on 200 m high Grassy Hill. It faces east and picks up low waves generated by the Trades. The beach can only be reached on foot, or by boat. There are two walking tracks, one leading from the Grassy Hill Road down the slopes for 400 m to the beach, the other a 600 m walk over the hill from the Finch Bay car park. For this reason it is usually deserted. It is, however, worth the walk. It is bounded by high rounded granitic boulders which become progressively exposed at low tide. There is a dry, narrow beach at high tide, which is partly shaded by overhanging trees, while at low tide the low beach is as wide as it is long (Fig. 2.12a).

Swimming: This is a relatively safe beach, particularly at high tide. When the Trades are blowing, waves up to 1 m high break across a wide, shallow surf zone.

Surfing: There are waves when you have a strong Trade wind blowing, however they tend to be short and sloppy.

Fishing: The rocks on either side provide the best spots at high tide.

Summary: This is a picturesque beach and worth the walk if you want some solitude and some small waves to play in.

642 FINCH BAY

No.	Beach	Unpatrolled			
		Rating HT	LT	Type	Length
642	Finch Bay	1	2	UD+tidal ck	500 m

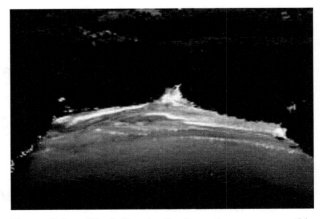

Figure 4.4   *Finch Bay is Cooktown's most accessible and popular beach. While waves are usually low, strong tidal currents flow out of Alligator Creek and the tidal shoals result in variable water depth off the beach.*

Finch Bay is 2 km by road from downtown Cooktown, and with vehicle access and a small car park, it is Cooktown's swimming beach. The beach is 500 m long and faces north-east. It is partly protected from the Trade winds by high, densely vegetated, granite headlands. Alligator Creek crosses the centre of the beach, with its tidal channel tending to meander across the southern half of the beach. The car park is beside the creek, with a short walk to the beach (Fig. 4.4). There is a narrow dry beach at high tide, fronted by sand flats up to 500 m wide at low tide. These are flat on the northern half of the beach, and cut by the tidal creek on the southern half. There is also a small climbing sand dune behind the northern end.

Swimming: This is the most popular swimming beach in Cooktown, however be careful near the tidal creek and channel, as the water is deeper and contains strong tidal currents. It is best to only swim on the northern half.

Surfing: Usually flat, with a low short wave when the Trades are strong.

Fishing: A popular spot to fish into the creek on a rising tide, or off the rocks at high tide.

Summary: Cooktown's favourite beach, even if it straddles Alligator Creek. It is the most accessible beach and offers some protection from the Trade winds.

643 QUARANTINE BAY

| No. | Beach | Unpatrolled | | |
		Rating HT LT	Type	Length
643	Quarantine Bay	2 3	cobble R +LTT-SF	1 km

Quarantine Bay is located 8 km by road from Cooktown and 4 km off the Cooktown Road. It is a 1 km long, northeast facing beach, lying in lee of the densely vegetated 100 m high Monkhouse Point, that protrudes 1 km seaward of the southern end of the beach (643). The point together with Mount Cook, which rises 415 m in lee of the beach, surround the beach with steep, tropical slopes. A road crosses a saddle toward the southern end, giving good access to the beach and a few beach houses that back the central and northern section of the beach. There are no facilities.

The beach is backed by dense vegetation. It has a steeper high tide beach that contains a mixture of sand and gravel, while at low tide a low gradient bar grades into wide tidal shoals toward the southern half. Polished granite boulders dot the beach, while the granite rocks of the headlands fringe each end.

Swimming: Best at high tide when there is relatively deep water off the beach. At low tide, wide, shallow sand flats front the beach.

Surfing: Usually only low wind waves and no surf.

Fishing: Best at high tide.

Summary: A quiet beach with good access but no facilities, and some interesting boulders strewn along the beach.

644 GOLF COURSE

| No. | Beach | Unpatrolled | | |
		Rating HT LT	Type	Length
644	Golf Course	2 3	R+LTT	3.5 km

On the south side of Monkhouse Point is a relatively straight, 3.5 km long, east facing beach (**644**) that runs down to the mouth of the Annan River. The Cooktown Golf Course occupies the low dunes behind the northern section of the beach, with some of the greens lying within 100 m of the beach. The beach can be accessed via the golf course, or along a 4WD track that runs along the base of Monkhouse Point to the northern end of the beach, and then runs down the length of the soft beach to the river mouth.

The beach is relatively narrow and steep at high tide, while at low tide waves break across a continuous narrow bar. A foredune and some vegetated blowouts back the beach, while the southern end protrudes into the river mouth as a 500 m long spit. In addition, an extensive tidal delta fronts the southern 1 km of beach, while the deep river channel with strong tidal currents lies off the tip of the spit.

Swimming: A relatively safe beach during normal low wind waves, with deeper water off the beach at high tide, and a small surf zone breaking over the bar at low tide. Be very careful at the river mouth because of the strong tidal currents and saltwater crocodiles.

Surfing: Usually only low wind waves, with a small break over the bar at mid to low tide.

Fishing: The river mouth is the most popular spot, while the beach can be fished at mid to high tide.

Summary: A relatively undeveloped beach frequented more by golfers and fishers than swimmers. It faces straight into the Trades and is very windy when they are blowing.

645 ANNAN RIVER (S)

No.	Beach	Rating HT LT	Type	Length
		Unpatrolled		
645	Annan River (S) 2	3	B+SF	600 m

On the south side of the **Annan River** mouth is a 600 m long beach and spit (645) that is tied to the northern slopes of Mount McIntosh. There is no road access to the beach which is best reached by boat down the river. The beach itself is relatively straight with a narrow curved spit at the northern tip, a low rocky outcrop at the southern end, and tidal sand shoals extending up to 1 km off the beach, which are exposed to varying degrees at low tide. The deep river channel lies immediately off the northern tip. The beach is backed by 300 m wide, low, open dunes and the mangrove-fringed Esk River, with joins the Annan at the mouth.

Swimming: You can only swim here at high tide as the sand flats are exposed at low tide. However be careful, particularly at the river mouth, as there is a deep channel with strong tidal currents and crocodiles.

Surfing: None.

Fishing: Best on the beach at high tide and day-long at the river mouth.

Summary: A low energy, though dynamic, beach that switches from a narrow high tide beach to a very wide, low tidal sand flat, together with the strong tidal flows in the river.

646-649 MT MCINTOSH

No.	Beach	Rating HT LT	Type	Length
		Unpatrolled		
646	Mt McIntosh (1) 1	2	B+SF	0.5 km
647	Mt McIntosh (2) 1	2	R+LTT	1.0 km
648	Mt McIntosh (3) 1	2	R+LTT	0.3 km
649	Mt McIntosh (4) 1	2	R+LTT	0.3 km

Mount McIntosh is a 135 m high, densely vegetated hill, whose crest lies just 300 m from the shore. Its northern and eastern slopes are fringed by four near-continuous sandy beaches, interspersed with numerous rocks and rocky reefs. The beaches and rocks occupy 2 km of shore and form an irregular boundary between the sandy Annan River mouth beaches to the north, and the sandy to muddy Walker Bay to the south. There is no vehicle access to any of the beaches, with boats providing the best access, particularly at high tide. The beaches are moderately protected by Grave Point and the shallow sand flats, resulting in waves averaging less than 0.5 m.

The first beach (**646**) begins immediately south of the southern Annan River sand spit and runs almost due south, curving to follow the trend of the backing slopes. It is a narrow high tide beach and fronted by irregular tidal flats, that extend up to 300 m to the east. The second beach (**647**) is more irregular, trends south-east and comprises a mixture of sand and numerous rocks, together with a few mangroves, fronted by 100 to 200 m wide rocks and sand flats. The southern two beaches (**648 & 649**) are pocket beaches, each 300 m long, facing north-east, and bordered by low rocky promontories, with more rocks scattered along the beach, and 100 m wide sand flats.

Swimming: All four beaches are more suitable for swimming at high tide, as at low tide they are fronted by shallow to bare sand flats.

Surfing: None.

Fishing: Only at high tide when water covers the sand flats.

Summary: Four near-continuous irregular beaches, backed by the rising slopes of Mount McIntosh, fronted by 100 to 300 m wide, shallow sand flats, with rocks scattered along the beach, and sand flats.

650 WALKER BAY

No.	Beach	Rating HT LT	Type	Length
		Unpatrolled		
650	Walker Bay	1 2	R+LTT	1.3 km

Walker Bay is a 2 km wide, north facing bay in lee of Grave Point. The more protected eastern two-thirds of the bay are fringed by mangroves, while in the slightly more energetic western third there is a 1.3 km long, north-east facing sandy beach (650). The beach runs south from the low rocky boundary with the southern Mount McIntosh beach (649), down to where it grades into mangroves and finally tidal flats. The beach receives low waves, less than 0.5 m high, and consists of a narrow high tide beach backed by dense vegetation, with a 50 m wide low tide bar, that increases in width to each end. A small creek drains across the northern end. There is no vehicle access to the beach and no facilities.

Swimming: There is deep, usually calm water off the beach at high tide, and a shallow bar at low tide.

Surfing: None.

Fishing: The rocks at the northern end provide access to slightly deeper water; otherwise the beach is better at high tide.

Summary: A natural, sandy beach running from the low, rocky headland down to the mangrove-fringed bay shore.

651   WALKER BEACH PT (N)

No.	Beach	Unpatrolled Rating HT	Unpatrolled Rating LT	Type	Length
651	Walker Pt (N)	1	2	R+LTT+ rocks	100 m

On the north side of **Walker Point** is a small, 100 m long beach bordered by irregular rock headlands, with a large rock cutting the beach in half. The beach is composed of sand and gravel, is relatively steep and faces due north, thereby affording protection from the Trade winds. There is a 4WD track to the beach which is used by fishers for camping and launching boats.

Swimming: There is deep water right off the beach at high tide, with shallow, rocky tidal flats at low tide.

Surfing: None.

Fishing: This is a popular spot to launch boats and to fish off the beach and adjacent rocks.

Summary: A relatively remote, though utilised, beach because of its sheltered location.

652-654   WALKER PT (S1-S3)

No.	Beach	Unpatrolled Rating HT	Unpatrolled Rating LT	Type	Length
652	Walker Pt (S 1)	1	2	UD	50 m
653	Walker Pt (S 2)	2	3	UD	200 m
654	Walker Pt (S 3)	1	2	UD	100 m

Walker Point is a finger-like, 20 m high headland that protrudes east for 300 m. It forms the northern boundary of a 1 km wide, open bay, within which are three small beaches occupying 350 m of the 1.5 km of rocky bay shoreline.

The northern beach (**652**) is a 50 m long pocket of sand lying beneath the southern side of the headland. It is awash at high tide and fronted by sand flats and reef at low tide.

The second beach (**653**) lies in the centre of the bay. It is longer (200 m), faces east and is backed by low, casuarina covered dune ridges. There is 4WD access to the beach, which is used at high tide for launching boats, while fishers also camp in amongst the casuarinas. The beach is composed of fine sand which, with waves averaging 0.5 m, produces a 100 m wide, low gradient intertidal beach, with coral reefs lying just below the low tide level.

The southern beach (**654**) is another 100 m long pocket of low sand flats, bordered by fringing rocks, and backed by a cleared open valley. It faces east, but is slightly protected by the southern headland, resulting in lower waves and a lower beach.

Swimming: All three beaches have very low gradients, which result in shallow water over the bar at high tide, and wide, bare sand flats at low tide, together with rocks and reefs off the flats.

Surfing: Only the main beach (653) receives the full force of the Trade wind waves which break across the shallow bar to produce a low, choppy break.

Fishing: The best shore fishing is at high tide off the rocks, and in a boat over the reef.

Summary: Three relatively natural beaches with 4WD access and space for camping at the main beach.

655   ARCHER PT (N2)

No.	Beach	Unpatrolled Rating HT	Unpatrolled Rating LT	Type	Length
655	Archer Pt (N 2)	2	3	UD	300 m

One kilometre north-west of Archer Point is the largest beach in this area (655). It is 300 m long and faces east into the Trades, resulting in waves averaging 0.5 m. The beach is composed of fine sand which produces a narrow high tide beach and a 100 m wide intertidal beach at low tide. It is bordered to the south by the rising slopes of 175 m high Mount Archer. A vehicle track leads to the rear of the beach, where there is a clearing in the casuarinas used for camping. Fishers also launch boats from the beach at high tide.

Swimming: This beach provides some low waves and small surf when the Trades are blowing, however watch the scattered rocks toward the northern end.

Surfing: When the Trades are blowing there is small choppy surf.

Summary: A picturesque beach with reasonably good access and space for camping, and relatively safe swimming.

656   ARCHER PT (N1)

No.	Beach	Unpatrolled Rating HT LT	Type	Length
656	Archer Pt (N 1)	1 2	B+SF	100 m

Archer Point is a conical 60 m high headland capped by an automatic lighthouse. Below the steep, grassy northern slopes is a 100 m long sand and gravel beach (656), fringed by rocks and boulders, and fronted by a small coral reef. There is a steep 4WD track down to the beach, which is used for camping and launching boats at high tide.

Swimming: Usually calm with best conditions at high tide. The reef is right off the beach.

Surfing: None.

Fishing: Used for launching boats, with the best shore fishing off the beach or rocks at high tide.

Summary: A small, accessible beach popular with local fishers because of its protected northern orientation.

657   ARCHER PT (WHARF)

No.	Beach	Unpatrolled Rating HT LT	Type	Length
657	Archer Pt (wharf)	2 3	R+LTT	200 m

One kilometre south of Archer Point is a smaller point from which extends a now disused jetty. It was used to service the Leigh Creek tin mine, located 8 km to the south. Today it forms the southern boundary of a 200 m long east facing beach (657). The beach is composed of cobbles at high tide, with sand lower down, and a sand and rocky bar exposed at low tide. During the Trade winds, waves up to 1 m break over the bar. The slopes behind the beach are cleared, and there is car access to the northern end where boats are launched at mid to high tide, and a road to the jetty running along the southern end of the beach.

Swimming: This is the most exposed beach in the Archer Point area and offers some low surf during the Trades, and deep water off the beach at high tide; however watch the numerous small rocks on the beach and bar.

Surfing: If there is any surf at Archer Point it will be biggest at this beach, with a reasonable beach break over the bar.

Fishing: This is a popular spot owing to the jetty, rocks and more energetic surf.

Summary: An accessible ocean beach offering historic remains of the jetty and the 'biggest' surf in the area.

658   ARCHER PT (WHARF S)

No.	Beach	Unpatrolled Rating HT LT	Type	Length
658	Archer Pt (wharf S)	1 2	R+reef	100 m

On the south side of the Archer Point jetty is a 100 m long, south-east facing pocket of sand, bordered by 20 m high rocky bluffs (658). The beach is backed by a small foredune and grassy slopes, and fronted by a small fringing coral reef. There is car access right to the beach.

Swimming: A nice spot to snorkel over coral at high tide.

Surfing: None.

Fishing: The reef can be fished at high tide from the beach or rocks.

Summary: A small open pocket of sand and accessible reef.

659   ARCHER HILL (S)

No.	Beach	Unpatrolled Rating HT LT	Type	Length
659	Archer Hill (S)	1 2	R+SF	250 m

The southern slopes of **Archer Hill** run down to an open, east facing bay. Along the northern end of the bay is a 250 m long strip of shallow sand, protected from the waves by a fringing coral reef. The protection is sufficient to permit mangroves to occupy the shore for 1.5 km south of the beach. The Archer Point Road runs just past the northern end of the beach where there are two parking areas. The beach (659) initially consists of a narrow high tide beach fronted by 100 m wide tidal sand flats, which give way to wider tidal flats and then mangroves.

Swimming: Only possible at high tide, however crocodiles may be present so beware.

Surfing: None.

Fishing: Best at high tide in a boat when the reef and mangroves can be fished.

Summary: A very accessible beach right off the road; mostly suitable for walking on the flats at low tide, or fishing at high tide.

660 WALSH BAY (N)

No.	Beach	Unpatrolled		Type	Length
		Rating HT	LT		
660	Walsh Bay (N)	1	2	R+SF+reef	400 m

Walsh Bay is an open, east facing 2 km wide bay. Its shoreline, however, is protected by 500 m wide reef flats and reef which lower waves sufficiently for mangroves to occupy much of the central and southern shores of the bay. Only along the northern 400 m of the bay are the waves high enough to maintain a narrow high tide beach (660), backed by dense tropical vegetation and fronted by the reef flats.

Swimming: You can only swim off the beach and over the reefs at high tide, while at low tide the reefs can be accessed on foot.

Surfing: None.

Fishing: Only from the shore at high tide.

Summary: An isolated beach surrounded by rainforest-covered hills and fronted by reefs.

661 FORSBERG PT

No.	Beach	Unpatrolled		Type	Length
		Rating HT	LT		
661	Forsberg Pt	1	2	R+SF+reef	250 m

Forsberg Point lies at the eastern tip of a steep ridge that climbs steeply to 330 m granite-capped Bald Hill, 1 km to the west. On the south side of the point is a 250 m long, low energy beach (661), that grades into mangrove-covered tidal flats. The beach is backed by low, tree-covered flats, has a few scattered mangroves at its base, and is fronted by sand, then reef flats extending 500 m seaward. The flats lower wave height to about 0.1 m at high tide.

Swimming: Only suitable for a wade at high tide, or a walk out across the sand and reef flats at low tide.

Surfing: None.

Fishing: Best at high tide, especially from a boat over the reef flats.

Summary: An isolated, low energy beach surrounded by tropical vegetation.

662 WHALEBONE

No.	Beach	Unpatrolled		Type	Length
		Rating HT	LT		
662	Whalebone	1	2	R+SF+reef	500 m

Whalebone beach is a 500 m strip of sand lying immediately north of Obree Point. The beach (662) is backed by densely forested slopes that rise over 500 m. The tropical vegetation overhangs the low energy, narrow high tide beach, while at low tide the gently sloping beach and sand and reef flats extend 250 m out to the edge of the reef.

Swimming: The reef and sand flats are covered at high tide, but exposed at low tide.

Surfing: None.

Fishing: Best from a boat at high tide over the reef edge.

Summary: An isolated beach with no land access, backed by dense vegetation and fronted by a fringing coral reef.

663, 664 CEDAR BAY

No.	Beach	Unpatrolled		Type	Length
		Rating HT	LT		
663	Cedar Bay (N)	2	3	R+LTT	3.5 km
664	Cedar Bay (S)	2	2	R+LTT	700 m

Cedar Bay is an open, 7 km long, east facing bay, contained in an amphitheatre of steep, densely vegetated slopes rising to over 600 m within 2 km of the shore. The bay contains two beaches; the main northern beach (**663**) is 3.5 km long and runs south in a crenulate fashion from the mouth of the small Ashwell Creek to the headland below 590 m high Mount Finlay.

On the southern side of this headland is the 700 m long southern beach (**664**), a straight beach wedged in below slopes that rise steeply to Mount Ramsay. Three pockets of fringing reef lie along the northern beach with open water in between, while the southern beach has no reef.

Swimming: Both beaches have relatively deep water off the beach at high tide, with alternating reef and open water along the northern beach.

Surfing: The beaches receive low wind waves that produce some low choppy surf.

Fishing: Best off the northern beach in the deeper water beside the reefs and at Ashwell Creek mouth.

Summary: A relatively long bay with two stunning beaches, both backed by lush vegetation overhanging the beach, and fronted in part by coral reefs.

665 WEARY BAY

No.	Beach	Unpatrolled		Type	Length
		Rating			
		HT	LT		
665	Weary Bay	2	3	R+LTT	8.5 km

Weary Bay is an open, east facing, 9 km long bay at the mouth of the Bloomfield River. Two hundred and eighty metre high Rattlesnake Point forms the northern boundary with the river mouth and 270 m high Collins Hill borders the southern end. Most of the bay shoreline is occupied by 8.5 km long Weary Bay Beach (665), which extends south from Rattlesnake Point to the river mouth. The small Fritz Creek drains across the northern end, with the mobile Bauer Inlet usually crossing the centre of the beach. The river is 200 m wide, with tidal shoals that extend 1 km seaward of the beach.

The beach is backed by a low sand barrier and swamp up to 500 m wide, then the more extensive Wyalla Plain which extends up to 3 km inland. There is road access to the southern end of the beach, which is located 1 km from the Cooktown Road and 2 km from the small settlement of Ayton. The beach receives waves averaging 0.5 m, but these can rise to over 1 m during strong Trade winds. The waves produce a moderately sloping high tide beach fronted by a low, shallow 100 m wide bar that runs the length of the beach.

Swimming: There is low surf along the beach, which breaks across the shallow bar. Best at mid to high tide. When waves exceed 1 m rips flow across the bar, particularly at low tide. Do not swim at the river mouth owing to the strong tidal currents and possible crocodiles.

Surfing: During strong Trade winds, waves up to 1 m break across the bar as a sloppy beach break.

Fishing: Best at the creek and river mouth.

Summary: A long, low beach just off the Cooktown Road. It offers many kilometres of natural beach and the chance of a few usually low waves breaking across the bar.

Cape Tribulation National Park
Area: 16 965 ha
Bloomfield River to Daintree River
Beaches: 666 to 691 (25 beaches)

Public camping: Noah Beach (677) and Thornton Beach (680)
Private camping:Cape Tribulation (673) and Myall Creek (675)

For information contact:

The Ranger	QPWS
Cape Tribulation National Park	Johnston Road
PMB 10 PS 2041	PO Box 251
Mossman Qld 4873	Mossman Qld 4873
Phone: (07) 4098 0052	Phone: (07) 4098 2188
Fax: (07) 4098 0074	Fax: (07) 4098 2279

---

666 WEARY BAY (S)

No.	Beach	Unpatrolled		Type	Length
		Rating			
		HT	LT		
666	Weary Bay (S)	1	2	TSF	1.2 km

Collins Hill rises to 270 m immediately south of the Bloomfield River mouth. Beginning on the south side of the river mouth and running along the base of the steep slopes is an irregular, thin ribbon of sand that extends for about 1.5 km. The narrow high tide beach (666) is fronted by up to 1 km wide river and tidal shoals, with occasional mangroves growing along the beach. There is a small tourist resort at the southern end, that has an access jetty, but no vehicle access.

Swimming: Only possible for a wade at high tide.

Surfing: None.

Fishing: Best at high tide amongst the mangroves.

Summary: A narrow beach wedged in between the steep lush slopes of Collins Hill and the Bloomfield River mouth shoals.

---

667   BEACH 667

No.	Beach	Unpatrolled		Type	Length
		Rating			
		HT	LT		
667	Beach 667	2	3	R+LTT+reef	1.4 km

This 1.4 km long, north-east facing beach (667) lies at the base of 200 m high, steep, vegetated slopes. The beach is crenulate alongshore owing to extensive fringing coral reefs and a small headland that truncates the northern end of the beach at high tide. The beach is moderately steep, with deep water off the beach at high tide, while at low

tide two patches of coral reef extend up to 1 km offshore. There is a residence on the slopes overlooking the southern end of the beach, but otherwise no development and no vehicle access.

Swimming: There is deeper water between the reefs and at high tide.

Surfing: Usually none.

Fishing: An excellent location with access to the reefs from the beach or off the beach by boat.

Summary: A moderately sheltered beach nestled between lush tropical slopes and coral reefs.

668   COWIE BAY

| No. | Beach | Unpatrolled | | |
		Rating HT LT	Type	Length
668	Cowie Bay	1 2	R+LTT	1.7 km

Cowie Bay is a 1.7 km long, east facing beach (668) extending south from 60 m high Cowie Point to a low, reef fringed headland. Melissa Creek drains across the northern end of the beach, and Bind Creek across the southern end, with fringing reefs adjacent to both headlands. The beach lies 1 km off the Cooktown Road, however there is no formal vehicle access. The beach is backed by a low, 200 m wide sand barrier, and fronted by a gently sloping 100 m wide bar. Waves averaging about 0.5 m break across the bar as a low choppy surf.

Swimming: This beach offers a low surf, breaking across a shallow bar.

Surfing: Usually low choppy waves, with waves up to 1 m during strong Trade winds.

Fishing: Best at high tide at each of the creek mouths and off the reef flats.

Summary: A natural beach offering relatively safe swimming, plus the creeks and reefs for fishing.

669   DONOVAN PT (N)

| No. | Beach | Unpatrolled | | |
		Rating HT LT	Type	Length
669	Donovan Pt (N)	1 2	B+SF	500 m

To the north of **Donovan Point** is a 2 km wide, broad, V-shaped embayment, containing a 500 m long beach (669) in the V. The entire beach and bay system is fronted by a near-continuous fringing coral reef, extending in places 1 km from the beach. The beach consists of a narrow high tide fringe of sand, fronted by intertidal sand flats up to 600 m wide, and then the fringing reef. Because of the reef, waves are usually very low and mangroves are scattered along the central part of the beach. The beach lies just off the Cooktown Road, with easy vehicle access to the back of the beach.

Swimming: There is shallow water over the tidal flats at high tide, while at low tide the waves lap over the low tide flats and reef.

Surfing: None.

Fishing: Best at high tide from the beach, and in a boat over the reef.

Summary: A very accessible beach just off the road.

670   DONOVAN PT (S)

| No. | Beach | Unpatrolled | | |
		Rating HT LT	Type	Length
670	Donovan Pt (S)	1 2	B+SF	2 km

To the south of **Donovan Point** is a gently curving 2 km long beach (Fig. 1.4). The northernmost section lies just below Cooktown Road and is the most accessible and most exposed section, with waves usually breaking across the beach. Immediately to the south coral reefs fringe the beach (670), lowering waves and producing extensive intertidal sand flats, the central section of which are scattered with mangroves.

Swimming: The northern 300 m long section is the most accessible and most suitable for swimming, as sand flats front the beach to the south.

Surfing: Only low sloppy waves breaking across the northern beach section.

Fishing: The northern end is best for both beach and rock fishing, while a boat is needed to fish the reef.

Summary: An accessible beach right next to Cooktown Road.

671 EMMAGEN CK

No.	Beach	Unpatrolled		Type	Length
		Rating HT	LT		
671	Emmagen Ck	2	3	R+LT+reef	400 m

Emmagen Creek is a permanently open creek that has built a cobble delta that protrudes 100 m from the shore. On the southern side of the delta is a 400 m long sandy beach (671), that is backed by the densely vegetated creek floodplain, with trees hanging over the beach, and fronted by a fringing coral reef. The beach faces east and receives waves averaging 0.5 m at high tide which produce a low surf, while sand and reef flats are exposed at low tide. The Cooktown Road runs 400 m west of the beach, however there is no vehicle access.

Swimming: A nice spot for a swim or snorkel over the reef at high tide.

Surfing: Only low waves at high tide.

Fishing: Best off the delta, either into the surf or creek.

Summary: An isolated tropical beach with the interesting cobble creek delta forming the northern boundary.

672 PILGRIM SANDS

No.	Beach	Unpatrolled		Type	Length
		Rating HT	LT		
672	Pilgrim Sands	1	2	R+LT+reef	1.5 km

Pilgrim Sands is the name of a small tourist settlement and private camping area in lee of this beach. The access road reaches the beach (672) on the north side of a cobble and mangrove-fringed creek mouth. The beach extends north for 1.5 km from the small creek. It is narrow at high tide, wedged in between steep, vegetated slopes and a fringing coral reef, with reef increasing northward. At low tide the beach widens and the reef flats are exposed.

Swimming: Best at high tide, especially if you want to snorkel over the reef.

Surfing: Usually none.

Fishing: Best over the reefs and at the creek mouth at high tide.

Summary: A popular, accessible beach; nice for a walk

between the tropical forest and the reef, both just metres apart.

673 CAPE TRIBULATION BEACH

No.	Beach	Unpatrolled		Type	Length
		Rating HT	LT		
673	Cape Tribulation	1	2	B+SF	700 m

Cape Tribulation is a destination for most travellers driving north of Cairns. The cape was named by Captain Cook in June 1770, following the Endeavour's grounding on a coral reef off the cape. As Cook wrote, he called the "North point Cape Tribulation, because here began all our Troubles". The cape itself is part of the Cape Tribulation National Park, with freehold land to either side. The main road parallels the back of the beach, with a car park right behind the beach.

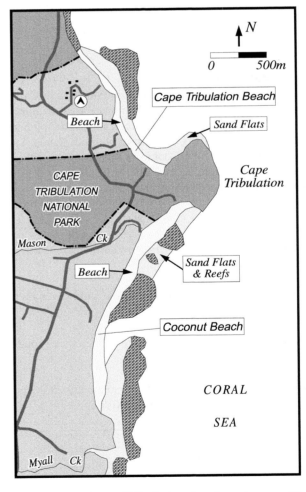

Figure 4.5 *Cape Tribulation has beaches to either side. To the north is the protected, very low energy cape beach which faces east to north-east. On the southern side is the longer, more exposed Coconut Beach; popular for swimming and snorkelling over the fringing reefs.*

There are beaches either side of the cape; the northern called Cape Tribulation Beach (673), the longer southern called Coconut Beach (Fig. 4.5). Cape Tribulation Beach is tucked in lee of the cape, faces north-east and is 700 m long. It is a low, flat beach, narrow at high tide and up to 300 m wide at low tide. Waves are usually low, with a few mangroves growing out along the side of the cape. Dense tropical vegetation surrounds the beach and provides excellent shade. The northern end of the beach stops amongst the rocks, reef and mangroves at the mouth of Pilgrim Sands Creek, while the southern end abuts the cape.

Swimming: This is a popular spot to wade, however it's often too shallow to swim, except on high tide. Best to stay on the sand and clear of the rocks at either end.

Surfing: None.

Fishing: Best at high tide, either off the rocks around the cape or at the creek mouth.

Summary: A lovely, unspoiled tropical beach, where the forest does meet the sea. More suited for walks and a wade than swimming.

674, 675 COCONUT BEACH

No.	Beach	Unpatrolled			
		Rating		Type	Length
		HT	LT		
674	Coconut	1	2	R+reefs	2.5 km
675	Myall Creek	1	2	R+reefs	2 km

On the southern side of Cape Tribulation are two beaches, totalling 4.5 km in length and separated by Myall Creek. Both beaches have low gradient, narrow, high tide beaches backed by palm trees and tropical vegetation, and fringed with patches of coral reef.

The first beach (674), **Coconut Beach,** runs almost due south from Cape Tribulation and the mouth of the meandering Mason Creek, for 2.5 km to the cobble and mangrove-fringed mouth of Myall Creek. The narrow high tide beach winds its way south, as both the creeks and three areas of fringing reefs produce undulations in the shoreline. The beach protrudes seaward in lee of the reef, while in the embayments there is deeper water and a sandy bottom off the beach. This is the more popular of the cape's beaches, owing to the reef and deeper water. Many travellers see their first coral reef at this beach. The reef runs right to the shore in places and can be reached by snorkelling off the beach.

The southern beach, **Myall Creek** beach (675), is 2 km long, faces south-east and has a more continuous fringing coral reef. Both beaches have good access, being located just east of the Cooktown Road, and are very popular spots for swimming and viewing the reef. There is a private camping area at Myall Creek.

Swimming: The northern beach is more popular for swimming, owing to less coral, while both are excellent for snorkelling over the reef.

Surfing: None at the beach, while waves do break over the reef.

Fishing: The northern beach has some channels between the reefs, while the reefs can be fished from the beach at high tide.

Summary: Two popular beaches, particularly for those wanting to get into the water to swim or see the reef.

676-678 NOAH HEAD, NOAH CREEK

No.	name	Unpatrolled			
		Rating		Type	Length
		HT	LT		
676	Noah Head	2	3	R+LTT	100 m
677	Noah Creek (N)	2	3	R+LTT	1.6 km
678	Noah Creek (S)	2	3	R+LTT	2.3 km

Noah Head rises over 100 m and protrudes sufficiently to separate the Myall Creek beach from those surrounding Noah Creek. Between the head and lying either side of Noah Creek are three exposed beaches. The Cooktown Road skirts close to all three, providing good access, with a picnic and public camping area located just south of the head in lee of Noah Creek (north) beach (677).

The three near-continuous beaches face slightly south of east into the predominate south-east Trade winds and waves, which result in waves averaging 0.5 m and more during strong Trade winds. The waves maintain a wide, low gradient beach with a low surf zone breaking across the shallow bar at high and low tide.

The northernmost beach (**676**) is a 100 m pocket of sand cut off from the main beach at high tide by a spur of Noah Head, though readily accessible at low tide. The main beach (**677**) is 1.6 km long and extends from the spur down to Noah Creek, which drains permanently across the beach. On the south side of the creek is the 2.3 km long southern beach (**678**) which ends below 450 m high Table Mountain.

Swimming: These three beaches offer some of the more accessible and potentially large surf in the region. They are relatively safe, though rips can form when waves exceed 0.5 m, particularly at low tide.

Surfing: If there is any surf on the coast it will be breaking along these three beaches.

Fishing: All three offer good fishing, with rocks at the ends of all three beaches and Noah Creek located in the centre.

Summary: Three very accessible beaches, with the main beach offering picnic and camping areas, and a chance of some low surf.

679 BOUNCING STONES

| No. | Beach | Unpatrolled | | Length |
		Rating HT LT	Type	
679	Bouncing Stones	2 3	cobble R+LT	100 m

Bouncing Stones beach (679) is a popular tourist stop - not so much to swim, but when the tide is high and the waves breaking, to see if the stones that compose the steep high tide beach are in fact bouncing. The beach lies right next to the Cooktown Road and is very accessible. However, to preserve the stones and protect the beach, which has important aboriginal significance, access to the beach is restricted and moving the stones is banned.

Swimming: Swimming is not permitted at the beach because of the restriction.

Surfing: Waves break across a sandy bar at low tide, however like swimming, surfing is also prohibited.

Fishing: Not permitted.

Summary: A popular beach to view, however beach access is essentially prohibited in the interest of preserving this small but fascinating beach.

680 THORNTON

| No. | Beach | Unpatrolled | | Length |
		Rating HT LT	Type	
680	Thornton	1 2	R+LTT/SF	1 km

Thornton beach (680) is situated on the northern shore of Alexandria Bay and on the northern side of Cooper Creek mouth. The Cooktown Road runs just behind the beach, with a shop and kiosk on the road, a public camping area and several vehicle access tracks to the beach. This popular beach faces south-east and receives some surf along the northern end, however the central and southern part of the 1 km long beach is protected by Stuck Island and the tidal flats of Cooper Creek. The beach is wide and flat, with high tide almost reaching the trees, while at low tide a 200 m wide bar and sand flats front the beach.

Swimming: Best toward the more energetic northern end, while the creek mouth should be avoided owing to the possibility of crocodiles.

Surfing: Chance of some low surf at the northern end.

Fishing: A very popular spot, particularly Cooper Creek and offshore around Stuck Island.

Summary: A very accessible and popular beach for fishing, picnicking and swimming.

681 ALEXANDRIA BAY

| No. | Beach | Unpatrolled | | Length |
		Rating HT LT	Type	
681	Alexandria Bay	2 3	R+LT	3 km

Alexandria Bay is an open bay lying between the northern Table Mountain and the southern Bale Hill. A 3 km long beach (681) fronts a 2 km wide swamp that partially fills the bay. The beach faces east and is bordered by Cooper Creek in the north and McKenzie Creek to the south. The creeks and backing swamp prohibit vehicle access, with boat being the best form of access. The beach receives moderate waves and usually has a narrow high tide beach fronted by a 200 m wide shallow bar, across which the waves break at low tide. The beach has been experiencing some erosion, which is undercutting and exposing the tall melaleuca trees that back the beach. The beach also widens toward each of the inlets.

Swimming: This is a relatively safe beach with low waves and a shallow, gently sloping bar. However stay clear of the inlets owing to crocodiles.

Surfing: There is moderate surf during good Trade Wind conditions that breaks across the shallow bar.

Fishing: Best in the two inlets and creeks.

Summary: An isolated beach offering good fishing, a quiet swim and possibly a surf.

682 ALEXANDRIA BAY (S)

No.	Beach	Unpatrolled			Length
		Rating		Type	
		HT	LT		
682	Alexandria Bay (S)	1	2	R+LT	700 m

On the north side of Bale Hill, just east of the McKenzie Creek mouth, is a narrow strip of sand (687) running in a north-facing arc for 700 m along the foot of the hill. It is very protected and fronted by shallow tidal flats. There is private road access to the western McKenzie Creek end, but no public access other than boat.

Swimming: Not recommended, owing to the proximity of the creek and possibly crocodiles, and the shallowness of the water over the tidal flats.

Surfing: None.

Fishing: Best at high tide in a boat off the tidal flats and tidal shoals of adjoining McKenzie Creek.

Summary: A relatively narrow, low energy, isolated beach.

683 BAILEY PT

No.	Beach	Unpatrolled			Length
		Rating		Type	
		HT	LT		
683	Bailey Pt	2	3	R+LT	100 m

Bailey Point lies 1 km east of 280 m high Bailey Hill. The small pocket beach (683) is situated immediately south of the tip of the point. It is 100 m long and faces east, with protruding rocks bordering each end as well as some large rocks across the centre of the beach. The densely vegetated slopes of the hill back the beach. There is no vehicle access.

Swimming: This is a small, little used beach which does provide some interesting rocks and rock reefs for viewing.

Surfing: Usually only a low, choppy beach break when the Trades are blowing.

Fishing: Best off the rocks at high tide.

Summary: An isolated little beach perched beside the prominent Bailey Point.

684-686 COW BAY

No.	Beach	Unpatrolled			Length
		Rating		Type	
		HT	LT		
684	Cow Bay (N)	2	3	R+LT	200 m
685	Cow Bay (mid)	2	3	R+LT	500 m
686	Cow Bay (S, main)	2	3	R+LT	1 km

Cow Bay is an open, east facing bay, containing three beaches separated by rocky spurs at high tide, but continuous at low tide. The main beach (686) is accessible via the Cow Bay Road and there is a car park and picnic area behind the centre of this beach. Trees hang over the back of the beach, providing shade as well as an attractive backdrop. The adjoining two northern beaches (684, 685) must be accessed on foot or by boat. All three beaches face east and receive waves averaging 0.5 m, which combines with the fine beach sand to maintain a wide, low gradient beach across which the surf breaks when the Trades are blowing. Besides the rocky spurs and boundary headlands, there are patches of coral reef off the middle and main beaches.

Swimming: All three provide a wide, low gradient beach with usually low surf and no rips.

Surfing: Best when there are good Trades blowing, which produce a sloppy 0.5 to 1 m high beach break.

Fishing: Best off the rocks at high tide.

Summary: The main beach is very accessible and relatively popular, being just 5 km off the Cooktown Road and with development taking place on the adjoining land.

687, 688 BLACK ROCK

No.	name	Unpatrolled			Length
		Rating		Type	
		HT	LT		
687	Black Rock (N)	2	3	R+LT	150 m
688	Black Rock (S)	2	3	R+LT+reef	200 m

Black Rock reef lies 600 m east of a 300 m high, densely vegetated spur. On either side of the spur are two small beaches. The northern beach (687) is 150 m long, faces north-east, has a small headland as its northern boundary and high rocky slopes to the south, together with some rocks scattered across the beach. The southern beach (688) faces east, is surrounded on three sides by steep vegetated slopes, and is fronted by a coral reef up to 100 m wide.

Swimming: Both beaches receive low waves and are relatively safe, with the southern beach also offering the reef for viewing.

Surfing: Usually none.

Fishing: Best off the rocks and over the reef at high tide.

Summary: Two small, attractive, relatively isolated beaches which can best be accessed by boat.

689 MT ALEXANDRIA

No.	Beach	Unpatrolled		Type	Length
		Rating HT	LT		
689	Mt Alexandria	2	3	R+LT	1.5 km

Mount Alexandria is 500 m high and slopes steeply to the east, where it is fringed by a 1.5 km long beach (689) bounded by headlands at each end, together with a patch of reef off the northern end. There is a camp, but no access to the beach, other than by boat. The beach faces east and receives waves averaging 0.5 m. At high tide they surge up the steep, tree-lined beach, while at low tide a flat, 100 m wide bar is exposed.

Swimming: Best at high tide.

Surfing: Usually a low break across the bar when the Trades are blowing.

Fishing: Better at high tide and off the rocks at each end.

Summary: An isolated beach with a dramatic green backdrop of Mount Alexandria. Used by locals for camping and visited by boaters.

690 CAPE KIMBERLEY(N)

No.	Beach	Unpatrolled		Type	Length
		Rating HT	LT		
690	Cape Kimberley (N)	2	3	R+LT+ rocks	100 m

This beach (690) is a narrow, 100 m long strip of sand and rock pockets wedged in below the 100 m high slopes on the north side of Cape Kimberley. It is awash at spring tide and fronted by a rock-strewn bar at low tide. There is no access other than by boat.

691 CAPE KIMBERLEY

No.	Beach	Unpatrolled		Type	Length
		Rating HT	LT		
691	Cape Kimberley	2	3	R+LT	3.5 km

Cape Kimberley forms the northern boundary of the 12 km wide bay into which the Daintree River flows. Between the river mouth and the cape is the 3.5 km long Cape Kimberley beach (691). The beach faces south-east, directly into the Trades, however Snapper Island 2.5 km offshore and the river mouth shoals lower the waves along parts of the beach. The beach is accessible by the 10 km long Cape Kimberley Road that runs out to the cape from the Daintree ferry, skirting the extensive floodplain and swamp that backs the beach. There is a resort and camping area at the northern end of the beach, together with public parking and facilities, making this a relatively popular beach. The more exposed parts of the beach receive waves averaging 0.5 m, which break across a wide, low gradient beach that is up to 200 m wide at low tide. Toward the southern end, the Daintree River mouth shoals extend up to 1 km off the beach, substantially lowering the waves and widening the low tide beach.

Swimming: Best at the accessible northern end. Stay clear of the river mouth owing to strong tidal currents and possibly crocodiles.

Surfing: Also best at the northern end, where the waves are larger and break across the low bar as a sloppy beach break.

Fishing: Best fishing is off the rocks at the northern end, or into the river mouth in the south.

Summary: A very accessible and relatively popular beach with facilities for tourists and the public.

692 WONGA

No.	Beach	Unpatrolled		Type	Length
		Rating HT	LT		
692	Wonga	2	3	R+LT	10.5 km

Wonga beach (692) is a 10.5 km long, east facing beach that runs in a sinuous curve from Rocky Point north to the mouth of the Daintree River. It occupies part of the wide Trinity Bay (named by Captain Cook on the day of the same name); he commented on its shallow sea floor.

The southern 4 km are backed by houses and a caravan park, all located on a 1 km wide sand plain (barrier) behind the beach. There is a beach boat ramp in front of the park. The beach is fringed by casuarina trees and has a moderately steep and narrow high tide beach, which at low tide is fronted by sand flats more than 100 m wide, particularly in the south. The Daintree runs behind the northern end of the beach, which is undeveloped and slowly curves around in lee of the Daintree River mouth shoals. The shoals extend up to 2 km east of the beach.

Swimming: It is best to swim at high tide when there is deeper water over the bar. Stay clear of the Daintree mouth, owing to the strong tidal currents and possible presence of crocodiles.

Surfing: Usually a low beach break; best at mid to low tide.

Fishing: Best at high tide, in a boat over the reef off the beach, or in the river mouth.

Summary: A long, accessible beach with tourist and public facilities.

693   ROCKY PT (DAYDREAM)

No.	Beach	Unpatrolled			
		Rating HT LT		Type	Length
693	Rocky Pt (Daydream)	2	3	R+LT+ rocks	2.2 km

Rocky Point (Daydream) forms the southern boundary of its long neighbour, Wonga Beach. The point is the southern extent of the Dagmar Range and elevations of 100 to 200 m are reached within a few hundred metres of the coast. The Daintree Road is forced to hug the shoreline. Immediately below the 2 km long, winding point and road is a narrow, crenulate, sandy high tide beach (693), interspersed with rocks and fronted by a wider low tide beach and rock flats. Toward the southern end is a good concrete boat ramp and a groyne to shelter it from the Trade winds and waves.

Swimming: This beach is relatively safe and best at high tide, however be careful of the numerous rocks.

Surfing: Usually none, however waves can run along the rock flats in strong Trades.

Fishing: Good at high tide over the rocks and flats, however most fishers use the ramp to go outside.

Summary: A highly visible beach right below the main road, best suited for viewing rather than recreating.

694-696   ROCKY PT, NEWELL, COOYA

No.	Beach	Unpatrolled			
		Rating HT LT		Type	Length
694	Rocky Pt (S)	2	3	R+LTT	2.7 km
695	Newell	2	3	R+LTT	2.0 km
696	Cooya	1	2	R+SF	2.0 km

Between Rocky Point and Island Point is an 11 km wide, open bay. For the first 7 km south of Rocky Point is a strip of sandy beach cut by Saltwater Creek and the Mossman River into three beaches: Rocky Point (south), Newell Beach and Cooya. Increasing protection from Island Point, toward the southern end of Cooya Beach, permits mangroves to grow along the shoreline for the next 7 km to Port Douglas.

Rocky Point (south) beach (694) is 2.7 km long and extends nearly due south from the point to Saltwater Creek, which maintains a 50 m wide inlet. There is access to the beach from the Daintree Road, that skirts the northern end of the beach at Rocky Point, otherwise there are no facilities, as sugarcane farms back the beach. The beach receives waves averaging 0.5 m and is moderately narrow and steep at high tide, with a 100 m-plus low tide sand bar.

Newell beach (695) is a 2 km long, east facing beach that is bordered by Saltwater Creek in the north and the Mossman River in the south. It lies 2 km east of the Daintree Road, with a road running out to the Palm Beach settlement that incorporates a store, caravan park, public parking and facilities. The beach is narrow and steep at high tide, but has a wide, shallow sand bar at low tide.

Cooya beach (696) lies on the south side of the Mossman River and extends to the south for 2 km before the low energy beach gives way to mangrove-covered tidal flats. The beach lies 4 km east of Mossman and a road runs out to the Cooya settlement, which includes a nice foreshore reserve and facilities, but no store. The beach receives low waves and has a steep, narrow high tide beach, while at low tide the sand flats extend up to 500 m off the beach.

Swimming: You can only swim here at high tide owing to the wide tidal flats.

Surfing: None, except during cyclones.

Fishing: Best up at the river mouth, or down amongst the mangroves in a boat at high tide.

Summary: This is the closest beach to Mossman; it has good public beachfront facilities, with usually low to calm waves.

Captain Cook Highway

Mossman to Cairns
length 95 km
beaches 697-724

The Captain Cook Highway is one of the world's great ocean drives. It winds along the coastal plain and cliffhanging slopes from Mossman to Cairns, only a road distance of 60 km, but in so doing passes within a whisker of several beaches and presents stunning vistas of beach after beach. It is best to do this drive from north to south so that you are on the beach side of the road, which provides better views, easier parking and no road crossing if you get out for a photo or swim.

697 FOUR MILE BEACH

PORT DOUGLAS (SLSC)

Patrols:
Surf Life Saving Club:	July to May
Lifeguard on duty:	all year
Stinger enclosure:	November to May

No.	Beach	Rating		Type	Length
		HT	LT		
697	Port Douglas (Four Mile)	2	3	R+wide LT	4 km

Four Mile Beach (Port Douglas) (697) is one of the best known and most popular beaches of northern Australia. What was a sleepy coastal town in the 1970's became an international tourist destination by the late 1980's. Three major resorts back the beach, together with golf courses and a marina in the town. The beach runs due south for 4 km and is also known as Four Mile Beach. It begins at the base of prominent, 70 m high Island Point and ends amongst the rocks, reefs and mangroves of the Mowbray River mouth. For the most part the beach is very low, flat, and at low tide up to 200 m wide. The sand flats widen toward the river mouth (Fig. 4.6).

The northern end of the beach, just 1 km from the town centre, is the most accessible and most popular. It is also the site of the Port Douglas Surf Life Saving Club (Fig. 4.7), founded in 1984, that patrols the beach as well as maintaining a stinger enclosure. Down the beach is the large Mirage Resort which provides three beach access tracks. Further access tracks are available through the residential development along the southern half of the beach, including one at the sailing club.

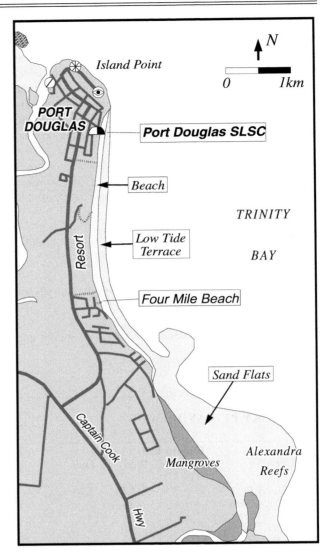

Figure 4.6 *A map of Port Douglas and adjoining Four Mile Beach. Port Douglas SLSC is located at the northern end.*

Figure 4.7 *Four Mile Beach at Port Douglas. The surf life saving club and stinger net are located in the centre of this view.*

Swimming: Generally better at high tide, when it's deeper and not such a long walk to the water. Rips are rare, except during strong Trade winds and accompanying waves. During the stinger season (November to May) a net enclosure is located in front of the surf lifesaving club.

Surfing: Usually low, short waves, with the beach more popular for wind surfing, particularly when the Trades are blowing.

Fishing: It is flat off the beach and most fishers use the northern headland, the town jetties or head offshore in a boat.

Summary: A popular tourist destination and well worth a visit. There is plenty of space and the added safety of a patrolled beach and stinger enclosure.

698, 699   DOUGLAS BEACH, ALEXANDRIA REEFS

| No. | Beach | Unpatrolled | | | |
		Rating HT	LT	Type	Length
698	Douglas Beach	1	2	TSF	2.5 km
699	Alexandria Reefs	1	2	TSF	3.5 km

Between the southern end of Port Douglas (Four Mile) Beach and Yule Point are 6 km of crenulate tidal flats, which lie either side of the small Mowbray River mouth. To the north of the river mouth the tidal flats are known as Douglas Beach, while to the south they are named Alexandria Reefs, for the adjacent coral reefs. The reefs and Yule Point lower average waves to a few centimetres at the shoreline.

Douglas Beach (698) faces north-east and lies 1.5 km off the Cook Highway, with a road off the Port Douglas Road leading to houses behind the northern end of the beach. Apart from the few houses there are no facilities. The beach in lee of **Alexandria Reefs** (699) extends from the river mouth to the north side of Yule Point, where the Cook Highway runs right over the very southern end of the beach, providing a view of the mangroves and tidal flats. Both beaches have narrow high tide beaches fronted by sandy tidal flats up to 1.5 km wide, and then fringing coral reef. There are pockets of mangroves along the beaches, as well as dense mangroves either side of the river mouth.

Swimming: Neither beach is really suitable for swimming, owing to the shallow tidal flats.

Surfing: None.

Fishing: Best at the river mouth and over the reefs at high tide.

Summary: Two little used beaches, owing to the restricted access and dominance of tidal flats.

700   PEBBLY

| No. | Beach | Unpatrolled | | | |
		Rating HT	LT	Type	Length
700	Pebbly	2	3	R+LT	1.5 km

Pebbly beach (700) is a curving, north-east facing beach lying between 100 m high Yule Point and 40 m high White Cliffs. The Cook Highway skirts the northern half, providing excellent views and access. However there are no facilities at the beach. The beach is named after the shingle and cobbles which compose the steep high tide beach, that grades into sand lower down the beach and on the bar. Removal of the cobbles is prohibited. Rocks outcrop at either end and there are some rock flats on the bar.

Swimming: This beach receives moderate waves and usually has a narrow surf zone that is best for swimming at mid to high tide.

Surfing: The waves break across a narrow bar, producing a low, sloppy beach break.

Fishing: Best off the rocks at high tide.

Summary: A very accessible beach right by the highway.

701   WHITE CLIFFS

| No. | Beach | Unpatrolled | | | |
		Rating HT	LT	Type	Length
701	White Cliffs	2	4	R+rock flats	200 m

On the north side of the White Cliffs point are two pockets of high tide sand fronted by intertidal rock flats. The beach (701) is backed by 40 m high cliffs and natural vegetation, then farms, with the only access being on foot from the southern end of Pebbly Beach. While the point offers an interesting walk and some good fishing spots at high tide, the beach is unsuitable for swimming due to the numerous rocks.

702 OAK

		Unpatrolled		
No.	Beach	Rating HT LT	Type	Length
702	Oak	2 3	R+LT	1.6 km

Oak Beach is a straight, east facing, 1.6 km long beach (702) running almost due south from White Cliffs. The Cook Highway clips the southern end, where there is a car park and access to the beach. The remainder of the beach is backed by farms, with a creek draining across the centre of the beach and the rocks of White Cliffs protruding 100 m off the northern end. The beach is composed of a mixture of sand and cobbles, with steep, gravel cusps at high tide, fronted by a moderately steep beach and 50 m wide bar. Waves average 0.5 m and break across the bar at mid to low tide.

Swimming: This is a relatively safe beach for swimming when waves are low, however there can be rips and a drag along the beach when waves exceed 0.5 m.

Surfing: Usually a low, sloppy break over the bar.

Fishing: Best off the rocks at either end, and the creek mouth, when open.

Summary: A highly visible beach from the highway, with good access at the southern end, but no facilities.

703 LITTLE REEF

		Unpatrolled		
No.	Beach	Rating HT LT	Type	Length
703	Little Reef	1 2	R+LT+reef	700 m

Little Reef beach is named after the patch reef that lies immediately off the centre of the beach. The presence of the reef lowers waves in the centre and has caused the shoreline to protrude out to form a cuspate foreland. The beach (703) is backed by the Cook Highway, with limited parking and access via a car park behind the centre of the beach. Steep slopes and rocky shorelines border the beach, while the centre is protected by the reef. The beach is moderately sloped and is up to 80 m wide at low tide.

Swimming: Best at high tide, for both a swim and snorkel over the reef.

Surfing: Usually only a low shorebreak at high tide.

Fishing: Best at high tide, when the rocks and reef can be fished.

Summary: A very visible beach providing easy access to the reef.

704, 705 TURTLE CK

		Unpatrolled		
No.	Beach	Rating HT LT	Type	Length
704	Turtle Ck	2 3	R+LT	200 m
705	Turtle Ck (S)	2 3	R+LT	150 m

Turtle Creek is a steep, small creek that drains across Turtle Beach. The beach (704) and its adjoining southern neighbour (705) sit at the base of steep, vegetated slopes that rise 400 m to Arnold Knob 2 km to the south. The Cook Highway is cut into the slopes 60 m above the beaches, providing views of both beaches to passing motorists. There is limited highway parking and a steep descent to the southern beach, while Turtle Creek Beach is occupied by a private resort and is not accessible to the public, except by boat.

Both beaches are relatively short, bordered by rocky shorelines, with two clumps of rock dividing Turtle Creek Beach. They have relatively steep, narrow, cusped high tide beaches fronted by narrow, continuous bars. Sand moving around from the southern beach can form an offshore bar off the southern end of the main beach. Waves average 0.5 m, breaking across the bar at low tide and surging up the beach face at high tide.

Swimming: Turtle Creek Beach is popular with resort guests, while the public can scamper down to the southern beach for a swim.

Surfing: Usually a low, narrow beach break.

Fishing: Best off the beach or rocks at high tide.

Summary: Two pockets of sand, with the southern beach remaining little used, while the main beach houses a small resort.

706 WANGETTI

		Unpatrolled		
No.	Beach	Rating HT LT	Type	Length
706	Wangetti	2 3	R+LT	4 km

Wangetti beach (706) is a 4 km long, east-north-east facing beach, running in a near-straight line from Slip Cliff

Point in the north to Red Cliff Point in the south. Most tourists to the coast stop at the large car park and lookout on the northern point, that provide a spectacular view of the entire beach. While the beach is highly visible, there is only parking and access at either end. The highway runs a few hundred metres inland from the rest of the beach and there are no facilities.

The beach receives waves averaging 0.5 m, that break across a narrow, continuous bar at low tide and surge up the moderately steep beach at high tide. Hartley's Creek drains across the northern section, causing the beach to protrude slightly and form a small lagoon in behind the beach. The southern end is eroding and exposing a cobble basement to the otherwise sandy beach.

Swimming: A relatively safe beach under average wave conditions, however higher waves will produce a northward drag along the beach and occasional rips.

Surfing: Usually a low, narrow beach break; best off the protruding creek mouth.

Fishing: The rocks at each end offer the best vantage point, while the narrow bar can be fished at high and low tide.

Summary: One of Australia's most photographed, yet little used, beaches. Great for the tourists to see and for those who want a long, quiet beach.

707, 708   RED CLIFF PT, BEACH 708

No.	name	Unpatrolled			
		Rating HT	LT	Type	Length
707	Red Cliff Pt (S)	3	4	R+rocks	500 m
708	Beach 708	2	3	R+rocks	500 m

At the southern end of Wangetti Beach, the Cook Highway again takes on the steep slopes of the Macalister Range. The ranges rise nearly 500 m within 1.5 km of the coast, and require the highway to hug the lower slopes and skirt the narrow beaches. Just east of Red Cliff Point is a 500 m patch of sand and numerous boulders and cobbles, that comprises a very accessible and visible beach (**707**), but not one suited for safe swimming. Two kilometres further east is a second beach (**708**) of a similar nature, only with a western pocket of sand. Both beaches receive waves averaging less than 0.5 m and have a steep, sand-cobble high tide beach fronted by rocky and sandy patches at low tide.

Swimming: While waves are usually low along these beaches, the prevalence of rocks on the beach and in the

surf make them generally unsuitable for safe swimming.

Surfing: During strong Trades there are some right-handers running along the steep, rocky shore.

Fishing: Reasonably popular spots owing to the easy access and rocky vantage points.

Summary: Two highly visible beaches worth a photo stop, but with no facilities and unsuitable for safe swimming.

709-711   SIMPSON PT

No.	name	Unpatrolled			
		Rating HT	LT	Type	Length
709	Simpson Pt (W)	2	3	R+LT	600 m
710	Simpson Pt (S 1)	2	3	R+LT	500 m
711	Simpson Pt (S 2)	2	3	R+LT	400 m

Simpson Point is a prominent spur where the coast turns nearly 90 degrees and trends south-east toward Ellis Beach. The point is a jumble of rocks and boulders (Fig. 3.1d), with the highway cutting across just behind the rocks. To either side are three sandy beaches, each bordered by rocks (**709, 710, 711**). All three lie right next to the highway, which provides both excellent views and easy access, although there are no facilities.

The beaches receive waves averaging less than 0.5 m, which combine with the coarse sand to build relatively steep, narrow high tide beaches, fronted by a narrow, continuous bar. Waves surge up the beach at high tide and break over the bar at low tide. There are rocks at the ends of all three, as well as scattered in the surf, while the two southern beaches are partially broken by rocks extending across parts of the beaches.

Swimming: Three very accessible and relatively safe beaches, with deeper water off the beaches at high tide and a little surf at low tide.

Surfing: Usually small wind waves that break across the bar at mid to low tide.

Fishing: Best off the rocks at the ends and scattered along the beaches, as well as along the steep beach at high tide.

Summary: Three attractive, very accessible, small beaches offering a nice place to stop for a swim, photo or to fish.

712 ELLIS BEACH (SLSC)

Patrols:				
Surf Lifesaving Club:			July to May	
Lifeguard on duty:			November to May	
Stinger net:			November to May	
No.	Beach	Rating	Type	Length
		HT LT		
712	Ellis	1 2	R+LT	2.5 km

Ellis Beach (712) begins where the Cook Highway, 20 km north of Cairns, really meets the Coral Sea. Rounding Buchan Point, the highway parallels the entire length of the 2 km long beach, continuing on past northern Simpson Point. Steep, forested slopes rise up to 600 m within 1 km of the highway, while a soft, sandy beach lies immediately east of the highway (Fig. 4.8).

The beach faces north-east and runs essentially straight for 2.5 km from the northern side of Buchan Point up to a group of rocks. There is a pocket sand and cobble beach at the junction with the point, otherwise it is an essentially steep, sandy beach. The Ellis Beach caravan park straddles the highway in the southern half of the beach, with the centre occupied by the Ellis Beach Surf Life Saving Club, a kiosk and store. There is good access and parking for most of the length of the beach.

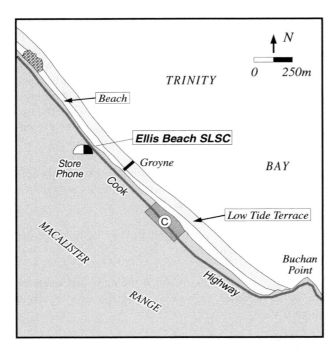

Figure 4.8   *Ellis Beach extends in a north-west direction from Buchan Point, with the Cook Highway paralleling the back of the beach. The surf club patrols the mid section of the beach, near the small groyne.*

The beach is composed of medium sand which produces a soft, moderately steep high tide beach, fronted by a low gradient, continuous, 50 m wide bar, that is exposed at low tide. Waves average 0.5 m, only increasing in height during strong south-easterlies. A small rock groyne crosses the beach in the patrol area (Fig. 4.9).

Figure 4.9   *Ellis Beach and surf club is wedged in below the steep slope of the Macalister Range. The club house and patrol area lie adjacent to the small rock groyne visible in this view.*

Swimming: The surf lifesaving club patrols the centre of the beach just across the highway from the club house. It is best to swim here both for safety and for access to the adjacent amenities.

Surfing: There is usually a low, sloppy surf, particularly at mid to low tide.

Fishing: You can fish off the steep beach at high tide, however at low tide you will need to walk out onto the bar to cast across the narrow surf zone.

Summary: This is Cairns northernmost patrolled beach and also its most accessible from the highway. An attractive beach wedged between the sea and the mountains, with sufficient facilities for the visitor or camper.

713, 714 BUCHAN PT, PALM BEACH PT

		Unpatrolled		
No.	Beach	Rating	Type	Length
		HT LT		
713	Buchan Pt	2 3	R+LT	1.2 km
714	Palm Beach Pt	2 3	R+rocks	50 m

The view of **Buchan Point** is the first coastal vista that drivers sees north of Cairns, and consequently there is a parking area by the highway where many tourists stop to record the view. The beach (713) faces north-east and runs for 1.2 km from 20 m high Buchan Point down to where the highway first reaches the coast. The beach is moderately steep and narrow at high tide. The usually low waves surge up the high tide beach and break across a narrow, continuous bar at low tide. There are several houses at the southern end of the beach but no public facilities.

The **Palm Beach Point beach** (714) is a 50 m long pocket of sand and rocks on the northern side of the point. It can be reached on foot down a steep slope from a walking track that skirts the top of the 40 m high point.

Swimming: Buchan is a relatively safe beach under normal low waves, with some low surf at low tide. The point beach is isolated and rocky.

Surfing: Best in strong Trades at mid to low tide, when a sloppy beach break is produced.

Fishing: Both beaches offer good rock fishing spots.

Summary: Buchan is highly visible as the first beach seen from the highway north of Cairns, while the point beach is only known to locals.

715 PALM COVE (CAIRNS SLSC)

Patrols (Palm Beach):
  Surf Life Saving Club: July to May
Patrols (Palm Beach, Clifton, Kewarra):
  Lifeguard on duty: all year
  Stinger net: November to May

No.	Beach	Rating HT	LT	Type	Length
715	Palm Cove	2	3	R+LT	5.5 km

Palm Cove Beach (715) lies at the northern end of a 5.5 km long beach that runs south from Buchan Point, where it faces east, through to **Clifton Beach** and on to **Kewarra Beach,** where it faces north-east (Fig. 4.10). Deep Creek drains across the beach between Clifton and Kewarra. Kewarra ends in the south at Taylor Point. Twin-peaked Double Island is located 1 km off Buchan Point.

The Cairns Surf Life Saving Club, founded in 1924, is located on the beachfront 500 m south of Buchan Point (Fig. 4.11). The beach is located 1 km off the Cook Highway and, besides the surf lifesaving club, has most facilities for visitors and tourists, including a variety of tourist accommodations. There is also a boat ramp and fishing jetty just south of the point. The road parallels the beach, with ample parking and a foreshore reserve between the road and beach.

The beach is composed of relatively coarse sand which, together with the waves averaging just under 0.5 m, produces a soft, steep high tide beach, and a 40 m wide attached bar at low tide (Fig. 4.12). Waves surge up the steep beach at high tide and break across the bar at low tide.

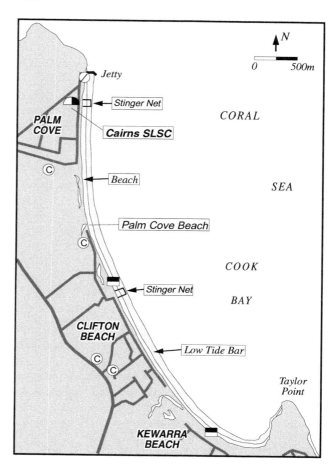

Figure 4.10   *Cairns SLSC is located toward the northern end of Palm Cove Beach, and during the summer months operates a stinger net. The 5.5 km long beach is also the site of Clifton and Kewarra Beaches, which are both patrolled by lifeguards.*

Swimming: This is a relatively safe beach under normal low waves. However, waves and currents increase during strong south-easterlies. The three patrolled beaches and net enclosures offer the safest bathing, particularly during the stinger season.

Surfing: Usually low, sloppy waves, best at mid to low tide and when there is a bit of south-east wind to pick up the waves.

Fishing: The jetty and northern rocks are the most popular spots, with fishing off the beach best at high tide.

Summary: One of Cairns' longer and more popular beaches, both with locals and the increasing number of tourists.

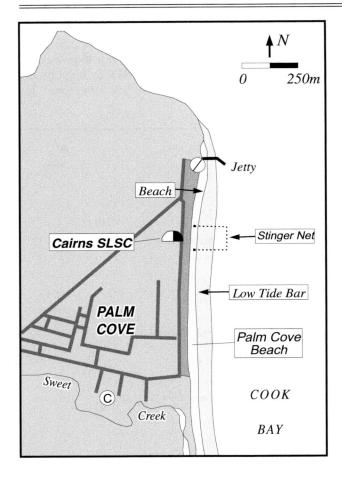

Figure 4.11   *Detailed map of Palm Cove, showing location of seasonal stinger net in front of Cairns SLSC.*

Figure 4.12   *View of Buchan Point and Palm Cove Beach, site of Cairns Surf Life Saving Club. The club and patrol area are located just south of the jetty.*

716 TRINITY BEACH

Patrols:
  Lifeguard on duty: all year
  Stinger net: November to May

No.	Beach	Rating HT	LT	Type	Length
716	Trinity	2	3	R+LT	1.5 km

Trinity Beach is a popular, well developed beach located 2 km east of the Cook Highway, approximately 15 km north of Cairns. The 1.5 km long beach faces north-east and is bounded by 215 m high Earl Hill to the south and 60 m high Taylor Point to the north (Fig. 4.13). Trinity Beach shopping centre is located behind the centre of the southern half of the beach. In addition, facilities for both locals and tourists are available along the beachfront, which includes a continuous foreshore reserve, parking and backing road.

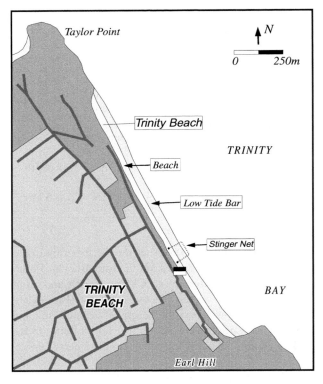

Figure 4.13   *Trinity Beach is a straight, 1.5 km long beach with the patrol area and stinger net located adjacent to the main shopping area.*

The beach is composed of coarse sand that produces a steep, narrow beach at high tide with little surf (Fig. 2.14a), and a wider, lower gradient, exposed, 40 m wide bar at low tide, with a wider surf. Waves average 0.5 m, picking up during strong south-easterly winds.

Lifeguards patrol the central beach area where the stinger enclosure is located.

Swimming: It is safest to swim in the central patrolled area, and in the enclosure during stinger season.

Surfing: Usually low waves that break better at mid to low tide, and are a little higher during strong south-east winds.

Fishing: Best at high tide off the beach, and off the rocks at each end.

Summary: A well developed and popular beach with all facilities, plus lifeguards and a stinger enclosure.

717, 718 MOON RIVER, YORKEYS PT

No.	Beach	Unpatrolled		Type	Length
		Rating HT	LT		
717	Moon River	2	3	R+LT	800 m
718	Yorkeys Pt	1	2	R	100 m

Moon River is a tidal creek that drains the Half Moon Creek Wetland Reserve, that extends for 2.5 km south of the river mouth. The beach has been built across the creek mouth between Earl Hill to the north and 70 m high Yorkeys Knob to the south. A harbour-marina was constructed in 1992 immediately south of the river mouth, and its two harbour walls have substantially stabilised the creek. The marina is backed by a resort and golf course, while there is a canal estate off the creek, behind the main part of the beach.

Moon River beach (717), now truncated by the marina, is 800 m long, faces north-east and receives waves less than 0.5 m high. To the north they form a steep beach and narrow bar; however at the creek mouth, tidal shoals extend up to 300 m off the beach. There is vehicle access to the marina and a large car park, however the main beach has a road to the canal estate, then a walking track to the beach.

On the south side of the harbour, the wall has caused sand to accumulate between it and Yorkeys Point, forming a 100 m long, north facing pocket beach (**718**). It lies right next to the seawall, marina car park and boat ramp.

Swimming: The best spots are the northern end of the main beach and the little pocket beach, as both offer deeper water off the beach and usually low waves.

Surfing: Usually none.

Fishing: Best off the harbour walls, Yorkeys Point, and at the creek mouth.

Summary: A very accessible marina, but more difficult to access main beach.

719 YORKEYS KNOB

Patrols:				
Lifeguard on duty:		November to May		
Stinger net:		November to May		
No. Beach	Rating HT	LT	Type	Length
719 Yorkeys Knob	1	2	R+LT	1.7 km

Yorkeys Knob is a 70 m high headland which forms the northern end of the beach of the same name (719). The beach runs toward the south-east for 1.5 km where it encounters the mouths of Yorkeys and the larger Richters Creeks, and their associated tidal shoals (Fig. 4.14). Yorkeys Knob settlement, including two caravan parks, parallels the northern half of the beach, with houses also spreading up onto the headland. There is excellent beach access for the length of the beach.

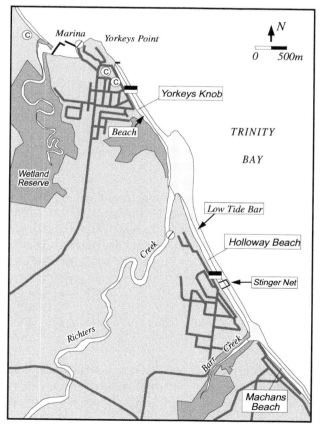

Figure 4.14 *Yorkeys Knob Beach abuts Yorkeys Point, and extends south-east to Richters Creek, beyond which is Holloway Beach and then Machans Beach. Both Yorkeys and Holloway are patrolled by lifeguards.*

This attractive, tree-lined beach is composed of medium sand, that has built a steep high tide beach, while at low tide, low sand flats are exposed. The sand flats increase dramatically in width toward the southern creek mouths. A rock groyne is located at the north end and extends out from the Knob. This was constructed in 1959-60, following severe beach erosion during the 1956 cyclone. Sand has built up on the southern side of the groyne providing a degree of protection for the road and houses. However, a rock seawall is still present along most of the developed beach section.

Swimming: Best at high tide in the northern patrolled area. Sand flats are exposed at low tide, while closer to the creek mouths there are strong tidal currents.

Surfing: Usually only small wind waves that are best at mid tide. Waves reach 1.5 m during strong Trade winds, with best surfing and windsurfing toward the northern end.

Fishing: The rock groyne is popular at high tide, while the creek mouth can be fished from the banks or a boat.

Summary: A quiet beach 5 km off the highway; favoured by locals and campers.

720 HOLLOWAY BEACH

Patrols:				
Lifeguard on duty:			November to May	
Stinger net:			November to May	
No.	Beach	Rating HT LT	Type	Length
720	Holloway	1 2	R+LT	2.5 km

Holloway Beach (720) is one of three beaches that form the northern shoreline of the Barron River delta. The beach is very much part of the delta, with Richters and Barr Creeks forming the northern and southern boundaries of the 2.5 km long beach. The beach and backing settlement is, in fact, surrounded by creeks and mangrove flats associated with the delta. The beach lies 5 km off the highway and is backed for the most part by residential development, with good road access the length of the beach.

The beach faces north-east and is composed of medium sand that produces a relatively steep high tide beach, fronted by a low tide bar. The bar grades into the very extensive intertidal flats toward the two creek mouths.

Swimming: The best swimming is at mid to high tide, as the sand flats are exposed and very shallow at low tide. A stinger enclosure operates during the summer season.

Surfing: Only small wind waves, better at mid tide.

Fishing: The creek mouths are the most popular areas, as well as the beach at high tide.

Summary: A narrow high tide beach, backed by palm trees, with good access for the length of the beach.

721 MACHANS

No.	Beach	Unpatrolled Rating HT LT	Type	Length
721	Machans	2 3	R+LT/SF	2.5 km

Machans Beach (721) is located immediately north of the main Barron River mouth and is part of the Barron River delta. Consequently, while it is a relatively low energy beach with waves averaging less than 0.5 m, it has a very dynamic shoreline owing to changes in the river mouth. The beach is 2.5 km long, with a road and beachfront houses backing most of the beach. In order to protect the houses from shoreline changes, the beach has largely been replaced or backed with a rock seawall. The northern end of the beach has a relatively narrow bar, which toward the southern end is replaced by tidal flats up to 1 km wide. The beach is occasionally breached by Redden Creek, with the southern half called Redden Island and the beach South Machan Beach. However the creek may close for years at a time, forming one continuous beach.

Swimming: Best toward the northern half away from the shallow tidal flats. Beware of crocodiles near the river mouth.

Surfing: Usually none.

Fishing: Best in Barr Creek, when open, and at the river mouth.

Summary: A long, narrow beach that has been overdeveloped, resulting in the need for the seawall.

722, 723 BARRON, ELLIE PT

No.	Beach	Unpatrolled Rating HT LT	Type	Length
722	Barron	1 2	B+SF	2 km
723	Ellie Pt	1 2	B+SF	1 km

For 4 km south of the Barron River mouth and delta, the shoreline protrudes over 2 km seaward as a result of sediment deposited by the river. In fact, the main river mouth used this part of the delta until a switch in river channels in 1939. Much of this area is low-lying tidal flats with extensive mangroves, with thin strips of beach fronted by tidal flats extending a further 1 km out into Trinity Bay.

There are two beaches along this dynamic section; a 2 km strip of sand running around Casuarina Point called **Barron** beach (722), and a 1 km strip around **Ellie Point** (723). Both are low-lying tidal flats and are bordered by the Barron River and Airport Creek. Neither beach has land access, and both are subject to shoreline changes during and after river floods.

Swimming: Unsuitable, owing to the extensive tidal flats and possible mangroves.

Surfing: None.

Fishing: Best in the adjoining river and creeks.

Summary: Two little used strips of sand, close to Cairns, but located in the tidal flats.

724 CAIRNS

No.	Beach	Unpatrolled		Type	Length
		Rating			
		HT	LT		
724	Cairns	1	1	TSF	1.5 km

Cairns is largely built on a low, sandy beach system which, fed by sands from the Barron River, has built out up to 4 km into Trinity Bay. Today, the seaward end of the system is fringed by mangroves to the north, bordered by wide Trinity Inlet to the south, while along the Cairns Esplanade is 1.5 km of high tide beach and another 1 km of seawall. The beach (724) is somewhat artificial, as there has been extensive reclamation of the mangroves and tidal flats, and placement of sand to produce a recreational area along the wide foreshore reserve that runs the full length of the Esplanade.

The beach is a steep, narrow strip of sand at high tide, and tidal flats up to 1.5 km wide at low tide. It has excellent access and parking along the full length and numerous facilities in the backing reserve.

Swimming: This is not a popular beach for swimming, owing to the shallow tidal flats. Best at high tide.

Surfing: None.

Fishing: Only at high tide.

Summary: Cairns' gateway beach, however it's really a converted tidal flat which, none the less, has a magnificent foreshore reserve and provides good access to, and view of, Trinity Bay.

725-729 GIANGURRA, SECOND BEACH, BROWN BEACH, KOOMBAL, SUNNY BAY

No.	Beach	Unpatrolled		Type	Length
		Rating			
		HT	LT		
725	Giangurra	1	1	B+SF	800 m
726	Second Beach	1	1	B+SF	500 m
727	Brown Bay	1	1	B+SF	500 m
728	Koombal	1	1	B+SF	500 m
729	Sunny Bay	1	2	R+LT	200 m

Opposite Cairns is the Murray Prior Range, that rises to over 500 m. On the Cairns side of the range are a series of low energy beaches and tidal flats, while the Yarrabah Aboriginal Community and several more beaches lie to the east. The first five beaches occupy the 6 km of steep shore between the mangroves of Cairns Harbour and False Cape. They all face north-west and are well protected from most waves by their orientation and extensive tidal flats.

The first four beaches between Bessie Point and Koombal are serviced by the Yarrabah Road, and coincide with the small foreshore settlements of Giangurra, Lyons Point and Koombal. **Giangurra** (725) has an 800 m long high tide beach fronted by 1 km of tidal flats. Either side of Lyons Point and in front of Koombal are three near-continuous, but irregular, beaches (**726, 727, 728**), each about 500 m long, and fronted by 500 m wide tidal flats. Finally at Sunny Bay, in lee of False Cape, is a 200 m long, steep, sandy beach (**729**) with large boulders and fronted by a narrow bar/tidal flat. It has no road access.

Swimming: At all beaches it is only possible to swim at high tide, owing to the shallow tidal flats.

Surfing: None.

Fishing: Only at high tide from the beaches.

Summary: A series of quiet, low energy beaches, right opposite Cairns but a world away.

730-733 MISSION BAY BEACHES

No.	Beach	Unpatrolled		Type	Length
		Rating			
		HT	LT		
730	False Cape (E)	1	2	R+LT	200 m
731	Yarrabah	1	1	B+SF	3.3 km
732	Bulburra	1	1	B+SF	500 m
733	Palm Beach	1	1	B+SF	300 m

Mission Bay is a 7 km wide, north facing bay bordered by False Cape and Cape Grafton. Cook named Cape Grafton in 1770 and noted in the bay several fires and some (aboriginal) people. He anchored 3 km offshore owing to the shallow nature of the bay. Later a mission was established in the bay, and today it is the site of the Yarrabah Aboriginal Community. It is therefore a region of long and continuous aboriginal occupation.

The settlement is located along the western end of Yarrabah Beach, at the end of the Yarrabah Road. It is the only one of the four bay beaches that can be reached by road. There is a small, exposed beach (**730**) out on the east side of False Cape, that consists of a high tide strip of sand and large granite boulders. The main **Yarrabah Beach** (731) is 3.3 km long, faces north and consists of a narrow, low energy high tide strip of sand fronted by tidal flats up to 1.5 km wide. It grades into mangroves at its western end.

Bullburra (732) and **Palm** (733) beaches occupy 500 m and 300 m breaks in the mangroves toward the centre and eastern end of the bay. They have no road access and can only be reached by boat at high tide. Both are fronted by 500 m wide tidal flats.

Swimming: It is only possible to swim at high tide owing to the wide, shallow tidal flats.

Surfing: None.

Fishing: Again, only at high tide from the beaches.

Summary: All these beaches are on Aboriginal land and a permit is required to visit the community and its shoreline.

734, 735 TURTLE BAY, LITTLE TURTLE BAY

No.	Beach	Unpatrolled		Type	Length
		Rating			
		HT	LT		
734	Turtle Bay	1	2	R+LT	800 m
735	Little Turtle Bay	1	2	R+reef	100 m

Turtle Bay is an open, hook-shaped bay 4 km south of Cape Grafton. It faces north to north-east and houses a steep, 800 m long beach (734) which, toward its northern end, is broken by large granite boulders into two smaller beaches. The bay provides good anchorage just off the beach, particularly in the more protected southern corner.

Little Turtle Bay (735) lies 1 km further south and faces south-east into the Trades. It is bordered by two small headlands, is protected by a fringing coral reef and consists of the reef and a 100 m long high tide beach. Both beaches are only accessible by boat.

Swimming: Turtle Bay offers a nice, steep beach with deep water right off the beach, particularly at high tide. Little Turtle Bay has the reef running for 200 m off the beach.

Surfing: There are only waves breaking over the Little Turtle Bay reef at high tide.

Fishing: Both beaches have good vantage points at high tide, and from the rocks all day.

Summary: Two nice, steep, sandy beaches, with Turtle Bay the best for boats and Little Turtle Bay offering its small reef.

736, 737 KING BEACH

No.	Beach	Unpatrolled		Type	Length
		Rating			
		HT	LT		
736	King Beach (N)	2	3	R+LT	100 m
737	King Beach	2	3	R+LT	3 km

King Beach is a 3 km long, south-east facing beach located in Wide Bay, in lee of Fitzroy Island 2.5 km offshore. The beach is bounded by prominent, 200 to 300 m high headlands that extend more than 1 km seaward. The main beach (737) is backed by extensive vegetated coastal dunes, with a 10 m high foredune behind the beach. At the junction of the northern headland and the beach is a small, 100 m long pocket beach (736), bordered by scattered granite boulders. Between the two beaches is a previously occupied clearing. There is, however, no vehicle access to the beach. The main beach is composed of coarse sand and consequently is relatively narrow at high tide, with a steep beach face, fronted by a 50 to 100 m wide continuous bar. During the Trades, waves 0.5 to 1 m high break across the bar at mid to low tide.

Swimming: Both beaches have relatively deep water

off the beach at high tide and a surf at low tide during the Trades. When waves exceed 0.5 m rips may flow across the surf, so be careful.

Surfing: Only during the Trades when there is a sloppy break across the bar, which is best at mid to low tide.

Fishing: Reasonably good beach fishing at low tides, with rocks at each end.

Summary: Two isolated, natural beaches, best accessed by boat and used for camping and fishing by the local community.

CAIRNS TO HINCHINGBROOK ISLAND (Fig. 4.15)

Length of Coast:	202 km
Beaches:	87 (beaches 740-826)
Islands:	Dunk Island (2 beaches)
Surf Life Saving Clubs:	Etty Bay, Mission Beach
Lifeguards:	Bramston Beach
Major towns:	Innisfail, Mission Beach region

Regional Map 2: Cairns to Hinchinbrook Island

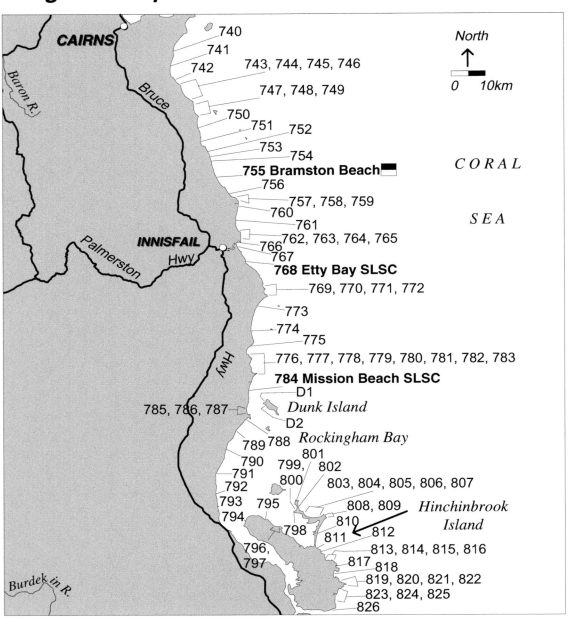

Figure 4.15   *Regional Map 2-Cairns to Hinchinbrook Island, beaches 740 to 826. Beaches with Surf Life Saving Clubs indicated in bold, beaches with lifeguard patrol only indicates by flag.*

738-740 OOMBUNGHI

No.	Beach	Unpatrolled		Type	Length
		Rating			
		HT	LT		
738	Oombunghi	2	3	R+LT	4.5 km
739	Oombunghi (S 1)	2	3	R+LT	1.0 km
740	Oombunghi (S 2)	2	3	R+LT	600 m

Oombunghi Beach lies 9 km south of the Yarrabah settlement and is connected to it by a gravel road. The road reaches the southern end of the beach and runs on along the coast to the southern two beaches. There are scattered houses and shacks along the road and on all three beaches. There is also a vehicle track to the northern end, and a few houses.

Oombunghi Beach (738) is 4.5 km long and faces southeast, into the Trades, and is bordered by Deception Point to the north and lower rocks to the south. The beach has a moderately steep high tide beach, owing to the medium to coarse sand, fronted by a continuous bar 50 to 100 m wide. Waves surge up the beach at high tide and break across the bar at mid to low tide. There is a small creek draining across the beach toward its southern end, with several beachfront houses between the creek mouth and the southern boundary rocks.

On the south side of the rocks is the first southern beach (**739**). It is 1 km long, faces east and has finer sand than the main beach. The sand and slightly lower waves produce a lower, wider beach. There are also low rocks bordering each end and in the centre of the beach. The road runs behind the beach, with a few houses scattered along the foreshore.

The second southern beach (**740**) has even finer sand and a very low, 100 m wide beach at low tide. Rocks border the ends, as well as a rocky islet just off the beach that is connected by sand at low tide. The road ends at this beach, where there are a few houses, then continues on as a 4WD track to the mouth of Saltwater Creek.

Swimming: All three beaches are used by the locals for swimming and fishing. The main beach has deep water off the beach at high tide and surf at mid to low tide, while the two southern beaches have lower surf breaking across the low beach and bar.

Surfing: Waves up to 1 m high break across the bars during the Trades.

Fishing: Best at high tide and off the many rocks that border these beaches.

Summary: Three relatively natural tropical beaches that provide beachfront living and surf for the Yarrabah community.

741 SALTWATER CK

No.	Beach	Unpatrolled		Type	Length
		Rating			
		HT	LT		
741	Saltwater Ck	2	3	R+LT	6 km

Saltwater Creek beach (741) is a 6 km long, east facing beach, that remains untouched by development, other than a fishing camp at the southern end. The beach is bordered by the rock-impounded Saltwater Creek entrance at the northern end, and low rocks at the southern end. It is backed by densely vegetated, low sand ridges, then steep slopes rising several hundred metres to the crestline of the Malbon Thompson Range. The entire area is part of the Yarrabah Aboriginal Community, and there are no access roads to the beach.

The beach itself is composed of coarse sand derived from the surrounding granite. This sand, coupled with waves averaging 0.5 m, have built a steep and narrow beach at high tide (Fig. 4.16), fronted by a 50 m wide bar at low tide. During high wave conditions, small rips, spaced every 100 m, cut across the bar, producing up to 60 rips along the beach. In places the beach has eroded back into the overhanging vegetation, causing it to fall on the beach.

Figure 4.16 *Saltwater Creek beach, a steep narrow high tide beach, with waves just breaking on the low tide bar.*

Swimming: A reasonably safe beach under normal low waves, with deep water off the beach at high tide and some small surf at low tide. However when waves exceed 0.5 m, rips cut across the surf zone.

Surfing: Only during strong Trades, when waves may reach a metre and break across the bar at mid to low tide.

Fishing: Best at high tide and following bigger seas,

when there are rip holes along the beach.

Summary: An isolated beach on aboriginal land and only accessible by boat.

742, 743   BEACH 742-743

		Unpatrolled		
No.	Beach	Rating HT LT	Type	Length
742	Beach 742	2 3	R+LT+rocks	2.5 km
743	Beach 743	2 3	R+LT+rocks	600 m

At the southern end of Saltwater Creek Beach, the steep, vegetated slopes reach the sea and the coast trends south-east. Running along the base of the ranges is a strip of sand, interspersed with small, rocky outcrops. Beaches 742 and 743 occupy the first 3 km of the rocky coast. Beach **742** consists of six near-continuous high tide strips of sand, separated by small, rocky headlands of varying sizes, with some rocks scattered along the beaches and exposed at low tide. A larger headland separates it from Beach **743**, which consists of two small pocket beaches, totalling about 600 m in length. Neither beach has any access or development.

The two beaches are composed of the coarse, granitic sand that produces a steep high tide beach, fronted by a flatter low tide beach about 30 m wide, with rocks outcropping along the beaches. If approaching the beaches by boat at high tide, beware of the submerged rocks.

Swimming: Two relatively safe beaches, with usually low waves. However, care must be taken of the numerous rocks, particularly at high tide.

Surfing: Only a small break across the bar at mid to low tide.

Fishing: Numerous good spots on the many rocks along the beaches and headlands.

Summary: Two natural beaches backed by steep, densely vegetated slopes and fringed by rocks.

744   BEACH 744

		Unpatrolled		
No.	Beach	Rating HT LT	Type	Length
744	Beach 744	2 3	R+LT	3 km

Beach 744 is an east facing, 3 km long beach, that has built out 200 to 300 m from the backing steep slopes of the Malbon Thompson Ranges, which rise to 1 000 m within 2.5 km of the beach. There is no access to the beach, and today dense vegetation covers the backing sand ridges and steep slopes.

The first 500 m of the beach are intersected by a rocky point, and numerous rocks outcrop along the beach. For the next 2.5 km it is a continuous, steep, sandy beach, with vegetation hanging over the high tide beach. A 50 m wide, continuous bar is exposed at low tide.

Swimming: There is deep water off the main beach at high tide, while low waves break across the bar at mid to low tide.

Surfing: Only low waves breaking across the narrow bar.

Fishing: Best at high tide, although you can cast across the bar at low tide. There is a rough fishing camp on Beach 744.

Summary: An isolated beach surrounded by dense tropical vegetation.

745-749   BEACH 745-749, PALMER PT (N)

		Unpatrolled		
No.	Beach	Rating HT LT	Type	Length
745	Beach 745	2 3	R+LT+rocks	1 km
746	Beach 746	2 3	R+LT+rocks	2 km
747	Beach 747	2 3	R+LT+rocks	1.5 km
748	Beach 748	2 3	R+LT	2 km
749	Palmer Pt (N)	2 3	R+LT+rocks	500 m

Between Beach 745 and Palmer Point is a 7 km section of coast that runs around the base of a 500 m high spur of the Malbon Thompson Range. Near-continuous beaches, interrupted by granite outcrops and small headlands, fringe the spur. All five beaches (745-749) are composed of coarse granitic sand and all have steep high tide beaches, with 30 to 50 m wide low tide bars. What differentiates the beaches is their integrity between the rocks, their orientation and degree of rock outcrop along the beach. There is no land access to any of the beaches and no facilities, although some are used for fishing camps.

Beach **745** is 1 km long and faces north-east, with rocks outcropping at either end and along much of the northern half. There is a small, rocky headland at the southern end.

Beach **746** consists of six scallops of sand, extending

for a total of 2 km along the base of the steep slopes, and facing north-east. Each sand patch is bordered by low, granite outcrops and all have varying degrees of rocks on the beach and bars.

Beach **747** is a more east facing beach, being located on the south side of a turn in the spur. The rocks of the spur give way sufficiently for three near-continuous strips of sand, totalling 1.5 km.

Beach **748** is the straightest and longest continuous strip of sand, 2 km in length and facing nearly due east. It is backed by a narrow valley and there are some low sand ridges behind the beach, while the small valley creek drains across the northern end and ponds a small lagoon behind the beach.

Beach 749 is partially protected by **Palmer Point**. It is 300 m long, faces north-east and is backed by steep, densely vegetated slopes.

Swimming: The best beaches for swimming are those that are more rock-free, namely Beach 746, which is backed by an open valley rather than steep slopes. Be careful of the numerous rocks, that may be submerged at high tide.

Surfing: Usually low waves breaking across the bar, with larger waves during strong Trade winds.

Fishing: The local aboriginal community has a number of fishing camps on these beaches, attesting to the possibility of some good spots. The beaches have deep water at high tide and there are numerous rocks and small headlands to fish from. Be careful of the many rocks if fishing from a boat.

Summary: A series of beautiful, natural beaches mixed with numerous granite rocks, and all backed by dense tropical vegetation.

750, 751   WOOLANMARROO

No.	Beach	Unpatrolled		Type	Length
		Rating HT	LT		
750	Flirt Pt (Woolanmarroo)	2	4	R+LT+inlet	6.5 km
751	Constantine Pt (Woolanmarroo S)	2	4	R+LT+inlet	3 km

There is a breech in the Malbon Thompson Range between Palmer and Bramston Points. It has been carved by the Mulgrave and Russell Rivers, that converge and flow through the breech at Mutcheno Inlet. On either side of the unstable inlet are two sandy barriers and fronting beach systems, with an extensive tidal delta and shoals at the inlet.

Between Palmer Point and the inlet is 6.5 km long **Flirt Point beach** (750, also known as Woolanmarroo) and the southern end of the Yarrabah Aboriginal Land. The beach begins right at the small Palmer Point and runs south for 4 km before it encounters the inlet shoals, which extend up to 1.5 km seaward. The shoals and 200 m wide inlet are very unstable. The beach consists of a steep high tide beach, fronted by a continuous, 50 m wide bar; however at the inlet, the bar gives way to tidal shoals that extend well seaward at low tide.

On the south side of the inlet is the even more unstable **Constantine Point**, a 1 km long, narrow spit, which then becomes a more stable beach (751) that extends for a further 2 km to the protected lee of Bramston Point. Unfortunately a number of houses have been built on the spit and all are in danger of being eroded as the spit continues its natural oscillations. A few makeshift seawalls have been built in front of the houses. Both beaches are only accessible by boat, and apart from the houses there are no facilities.

Swimming: The best swimming is well away from the tidal inlet, that has strong tidal currents.

Surfing: Usually only low waves breaking across the narrow bar.

Fishing: The inlet is a popular spot, either fished directly from the shore or more commonly by boat.

Summary: Two unusual beaches framing a dynamic inlet, and backed by a narrow breech in the ranges.

752-754   BRAMSTON

No.	Beach	Unpatrolled		Type	Length
		Rating HT	LT		
752	Bramston Pt	2	3	R+LT+rocks	3 km
753	Bramston Pt (S)	2	3	R+LT+rocks	800 m
754	Bramston (N)	2	3	R+LT	600 m

Bramston Point forms a major promontory. The backing slopes rise to 560 m and its densely vegetated slopes run right down to the shoreline, where the tropical vegetation overhangs the steep, narrow beaches that fringe the base. It is composed of granitic rocks, and large granite boulders and rocks border all beaches, breaking some of the beaches into a series of smaller sandy pockets. There is no vehicle or foot access to these beaches and no facilities.

Bramston Point beach (752) begins on the south side of the point. It faces north-east and consists of a 3 km series of nine pockets of sand, each separated from the next by clumps of granite rocks and boulders. In addition, rocks outcrop on most of the beaches. The beaches are all relatively steep, with a narrow high tide beach overhung with trees, and a 50 m wide low tide bar, that usually has some rocks.

Bramston Point (south) (753) is a straighter, 800 m long beach that faces more easterly. It has a small creek that is impounded behind the northern end, forming a small lagoon. The creek crosses the beach next to some large boulders. It has a steep beach face and a continuous bar along the base of the beach, across which waves break at mid to low tide.

Bramston Beach (north) (754) lies around a small headland from Bramston Beach. It is 600 m long, faces east and has a 50 to 100 m wide low tide bar. There are the remnants of a clearing on the steep slopes behind the beach, but otherwise no development. This beach can be reached on foot from the track that leads to the mouth of Wyvuri Creek, at the northern end of Bramston Beach.

Swimming: All three beaches have relatively deep water off the beach at high tide, and waves breaking across the bar at mid to low tide. Rips only occur when waves exceed 0.5 m. Be careful of the many rocks, that may be submerged at high tide.

Surfing: Only sloppy beach breaks across the bar at mid to low tide.

Fishing: There are numerous vantage points along these beaches, both on the beaches at high tide and on the many rocks, as well as the small creek on Bramston Point (south) Beach.

Summary: Three natural beaches that are really only accessible by boat, all offering dense tropical vegetation literally overhanging the sea.

755 BRAMSTON BEACH

Patrols:				
Lifeguard on duty:			Christmas & Easter	
Stinger net:			November to May	
No. Beach	Rating HT LT		Type	Length
755 Bramston	1	2	R+LT	8.5 km

Bramston Beach (755) is located 20 km north of Innisfail, and lies 18 km east of the Bruce Highway. There is a small holiday settlement, a store, motel and beachfront camping area, all located toward the southern

end of the 8.5 km long beach. The beach faces east-north-east, with prominent Rocky Point forming the southern boundary, and Wyvuri Creek, which exits against some granite rocks, at the northern boundary (Fig. 4.17). Luxuriant tropical vegetation fringes the back of the beach. The main road reaches the beach at the groyne and camping area, with good access to the southern end and Rocky Point, and a rough track leading up to the northern end and Wyvuri Creek mouth.

Waves average about 0.5 m and produce a moderately steep high tide beach, fronted by a low, flat low tide bar (Fig. 4.18). The main part of the beach, adjacent to the camping area, has a rock groyne that has trapped the sand to build a wide, steep beach to the south, with a low, flat, eroding beach to the north. Toward Rocky Point, wider tidal flats replace the bar. The small Joyce Creek flows out against the point, with a breakwater partly protecting the creek, so as to allow small boats to reach the boat ramp inside the creek mouth.

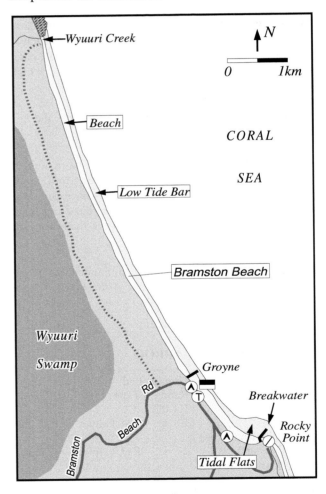

Figure 4.17   *Eight kilometre long Bramston Beach is bordered by Wyvuri Creek in the north and low Rocky Point in the south. The main settlement, caravan park and patrolled area are located immediately south of the groyne.*

Figure 4.18   *Bramston Beach shown at low tide.*

Swimming: A relatively safe beach with best swimming at mid to high tide, when the bar and tidal flats are covered. The patrolled area is immediately in front of the camping area and a stinger net is operated during the summer.

Surfing: Usually low waves breaking across the bar. Strong south-easterlies are required to produce larger wind waves.

Fishing: The southern Joyce Creek and breakwater is the most accessible and popular spot, together with the main groyne and beach at high tide.

Summary: A nice, quiet spot all drenched in tropical vegetation, with limited tourist facilities, a stinger enclosure, a long beach and interesting creeks and headlands.

---

756   ROCKY PT

No.	Beach	Unpatrolled		Type	Length
		Rating HT	LT		
756	Rocky Pt (S)	2	3	R+LT	4 km

On the south side of the low **Rocky Point** is a 4 km long, north-east facing beach (756) that extends down to the protected lee of Cooper Point, capped by 200 m high Mount Cooper. There is access to the north end of the beach from Bramston Beach, which leads to a few houses set behind the beach. There is no other development and no facilities.

The beach has a moderately steep high tide beach, fronted by a continuous, 50 m wide bar, which widens considerably toward Cooper Point, resulting in shallow shoals extending several hundred metres off the southern end of the beach.

Swimming: Best at high tide, particularly in the south.

Surfing: Usually a low break over the bar at mid to low tide, however during strong Trades, some spilling waves break across the southern shoals at high tide.

Fishing: Better at high tide, particularly in the south.

Summary: A long beach that is beginning to be developed, offering plenty of sand and the low tide bar and shoals.

Ella Bay National Park

Beaches: 757-760, 761

Ella Bay National Park occupies the shore and backing land at both ends of Ella Bay, with freehold land in between.

757-759   COOPER PT

No.	Beach	Unpatrolled		Type	Length
		Rating HT	LT		
757	Cooper Pt (1)	1	2	B+SF	400 m
758	Cooper Pt (2)	2	3	R+LT	700 m
759	Cooper Pt (3)	2	3	R+LT	100 m

Cooper Point protrudes 2.5 km to the east from the southern end of Rocky Point Beach. Along the base of its steep, densely vegetated slopes is a strip of sand and metamorphic rocks that comprises the three beaches. The three all lie within Ella Bay National Park and are undeveloped, with no vehicle or foot access and no facilities.

Cooper Point (1) beach (**757**) is 400 m long, faces almost due north and receives low waves averaging about 0.3 m. It has a steep, narrow high tide beach covered with numerous rocks and boulders, and tidal shoals off the beach.

Cooper Point (2) (**758**) is a curving, 700 m long, north-east facing beach, bordered by large rock outcrops, together with clumps of rocks along and off the beach. It receives slightly higher waves, has a steep beach face and a shallow low tide bar grading to shoals.

Cooper Point (3) (**759**) is a small, 100 m long, east facing beach on the south-eastern tip of the point. Large fingers of metamorphic rock border the steep beach, while at low tide a 50 m wide bar is exposed.

Swimming: On all three beaches it is better to swim at high tide, as there are shallow shoals and bars exposed at low tide, plus the many rocks. Beware of the rocks that may be submerged at high tide.

Surfing: During strong Trades, there is a chance of a wave running along the north side of Cooper Point and breaking over the low tide shoals.

Fishing: Best at high tide off the beaches or on the many rocks.

Summary: Three natural beaches backed by the steep, tropical slopes of Mount Cooper.

760, 761   ELLA BAY

No.	Beach	Unpatrolled Rating HT	LT	Type	Length
760	Ella Bay (N)	2	3	R+LT	2.8 km
761	Ella Bay	2	3	R+LT	4.8 km

Ella Bay is an open, 9 km long, east facing bay, lying between Cooper and Heath Points. The bay is backed by an amphitheatre of ranges rising 300 to 400 m, with the former valley largely filled with sediments of the extensive Ella Swamp and the two Ella Bay beaches. The sediments have been derived from the large Johnstone River a few kilometres to the south, with the sands and mud driven northward and into the bay by waves and strong tidal currents. Ella Bay National Park backs the northern and southern ends of the bay and encompasses much of the densely vegetated lowland and eastern slopes of the backing Seymour Range.

Ella Bay (north) Beach (760) runs for 2.8 km from Cooper Point to a small unnamed headland. Cooper Creek runs out across the northern end of the beach hard against the rocks of the point. The beach is backed by a low, 500 m wide sand barrier, then the swamp. There is no land access to the beach and no development. The beach is steep at high tide, with waves surging against the beach face, while at low tide a 50 m wide bar causes the waves to break off the beach.

Ella Bay Beach (761) runs from the small headland down to the protected lee of Heath Point. It is 4.8 km long and similar to the north beach, only with a narrow strip of sand separating it from the swamp. A small creek occasionally drains across the centre of the beach. There is a gravel road around Heath Point to the southern end, where there is a car park and camping area. The high tide beach is steep, while the low tide bar and shoals increase in width toward Heath Point.

Swimming: Better at mid to high tide when the bar is covered.

Surfing: Usually a low break across the bar at mid to low tide.

Fishing: The beaches can be fished at high tide, as there are extensive shoals exposed at low tide at the accessible southern end.

Summary: Two relatively natural beaches in a tropical setting, with essentially no development but good access to the southern beach.

762-764   HEATH PT

No.	Beach	Unpatrolled Rating HT	LT	Type	Length
762	Heath Pt (1)	1	2	R+tidal shoals	600 m
763	Heath Pt (2)	1	2	R+tidal shoals	400 m
764	Heath Pt (3)	2	3	R+LT	200 m

Heath Point is a 200 m high spur of metasediments (schists) that forms the southern end of Ella Bay. The road from Flying Fish Point winds around the base of the point to give access to Ella Bay, with a short climb required to reach the three beaches at the base of the point.

The beaches comprise a near-continuous, crenulate strip of sand that runs from the southern end of Ella Bay out to the eastern tip of the point. The high tide beaches are all steep, with extensive tidal shoals off the first two beaches. Sand for the shoals, probably originating from the Johnstone River, is moved by tides and waves around Heath Point. The mobile field of sand shoals slowly moves along the beaches into Ella Bay. As a result of this movement the width of the beaches and shoals varies from year to year.

Heath Point (1) Beach (762) is 600 m long, faces northeast and is fronted by the tidal shoals. It is easily accessible on foot from the Ella Bay Road and car park.

Heath Point (2) (763) is 400 m long, faces slightly more northerly and has to be reached by climbing down from the road. Both beaches have a steep high tide beach, with tidal shoals extending 200 to 300 m off the beach.

Heath Point (3) (764) is located on the eastern tip of the point. It is 200 m long and faces east. As a consequence it receives higher waves, has a steep beach and a narrower low tide bar.

Swimming: Swimming is only possible at mid to high tide because of the shallow tidal shoals. However, watch for submerged rocks off the beaches.

Surfing: During strong Trades, waves wrap around Heath Point and break over the tidal shoals, producing sloppy, spilling breakers. Otherwise there is usually no surf.

Fishing: Only at high tide on the first two beaches, or off the rocks on the point beach.

Summary: Three crenulate beaches lying below the Ella Bay Road, but a little difficult to access. A nice spot for a walk from Ella Bay to Heath Point.

765-767 FLYING FISH PT, COCONUT BAY, CONQUETTE PT

No.	Beach	Unpatrolled Rating HT	LT	Type	Length
765	Flying Fish Pt	1	2	R+tidal shoals	3 km
766	Coconut Bay	1	2	B+SF	1 km
767	Conquette Pt	1	2	R+tidal shoals	1.5 km

The Johnstone River, one of the larger in north Queensland, enters the sea at Gladys Inlet. It has a 300 m wide entrance and tidal shoals extending up to 1 km seaward of the inlet. There are three beaches associated with the river mouth - Flying Fish Point to the north, Coconut Bay Beach in the mouth, and Conquette Point Beach forming the southern entrance.

The town of Innisfail is located 5 km west of the river mouth. A road runs out from Innisfail along the north side of the river, past the back of Coconut Bay to Flying Fish Point. On the south side, the Conquette Point Road runs along the southern side of the river to within 400 m of the beach.

Flying Fish Point beach (765) is 3 km long, extending straight south from Heath Point to Flying Fish Point; a small, rocky headland that forms the river's northern head. There is a small settlement behind the southern end of the beach, including a store and limited facilities. The beach is relatively steep at high tide, while at low tide the Johnstone River tidal shoals extend several hundred metres off the beach. Changes in the width of the beach, caused by the river fluctuations, have resulted in wooden groynes being built across the beach, which have succeeded in producing an ugly beach, if nothing else.

Coconut Bay beach (766) is set just inside the river mouth. It is 1 km long and faces south-east. The Flying Fish Point Road runs along the back of the narrow beach, with beachfront houses along the length of the beach. The beach is fronted by low, flat tidal and river flats and the river channel, which clips the southern end of the beach.

Conquette Point Beach (767), on the southern side of the river, has to date been spared from development and, while the beach also oscillates with the river, apart from

a few trees falling in, no one is concerned. This beach has finer sand and a lower, wider beach face, with tidal shoals extending over 1 km off the beach that are exposed at low tide. There is also a small creek, that houses a few mangroves, draining across the southern corner of the beach.

Swimming: You can only swim on these beaches at mid to high tide because of the extensive tidal shoals and flats. Be careful at Coconut Bay and near the river mouth, as strong tidal currents persist and crocodiles may be about.

Surfing: Usually none, apart from spilling waves over the tidal shoals during strong Trades.

Fishing: Everyone heads for the river mouth, which can be fished from all three beaches and in boats, with a boat ramp at Flying Fish Point.

Summary: Three low energy but dynamic beaches, owing to their location next to the Johnstone River. Its floods and sediments play a major role in the nature and stability of these beaches.

Etty Bay National Park

Beaches: none

Etty Bay National Park is a small coastal park backed by the steep, densely vegetated slopes of the Moresby Range.

---

768 ETTY BAY (SLSC)

Patrols:				
Surf Life Saving Club:		July to May		
Lifeguard on duty:		Christmas & Easter		
Stinger net:		November to May		
No.	Beach	Rating HT LT	Type	Length
768	Etty Bay	1 2	R+LT	700 m

Etty Bay is the surfing beach (768) for the town of Innisfail. It is located 15 km by road from Innisfail and 7 km from the Bruce Highway, with the final kilometre climbing 100 m over the coastal Moresby Range before descending into the picturesque little bay. The bay is a small indentation in the range, sufficiently large to contain the 700 m long, east facing beach, and with enough flat land for a road, the Etty Bay Surf Life Saving Club and a small caravan park, running into a narrow northern valley (Fig. 4.19). This has long been a popular beach, with the surf lifesaving club founded in 1935.

The beach is backed by a narrow, tree-shaded reserve, then steep, forested slopes rising up to 250 m behind the beach. The surf lifesaving club, a store and toilets sit behind the middle of the beach. The beach is composed of fine sand that produces a gently sloping beach. At high tide it is less than 50 m wide, however at low tide a shallow bar over 100 m wide is exposed. Waves average 0.5 m and spill across the shallow bar. There are rocks on the beach and in the surf at the southern end (Fig. 4.20), with a small, rocky promontory crossing the northern end that divides the beach at high tide. The stinger net (Fig. 3.7b) is located in front of the surf lifesaving club.

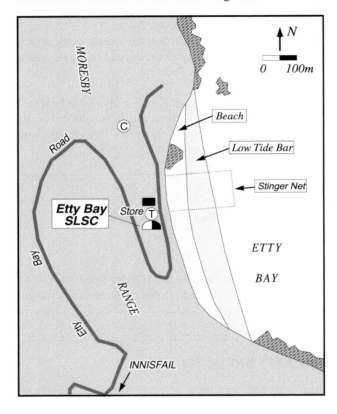

Figure 4.19 *Etty Bay is a picturesque, 700 m long beach backed by the steep Moresby Range. Backing the beach is a caravan park and surf club, with a seasonal stinger net in front of the club.*

Swimming: This is a relatively safe beach, with a usually shallow surf zone. Strong south-easterlies build bigger waves which both widen the surf zone and produce rips, particularly against the northern rocks. So it is best to stay in the centre of the beach, in the stinger net area.

Surfing: Low waves spill across the surf zone, with higher waves accompanying strong south-easterlies.

Fishing: Owing to the low beach slope, it is best to fish at high tide, and from the rocks at each end.

Summary: One of the more attractive beaches on the Queensland coast, with sufficient amenities for day visitors, plus the caravan park for travellers.

Figure 4.20 *Picturesque Etty Bay Beach is backed by steep, vegetated slopes and bordered by rocky headlands. This view along the beach shows the stinger net equipment and typical low surf, and boundary rocks.*

769, 770   ROBINSON

No.	Beach	Unpatrolled		Length
		Rating HT LT	Type	
769	Robinson	3 4	R+LT+rocks	500 m
770	Robinson (S)	4 4	R+LT+rocks	250 m

Robinson Beach is located 2 km south of the entrance to Mourilyan Harbour. It consists of two pockets of sand wedged in at the based of 100 m high Esmeralda Hill, to the south of Hayter Point. Dense tropical vegetation extends right to the back of the beaches, and the main northern beach is backed by the narrow Seaforth valley that links it with Mourilyan Harbour. There is no development on either beach and no vehicle access.

The main beach (**769**) is 500 m long, faces east and receives waves averaging 0.5 m that surge up a moderately steep and narrow beach at high tide, while at low tide they break across an 80 m wide low tide bar. The beach is bordered by rock-strewn headlands, with a substantial rock field in the middle of the beach, as well as some rocks scattered across the bar. The presence of the rocks plus the occurrence of rips against the rocks induce additional hazards on this otherwise attractive beach.

The southern beach (**770**) lies on the south side of the rocks that separate it from the main beach. The beach itself if 250 m long and consists of a steep high tide beach and a wider, flatter low tide bar. However, rocks completely dominate both ends as well as almost enclosing the beach in rock reefs and a sea stack. As a result, waves are lower at the beach; however it is extremely hazardous to swim in the surf owing to the rocks and associated rips.

Swimming: These beaches are only suitable under calm conditions, when there is good diving around the rocks. Any waves produce both hazardous turbulence around the rocks and rock-controlled rips.

Surfing: Only on the northern beach away from the rocks, when waves exceed 0.5 m.

Fishing: There are excellent locations off the numerous rocks and reefs on both beaches.

Summary: Two small, natural, tropical beaches. Best to visit in calm conditions, owing to the presence of rocks on and off the beaches.

Cowley Army Training Area
Beaches 771-773

771, 772 BROWNS

No.	Beach	Unpatrolled		
		Rating HT LT	Type	Length
771	Browns	2 3	R+LT	1.6 km
772	Browns (S)	1 1	UD	300 m

Double Point is an 80 m high, densely vegetated headland lying at the northern end of Cowley Beach. The point was an island when the coast was flooded by rising sea level 6000 years ago. It has since been joined to the mainland, as the beaches to either side built out over 3 km. On the north side of the point are two beaches, linked by sand, but separated at high tide by a small, 25 m high, rocky point linked by a sandy tombolo. Both beaches are part of an army training reserve and are off-limits to the public beyond the high water mark. There is no development on either beach and no access tracks.

The main **Browns** beach (**771**) is 1.6 km long, faces east and is bounded by a 40 m high northern headland and the southern tombolo. It is backed by the densely vegetated, 2 km wide, low beach-foredune ridge plain. It receives waves averaging 0.5 m, that break across a 100 m wide bar at low tide and surge across a narrow surf zone and up the narrow tree-lined beach at high tide. Rips are usually absent.

The smaller southern beach (**772**) lies immediately behind the tombolo and faces north-east. It is well protected by Double Point and the tombolo, and waves average 0.3 m. As a result the beach is low, with a 250 m wide

low tide bar and a few low mangroves scattered across the high tide area. At high tide waves lap against the narrow high tide beach, at mid tide they break across the bar, while at low tide the bar is exposed between the point and tombolo.

Swimming: Both beaches are relatively safe under normal low wave conditions, with rips only occurring on the northern beach when waves exceed 1 m. Swimming is best at mid to high tide because of the shallow bars.

Surfing: During strong Trades and higher waves there is a protected break over the southern bar at mid to high tide, while waves greater than 0.5 m will produce a break across the northern beach bar at mid tide.

Fishing: Best off the rocks at high tide.

Summary: Two undeveloped beaches bordering an army reserve, that offer some protection in the south, but with entry prohibited beyond high water.

773 COWLEY

No.	Beach	Unpatrolled		
		Rating HT LT	Type	Length
773	Cowley	2 3	R+BR	7.5 km

Cowley beach (**773**) is a 7.5 km long, exposed, southeast facing beach. It is located between Double Point and the shifting mouth of Liverpool Creek. The small settlement of Inarlinga is located 1 km north of the creek mouth, with a small army camp immediately to the north of the settlement. The settlement consists of a few houses, with a concrete boat ramp across the beach. The beach can be reached by turning off at Cowley on the Bruce Highway and driving 7 km to the coast. The army has cleared a runway behind the northern half of the beach, but otherwise it is undeveloped.

The beach fronts one of the most substantial beach-foredune ridge plains on the Queensland coast. Over the past 6000 years, up to 60 low, shore-parallel sand ridges have built the beach 3 km seaward, in the process joining Double Point to the mainland. This process is an indication of the high level of wave energy, an abundant sand supply and the sand-trapping effect of Double Point. The entire area should be preserved as a geomorphological site because of its significance and beautifully preserved record of shoreline evolution. There is also a small beach located in an open seacave on the south side of the point.

Today the long beach receives waves averaging just over 0.5 m and often higher. These maintain a narrow, moderately steep high tide beach and an 80 m wide bar that, during and following higher waves, is cut by rips approximately every 100 m, with up to 60 rips occurring along the beach. Toward the creek mouth the bar widens into a series of tidal sand shoals.

Swimming: The beach is moderately safe under normal low waves, however beware of rip holes and currents, particularly towards low tide and if waves exceed 0.5 m. Strong tidal currents also flow through the southern creek mouth.

Surfing: This beach picks up all southerly waves and usually has a beach break which increases in height with the Trades.

Fishing: This is a popular beach for both launching boats during calm periods and driving the beach in 4WDs to fish the rip holes and southern creek.

Summary: A long beach largely controlled by the army, with public access to the small Inarlinga settlement and the beach via 4WD.

Kurrimine National Park

Beach 774
Kurrimine National Park is a small coastal park extending for 3 km south of the southern side of Liverpool Creek. It encompasses the northern half of Murdering Point Beach, the densely vegetated backing beach and foredune ridge plain.

774-776   KURRIMINE

No.	Beach	Unpatrolled		Type	Length
		Rating HT	LT		
774	Murdering Pt	1	1	R+SF	7 km
775	Kurrimine	1	1	R+SF	3.5 km
	Stinger net:	November to May			
776	Maria Creek	1	1	R+SF	1.5 km

Between Liverpool Creek and Muff Creek, 11 km to the south, are three lower energy beaches dominated by the presence of extensive fringing coral reefs and three creek mouths. Only the central **Kurrimine** Beach is accessible via the Murdering Point Road. Kurrimine has a beachfront location and offers a range of caravan parks, a resort and limited shopping and tourist facilities, as well as a concrete beach boat ramp. There is no road access to the northern Murdering Point Beach, nor to the southern Maria Creek Beach although there is a boat ramp located in Muff Creek, which forms the southern boundary of the beach.

Murdering Point beach (774) is 7 km long, faces east and extends from the reef-protected, though dynamic, mouth of Liverpool Creek, south for 1 km to a small, reef-formed, cuspate foreland, then essentially due south to Murdering Point. The point itself is a low, sandy, substantial, cuspate foreland formed in lee of the Sisters Islands and the extensive King Reef complex, that surrounds the point and extends up to 5 km offshore. As a result of the reef, the beach is calm at low tide and fronted by extensive sand and reef flats, with only low waves reaching the steep beach at high tide. The shoreline, however, remains dynamic and large sections are being eroded back into tall melaleuca trees. The northern half of the beach is part of Kurrimine National Park.

Murdering Point is connected with two shipwrecks. A longboat from the 'Maria', wrecked on Bramble Reef, landed at the creek in 1871, and the 'Riser' was wrecked on King Reef in 1878. The crew of the 'Riser' and some of the longboat crew were killed by aborigines.

Kurrimine Beach (775) is a popular holiday destination, offering the above facilities and a relatively safe beach. The beach is approximately 3.5 km long, running south-west from Murdering Point to the low, long, narrow spit that forms the northern side of Maria Creek mouth. Housing development backs all the beach to the spit, while the length of the spit varies depending on the location of Maria Creek. The beach is calm at low tide and fronted by sand flats and ridges, with low waves lapping against the steep high tide beach. A fixed stinger net is operational during the stinger season, which is usable at mid to high tide. This beach is also experiencing erosion, particularly around the boat ramp and adjoining caravan park.

The southern **Maria Creek** beach (776) is a low wave energy, but dynamic, beach that varies considerably in shape and length, depending on the nature of the creek mouth and adjacent northern spit. It can be up to 1.5 km long and usually consists of a narrow high tide beach, fronted by low, 500 m wide, intertidal sand flats that consist of several low ridges and runnels. In addition, Liverpool Creek drains across the northern end of the flats and the smaller Muff Creek cuts across the southern end of the beach, stranding the southern corner at high tide.

During the wet season, **Maria Creek** delivers such volumes of freshwater as to turn the adjoining sea fresh. In the late 19th century, small sailing ships used to stop off at Maria Creek in the wet season to replenish their water casks directly from the sea.

Swimming: Kurrimine is the most popular location, providing good access and relatively safe conditions usually free of waves. However, sand flats are exposed at low tide, so time your swim between mid and high tide.

Use the stinger enclosure during the stinger season. Beware of the tidal currents at the Liverpool, Maria and Muff Creek mouths.

Surfing: None, owing to the protecting reefs and tidal flats.

Fishing: Kurrimine and the three creeks are all popular locations. The beach can only be fished at high tide, while the creeks are popular from boats or the banks.

Summary: Three lower energy, reef-protected beaches and creek mouths, with Kurrimine being the centre of attention and all facilities.

777, 778   GARNER

No.	Beach	Unpatrolled		Type	Length
		Rating HT	LT		
777	Garner (N)	1	2	B+SF	300 m
778	Garner	1	2	B+SF	1.2 km

Garner Beach consists of two near-continuous pocket beaches tucked in between two 60 m high bushy headlands. A road off the Bingal Bay Road runs right to the back of both beaches, servicing a few houses and a shady beachfront reserve that is used for camping; otherwise there are no facilities.

The northern beach (**777**) is 300 m long and bordered by low, rocky outcrops. It faces north-east and is protected by sand flats, then coral reef. It is bare sand at low tide, with only low waves at high tide lapping against the moderately steep high tide beach. The main southern beach (**778**) is 1.2 km long and swings around to face north-east, then north. It is even more protected, with 300 m long low tide sand flats off the beach, then the reef, with rocky outcrops at each end. The more protected southern headland is fringed by mangroves.

Swimming: Both beaches receive very low waves and are best at high tide when the reef, rock and sand flats are covered.

Surfing: None.

Fishing: Best at high tide off the beach, or in a boat over the reef and rock flats.

Summary: One of the earliest settled parts of this coast and still relatively undeveloped.

779-781   BROOKS, BINGAL, BICTON

No.	Beach	Unpatrolled		Type	Length
		Rating HT	LT		
779	Brooks	2	3	R+BR	700 m
780	Bingal Bay	2	3	R+BR	800 m
781	Bicton Hill	1	2	R+LT	100 m

Bingal Bay is an open, east facing bay, occupying the 5 km of coast between Garner Beach and Clump Point (Fig. 4.21). The Bingal Bay Road reaches the coast at Bingal Bay Beach, while all six beaches in the bay are accessible by vehicle. There are some shops at Bingal Bay, but otherwise limited tourist facilities. There are three beaches in the northern part of the bay - Brooks, Bingal Bay and Bicton Hill.

Brooks beach (779) can be reached via a steep gravel road off the Garner Beach Road. It leads to a few houses and provides 4WD access to the beach, which is used by locals to store and launch boats. There is limited parking and no facilities. The beach is 700 m long, bordered by a 40 m high headland with fronting rock and coral reef, and 60 m high Ninney Point in the south. It faces east and receives waves averaging 0.5 m which, when large, can cut up to 10 rips across the low tide bar.

Bingal Bay beach (780) lies on the south side of Ninney Point and runs for 800 m down to the rocky foreshore that fronts Clump Point National Park. It faces east and, like Brooks, can be cut by rips every 100 m during and following higher waves. The Bingal Bay Road runs behind the southern half of the beach and there is a shady camping area in the foreshore reserve. A small creek drains across the middle of the beach and there is an old rock jetty and concrete boat ramp at the southern end. The high tide beach is relatively steep, narrow and backed by coconut trees, while at low tide the bar and rip channels (if present) are exposed.

Bicton Hill beach (781) is a 100 m long pocket of high tide sand, fronted by more rocks than sand at low tide. It is backed by the Bingal Bay Road, with limited street parking.

Swimming: Bingal Bay Beach is the best known, most accessible and by far one of the most popular beaches in the area. It offers relatively safe swimming that is best at high tide, with a chance of usually low surf and a few rips at low tide.

Surfing: Both Brooks and Bingal Bay receive usually low surf, that increases with the Trades; usually best at mid to low tide.

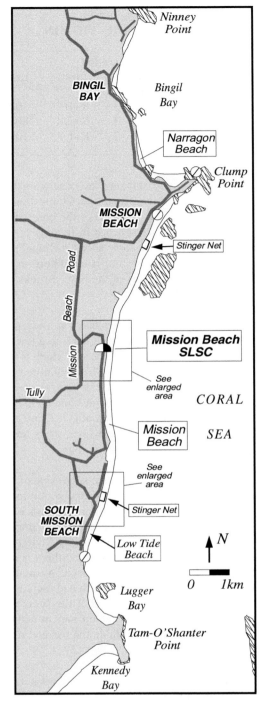

Figure 4.21 *The 20 km of coast between Ninney and Tam O'Shanter points contains some of Queensland's most scenic beaches, including Bingil Bay, the Mission beaches and Lugger Bay.*

Fishing: All three beaches offer rocks at each end, plus the chance of rip holes on Brooks and Bingal Bay.

Summary: Bingal Bay is one of Queensland's better known beaches, made famous by Prime Minister Harold Holt, who used to holiday there in the 1960's. All three beaches are attractive, tree-lined and backed by steeply rising vegetated slopes.

Bingal Bay was settled by the four Cullen brothers in 1882, followed by the Garners (1889) and Unsworths. The now idyllic tourist locations were then a scene of hard work - clearing the land, planting crops, waiting for ships to market their crops, and hoping they were bypassed by devastating cyclones. Then World War I and the associated lack of shipping transport, and the 1918 cyclone, all but crushed their early attempts. Transport by ship continued well into the 20th century, as the locals lobbied for 54 years before a road reached the settlement. The official opening of the road between El Arish and Bingal Bay occurred in July 1936.

782, 783 NARRAGON, CLUMP PT

No.	Beach	Unpatrolled		Beach type	Length
		Rating HT	LT		
782	Narragon	2	3	R+LT	1.1 km
783	Clump Point	1	2	TSF	200 m

Clump Point is a low, 1.5 km long finger of rock, reef and mangroves that forms the southern end of Bingal Bay and the northern boundary of Mission Beach. Lying to the north of the point are the very low energy Clump Point Beach and the longer Narragon Beach.

Narragon beach (782) is a 1.1 km north-north-east facing beach lying on the northern side of protruding Clump Point. The Bingal Bay Road runs the length of the beach and, because of its protected nature, there is a jetty toward the southern end of the beach that services the boats and ferries that transport tourists to Dunk Island and the reef. The beach receives waves averaging less than 0.5 m that surge up the high tide beach, while at low tide an 80 m wide low tide bar is exposed.

Clump Point beach (783) lies on the north side of the point, tucked in against the road. It faces north and is essentially a narrow high tide beach, fronted by a 500 m wide reef flat. There are a boat ramp and jetty off the tip of the point, that service the boats moored off the end of the flats.

Swimming: Narragon is a relatively safe beach, with usually calm to low waves and no rips, while Clump Point is unsuitable for swimming.

Surfing: None.

Fishing: Best off the jetty or rocks.

Summary: Narragon is a very popular tourist stop, used more for the jetty and boat trips than the beach.

784 MISSION BEACH
WONGALING BEACH
(MISSION BEACH SLSC)
SOUTH MISSION BEACH (LIFEGUARD)

Patrols:

Surf Lifesaving Club:	July to May	
Lifeguard on duty:	Christmas & Easter	
Stinger net:	November to May	

No.	Beach	Rating HT LT	Type	Length
784	Mission	1 2	R+LT	9.5 km

Mission Beach was settled by pioneer farmers in the 1880's, with the name deriving from an aboriginal mission established at South Mission Beach in 1912. The mission no longer operates and the farmers are being increasingly replaced by holiday and residential development, spreading 15 km from Bingal Bay in the north to South Mission Beach in the south. While the beaches are more than 16 km from the Bruce Highway and 20 km from Tully, the beautiful tropical beaches have long been popular for swimming and surfing. The original surf lifesaving club was founded at Bingal Bay in 1935 and, after some ups and downs, reformed at the present Mission Beach site in 1983.

Mission Beach (784) is 10 km long and faces essentially due east. Some coral reefs lie off the northern 3 km of beach, while Dunk Island, sighted and named by

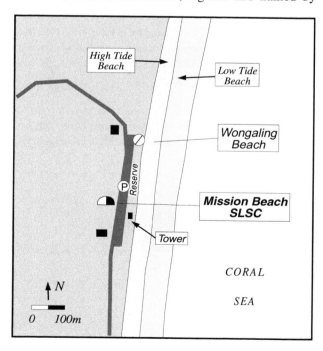

Figure 4.22   *Mission Beach Surf Club is located on the central Wongaling section of the 10 m long beach.*

Captain Cook, lies 5 km offshore and dominates the horizon off the southern part of the beach. Three gentle undulations in the shoreline are produced by waves moving around both Duck Island and the reefs. The beach begins at Clump Point, where it is known as Mission Beach and backed by an extensive foreshore reserve and shopping centre. Two kilometres on it becomes Wongaling Beach, which is the site of the Mission Beach Surf Life Saving Club (Fig. 4.22). The southern 3 km are known as South Mission Beach (Fig. 4.23).

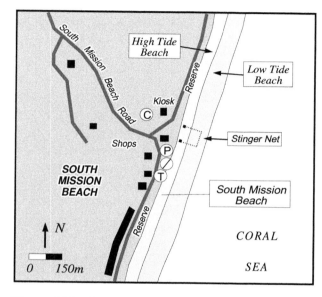

Figure 4.23   *South Mission Beach is patrolled by a lifeguard and is the location of the only stinger net on the long beach.*

The long beach has three main access points - at Mission, Wongaling and South Mission Beaches. These are associated with shopping centres at Mission and Wongaling, and a wide range of tourist facilities at all three beaches. Most of the beach is backed by a foreshore reserve, that provides good access and shade under the trees.

The beach is somewhat sheltered by Dunk Island and reefs, which results in low waves usually less than 0.5 m. These interact with the generally fine to medium sand to produce a 50 m wide high tide beach, with a narrow surf and a 100 m wide, flatter low tide beach, and wider surf zone (Fig. 4.24). Rips are usually weak or absent, however strong south-easterlies produce a northerly current along the beach. A strong northerly current also runs along the beach on outgoing spring tides.

Swimming: This is a popular but rarely crowded beach, owing to its length. It is best to swim at the main access points at Mission Beach, near the stinger enclosure, at the surf lifesaving club at Wongaling Beach, and at the lifeguard controlled stinger enclosure at South Mission Beach.

Surfing: Low, spilling waves dominate the entire beach, with larger waves during stronger south-easterlies.

Fishing: Fishing is better at high tide because of the shallow beach and surf zone; otherwise by wading out onto the bar at low tide. The rocks at either end are also fronted by shallow water, so most fishers use the beach boat ramps at Mission and the southern end of South Mission Beaches to access the reefs.

Summary: A long beach providing a wide range of facilities, with generally good access, shady trees over the top of the beach, usually low waves and relatively safe swimming conditions.

Figure 4.24 *A view of South Mission Beach towards low tide, looking north toward the patrol area.*

785, 786 LOVERS

No.	Beach	Unpatrolled		Type	Length
		Rating HT	LT		
785	Lovers (1)	2	3	R+LT+rocks	300 m
786	Lovers (2)	2	3	R+LT+rocks	100 m

At the southern end of Mission Beach, just past the boat ramp, rocks begin to litter the beach as 80 m high, densely vegetated, rocky slopes reach the coast. For 500 m these slopes separate Mission Beach from Lugger Bay. Along the base of the slopes are two small sand and rock beaches, known collectively as Lovers Beach. They are accessible from the Kennedy Track - a walking track that begins at the boat ramp and runs around the shore for 3.5 km to the Kennedy Bay picnic area.

The main **Lovers Beach** (785) lies 200 m south of the boat ramp and is separated from South Mission Beach by 100 m of rocks and boulders. The beach itself is about 300 m long, faces east, receives waves usually less than 0.5 m high and consists of a high tide sand beach fronted

by a sand and rock low tide beach. The second, smaller Lovers Beach (**786**) is another 300 m further on, beyond some more substantial rock platforms and rocks, and is just 100 m long, consisting of a similar high tide sand beach and low tide rocks.

Swimming: Care should be taken because of the rocks around these two beaches, especially at low tide.

Surfing: None.

Fishing: Popular spots because of the rock platforms, particularly between the two beaches, and with relatively easy access.

Summary: Two beaches viewed by those walking the Kennedy Track; mainly used for sunbaking rather than swimming.

787 LUGGER BAY

No.	Beach	Unpatrolled		Type	Length
		Rating HT	LT		
787	Lugger Bay	1	1	B+SF	1.5 km

Lugger Bay (787) forms the southern end of the Mission Beach system. It is a 1.5 km long bay tucked in lee of Tam O'Shanter Point. It initially faces east, but curves right around to finally face due north. It is a protected, low energy beach consisting of a narrow high tide beach fronted by low tide sand flats that widen to 300 m in the southern corner. The beach is backed by a low, densely vegetated foredune, then cleared farm land, with the small Mije Creek exiting in the southern corner. A footbridge, part of the Kennedy Track, crosses the creek and provides access to Tam O'Shanter Point.

Swimming: This is a relatively safe beach, as waves are usually very low to calm and rips absent. However, you can only swim at high tide as the sand flats are exposed at low tide.

Surfing: None.

Fishing: Only at high tide and in the creek.

Summary: A lovely, curving beach; part of the Kennedy Track and well worth the walk.

788 KENNEDY BAY

No.	Beach	Unpatrolled		Type	Length
		Rating HT	LT		
788	Kennedy Bay	2	3	R+BR	3.3 km

Kennedy Bay is the site where, in 1848, Edmund Kennedy and his party began their epic and fateful trek to Cape York. While not part of Edmund Kennedy National Park, located a few kilometres to the south, the bay bears his name and it is accessible from Mission Beach along the delightful 3.5 km long Kennedy Track. The track skirts Lovers Beach, Lugger Bay and Tam O'Shanter Point to reach the northern end of the bay, with a picnic area provided in the shade behind the beach. This is the only land access to the otherwise relatively remote and natural beach.

The beach (**788**) is 3.3 km long and initially faces south-west in lee of the protection afforded by the southern tip of Tam O'Shanter Point. Such is the protection that sand flats and a few mangroves front the northern end of the beach. The beach then swings around to face south-east into the Trade winds and waves, which average over 0.5 m. These produce a 150 m wide low tide bar, usually cut by rips every 100 m at low tide, with a trough and second bar paralleling much of the main part of the beach. Toward the southern end, the bars give way to the tidal shoals and currents of the Hull River, that extend for 500 m off the river mouth.

Swimming: This can be a hazardous beach when the Trades are blowing and waves exceed 0.5 m. Surf and strong rips are present, particularly toward low tide. In addition, there are strong tidal currents at the Hull River mouth. The safest swimming is at the northern, more protected, end.

Surfing: There is a chance of surf here, as it is one of the more exposed beaches in the area. However it requires a walk from South Mission Beach.

Fishing: Like the surf, the rips also produce good fishing holes along the beach, plus the deep river mouth at the southern end. However it also requires a walk of a few kilometres.

Summary: A dynamic, natural beach, much the same today as when Kennedy saw it 150 years ago. Well worth the walk, but take care if swimming as there may be strong rips.

D1, D2 DUNK ISLAND

No.	Beach	Unpatrolled		Type	Length
		Rating HT	LT		
D1	Pallon	1	1	B+SF	1.4 km
D2	Brammo Bay	1	1	B+SF	1.2 km

Dunk Island lies just 4 km off South Mission Beach and with a ridge line of over 200 m high, it is clearly visible from the mainland. Captain Cook sighted and named this 'Tolerable high island' in June 1770. A regular ferry service now operates from Narragon Beach to the island, accommodating day-trippers as well as longer term guests. Most of the island is part of the 730 ha Dunk Island National Park. However, the western corner has been partly cleared and includes a 500 m long, low, sandy, cuspate foreland. This area houses a tourist resort and landing strip. There are two beaches on either side of the foreland - Pallon on the south and Brammo Bay on the north.

Pallon beach (D1) is 1.4 km long and faces south-west toward Lugger Bay. It is shielded from most waves by the island and fringing reefs, and at high tide calm to low wave conditions usually prevail. At low tide up to 1 km of sand and reef flats are exposed.

Brammo Bay (D2) on the north side of the foreland is even more protected, owing to its northerly aspect. Consequently the island jetty and settlement, including the resort, are located along its 1.2 km long sandy shore. The beach is relatively narrow at high tide, while at low tide sand and reef flats are exposed, particularly toward the centre and eastern end. The jetty is located at the western end where the water is deeper.

Rockingham Bay was sighted by Captain Cook in June 1770, after he rounded Cape Sandwich on the northern tip of Hinchinbrook Island. He noted the 'fine, large bay' and named it after a recent English Prime Minister.

The bay extends from Dunk Island and Tam O'Shanter Point in the north to the northern shore of Hinchinbrook Island in the south, a distance of 25 km. Its shoreline south of the point is dominated by the Tully River and several smaller creeks, as well as intervening low, prograding beach ridge plains, backed by extensive mangrove swamps. Cardwell lies at the south-western limit of the bay, where it merges with the northern end of the Hinchinbrook Channel.

Swimming: Both beaches are relatively safe, with usually calm to low wave conditions. Best swimming and snorkelling over the reefs is at high tide, with the flats exposed at low tide.

Surfing: None.

Fishing: The jetty is a favourite spot, providing access to the only deep water close to shore; otherwise use a boat to go out over the reef at high tide.

Summary: Dunk Island was immortalised by the beach-combing author E. J. Banfield, who lived on the island in the early 20th century. It remains a primarily natural environment, now preserved by its largely national park status.

789 GOOGARRA

		Unpatrolled			
No.	Beach	Rating		Type	Length
		HT	LT		
789	Googarra	2	3	R+BR	4.5 km

Googarra beach (789) is best known for the two settlements at either end - Hull Heads at the northern tip on the Hull River inlet, and Tully Heads at the southern tip on the Tully River mouth. The two settlements lie just over 10 km in from the Bruce Highway and about 15 km from Tully. There is a caravan park at Tully Heads and another at the junction of the two roads to Tully Heads. Both have boat ramps and are popular fishing and holiday destinations.

The beach linking the two settlements is 4.5 km long, faces south-east and is bordered at each end by the dynamic Hull and Tully River mouths and their associated tidal channels and extensive tidal shoals, that extend a few hundred metres seaward. Away from the tidal shoals, the beach is exposed to waves averaging 0.5 m that produce a moderately steep high tide beach, fronted by a near-continuous low tide bar. Beyond the low tide bar are one to two more outer bars, that are maintained by periods of higher waves accompanying strong Trade winds.

Swimming: The safest swimming is in the areas away from the river mouths. There are strong currents and deep and variable channels associated with both entrances, as well as a possibility of crocodiles and stingers.

Surfing: Chance of a surf over the bars away from the river mouth, especially if the Trades are blowing strong.

Fishing: Both settlements are popular fishing spots, offering creek, river, boat and beach fishing.

Summary: Two small settlements linked by a 4.5 km long beach and framed by wide river shoals. The beaches are more popular with the Tully region locals than passing tourists.

Edmund Kennedy National Park

Area:	6 039 ha
Beaches	791-793
Length of coast:	16 km

Edmund Kennedy National Park is a coastal park containing 16 km of sandy shore, backed by densely vegetated beach ridges and then mangrove swamps that fringe parts of the Tully River and Dallachy, Ten Mile, Wreck and Meunga Creeks. The only access is at the southern Meunga Creek, where there is also a camping area and boat ramp.

For information:
Queensland Department Environment & Heritage
Bruce Highway
PO Box 74
Cardwell Qld 4816
phone: (07) 49 66 8601

For information:
Queensland Department Environment & Heritage
Bruce Highway
PO Box 74
Cardwell Qld 4816
phone: (07) 49 66 8601

790-793 EDMUND KENNEDY NATIONAL PARK

		Unpatrolled			
No.	Beach	Rating		Type	Length
		HT	LT		
790	Tully-Murray R	2	3	R+LT+bars	6.5 km
791	Murray R-Dallachy Ck	2	3	B+SF	8 km
792	Dallachy-Wreck Ck	2	3	B+SF	3 km
793	Wreck-Meunga Ck	2	3	B+SF	4.2 km

South of the Tully River is a 25 km section of essentially undeveloped beaches and tidal creeks. It contains four beaches bordered by five creeks. They are all backed by extensive low, sandy beach and foredune ridge plains that extend up to 5 km inland. Due to the low, wet, sandy nature of the soils the land was not cleared for farming and now the three southern beaches and backing plains are part of the Edmund Kennedy National Park. The only vehicle access to the park is at the southern end at Meunga Creek where there is a camping area and creek boat ramp, plus a resident crocodile.

The beaches begin on the southern side of the Tully River. The first (**790**) is 6.5 km long, faces east-south-east and runs from Tully River to Murray River. It consists of extensive tidal shoals off the river mouths, a narrow high tide beach fronted by a wide low tide bar and additional outer bars at low tide.

Beach **791** runs for 8 km from the Murray River to the smaller, winding Dallachy Creek that drains the extensive backing sand plains and swamps. It has a narrow high tide beach, wide tidal shoals in the north that grade into multi-ridge sand flats for 3 km, then a normal beach and low tide bar and finally the Dallachy Creek tidal shoals.

Beach **792** is 3 km long, faces east and extends from Dallachy Creek to Wreck Creek. Due to its shorter length and lower wave energy, its low tide environment is completely dominated by the tidal shoals and ridged sand flats that extend up to 1 km off the beach.

Beach **793** is a straight, east facing, 4.2 km long beach running between Wreck and Meunga Creeks. It is bordered by extensive creek tidal shoals, with 1 km wide tidal and sand flats fronting the entire beach, and is backed by a narrow high tide beach.

Swimming: The safest swimming is well clear of the creeks and their strong tidal currents on the central parts of these beaches. You can only swim at mid to low tide because of the extensive tidal shoals and flats.

Surfing: Usually low waves at high tide and nothing but sand flats at low tide.

Fishing: Best in the creek and inlet at mid to high tide, with a boat ramp at Meunga Creek.

Summary: Four undeveloped and natural beaches that receive little attention despite their close proximity to the highway.

---

794 CARDWELL

No.	Beach	Unpatrolled Rating HT	LT	Type	Length
794	Cardwell	1	2	R+mud flats	6.5 km

Cardwell is a small town of 1500 people that straddles the Bruce Highway and fronts the magnificent Hinchinbrook Channel and Island. While Cardwell was settled in 1864, apart from sugarcane farms, it has avoided substantial development, with tourism likely to lead it into the next century. The town has most tourist facilities and provides the gateway to Hinchinbrook Island as well as the backing hinterland.

Cardwell Beach (794) is well protected by Hinchinbrook Island and is usually calm. Consequently

the beach consists of a steep, soft, 20 m wide high tide beach, fronted by mud flats at low tide. There are a jetty and boat ramp that are only useable at high tide, so watch the tides if swimming or boating. The beach faces east and is 6.5 km long, extending from the wide, sandy tidal shoals of Meunga Creek in the north to the low, mangrove-fringed mud flats of Oyster Point in the south. A major development has been proposed for Oyster Point, but is at present the subject of ongoing debate over its potential impact on the mangroves and seagrass meadows of the Hinchinbrook Channel.

Swimming: You can only swim at high tide, as it is a mud bath at low tide.

Surfing: None.

Fishing: Best at high tide off the beach or jetty, or from a boat in the channel. There is also a boat ramp at Meunga Creek.

Summary: Everyone driving along the Bruce Highway has to pass through Cardwell, however the few who stop do so to visit the surrounding features rather than the beach.

Hinchinbrook Island and National Park	
Area:	39 350 ha
Coast length:	115 km
Beaches:	795-826 (32 beaches)

Hinchinbrook Island was sighted by Captain Cook on 8 June 1770. He named the southern Halifax Bay, northern Rockingham Bay, and Point Hillock and Cape Sandwich, the easternmost and northernmost points of the island. He did not, however, name the island as he was unaware that it was separated from the mainland by the narrow Hinchinbrook Channel. The island was settled by timber cutters in the 1880's, with little permanent white settlement even to today. A mission was briefly set up in Missionary Bay in 1874 to attract the decimated local aborigines, but it was already too late. Tourism started in a small way in the 1930's, and today there is just one tourist resort on Hinchinbrook, hidden away on the northern tip at Cape Richards.

Most of the island was declared a national park in 1932 and today the entire island has that status. The island is 33 km long, up to 20 km wide and reaches a height of 1121 m at Mount Bowen, with Mount Diamantina and Mount Straloch both over 900 m. It contains 115 km of shoreline, half of it in the mangrove-fringed Hinchinbrook Channel (itself a national park) and Missionary Bay. The eastern half contains more energetic sandy beaches, including some sand dunes in Ramsay Bay and rocky shores. All in all, the island contains a tremendous variety of tropical environments, most in relatively pristine conditions, contained on a large island, yet only 1 km off the mainland.

For information:
Queensland Department Environment & Heritage
Bruce Highway
PO Box 74
Cardwell Qld 4816
phone: (07) 49 66 8601

795-797 WAGGON WHEELS, MISSIONARY BAY

| No. | Beach | Unpatrolled | | | |
		Rating HT	LT	Type	Length
795	Waggon Wheels	1	1	B+SF	700 m
796	Missionary Bay (1)	1	1	B+SF	300 m
797	Missionary Bay (2)	1	1	B+SF	100 m

Missionary Bay is a 20 km long, 5 km wide bay located on the northern side of Hinchinbrook Island. Its eastern half is filled with mangrove-covered mud flats up to 6 km wide. Only its 10 km long south-western shore is exposed to waves and even here, mangroves still dominate. In amongst the mangroves are three patches of sand - Waggon Wheels Beach and Missionary Bay Beaches 1 and 2. All three face north and are well protected from wind and waves, with only northerly winds producing low waves across the bay. All three consist of a narrow, sandy high tide beach, fronted by 200 to 300 m of sand to mud tidal flats. There is no vehicle access and no facilities at any of the beaches, with the only access being by boat at high tide.

Waggon Wheels beach (795) is 700 m long, with fringing mangroves and is backed by the densely vegetated slopes of 70 m high Mount Pitt. Beaches **796** and **797** lie adjacent to each other separated by a low headland; both are bordered by rocks with some fringing mangroves.

Swimming: You can only swim at these beaches at high tide, because of the tidal flats.

Surfing: None.

Fishing: Only at high tide from the beach.

Summary: Three natural beaches set in a low energy, mangrove dominated shoreline.

798-800 MACUSHA COVE, TWO SISTERS

| No. | Beach | Unpatrolled | | | |
		Rating HT	LT	Type	Length
798	Macushla Cove	1	1	B+SF	500 m
799	Two Sisters (1)	1	1	B+SF	400 m
800	Two Sisters (2)	1	1	B+SF	200 m

Kirkville Hills is an 80 m high and 3 km wide hillock located 3 km south of Cape Richards, the northern tip of the island. The base of the hills are exposed to the higher waves of Shepherd Bay to the east and the calmer conditions of Missionary Bay to the west. These three beaches fringe the western base of the hills. All three are in a natural state, with no development and no access other than by boat or foot.

Macushla Cove (798) lies on the south side of the hills and faces south. It consists of a narrow, sandy, 500 m long high tide beach, backed by the rising slopes of the hills, with rocks outcropping on the beach. It is fronted by sandy to muddy tidal flats, up to 200 m wide. To the east it grades into mangroves, while the western tip of the hills forms its western boundary.

Two Sisters beach (799) begins on the north side of the western tip, is 400 m long and is bordered at each end by rocky shore. It faces north-west and consists of a narrow high tide beach fronted by initially narrow tidal flats in the west, which rapidly widen to several hundred metres in the east.

The adjoining Two Sisters beach (**800**) is 200 m long, faces north-west and is wedged in between rocky shore to the west and the several hundred metre wide, muddy and mangrove-fringed shore that extends for 2 km to the base of Cape Richards.

Swimming: All three beaches are dominated by shallow tidal flats and are only suitable for swimming at high tide.

Surfing: None.

Fishing: Best at high tide from the fringing rocks and amongst the mangroves.

Summary: Three low energy strips of high tide sand, dominated by the mud and mangroves of Missionary Bay.

---

801 CAPE RICHARDS

| No. | Beach | Unpatrolled | | | |
		Rating HT	LT	Type	Length
801	Cape Richards	1	2	R+LT	300 m

Cape Richards is the northernmost tip of Hinchinbrook Island. It consists of a one square kilometre, 140 m high hillock, that is tied to the Kirkville Hills and the remainder of the island by a 200 m wide, 2 km long, sandy barrier. On the north side of the cape, facing north, is a 300 m long pocket beach (801). It is bordered by prominent rocky headlands, with a few large rocks on the beach, and is backed by the island's only resort, hidden in amongst the dense vegetation. A small jetty on the west side of the beach provides the only access to the resort.

The beach is composed of relatively coarse sand; together with the waves that average less than 0.5 m, this produces a narrow, moderately steep high tide beach, with a shallow bar at low tide. Resort guests launch a variety of watercraft from the beach.

Swimming: This is one of the best and safer swimming beaches on the island and can be used at high and low tide. However, be careful if swimming north of the rocks as strong tidal currents flow past the cape.

Surfing: Usually calm, with a low dumping shorebreak when waves are present.

Fishing: Best off the rocks at each end and the jetty on the west side of the resort.

Summary: This is the beach most visitors to the island stay and swim at. It is an attractive, rock-bound beach backed by the small and tasteful resort.

802, 803   SHEPHERD BAY (N)

No.	Beach	Unpatrolled			
		Rating HT LT		Type	Length
802	North Shepherd	2	3	R+LT	2 km
803	South Shepherd	2	3	R+LT	1.8 km

Shepherd Bay is an open, 8 km wide, north-east facing bay on the north-eastern tip of the island. It is bordered by Cape Richards and Cape Sandwich. The bay contains 11 km of shoreline, consisting of six beaches separated by rocky shore. The first two beaches (802 & 803) are the larger of the six and occupy half the bay shoreline. There is no development on the beaches and the only access is by boat or on foot from the Cape Richards resort.

North Shepherd Beach (**802**) faces east and runs for 2 km from the southern side of Cape Richards down to the eastern base of the Kirkville Hills. It receives waves averaging about 0.5 m and consists of a narrow high tide beach, fronted by a 200 m wide, low gradient low tide bar. It is backed by a 10 m high, 50 to 200 m wide foredune, which is in turn bordered by the mangroves and mud flats of Missionary Bay.

South Shepherd Beach (**803**) is located on the south side of the Kirkville Hills. It is 1.8 km long and faces north-east. It receives slightly lower waves and also has a narrow high tide beach and a low tide bar up to 200 m wide.

It is bordered by the rocky shore of the hills to the north, backed by both foredunes and a 40 m high hillock, with rocky shore to the south.

Swimming: Both beaches offer long stretches of clean sand and clear water. Swimming is better at mid to high tide. However, rips can form across the bar if waves exceed 0.5 m.

Surfing: Usually low, lapping waves, however high Trades can produce 0.5 m waves wrapping around the southern points and a small break over the bars.

Fishing: Best at high tide off the rocks at each end and along the beach.

Summary: Two long, natural beaches ideal for a beach walk and a relatively safe swim.

804-807   SHEPHERD BAY

No.	Beach	Unpatrolled			
		Rating HT LT		Type	Length
804	Shepherd Bay (1)	1	2	R+LT	300 m
805	Shepherd Bay (2)	1	2	R+LT	200 m
806	Shepherd Bay (3)	1	2	R+LT +rocks	200 m
807	Shepherd Bay (4)	1	2	R+LT	150 m

The southern shore of Shepherd Bay is predominantly rocky and backed by steep, densely vegetated slopes, rising to a ridge line up to 260 m high. Tucked into indentations in the 5 km of rocky shore are four small pocket beaches, totalling 850 m in length. They all receive low waves and are calm except during northern winds and waves. All four can only be reached by boat and have no development or facilities.

Beaches 804, 805 and 806 occupy the first kilometre of the shore. Beach **804** is 300 m long, faces north-east, has a few rocks on the beach and is bordered by low, rocky points. It is backed by dense vegetation containing a few palm trees. The beach is low and narrow at high tide, with a 50 m wide low tide bar. Beach **805** is similar, just 200 m in length, with more casuarinas backing the low beach and prominent granite boulders at either end. Beach **806** is also 200 m long, but contains less sand and is littered with large granite boulders, a result of steeper rocky slopes backing the beach.

Beach **807** is located 2 km further on toward Cape Sandwich. It consists of a 150 m long, north facing pocket of sand, bordered by rocky boulder fields, with a few mangroves growing in the eastern boulders. The beach is low and narrow at high tide, with a 50 m wide bar at low tide.

Swimming: All four beaches usually have low waves and relatively safe swimming conditions. However, watch the numerous boulders toward the ends and on Beach 806.

Surfing: Usually none, except during strong northerly winds.

Fishing: Excellent fishing over the rock reefs off the rocks.

Summary: Four lovely pockets of sand at the base of densely vegetated, craggy slopes.

808, 809   CAPE SANDWICH

| No. | Beach | Unpatrolled | | | |
		Rating HT	LT	Type	Length
808	Cape Sandwich (1)	2	3	R+LT	100 m
809	Cape Sandwich (2)	2	3	R+LT	500 m

Cape Sandwich was sighted and named by Captain Cook in June 1770. The cape consists of a craggy, 200 m high headland that protrudes east and forms the boundary between Shepherd Bay and the larger Ramsay Bay to the south. There is no access to the cape other than by boat or on foot. On the tip of the cape are two indentations, containing two pocket beaches.

Beach **808** is a small wedge of sand in a V-shaped indentation immediately below the tip of the cape. At high tide it is awash and consists of a boulder-covered high tide beach. At low tide, the beach lengthens to 100 m and exposes a low tide bar up to 80 m wide. Waves average 0.5 m, increasing in height during strong Trades. At these times a rip is generated against the northern rocks.

Beach **809** occupies a larger indentation and is 500 m long at low tide. It consists of a narrow high tide beach and low gradient, 200 m wide low tide bar which, during higher waves, also has a strong rip against the northern rocks and a smaller one against the southern rocks.

Swimming: Relatively safe conditions prevail when waves are low, however rips will form if there is any surf, particularly toward low tide.

Surfing: A chance at mid to low tide when there are moderate Trades blowing.

Fishing: Excellent access to deeper water off the granite boulder headlands that border each beach.

Summary: Two headland-fringed pockets of sand that almost disappear at high tide and become wide, low beaches at low tide.

810   RAMSAY BAY

| No. | Beach | Unpatrolled | | | |
		Rating HT	LT	Type	Length
810	Ramsay Bay	3	4	R+BR	8.5 km

Ramsay Bay is an open, east facing bay containing the longest and most energetic beach on the island, as well as in the region. The 8.5 km long beach also ties the northern rocks of Cape Sandwich to the island proper. In places it is only 100 m wide and on its western side is bordered by the 5 to 6 km wide mangrove-covered mud flats of Missionary Bay. The only access to the beach is by boat or on foot and there is no development.

The beach (810) faces east-south-east, aligning it into the prevailing Trade winds and waves. As a result, waves average over 0.5 m and are commonly over 1 m high. These produce a wide, energetic surf zone dominated by rips every 80 to 100 m (Fig. 4.25), with at times up to 100 perfectly formed rips along the beach. The beach itself consists of a prominent high tide berm and moderately steep swash zone, then a 100 m wide bar which on its outer edge houses the many rips. Behind the beach the Trade winds have built a continuous foredune, breached in places by blowouts reaching up to 70 m high. It is well worth climbing these to capture the contrasting view of the energetic sand beach on one side with the mud and mangrove dominated bay on the other side.

Figure 4.25 *The northern end of Ramsay Bay showing the wide beach at low tide, together with closely spaced rips in the surf.*

Swimming: This is the best surfing beach around, but comes with rips when waves are breaking, so use caution, particularly toward low tide.

Surfing: This is also the best beach to find rideable waves over the bar at mid to low tide.

Fishing: The rip holes act to improve the beach fishing, while there are also rocks at each end of the beach.

Summary: A magnificent long beach, with its more energetic surf, usually numerous rips and contrasting beach and mud-mangrove environments.

811 RAMSAY BAY (S)

No.	Beach	Unpatrolled Rating HT	LT	Type	Length
811	Ramsay Bay (S)	1	2	R+LT	200 m

At the southern end of Ramsay Bay Beach is a small, rocky headland, beyond which is a lower energy, low, 200 m long beach, called Ramsay Bay (south) (**811**). The beach curves around to face almost due north and is bordered on both sides by low, granitic headlands. Due to the low waves and finer sand, together with groundwater draining across the beach, it is almost awash at high tide, with a low, 50 to 100 m wide bar at low tide.

Swimming: Best at high tide when the bar is covered.

Surfing: None.

Fishing: Better off the rocks at high tide.

Summary: An isolated, low energy beach that provides a safer anchorage if exploring the adjoining Ramsay Bay Beach.

812 NINA

No.	Beach	Unpatrolled Rating HT	LT	Type	Length
812	Nina	1	3	R+BR	500 m

Nina Peak is a 300 m high, rocky crag that protrudes above the tropical surrounds. One kilometre due east of the peak is 500 m long, east facing **Nina Beach** (812). Viewed from the sea, the peak provides a dramatic backdrop to the beach. The beach is bordered by rock-fringed headlands and backed by a narrowing, 4 km long valley. A small creek that drains the valley flows out at the northern end of the beach, hard against the northern rocks.

Nina Beach receives waves averaging 0.5 m that surge against a narrow high tide beach, while at low tide a 60 m wide bar is exposed, usually with rips against the rocks and the possibility of two to three along the beach. The beach is backed by a low, densely vegetated foredune, including a few palm trees. There are no facilities and

the only access is by boat or on foot.

Swimming: Relatively safe under the usually low waves, however stay clear of the rocks when waves are breaking and watch for rips toward low tide.

Surfing: Usually a low, spilling break across the bar at mid to low tide.

Fishing: Best at high tide on the creek and off the rocks.

Summary: A natural, tropical beach surrounded by steep valley slopes.

813-816 LITTLE RAMSAY BAY, THE THUMB, AGNES ISLAND

No.	Beach	Unpatrolled Rating HT	LT	Type	Length
813	Little Ramsey Bay	1	3	R+LT	1.1 km
814	The Thumb	1	2	R+LT	150 m
815	Agnes Is (W)	1	2	R+LT	500 m
816	Agnes Is	1	2	R+LT +rocks	800 m

To the west of Agnes Island is an open, 3 km wide, north facing bay. It contains four beaches, all bordered by prominent headlands and backed by steeply rising, densely vegetated slopes, with the 1000 m high peak called The Thumb just 2 km to the west of the main beach. There is no development and no vehicle access to any of the beaches.

Little Ramsey Bay (**813**) is 1.1 km long, faces north-east, is bordered by low, rocky headlands and backed by a 4 km long, V-shaped valley that ends at 1100 m high Mount Bowen. The beach consists of a narrow, moderately steep high tide beach fronted by a low tide bar up to 80 m wide. Waves average less than 0.5 m, producing a 30 m wide surf zone at high to low tide. Rips are usually absent, however wave height increases northward along the beach and a rip may run out against the northern rocks. Two creeks cross the beach, at each end, with a few mangroves in the southern creek. A low casuarina-covered foredune backs the beach.

Beach **814** is a 150 m long, north facing pocket of sand, bordered by rocks at each end and at low tide fronted by rocks and a small coral reef. Steep, densely vegetated slopes back the beach.

Beach **815** is protected by two headlands that extend almost 500 m seaward of this 500 m long north-east facing beach. Within this well-protected embayment is a narrow high tide beach backed by a low foredune. It is fronted by a bar/sand flats, up to 100 m wide, and then a fringing coral reef. A small creek drains out across the western end of the beach with a few mangroves in the creek.

Beach **816** lies immediately in lee of Agnes Island and at low tide connects the island to the mainland. The beach faces essentially north-east and is predominantly sandy, with numerous rocks outcropping along and off the 50 m wide main beach, while in the southern corner a boulder beach dominates and links the island at low tide.

Swimming: These beaches all receive moderate protection form the Trades, with waves usually less than 0.5 m high and rips usually absent. They are best at high tide, with rocks and reefs causing some problems at low tide.

Surfing: Usually none.

Fishing: The best spots are usually in the creek mouths and off the rocks at high tide, or from a boat over the coral and rock reefs.

Summary: Four natural beaches offering a variety of scenery and usually low wave to calm conditions.

817   BEACH 817

No.	Beach	Unpatrolled		Type	Length
		Rating			
		HT	LT		
817	Beach 817	1	3	R+BR	400 m

Beach **817** is a pocket of sand surrounded on three sides by steep, grass-covered slopes that rise to 100 to 250 m. At the base of the slopes is the 400 m long beach and a low, densely covered foredune. The high tide beach is low and narrow, while at low tide an 80 m wide bar is exposed, usually with rips running out against the rocks and two along the beach.

Swimming: This is a very difficult to access and potentially hazardous beach, particularly when waves exceed 0.5 m and toward low tide, when up to four rips form in the surf.

Surfing: Waves average over 0.5 m and produce a beach break over the bars and rips.

Fishing: Best off the rocks at high tide, or in the rip holes at mid to low tide.

Summary: A relatively isolated beach sitting in its own little valley.

818   ZOE BAY

No.	Beach	Unpatrolled		Type	Length
		Rating			
		HT	LT		
818	Zoe Bay	1	2	UD	2.5 km

Zoe Bay occupies the front of a 2 km wide, 4 km long valley that has been largely infilled with sand and mud. The 2.5 km long beach (818) forms a spit toward the seaward end of the valley, with the high valley sides narrowing to a width of 1.5 m to form the valley and bay mouth. Behind the beach, the flat valley floor is occupied by a meandering tidal creek and extensive mangroves, while steep slopes, rising to several hundred metres, surround the valley.

The beach faces east and receives waves averaging 0.5 m. These have built the fine beach sands into a low, narrow high tide beach fronted by a very low gradient, 150 m wide, intertidal sand bar, with the low waves spilling across a 100 m wide surf zone (Fig. 4.26). The main valley creek maintains an inlet at the northern end of the beach and flows out across a 400 m wide tidal delta, that is also exposed at low tide. A smaller creek also drains across the very southern end of the beach.

Swimming: The safest swimming is on the main beach and clear of the two tidal creeks, where strong currents and deep channels are common.

Surfing: Usually low, spilling breakers that improve as the Trades pick up, particularly in the most protected southern corner.

Fishing: Best off the rocks and in the two creeks.

Summary: A moderately long, low beach surrounded by prominent headlands, two creeks and the extensive backing tidal flats and mangroves.

Figure 4.26 *The wide, low gradient Zoe Bay beach, backed by the high peaks of Hinchinbrook Island.*

819-821 HILLOCK PT (N)

| No. | Beach | Unpatrolled | | | |
		Rating HT	LT	Type	Length
819	Beach 819	1	2	R+LT	150 m
820	Beach 820	1	2	R (boulder)	100 m
821	Hillock Pt (N)	1	2	R (boulder)	100 m

Hillock Point is the most easterly tip of Hinchinbrook Island. It was sighted and named 'on account of its figure' by Captain Cook in June 1770. Between Zoe Bay and Hillock Point is 5 km of steep, rocky coast containing three small beaches. The largest is Beach **819**, a 150 m long, north-east facing pocket of high tide boulders, fronted by a high and low tide sand beach, and backed by steeply rising, vegetated slopes. The beach receives waves averaging about 0.5 m and is 10 m wide at high tide, while exposing a 50 m wide bar at low tide.

Beach **820** is a 100 m pocket of sand, shell and boulders; the latter delivered by the creek that descends steeply from the small backing valley. Beach **821** is the northern part of the boulder and sand beach that ties 70 m high Hillock Point to the mainland. The beach faces north and is predominantly boulders, with some sand at high tide.

Swimming: Only beach 820 is suitable for swimming, owing to the dominance of boulders and rocks at beaches 820 and 821.

Surfing: Usually none.

Fishing: All three offer good access to rocks and rock reef.

Summary: Three pockets of sand and boulders that are mainly viewed by passing boat fishers.

822-824 HILLOCK PT (S)

| No. | Beach | Unpatrolled | | | |
		Rating HT	LT	Type	Length
822	Hillock Pt (S)	1	3	R+rocks	100 m
823	Beach 823	1	3	R+BR	600 m
824	Beach 824	1	3	R+BR	200 m

To the south of Hillock Point the coast remains steep and predominantly rocky for 6 km, with only three beaches occupying a total of 900 m of the shore. The first beach **(822)**, lies 500 m west of Hillock Point and consists of a 100 m long, east facing pocket of sand fronted by a string of boulders.

Beach **823** occupies the front of a sloping valley. It is 500 m long, faces south-east into the Trades and is fringed by boulder beaches and 100 m high rocky headlands and slopes. The beach itself consists of a low, narrow high tide beach, with a 50 m wide low tide bar, along the edge of which are usually two to three rips, with another two permanent rips against the boulders.

Beach **824** is a 200 m long sand beach, fringed by boulder beaches and backed by grassy slopes that rise steeply to a 60 m high ridge line. The beach is almost awash at high tide, with a 50 m wide low tide bar and a permanent rip against the northern rocks.

Swimming: All three beaches have permanent hazards, including boulders and headland rips on the longer two beaches. Use care if swimming, particularly toward low tide when the rips are strongest.

Surfing: Chance of a break over the bar and rips on the two more sandy beaches.

Fishing: All three offer plenty of rocks to fish from, plus the permanent and beach rip holes.

Summary: Three exposed, sandy beaches, dominated by the surrounding slopes and headlands.

825, 826 MULLIGAN BAY, PICNIC BEACH

| No. | Beach | Unpatrolled | | | |
		Rating HT	LT	Type	Length
825	Mulligan Bay	1	3	R+BR	4 km
826	Picnic Beach	1	2	B+SF	2.8 km

The south-eastern corner of Hinchinbrook Island forms the northern boundary of the Hinchinbrook Channel. At its mouth the channel is 2 to 3 km wide, with tidal shoals extending up to 5 km seaward. The strong tidal currents and waves have built an extensive coastal plain up to 2 km wide and containing over 4 square kilometres of sandy sediments. The eastern side of the plain is called Mulligan Bay and the southern side, facing into the channel, is Picnic Beach.

Mulligan Bay beach (825) is 4 km long, faces east and extends from a 60 m high headland at the northern end down to sandy George Point in the south. It is backed by a sediment-filled valley behind the first 2 km of beach, that is drained by a small creek flowing out against the northern headland. An 80 m high knoll backs the centre of the beach, with a small creek draining across the beach on its southern side. Beyond is the wider coastal plain extending down to George Point.

The beach receives waves averaging over 0.5 m, that combine with the medium beach sand to form a moderately steep high tide beach, fronted by a 200 m wide low tide bar. The bar commonly has rips spaced every 100 m along its outer edge. At and beyond George Point, tidal shoals extend off the beach and lower the wave height.

Picnic Beach (826) begins on the southern side of George Point and gradually curves around into the channel to face south toward Lucinda on the opposite side of the channel. The beach is very dynamic owing to the influence of the tidal shoals and currents that move pulses of sand along the beach, causing it to vary considerably in size, shape and width.

Swimming: Mulligan Bay Beach is moderately safe at high tide, with a greater chance of rip currents and channels at low tide, together with tidal currents off both creeks. Picnic Beach has low waves, but strong tidal currents right off the beach, particularly at low tide.

Surfing: Mulligan Bay Beach usually has a beach break, that picks up with the Trades.

Fishing: The channel is a very popular spot, with the creeks and rip holes also offering some variety.

Summary: These two beaches right opposite Lucinda are relatively popular for day-trippers, as the name Picnic Beach suggests.

LUCINDA TO CAPE UPSTART (Fig. 4.27)

Length of Coast:	389 km
Beaches:	102 (beaches 827-929)
Islands	Magnetic Island (22 beaches)
Surf Life Saving Clubs:	Forrest Beach, Alma Bay, Picnic Bay, Ayr (Alva Beach)
Lifeguards:	The Strand (Townsville)
Major towns:	Lucinda, Ingham, Townsville, Ayr

Regional Map 3: Lucinda to Cape Upstart

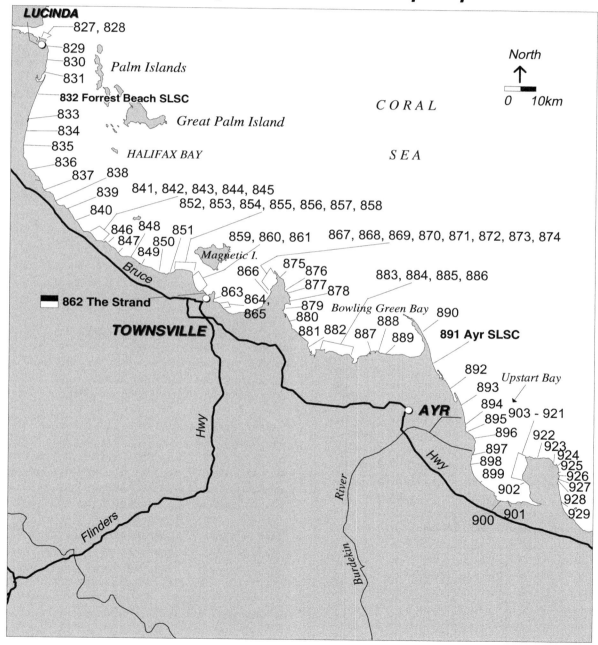

Figure 4.27   *Regional map 3   Lucinda to Cape Upstart, beaches 641 to 739. Beaches with Surf Life Saving Clubs indicated in bold, beaches with lifeguard patrol only indicates by flag.*

Halifax Bay is an open, 80 km long, north-east facing bay between Lucinda and Magnetic Island. The bay was named by Captain Cook, who described a 'large bay' between Point Hillock (on Hinchinbrook Island) and Cape Cleveland. Cook also named the southern Magnetic Island and noted the northern (Orpheus and Palm) islands. While the city of Townsville dominates the southern reaches of the bay, much of the bay shoreline is relatively undeveloped. The small town and sugar port of Lucinda sits at the northern extremity, with the Forrest Beach settlement 20 km to the south. Then there is essentially 45 km of undeveloped and often inaccessible shoreline, before reaching Balgal. Between Balgal and Townsville is 50 km of shore containing the small holiday settlements of Toomula, Toolaka and Saunders Beach.

827-829　DUNGENESS, LUCINDA

No.	Beach	Unpatrolled		Type	Length
		Rating			
		HT	LT		
827	Dungeness	1	1	B+SF	1.6 km
828	Lucinda	1	1	B+SF	700 m
829	Lucinda Pt	1	1	B+SF	4 km

Lucinda is a small town and sugar port located on the southern entrance to the Hinchinbrook Channel. The town sits on a low, sandy barrier that has accumulated at the entrance to the channel. It is backed by 3 km of mangrove-covered tidal flats and fronted by intertidal sand flats up to 2.5 km wide. Because of the overall shallowness of the coastline, the Lucinda jetty extends 5.3 km out to sea to reach water deep enough for the sugar ships. The town is located 7 km north-east of Halifax and 25 km from Ingham. It has a small shopping centre, a caravan park and facilities for boating and fishing.

The sandy barrier upon which the town rests contains three beaches - two in the channel and one facing the sea. The channel beaches both face north across the channel to Hinchinbrook Island. **Dungeness** Beach (827) is 1.6 km long and consists of a narrow, sandy high tide beach fronted by 200 to 300 m wide low tide sand flats. There are a few houses toward the western end and a boat ramp. A small creek crosses the beach toward the eastern end. **Lucinda** Beach (828) fronts the north side of the town and extends for 700 m east from the base of the long sugar jetty. It is backed by a rough, rocky seawall and narrow high tide beach, which is fronted by intertidal sand flats that increase in width from 100 to 300 m to the east. The eastern tip becomes a sinuous sandy spit called Lucinda Point.

Lucinda Point Beach (829) begins on the south side of the point and runs south in a crenulate fashion for 4 km to the mouth of Gentle Annie Creek. The narrow high

tide beach is backed by a sandy barrier occupied by houses for the first 2 km, before narrowing to a strip of sand between the shoreline and backing mangroves. Extensive sand flats, 2.5 km wide at the point, narrow to 1.5 km at the creek mouth.

Swimming: Swimming is only possible at mid to high tide, with extensive sand flats exposed at low tide. Stay well clear of the channel and the creek mouth, both of which have deep water and strong tidal currents, as well as possibly crocodiles.

Surfing: None.

Fishing: Lucinda is a very popular fishing town, with most visitors coming to fish the Hinchinbrook Channel and offshore reefs. The town has all facilities for fishing, including boat ramps at Dungeness and Lucinda.

Summary: Lucinda is well off the highway and mainly visited by locals and keen fishers. It offers quieter water and a beachfront town with sufficient facilities for the traveller.

830, 831　GENTLE ANNIE CK, TAYLORS BEACH

No.	Beach	Unpatrolled		Type	Length
		Rating			
		HT	LT		
830	Gentle Annie Ck	1	1	B+SF	4.2 km
831	Taylors Beach	1	1	B+SF	3.8 km

Gentle Annie Creek drains the extensive mangrove swamp on the south side of Lucinda. Between the creek mouth and Victoria Creek, 7 km to the south, is 8 km of beach, separated in the centre into Gentle Annie Beach and Taylors Beach by a small unnamed creek.

Gentle Annie beach (830) is 4.2 km long and consists of a narrow high tide beach, with 2 to 3 km of backing mangroves and beach ridges, and 1 to 1.5 km of intertidal sand flats. There is no development apart from an access track along a drain toward the southern end, and a few fishing shacks on the northern tip.

Taylors Beach (831) lies on the northern side of Victoria Creek mouth and is protected from waves by the northern spit of Forrest Beach. Waves are usually absent, however there is deep water in the tidal channel off the beach and strong tidal currents. The beach is backed by the small settlement of Cassady, that contains a few houses, a caravan park and a store. At the beach is a foreshore reserve containing a car park, picnic and play areas, an amenities block and a boat ramp.

Swimming: Both beaches are usually calm, however the wide, shallow tidal flats only permit swimming toward high tide. Take care at Taylors because of the strong tidal currents in the creek.

Surfing: None.

Fishing: Taylors Beach is a popular location, mainly with the local farmers and visiting fishers.

Summary: Gentle Annie is rarely visited, while Taylors is very accessible but 20 km off the highway and mainly used by locals.

832 FORREST BEACH

Patrols:				
Surf Life Saving Club:		September to May		
Lifeguard on duty:		Christmas & Easter		
Stinger net:		November to May		

No.	Beach	Rating HT LT	Type	Length
832	Forrest	2 3	R+BR	14.3 km

Forrest Beach (832) is the surfing beach for the town of Ingham. It lies 20 km east of the town and the Bruce Highway, and contains a small holiday settlement, spread for 4 km along the 14 km long beach. There is a store, hotel and caravan park at the beach, together with the Forrest Beach Surf Life Saving Club, founded in 1928. It is located in the wide foreshore reserve that backs the beach. The reserve also contains an amenities block, playground and picnic facilities (Figs. 4.28 & 4.29). The stinger net is directly in front of the surf lifesaving club, with a concrete boat ramp just to the north.

The beach lies toward the northern end of the open Halifax Bay and faces slightly south-east, exposing it to waves averaging between 0.5 and 1 m. These combine with the medium sized beach sand to build a 50 m wide high tide beach, fronted by a 50 m wide bar that is cut by small rip channels during and following waves greater than 1 m.

Swimming: A moderately safe beach at high tide, however care should be taken at mid to low tide when rips, if present, will intensify. Always swim in the patrolled/stinger net area.

Surfing: This is the best place to look for a wave in Halifax Bay, with beach breaks over the bar at mid to low tide.

Fishing: Beach fishing only, with rip holes common along the beach. There is a beach boat ramp that can be used at high tide during low wave conditions.

Summary: A beach popular with Ingham locals and an increasing number of travellers. It is well worth the drive out to the long attractive beach for a day trip, or for a quiet holiday.

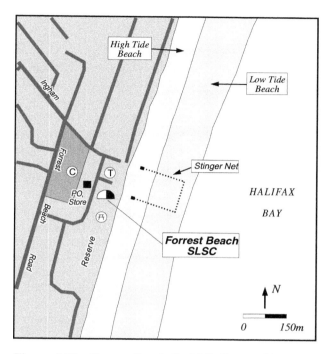

Figure 4.28 *Forrest Beach Surf Life Saving Club is located in a grassy foreshore reserve. The exposed beach usually has a low tide bar and occasional rips, and during the summer season is protected by a stinger net.*

Figure 4.29 *Forrest Beach, showing the surf club (centre) and the typical low tide bar drained by shallow rips.*

833-835 PALM CK, BRONTE, ORIENT-CATTLE

No.	Beach	Unpatrolled Rating HT	LT	Type	Length
833	Palm Ck	1	2	B+SF	2.5 km
834	Bronte	2	3	R+BR	4.5 km
835	Orient-Cattle Ck	2	3	R+BR	4.7 km

The central section of Halifax Bay consists of relatively undeveloped beaches. This is partly because the backing swamps make access difficult. South of Forrest Beach is 12 km of coast containing three beaches and one access track. This section of the bay is also more exposed, with east facing beaches like Forrest Beach.

The first beach is **Palm Creek** beach (833), which begins on the southern side of a small creek that separates it from Forrest Beach and extends for 2.5 km to the 500 m wide mouth of Palm Creek. Mangroves and tidal creeks back the narrow strip of sand and beach. The beach is straight and narrow at high tide, with a low tide system containing a few rips and more extensive sand ridges and shoals, as well as the tidal shoals at each end.

Bronte Beach (834) is 4.5 km long, running from Palm to Orient Creek mouths. There is also a small and often closed creek that crosses the middle of the beach. A circuitous, 15 km long 4WD access track runs across the backing swamps and sand ridges to the beach, to service a few shacks at the northern end. The beach consists of a narrow high tide beach and then a low tide bar that is usually cut by rips every 100 m.

Orient - Cattle Creek beach (835) is very similar to Bronte; 4.7 km long, backed by swamps and fronted by a rip-dominated low tide bar; however there is no vehicle access and no development.

Swimming: All three beaches usually have low to moderate waves, depending on the Trade winds. Swimming is safest toward high tide, as rips and holes become more prevalent toward low tide. Stay clear of the creek mouths as they contain deeper water and strong tidal currents.

Surfing: Chance of a break across the bars and rips at mid to low tide.

Fishing: These beaches offer the creeks and beach fishing into the rip holes.

Summary: Three isolated beaches, backed by extensive mangrove swamps, with only Bronte accessible by vehicle.

836-838 ELEANOR CREEK, CRYSTAL-OLLERA CK, OLLERA CK

No.	Beach	Unpatrolled Rating HT	LT	Type	Length
836	Eleanor-Crystal Ck	1	1	B+SF	9.5 km
837	Crystal-Ollera Ck	1	1	B+SF	8.5 km
838	Ollera Ck	1	1	B+SF	5.5 km

South of Cattle - Eleanor Creeks down to the mouth of Hencamp Creek at Balgal is 24 km of lower energy, northeast facing, crenulate shoreline, consisting of occasional creek mouths, narrow high tide beaches and extensive sand (and increasingly rocky) tidal flats. This section of coast is slowly being developed, with fishing shacks along the beach north of Crystal Creek and a newer development on the north side of Hencamp Creek.

The first beach (836) lies between **Eleanor** and **Crystal creeks**. It is 9.5 km long and initially consists of a narrow beach with extensive backing mangrove swamps. These narrow to the south and a vehicle track and holiday shacks parallel the southern 4 km of beach, to the mouth of Crystal Creek. The beach is initially fronted by the 1.5 km wide tidal flats of Eleanor Creek, with tidal and rock flats averaging 500 m to 1 km wide to the south. Where the rock flats dominate, the shoreline becomes increasingly crenulate. There is an extensive coral reef paralleling the beach offshore.

The **Crystal - Ollera Creek beach** (837) is 8.5 km long, with a highly crenulate shoreline, owing to variations in the 500 m wide rock and sand flats. Several small creeks also drain across the low beach and there are patches of mangroves along the low energy beach. A 4WD track runs 2 km to the beach from the Bruce Highway at Moongbulla railway siding. There is no development and no facilities at the beach, which is, however, used for beach boat launching.

Ollera Creek beach (838) extends from the mouth of Ollera Creek for 5.5 km to the small Hencamp Creek mouth. The narrow beach initially faces north-east and swings around to face north, then east, at Hencamp Creek. There is a new development at Hencamp Creek mouth, which can be reached by turning off the highway at the Kinduro railway siding.

Swimming: Because of the extensive shallow sand and rock flats, swimming is only possible at mid to high tide, when calm conditions are usually encountered.

Surfing: None.

Fishing: Best in the creeks and at the creek mouths, with Crystal Creek being the most accessible and also providing a boat ramp.

Summary: Three low energy beaches relatively close to the highway, that are slowly being developed for housing and recreation.

839, 840   HENCAMP CK, BALGAL

No.	Beach	Unpatrolled		Type	Length
		HT	LT		
839	Hencamp Ck	1	1	B+SF	1.5 km
840	Balgal	1	2	R+UD	9.5 km

Balgal is the first small coastal settlement south of Forrest Beach and lies at the entrance to Rollingstone Creek. North of the creek is a 1.5 km long, low energy beach (839), that is also crossed by the mouth of small **Hencamp Creek**. This beach consists of a east facing, relatively narrow and steep high tide beach, while low tide exposes sand and rock flats extending up to 500 m seaward of the creek mouths.

The main **Balgal beach** (840) lies 5 km from the highway and the small settlement parallels the northern kilometre of the beach. There is a store and caravan park as well as a boat ramp into the creek, which can only be used at mid to high tide. The beach runs south-east for 9.5 km to the rocks of Douglas Hill. There are three additional access roads to the central portion of the beach; two of these are off the Balgal Road, with housing development occurring at each. The beach is backed by a low, grassy foredune and receives waves averaging 0.5 m that maintain a 50 m wide, moderate gradient high tide beach and a 100 to 200 m wide, low gradient low tide bar. Waves spill across a wide surf zone at mid to low tide, while rips, when present, are usually weak.

Swimming: It is a relatively safe beach with a wide, low gradient surf and usually low waves.

Surfing: Usually only low, spilling breakers that pick up with the Trades.

Fishing: Best in the creek and over the rock reefs at high tide.

Summary: Balgal is a quiet little settlement with limited tourist facilities, but close enough to the highway to warrant a visit.

841-843   DOUGLAS HILL, TOOMULLA

No.	Beach	Unpatrolled		Type	Length
		HT	LT		
841	Douglas Hill	1	1	B+SF	200 m
842	Toomulla (Pt)	1	1	R+LT	100 m
843	Toomulla	1	2	B+SF	800 m

Douglas Hill is the first hill and bedrock outcrop in Halifax Bay south of Hinchinbrook Island. The 50 m high hill forms the southern boundary to Balgal Beach and around its 1 km base are two small beaches, while on its south side the small settlement of Toomulla is located between the hill and the wandering mouth of Saltwater Creek. Toomulla lies just 2 km from the highway, with vehicle tracks also leading to the hill and the southern end of Balgal Beach.

Douglas Hill beach (841) lies on the north side of the hill and consists of 200 m of sand that is continuous at low tide, but divided into three small pockets at high tide, all backed by the steep slopes of the hill. At low tide, sand waves extend 200 to 300 m off the beach. There is an access road to the very southern end of the beach and a small parking area.

Toomulla Point beach (842) is a 100 m long, north-east facing pocket of sand located immediately around a low, rocky point from the Douglas Hill Beach. A short access road and car park service both beaches. Rocks fringe either end and sand waves extend 200 m off the beach. Freehold land and some substantial houses overlook the small beach.

Toomulla beach (843) extends from the small, rocky boundary with the point beach for approximately 800 m to the mouth of Saltwater Creek. It is backed by the small holiday settlement of Toomulla, that has several houses, a large camping area and phone, but no store. The length and shape of the beach depends on the shifting entrance and sand bars at the creek mouth. The beach faces northeast and receives waves averaging less than 0.5 m. These break across 300 to 500 m wide sand and rock flats off the creek mouth. The high tide beach has a moderate gradient and is up to 50 m wide, backed by a low, casuarina-covered foredune and houses.

Swimming: All three beaches are very accessible and relatively safe, owing to the usually low waves. Swimming is better at mid to high tide when the flats are covered. Be careful at the creek mouth as it has strong tidal currents.

Surfing: Usually low, spilling waves.

Fishing: Best off the rocks at high tide and in the creek; otherwise in a boat over the rock reefs.

Summary: A quiet holiday settlement just off the highway; more a dormitory suburb of Townsville.

844-846   CASSOWAY CK, LEICHHARDT CK, CHRISTMAS CK

No.	Beach	Unpatrolled Rating HT	LT	Type	Length
844	Cassoway Ck	1	1	B+SF	3.5 km
845	Leichhardt Ck	1	1	B+SF	1 km
846	Christmas Ck	1	1	B+SF	3 km

Cassoway, Leichhardt and Christmas Creeks are three parallel creeks that have all delivered sand to the Halifax Bay shoreline, resulting in the beach building out on the order of several hundred metres. In between the abandoned beaches are mangrove-covered tidal flats. As a result of the low-lying nature of the land, there has been little development of this 6.5 km long section of the bay. There are vehicle access tracks leading to near the mouth of each of the creeks and a few shacks set back from the coast, but otherwise no major settlements or development.

Cassoway Creek beach (844) is 3.5 km long and runs from the southern side of Saltwater Creek mouth past the small mouth of Cassoway Creek to the lee of a sand spit extending north from Leichhardt Creek. The beach faces north-east and consists of a narrow high tide beach fronted by tidal flats that widen from 300 to 600 m near Leichhardt Creek. Waves are usually less than 0.5 m high.

Leichhardt Creek beach (845) is a dynamic, low sand spit that extends north from the mouth of Leichhardt Creek for up to 1 km. This is a low energy but highly dynamic strip of shore, that is influenced by both the creek and the waves. The low beach is fronted by sand and rock flats up to 600 m wide.

Christmas Creek beach (846) runs from the mouth of Christmas Creek as a 3 km long, well vegetated sand spit, that also increases in width to the north. The low beach faces north-east and is fronted by sand and rock flats up to 1 km wide, that are exposed at low tide.

Swimming: These three beaches are all difficult to access from the land and are little used. Waves are usually low and the beaches are relatively safe, however you can only swim toward high tide. Care must also be taken at each of the creeks, that can have strong tidal currents.

Surfing: Usually very low, spilling waves toward high tide.

Fishing: Best in the creeks or off the beaches at high tide.

Summary: Three low energy but relatively dynamic spits of sand that change shape and length from year to year.

847-849   TOOLAKEA, BLUEWATER, SAUNDERS

No.	Beach	Unpatrolled Rating HT	LT	Type	Length
847	Toolakea	1	1	B+SF	5.5 km
848	Bluewater	1	1	B+SF	1 km
849	Saunders	1	1	UD+tidal flats	5.5 km

Toward the southern end of Halifax Bay are the settlements of Toolakea, Bluewater and Jalloonda, separated by Bluewater and Deep Creeks. Toolakea is a narrow, 2 km long settlement at the southern end of Toolakea Beach, and lies 4 km from the highway. A road runs out from the small town of Bluewater, located on the highway, to Bluewater Beach, while Jalloonda is reached via a 6 km long road from the highway at Yabulu, that skirts the north side of 90 m high Mount Saunders. The only store in the three settlements is at Jalloonda.

Toolakea beach (847) is 5.5 km long, beginning at the mouth of Christmas Creek. It initially faces north-east, then swings around to face north at the settlement. It ends at the mouth of Bluewater Creek. The beach is backed by a low, casuarina-covered foredune, with a 50 m wide, moderately steep high tide beach, then sandy tidal flats that vary in width from a few hundred metres off the creek mouth to 200 m along the central portion of the beach.

Bluewater beach (848) is a changeable, 1 km long beach, owing to its location between two creek mouths. At times the low beach is fronted by a low sand spit, at other times it directly faces north-east into the waves. A row of houses and a low, grassy dune back the high tide beach, with sand flats merging into rock flats extending 200 to 300 m at low tide.

Saunders beach (849) is the beach for Jalloonda and the 2 km long settlement parallels the northern end of the 5 km long beach. It is bordered by Deep Creek to the north and the Black River in the south. One kilometre further on, the pipeline from the Yabulu nickel treatment plant crosses under the beach and runs 2 km out into the bay. The 100 m high chimney of the plant is a distinctive landmark. The beach faces north-east and consists of the low gradient high tide beach, fronted by tidal flats off the northern creek and river. The central portion has a 200 m wide, low gradient bar and the usually low waves spill across the wide, shallow surf zone.

Swimming: These three beaches all receive waves less than 0.5 m, with no waves at low tide. Swimming is best toward high tide and away from the creek mouths, which can have strong tidal currents. Dense casuarina trees provide shade behind the beaches.

Surfing: Usually none, other than low spilling waves toward high tide.

Fishing: Fishing is popular at all three settlements, including off the beach at high tide, in the creeks and in boats over the rock reefs.

Summary: Three essentially beach-holiday suburbs of Townsville, offering good access but limited facilities for the visitor.

850, 851   BLACK RIVER, BOHLE RIVER

No.	Beach	Unpatrolled Rating HT	LT	Type	Length
850	Black River	1	1	B+SF	4.5 km
851	Bohle River	1	1	B+SF	700 m

The Black and Bohle Rivers are two small rivers that enter the southern end of Halifax Bay. Lying toward the southern, more protected end of the bay, they receive low waves and have mangroves growing along parts of their shores. Black River is bordered by settlements on either side, with a road on its southern side running 7 km out to a growing coastal settlement that parallels the first kilometre of the beach. There is a second access road midway down the beach, and a vehicle track to the mouth of the Bohle River and Bohle River Beach.

Black River beach (850) is 4.5 km long, east-north-east facing and consists of a low, narrow high tide beach, fronted by tidal flats that are 500 m wide off the Black River mouth and increase to 2 km off the southern Bohle River mouth. A few mangroves fringe the northern end and increase in density to the south, where they are 500 m wide off the Bohle River.

The **Bohle River** beach is a 700 m long, low strip of north facing sand, situated on the eastern side of the river mouth. There is a 4WD track to the beach that is fringed by mangroves at each end and fronted by 1.5 km wide sand and mud flats.

Swimming: Only recommended and possible at high tide, owing to the wide, shallow tidal flats.

Surfing: None.

Fishing: Best in the rivers and at the river mouths.

Summary: Two low energy beaches with good access to the Black River mouth and dry weather access to the Bohle River mouth.

Cape Pallarenda Environmental Park

The Cape Pallarenda area was originally set aside as a quarantine area in 1915; the station was moved there, buildings and all, from West Point on Magnetic Island. The station was closed in 1973. During World War II the station was also used as an American Army Hospital, then in 1975 as temporary accommodation for the Australian Institute of Marine Science, until being declared an environmental park in 1986.

The park offers scenic coast and woodland walks and an interesting array of tropical flora and fauna.

For further information:
Queensland National Parks & Wildlife Service
(077) 21 2399

852-858   CAPE PALLARENDA PARK

No.	Beach	Unpatrolled Rating HT	LT	Type	Length
852	Beach 852	1	1	B+SF	400 m
853	Beach 853	1	1	B+SF	500 m
854	Shelly Creek (W)	1	1	B+SF	200 m
855	Shelly Beach	1	1	B+SF	3.1 km
856	Shelly Creek (E)	1	1	B+SF	400 m
857	Cape Pallarenda (1)	2	2	B+SF +rocks	250 m
858	Cape Pallarenda (2)	2	2	B+SF +rocks	100 m

The coast immediately north of Townsville has long been a relatively remote and little used area and today remains much the same. Between the mangrove-fringed Bohle River and Cape Pallarenda is 7 km of sandy and rocky coast, most of which is now contained within the Cape Pallarenda Environmental Park, with little development other than walking tracks. There are seven beaches along this low energy and generally north facing section of coast. There is limited 4WD access to some of the western beaches, with vehicle access to Cape Pallarenda and foot access to the remainder.

Beaches **852** and **853** lie 1.5 and 2 km respectively east of the Bohle River mouth and are backed by shrub-covered slopes that rise to a peak of 125 m. They are 400 and 500 m long respectively and both face essentially north-east. Their shorelines are composed of a mixture of narrow, sandy high tide beach, together with rocks and mangroves, while they are fronted by sandy tidal flats up to 1 km wide. There are vehicle tracks to the back of both beaches.

Shelly Creek (west) beach (854) is a 200 m long strip of sand lying between a low, rocky headland and the mouth of mangrove-fringed Shelly Creek, with a vehicle track reaching the headland. It faces north and is fronted by 500 m wide, sandy and shelly tidal flats.

Shelly Beach (855), as the name suggests, is a 3.1 km long ridge of shelly sand that is slowly growing to the west as a series of low, recurved spits. It faces north and is fronted by 1 to 2 km wide, sandy-shelly tidal flats. As the spit has developed, it has formed a 3 km long, mangrove-filled swamp between the spit and the slopes of the Many Peaks Range. The range is a 3 km long series of hills that rise to over 200 m.

Shelly Creek (east) beach (856) is a 400 m long, northeast facing strip of high tide sand lying at the base of the range, with numerous rocks and boulders along the beach. It is fronted by 2 to 3 km of sand, shell and rock tidal flats and can be accessed on foot via a track from Cape Pallarenda.

Cape Pallarenda (1) and (2) (beaches 857 & 858) are two small pocket beaches, 250 and 100 m long respectively. They lie on the north-east facing point of the cape and both are backed by vegetated slopes of the range that rise steeply to nearly 200 m. There is a car park just south of the southern beach and a walking track to both beaches.

Swimming: None of these beaches are really suitable for swimming, as they are tidal flats except at high tide. A number of the beaches also have numerous rocks.

Surfing: None.

Fishing: Only at high tide and best at the creek mouth.

Summary: Seven natural beaches lying within the environmental park and accessible along the walking tracks.

859-861   PALLARENDA, ROWES BAY, KISSING POINT

Patrolled

Roving IRB & 4WD patrols during stinger season (November to May)

No.	Beach	Rating HT	LT	Type	Length
859	Pallarenda	1	1	R+LT	2.5 km
860	Rowes Bay	1	1	R+LT/SF	4.2 km
861	Kissing Pt	1	1	B+SF	300 m

Pallarenda Beach and Rowes Bay are essentially one long beach cut in the middle by Three Mile Creek. Both beaches are backed by a 1 to 2 km wide, low, coastal plain. **Pallarenda Beach** (859) is 2.5 km long, faces east and receives low waves usually less than 0.3 m. These maintain a low, 50 m wide high tide beach fronted by 100 to 200 m wide intertidal sand flats. The northern end of the beach lies in the Environmental Park and has a number of facilities and picnic areas. Just south of the park entrance is a concrete boat ramp and a swimming enclosure, backed by the small Pallarenda residential settlement. The Cape Pallarenda Road runs right behind the beach, providing good access for the entire length, with a grassy foreshore reserve between the road and the beach.

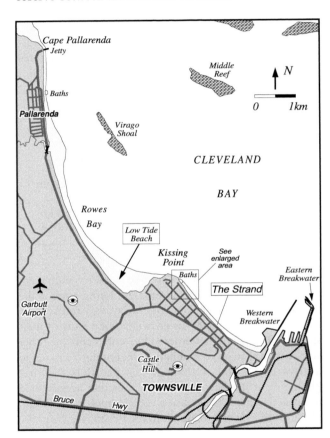

Figure 4.30 *A map of Townsville's regional beaches. They extend from Pallarenda in the north to Rowes Bay and The Strand, that fronts the city.*

On the south side of Three Mile Creek is the **Rowes Bay Beach** (860), which is 4.2 km long and faces east-north-east. It too is paralleled by a wide, grassy foreshore reserve, then the road. A small creek drains across the southern portion of the beach; just beyond which is the popular Rowes Bay caravan park. The tidal flats widen at each of the creek mouths, and toward the southern corner of the bay are up to 1 km wide.

Kissing Point Beach (861) is a 300 m long strip of sand that winds along the base of the 20 m high bluffs of

Kissing Point. There is residential development on the point and foot tracks down to the north facing, low energy beach, which is fronted by shallow tide flats that narrow toward the point.

Swimming: The best spot is at Pallarenda, with the added safety of the stinger net during summer. All beaches can only be used at mid to high tide owing to the low tide flats.

Surfing: Usually none.

Fishing: Best at high tide near the creek mouths.

Summary: Pallarenda and Rowes Bay are two of Townsville's more accessible and popular beaches, offering excellent access and plenty of space.

862 THE STRAND (TOWNSVILLE)

Lifeguard on duty:	all year		
Stinger net:	November to May		

No.	Beach	Rating HT LT	Type	Length
862	The Strand	**1** **1**	R+LT	2.7 km

The Strand is Townsville's main beach. A wide road of the same name parallels the back of the beach, with residential and tourist development occupying the entire length of the western side of the road. The beach is 2.7 km long and faces north-east toward Magnetic Island. It runs from northern Kissing Point south to the 1.5 km long western breakwater of Townsville harbour (Fig. 4.31). Between the road and the beach is a raised seawall, to protect the city from tropical cyclone storm surges, and a $29 million redevelopment of the foreshore reserve including a tree-lined walkway.

The entire reserve and beachfront underwent a major redevelopment in the late 1990's, which substantially improved beach access, parking and facilities, including the development of five protruding nodes containing a range of recreational facilities. These nodes also divide the once continous beach into six beach compartments between 200 m and 400 m in length. A new jetty extends seaward of the largest central node.

The beach (862) is protected by both the breakwater and offshore islands, including Magnetic Island. Waves are usually less than 0.5 m and at high tide surge against the narrow beach. At low tide there is a 50 m wide bar attached to the beach.

Swimming: The best swimming is at Kissing Point, either in the circular rock pool constructed at the point, with its own little beach, or immediately south of the pool, where there is a stinger net and lifeguard on duty (Fig. 4.32).

Surfing: There are usually no rideable waves. Very strong south-easterlies or a cyclone are required to have a surf at The Strand.
Fishing: Most fishers head for the Kissing Point rock pool, where they can fish off the walkway that runs around the pool. The beach is best at mid to high tide.

Summary: A well-developed beach backed by the city ofof Townsville. Waves are usually low to non-existent and swimming is relatively safe in the pool or stinger enclosures.

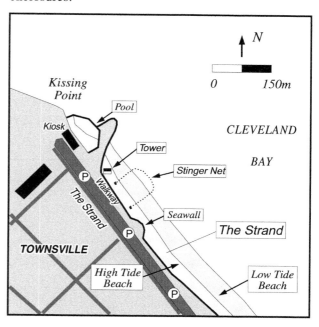

Figure 4.31 *A map showing the northern section of The Strand beach with its two swimming beaches and swimming pool on Kissing Point. They are patrolled 7 days per week, 365 days a year and have two stinger net enclosures in season.*

Figure 4.32 The Strand at Townsville has a steep beach face fronted by a low tide bar. This view also shows the stinger net cables.

863 MAGAZINE ISLAND

No.	Beach	Unpatrolled			
		Rating		Type	Length
		HT	LT		
863	Magazine Is	1	1	B+SF	200 m

Magazine Island is part of the port of Townsville and has been heavily modified by port development. Along the east side of the island, toward the mouth of Ross River, is a narrow, 200 m long strip of sand between the road and the creek. While the creek is used by boats for access to the bay, the beach is not recommended for swimming owing to the tidal flats adjacent to the beach, boat traffic and strong tidal currents in the creek that parallels the beach.

Magnetic Island

Magnetic Island is located 5 km off Townsville. It is serviced by a regular ferry from Townsville and many resi-

dents commute to Townsville to work. The island is approximately 50 km² (5184 ha) in area, with 40 km of coastline containing 24 beaches, many kilometres of steep, rocky shore and 5 km of mangroves on the protected western shore (Fig. 4.33). The island is fairly rugged, rising over 500 m at the central Mount Cook. The mountain is named after Captain Cook, who in June 1770 sailed past the island, naming it 'Magnetic Head or Island'. He was not sure of which since 'the compass did not traverse well when near it.' The island lies in a rain shadow area and, like Townsville, is relatively dry. There are four main areas of settlement, three on the southeast coast - at Picnic Bay where the ferry lands, a second at adjacent Nelly Bay, and a third at Geoffrey Bay, where a large marina is still under construction. The fourth is at the 3 km long Horseshoe Bay on the north coast, the largest bay on the island.

Most of the beaches are only easily accessible by boat, with less than half accessible by vehicle. The main swimming beaches are at Picnic Bay, Nelly Bay, Geoffrey Bay, Alma Bay and Arthur Bay, all on the south-east coast, and Horseshoe Bay on the north coast. There is a surf lifesaving club at Picnic Bay and the Arcadian Club at Alma Bay. The only stinger enclosure is at Picnic Bay.

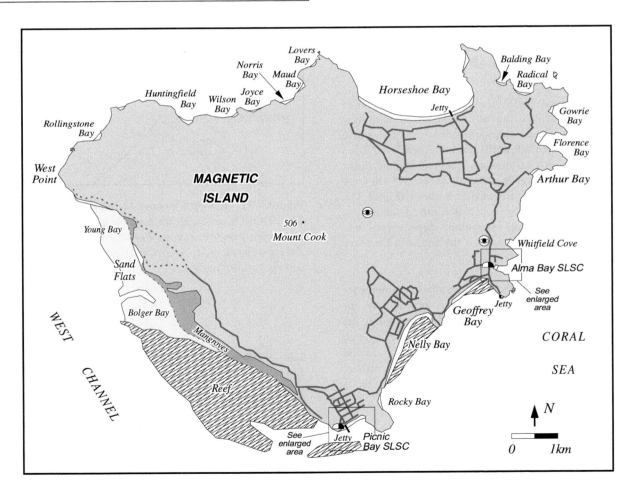

Figure 4.33   *A map of Magnetic Island showing the location of all island beaches.*

Magnetic Island National Park

Area: 2533 ha

Magnetic Island National Park occupies approximately half the island, including the high central peak of Mount Cook, and much of the north and north-eastern shores. The park contains a number of secluded beaches, 22 km of walking tracks and picnic areas at Balding and Radical Bays.

Adjacent to the park are also five areas of marine park, located along the north shore between Joyce and Lovers Bays, and on the south-eastern side at Balding and Radical Bays, and Florence and Geoffrey Bays.

For further information contact:
The Ranger
Hurst St, Picnic Bay, Magnetic Island
phone: (07) 78 5378

MT1-MT4 MAGNETIC ISLAND

No. Beach	Unpatrolled		Type	Length
	Rating			
	HT	LT		
MT1 Balding Bay	1	2	R+LT	200 m
MT2 Radical Bay	1	2	R+LT	250 m
MT3 Florence Bay	1	2	R+LT	200 m
MT4 Arthur Bay	1	2	R+LT	150 m

The north-east corner of Magnetic Island consists of granite hills rising nearly 200 m and a number of small bays and embayed beaches. The first five bays house four sandy beaches; Balding, Radical, Florence and Arthur Bays, while Balding, Radical and Florence Bays are also marine parks.

Balding Bay (MT1) is a small, semi-enclosed bay just over 100 m wide at the entrance, containing a curving, 200 m long, north facing beach. The national park bay can be reached via a walking track from Horseshoe or Radical Bays, but contains no facilities or development other than a picnic area. Protruding granite headlands protect the bay, which is a favoured anchorage for boats. The sandy beach has a moderate gradient and is 30 m wide at high tide and up to 70 m wide at low tide.

Radical Bay (MT2) lies immediately east of Balding Bay; it also faces north and is a more open, 250 m wide bay with prominent granite headlands. A road leads to the bay where there are a few houses and a small car park, but no facilities. The beach has a low to moderate gradient and varies in width from 20 m at high tide to up to 50 m at low tide.

Florence Bay (MT3) is an east facing, half kilometre deep, 200 m wide bay that contains both a sandy beach and a fringing coral reef. The entire bay is a marine park. There is vehicle access to the beach as well as a few houses and a small car park toward the southern end. The beach is 30 m wide at high tide, where it is partly shaded by casuarina trees, while low tide exposes sand flats over 50 m wide, with the reef fringing the northern half of the flats.

Arthur Bay (MT4) is a small, 150 m long bay containing the beach, a small fringing reef and granite headlands. The road to Radical Bay runs across the slopes behind the beach and there is a single house overlooking the southern end of the south-east facing bay. The beach is 20 m wide at high tide, widening to 50 m at low tide.

Swimming: Waves usually average less than 0.5 m at all four bays and swimming is relatively safe under these conditions. Just watch for the shallow coral reefs and the rocks adjacent to the headlands.

Surfing: Usually too small to worry about, with Florence and Arthur receiving slightly higher waves when the Trades are blowing; Florence is the best during strong south-easterlies.

Fishing: All four bays offer excellent opportunities for beach and rock fishing. Certain restrictions apply in the marine parks.

Summary: Four small but beautiful bays surrounded by steep, pine-covered, granite headlands and hills, together with clean, low energy, sandy beaches and some coral reefs.

MT5 ALMA BAY, ARCADIAN (SLSC)

Patrols:				
Surf Life Saving Club:		September to May		
Lifeguard on duty:		Christmas & Easter		
No. Beach	Rating		Type	Length
	HT	LT		
MT5 Alma Bay	1	2	R+LT	150 m

Alma Bay (MT5) is a very picturesque little bay, just 150 m in width at the beach. It faces due east and has two prominent granite headlands running out for 300 m on either end of the beach. Large granite boulders and rocks fringe the beach. The main island road runs past the back of the beach, with a foreshore reserve between the road and the beach (Figs. 4.34 & 4.35). The Arcadian Surf Life Saving Club, named after the backing settlement, was founded in 1928 and occupies the centre of the reserve.

The beach is often calm and only receives waves when the south-easterlies are blowing. The waves have built a gently sloping beach, that is 50 m wide at high tide and more than doubles in size at low tide. Rocks and boulders fringe each end of the beach and run out into the bay.

Swimming: This is a relatively safe beach, so long as you stay in the bay and close inshore. Be careful if swimming or snorkelling around the rocks as waves can surge and break over them.

Surfing: There is usually no surf. A strong south-easterly is required to have waves breaking in the bay.

Fishing: A popular spot to fish off the rocks. However be careful, as they are steep in places.

Summary: A very accessible and attractive little bay, with adequate facilities for the day-tripper, and a relatively safe place to swim.

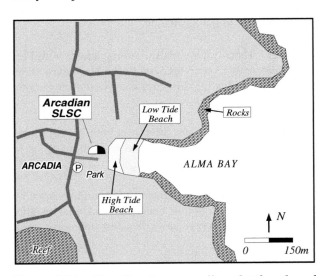

Figure 4.34   *Alma Bay has a small pocket beach and is backed by a grassy reserve that is the site of Arcadian SLSC.*

Figure 4.35   *Alma Bay, the site of the Arcadian Surf Life Saving Club, is a small, protected beach bordered by prominent granite rocks and headlands.*

MT6-MT8 GEOFFREY BAY, NELLY BAY, ROCKY BAY

No. Beach	Unpatrolled			
	Rating		Type	Length
	HT	LT		
MT6 Geoffrey Bay	1	1	R+coral reef	1 km
MT7 Nelly Bay	1	1	R+coral reef	1.75 km
MT8 Rocky Bay	1	2	R+LT+rocks	150 m

Between Bremner and Hawkings Points on the south-east coast of the island is a broad, 4 km wide bay containing the two larger Geoffrey and Nelly Bays, and the smaller Rocky Bay. All three are contained within prominent granitic headlands, with Hawkings Point rising to 90 m. The main island road from Picnic Bay to Horseshoe Bay runs along the shores of the two main bays and close to Rocky Bay, providing good access to all three.

Geoffrey Bay (MT6) is a 1 km long, south-east facing bay containing a straight, sandy, steep, narrow high tide beach, fronted by a continuous fringing coral reef that widens from 50 m in the west to over 300 m against the eastern Bremner Point. The entire bay is a marine park. The main road runs along behind the beach, and the Arcadia residential settlement extends for 500 m in from the eastern half of the bay. There is a narrow road out along Bremner Point to a small jetty and boat ramp. Waves are usually low at high tide and calm at low tide as the reef becomes exposed.

Nelly Bay (MT7) is similar to Geoffrey Bay; it is 1.75 km long, faces south-east and also has a fringing coral reef that widens from 10 m in the west to 300 m against the western Bright Point. The steep, narrow beach backs the reef. The main road skirts the western end of the beach before detouring through the backing Nelly Bay settlement. In the 1980's a major marina development was commenced and stalled at the eastern end of the beach. However, its prominent breakwaters now dominate this end of the beach, forming a man-made lagoon in their lee, into which drains Gustav Creek.

Rocky Bay (MT8) is a small indentation on the northern side of Hawkings Point. The main road is located on the 40 m high slopes at the northern end of the bay. There is steep access down the slopes to the beach. The bay contains a 150 m long, east facing beach that is almost awash and hidden amongst large granite boulders at high tide, while low tide reveals a 50 m wide bar. Due to its orientation, it receives waves averaging 0.5 m and is the highest energy beach on the island.

Swimming: Geoffrey and Nelly Bays are best for snorkelling over the reefs at high tide, with wading only possible at low tide. Rocky Bay offers the best chance for a small surf, however beware of the rocks at high tide.

Surfing: Only a chance at Rocky Bay when the Trades are blowing.

Fishing: Best over the reefs at high tide and off the rocks.

Summary: Three very accessible bays and site of much of the island's residential development as well as a budding large marina.

MT9 PICNIC BAY (SLSC)

Patrols:

Surf Life Saving Club:	September to May
Lifeguard on duty:	Christmas & Easter
Stinger net:	

No.	Beach	Rating HT LT	Type	Length
MT9	Picnic Bay	**1** **1**	R+LT+reef flats	700 m

Picnic Bay is the port of entry for Magnetic Island. The sole jetty extends 200 m out into the bay and receives the regular ferry from Townsville. The beach is backed by a foreshore reserve and road, and then the main island settlement of Picnic Bay.

The beach (MT9) is 1 km long and faces south-east. It is protected to the east by prominent, 90 m high Hawkings Point, that extends 1 km seaward of the beach. Waves are usually calm to low. At high tide the beach is only 20 m wide. As the tide falls, it widens to about 50 m and the extensive sand and reef flats are exposed. The flats extend several hundred metres off the eastern end of the beach, narrowing to 100 m off the western Nobby Head end (Fig. 4.36).

The Picnic Bay Surf Life Saving Club, founded in 1927, is located at the base of Nobby Head at the western end. The club maintains a stinger enclosure and a watch tower, both located up against Nobby Head. There is also a boat ramp crossing the beach, next to the jetty (Fig. 4.37).

Swimming: The best swimming is at high tide in the enclosure or, when safe, snorkelling over the reef flats. At low tide the bay is generally too shallow for comfortable swimming.

Surfing: None.

Fishing: The jetty is the most popular location, with deep water also accessible off Nobby Head.

Summary: The most visited and popular beach on the

island, with all the usual visitor and tourist facilities right on or behind the beach.

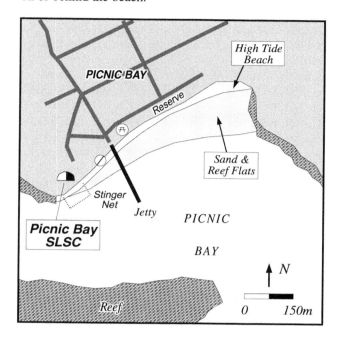

Figure 4.36   *Picnic Bay is the main settlement on Magnetic Island and the site of the main ferry jetty, Picnic Bay SLSC and a stinger net.*

Figure 4.37   *An aerial view of Picnic Bay. The surf club is located immediately behind the permanent stinger enclosure.   MT9*

MT10-MT12 COCKLE BAY, YOUNG BAY, WEST POINT

No.	Beach	Unpatrolled Rating HT LT	Type	Length
MT10	Cockle Bay	1 1	B+SF+reef flat	200 m
MT11	Young Bay	1 1	B+SF	400 m
MT12	West Point	1 1	R+LT	1.2 km

The south-western shore of Magnetic Island faces toward Townsville, just 5 km away. This is the low energy, protected side of the island and much of the 9 km of shoreline is dominated by tidal flats, coral reefs and mangroves. While the coast is freehold land, there is little development on this side of the island, with just one rough track running from Picnic Bay to the small settlements at Cockle and Young Bays and the north-western West Point.

Cockle Bay (MT10) lies 500 m west of Picnic Bay on the western side of 100 m high Sailors Rock. It is a low energy bay containing a low, narrow high tide beach fronted by sand, then coral flats up to 1 km wide. There is an open, 200 m section of beach at the southern end, with mangroves to either side. A few shacks back the beach, together with the vehicle track from Picnic Bay.

Young Bay (MT11) is a 2 km long bay largely fringed by mangroves. Toward the southern end of the bay is a 400 m long strip of sand fronted by 500 m wide sand flats, with mangroves fringing either side. There is a house on the beach and a vehicle track from the West Point track. Retreat Creek drains out at the northern end of the beach.

West Point (MT12) is the north-western tip of the island. On its southern side is a 1.2 km long, cuspate foreland that has built out 200 m from the backing steeper slopes, providing a flat portion of land upon which are several holiday shacks, all linked by the vehicle track to Picnic Bay. The beach is moderately steep and 30 m wide at high tide, while fronted by patches of reef and sand flats at low tide.

Swimming: West Point offers the best swimming, particularly at high tide; the two bays are shallow and are only accessible at high tide.

Surfing: None.

Fishing: Best at high tide owing to the shallow flats.

Summary: Three relatively remote and little visited island beaches.

MT13-MT16 ROLLINGSTONE BAY, LIVER PT, HUNTINGFIELD BAY, WILSON BAY

| No. | Beach | Unpatrolled | | | |
		Rating HT	LT	Type	Length
MT13	Rollingstone Bay	1	2	R+rock flats	70 m
MT14	Liver Point	1	2	R+boulders	50 m
MT15	Huntingfield Bay	1	2	R+LT	250 m
MT16	Wilson Bay	1	2	R+LT	60 m

The northern shore of Magnetic Island is the longest (12 km) and contains the greatest number of beaches (10). However, apart from Horseshoe Bay none of the beaches have vehicle access and are all located in the national park. However their isolation is part of the appeal of these small, but attractive and secluded, beach systems. They all tend to face north and are protected from the south-easterly winds, with waves averaging less than 0.5 m and often calm.

Rollingstone Bay (MT13) lies 1 km north-west of West Point. It is a small, 70 m long, north-west facing pocket of high tide sand and rock flats. The rock flats make interesting diving at high tide, but it is not really suitable for swimming or landing.

Liver Point beach (MT14) is similar, except all rock, only 50 m long and faces north. Again there is some good diving off the beach, while the beach is only suitable for walking.

Huntingfield Bay (MT15) is the first of the sandy beaches on the north shore. It is a 250 m long, north facing bay, that is backed by scrub-covered, rocky slopes that rise steeply to over 300 m. Large granite boulders fringe the back and sides of the beach, while the beach has a 20 m wide high tide beach, widening to 60 m at low tide. Waves average less than 0.5 m.

Wilson Bay (MT16) is a small, 60 m long, north-west facing beach contained in a curving granite bay, backed by steeply rising, wooden, granite slopes. The beach has a 10 m wide high tide beach backed by a few shady casuarinas, with a 50 m wide low tide beach and patchy coral reef right off the beach.

Swimming: Only Huntingfield and Wilson are suitable for swimming, with the added attraction of reef at Wilson. Waves are usually low to calm.

Surfing: None.

Fishing: Excellent at all beaches, with usually deep water off the beaches and rocks.

Summary: Four secluded beaches with Huntingfield and Wilson offering good anchorage.

MT17-MT20 JOYCE BAY, NORRIS BAY, MAUD BAY, LOVERS BAY

| No. | Beach | Unpatrolled | | | |
		Rating HT	LT	Type	Length
MT17	Joyce Bay	1	2	R+LT	50 m
MT18	Norris Bay	1	2	R+LT	100 m
MT19	Maud Bay	1	2	R+LT+reef	150 m
MT20	Lovers Bay	1	2	R+LT	30 m

The 2 km of coast between Joyce and Lovers Bays is all a marine park, while the backing slopes are part of the national park. This section of coast is well worth visiting, as it contains these four small, usually calm and picturesque bays and beaches.

Joyce Bay (MT17) is a 50 m long pocket of north facing sand, backed by steeply rising, wooded, granite slopes. Granite boulders both fringe and outcrop on the beach and in the small bay. The beach is almost awash at high tide and is 40 m wide at low tide. Waves are usually low to calm.

Norris Bay (MT18) is a 100 m long, north facing, open bay with the narrow high tide beach located at the base of steep, woody slopes. A 50 m strip of granite boulders separates it from Maud Bay, which contains a 150 m long, north-west facing beach (MT19) and fringing coral reef. The beach is partially lithified with beachrock and there is a small dune behind the beach hosting two fishing shacks, together with a small creek draining out at the eastern end.

Lovers Bay (MT20) is a tiny, 30 m pocket of sand wedged in below White Rock. It faces north-west and provides both anchorage and access to the rocks.

Swimming: All four beaches are suitable for relatively safe swimming under the usually low to calm wave conditions. Watch out for rocks and the reef at Maud Bay.

Surfing: None.

Fishing: Excellent fishing from the beach and adjoining rocky shores, as well as over the patches of coral reef just offshore.

Summary: Four small bays protected by marine and national parks and well worth a visit by boat or foot.

MT21, MT22　HORSESHOE BAY, WHITE LADY ROCK

Lifeguards on Duty:		September - May		
No.	Beach	Rating HT　LT	Type	Length
MT21	**Horseshoe Bay**	1　　2	R+LT	3.2 km
MT22	White Lady Rock	1　　1	R+reef	200 m

Horseshoe Bay is the largest bay on Magnetic Island and contains the longest and most popular beach. Located on the north-eastern end of the island, the bay is protected from the Trades and usually offers low waves and excellent swimming conditions. It is the place most visitors to the island head for. The main road from Picnic Bay ends

at the bay and the small settlement accommodates tourists as well as providing limited shopping facilities. A number of water-based tourist activities are available on the beach and out on the bay waters.

Horseshoe Bay beach (MT21) is 3.2 km long, curves around and faces roughly north. White Rock and White Lady Rock headlands protrude out 2 km at either end and provide substantial protection for the bay. The main settlement and tourist activities are located toward the eastern end of the bay, while there is road access to the entire bay shoreline. Two small creeks drain across the beach - one in the centre and one at the eastern end. The beach is 20 m wide at high tide, increasing to 50 m at low tide. There are usually sailing boats operating from, and moored off, the beach as well as a boat ramp across the beach at the settlement.

White Lady Rock beach (MT22) is a 200 m long, straight strip of west facing sand out on the western side of White Lady Rock. The narrow beach is fronted by a fringing coral reef that increases in width from 20 m in the south to 100 m at the rock. The land is in the national park, however the water off the beach is an oyster lease and usually off limits to the public.

Swimming: The island's most popular beach, usually offering calm conditions.

Surfing: None.

Fishing: Best off the beach at high tide.

Summary:

864, 865　ROSS RIVER, SANDLFY CK

No.	Beach	Unpatrolled			
		Rating HT　LT		Type	Length
864	Ross River (E)	1	1	B+SF	2.3 km
865	Sandfly Ck	1	1	B+SF	1.2 km

The eastern side of Ross River lies in stark contrast to the bustling Townsville on its western banks. South of the Ross River, the coast becomes increasingly protected by Cape Cleveland, and soon mangrove-covered tidal flats dominate the shore. The first 5 km do, however, receive sufficient low wave energy to maintain two low, irregular beach ridges and beaches, both fronted by wide tidal flats.

Ross River (east) (864) is a 2.3 km long, north-east facing, low, irregular beach ridge fronted by the tidal delta of the river, that extends up to 1.5 km off the beach and is exposed at low tide. There is a vehicle track from the Bruce Highway out to the beach, servicing a few fishing shacks located on the beach ridge.

Sandfly Creek beach (865) lies immediately to the east and consists of a similar low, low energy beach ridge fronted by sandy tidal flats up to 400 m wide. Sandfly Creek forms the eastern boundary, beyond which are 20 km of mangrove-fringed shore, all the way to Launs Beach at the base of Cape Cleveland.

Swimming: Only recommended at high tide.

Surfing: None.

Fishing: Best in the creeks and at high tide.

Summary: Two relatively isolated, low energy beaches, tidal flats and mangroves, in stark contrast to their Townsville neighbour.

Bowling Green Bay National Park

Beaches 875-877, 879-890

Bowling Green Bay National Park is a large national park with an area of 55 300 ha that occupies much of the southern shore of Cleveland Bay and Cape Bowling Green, and has a shoreline distance of 130 km. The park contains the extensive, low energy, mangrove-fringed shores of the bay and the western side of Bowling Green, as well as the more energetic, east facing beaches between Cape Cleveland and Cape Woora, and 22 km long Cape Bowling Green. The park extends inland to encompass much of the backing low gradient tidal flats, as well as 1 342 m high Mount Elliot.

For further information:

The Ranger
Bowling Green Bay National Park
PO Box 1954
Townsville Qld 4810
phone: (077) 78 8203

866-871   **BOWLING GREEN BAY NATIONAL PARK**

No.	Beach	Unpatrolled Rating HT	LT	Type	Length
866	Launs Beach	1	1	B+SF	50 m
867	Beach 867	1	1	B+SF	200 m
868	Beach 868	1	1	B+SF	40 m
869	Beach 869	1	1	B+SF	50 m
870	Beach 870	1	1	B+SF	100 m
871	Beach 871	1	1	B+SF	100 m

The eastern shore of Cleveland Bay consists of a 7 km long, rocky and sandy shore that ends at Cape Cleveland. Both the cape and the bay were named by Captain Cook in June 1770. Townsville lies 20 km across the bay on its western shore, with mangroves occupying much of the southern shore in between. The mangroves finally end at the eastern extremity of Launs Beach, where the shore also turns to a more northerly direction. For the first three kilometres north from Launs Beach, the shore alternates between six small, generally undeveloped, sandy beaches and the granitic rocks of Mount Cleveland; located 3 km to the east and reaching a height of 560 m.

Launs beach (866) is a 50 m pocket of sand at the far eastern end of a 1 km long, mangrove-fringed beach ridge. The mangroves may have been cleared for access. The eastern end of the beach abuts rocky slopes, and the hills behind provide a dramatic backdrop. Tidal flats extend up to 1.5 km off this very low energy, protected beach.

Beach **867** is a 200 m strip of sand around the first headland from Launs Beach. It consists of a low, narrow, west facing beach backed by a small, grassy beach ridge, with a small, mangrove-fringed creek draining out across the southern end, together with mangroves in the northern corner of the beach. The beach is bounded and backed by steeply rising granite slopes, with a few large boulders on and off the beach.

Beaches **868, 869** and **870** are three small, north-west facing pockets of sand; 40 m, 50 m and 100 m long respectively; lying along the base of a 1 km long headland. The three are separated by large granite boulders and backed by steeply rising, wooded slopes. There are a few shacks on beaches 869 and 870. Each beach has a small high tide beach, a 30 m wide low tide beach and tidal flats extending 300 to 400 m into the bay.

Beach **871** is a 100 m long sliver of sand lying between the base of a 120 m high slope and the tidal flats, with a large granite outcrop occupying the centre of the beach.

Swimming: All six have nice, small, sandy beaches at high tide, while extensive tidal flats are exposed at low tide.

Surfing: None.

Fishing: Only at high tide.

Summary: Six isolated beaches that can only be reached by boat at high tide.

accessible by boat at high tide.

872-874   LONG BEACH, RED ROCK BAY, CAPE CLEVELAND

No.	Beach	Unpatrolled Rating		Type	Length
		HT	LT		
872	Long Beach	1	1	B+SF	1.5 km
873	Red Rock Bay	1	1	B+SF	250 m
874	Cape Cleveland	1	1	R+coral reef	200 m

The 5 km of coast south of the western side of Cape Cleveland arches around in a broad, west facing bay, with a shore consisting of prominent granite slopes and headlands and three beaches; the longest, Long Beach, is 1.5 km in length. There is no access to these beaches other than by boat at high tide or on foot from the Cape Ferguson Road 20 km to the south.

Long beach (872) faces north-west and swings around in a gentle arc between two granite headlands. The beach is 10 to 20 m wide at high tide, fronting a 1.5 km long, low, grassy beach ridge, with extensive tidal flats extending up to 1 km off the beach, that are exposed at low tide. In addition, mangroves fringe the northern end of the low energy beach. There are a few shacks at the southern end, with two small creeks draining across either end of the beach.

Red Rock Bay beach (873) occupies a small, 250 m wide embayment, with granite rocks bordering each end. Mangroves grow along the rocky shore, as well as a few on the northern end, where a small creek drains across the beach. The beach is 10 m wide at high tide, widening to 30 m at low tide, where it merges with tidal flats extending another 300 m into the bay. The beach is backed by cleared, grassy slopes with a shack located on the slope at the southern end.

Cape Cleveland beach (874) is a 200 m long pocket of sand located immediately below the western side of the cape. The Cape Cleveland Lighthouse and some houses sit atop the 40 m high cape, and the beach is used to service the lighthouse. It has a concrete boat ramp that is useable at high tide, together with some storage sheds and a vehicle track up to the lighthouse. The beach itself consists of a 20 m wide high tide beach fronted by a small fringing coral reef, that widens toward the northern end.

Swimming: All three beaches provide good swimming conditions at high tide, with Cape Cleveland also offering the small reef.

Surfing: None.

Fishing: Best off the beach and rocks at high tide.

Summary: Three protected beaches that are only

875-878   TWENTY FOOT, PARADISE BAY, BRAY ISLAND, TURTLE BAY

No.	Beach	Unpatrolled Rating		Type	Length
		HT	LT		
875	Twenty Foot	2	3	R (boulder)	150 m
876	Paradise Bay	1	2	R+LT	1.2 km
877	Bray Island	1	2	R+LT	600 m
878	Turtle Bay	1	2	R+LT	2 km

Mount Cleveland is surrounded by 25 km of predominantly rocky shore, linked to the mainland by extensive tidal flats. On its eastern shore between the cape and Cape Ferguson, 14 km to the south, is a rocky shore containing four beaches, the longest being 2 km in length. Only the southern Turtle Bay is accessible by road and since 1985 has housed the Australian Institute of Marine Science. The first three can only be reached by boat or on foot over the steep backing terrain.

Twenty Foot beach (875) is a 150 m long, north-east facing boulder beach occupying the steep valley on the western side of which is Red Rock Bay. The beach is relatively steep, with boulders exposed all the way from high to low tide. Rocks and steep slopes border the beach. Twenty Foot Rock is a small islet located 500 m directly east of the beach.

Paradise Bay (876) is a straight, 1.2 km long, sandy beach facing east-north-east. It is backed by a broad, sloping valley and bordered by prominent, 100 m high granite headlands, all of which converge on 560 m high Mount Cleveland 2 km to the south. The beach is composed of fine sand which, with waves averaging 0.5 m, maintains a low to moderately steep beach. It is backed by a low, grassy and casuarina-covered foredune, and fronted by a narrow, continuous bar. The beach swings around at the very southern end where there is a safer anchorage, a small creek draining across the beach and a natural rock pool in amongst the large boulders.

Bray Island beach (877) is a 600 m long, relatively steep beach composed of coarse sand, wedged in between sloping, rocky headlands that rise to 100 m. It has a high berm backed by a low, casuarina-covered foredune. In the southern corner is a small creek containing a few mangroves. Bray Island lies 500 m off the southern tip of the beach.

Turtle Bay beach (878) is the longest of the cape's beaches, running for 2 km almost due south from 80 m high Cape Woora to 56 m high Cape Ferguson. The Australian Institute of Marine Science is located on gently sloping ground behind the beach, with the institute's small harbour and boat ramp in lee of the cape. The beach is composed of coarse sand and slopes steeply to a 50 m wide low tide bar. A small creek is usually dammed behind the middle of the beach. Waves average 0.5 m and surge up the steep beach at high tide, while breaking across the bar at low tide. There are large granite boulders at the southern end and scattered rock reefs off the southern end, including prominent Bald Rock.

The shoreline and adjacent sea between Paradise Bay and Chunda Bay, west of Cape Ferguson, is part of the Cape Ferguson Marine Park, with the section between Cape Woora and Chunda Bay reserved for scientific research and off limits to the public.

Swimming: Turtle Bay, otherwise known as AIMS Beach, is by far the most popular beach in the area and is used by workers at the institute, both for scientific research and recreation. It and Paradise Bay and Bray Island beaches both have deep water off the beaches at high tide, with surf over the low tide bars. Rips are usually absent, but do form when waves exceed 0.5 m. Twenty Foot Beach is rocky and unsuitable for swimming.

Surfing: Chance of a surf during the Trades at Paradise Bay, Bray Island and Turtle Bay, particularly at mid to low tide.

Fishing: Banned from Turtle Bay but permissible north of Cape Woora, although some restrictions apply.

Summary: Four more energetic beaches; the southern three are relatively natural, sandy beaches, bordered by prominent headlands and with Mount Cleveland as a backdrop.

Cape Bowling Green Bay

At Cape Ferguson the rocky shoreline gives way to the low sand, then mud, flats of the long, curving shore of Bowling Green Bay. While Cape Bowling Green lies 35 km directly east of Cape Ferguson, the linking shoreline occupies 75 km; the first 40 km is more sand than mud, while the eastern 35 km is dominated by wide, mangrove-fringed mud flats. There are eleven beaches located along the sandy section, all separated and at times dominated by numerous tidal creeks. There is road access to only two of the beaches, Big Beach and Cungulla. All the others lie as strips of sand backed by extensive high tide flats.

879-882   CHUNDA BAY, SALMON CK, BIG BEACH, CUNGULLA

| No. | Beach | Unpatrolled | | | |
| | | Rating | | Type | Length |
		HT	LT		
879	Chunda Bay	1	1	B+SF	1.6 km
880	Salmon Ck	1	1	UD	5 km
881	Big Beach	1	2	UD	7.2 km
882	Cungulla	1	1	B+SF	2.3 km

South of Cape Ferguson is 15 km of sandy shore, ending at the 1 km wide mouth of the Haughton River. The beaches all face east, turning slightly north-east toward the river. A 10 km long road from the highway at Clevedon runs out to the southern end of Big Beach and Cungulla, both of which have scores of shacks and holiday houses.

Chunda Bay lies tucked in behind Cape Ferguson and although east facing, is somewhat protected by the cape and extensive tidal flats. The bay's beach (**879**) is 1.6 km long with small tidal creeks entering the bay at either end. The high tide beach is narrow and eroding, with numerous casuarina trees falling onto the beach. It is fronted by 300 to 400 m wide, sandy tidal flats that are exposed at low tide and dissected by the two tidal creek channels.

Salmon Creek beach (880) is a 5 km long, east-southeast facing beach that is bordered by an unnamed northern creek and Salmon Creek to the south. Both Salmon Creek Beach and adjoining Big Beach are backed by up to 6 km of 60 low beach ridges, evidence of shoreline progradation over the past 6 000 years. The beach is composed of fine sand and consists of a 200 m wide, low gradient intertidal beach fronted by a low surf zone up to 100 m wide. A 5 m high foredune covered by spinifex, ipomoea and casuarina backs the beach.

Big Beach (881) is the biggest of the four at 7.2 km long. It faces north-east and runs from Salmon Creek to the wide mouth of the Haughton River. There is a road out to the houses and shacks at the southern end. The beach is backed by a low, casuarina-covered foredune and fronted by a wide, intertidal beach and 100 m wide surf zone, indicative of the overall low gradients.

Cungulla beach (882) lies on the northern shore of the Haughton River mouth and is a narrow, curving, east facing high tide beach, fronted by tidal flats up to 2 km wide that end at the deep river channel. An arm of the channel sweeps past the northern end of the beach, providing good boat access to the river. The entire beach is backed by a housing and shack settlement.

Swimming: Swimming is best at mid to high tide when

there is some depth in the water, otherwise it can be a long way to very shallow water at low tide. Rips are usually absent, unless strong Trades produce waves greater than 0.5 m. Be careful near the creek and river mouths, where there are strong tidal currents.

Surfing: Only low, spilling breakers on Salmon Creek and Big Beaches, that are best at mid to low tide.

Fishing: The creeks and rivers are the most popular spots, usually fished from a boat at high tide.

Summary: Four relatively isolated beaches with good access to Big Beach and Cungulla, but no facilities.

883-889 CONNORS ISLAND - HUGHES CK

No.	Beach	Unpatrolled		Type	Length
		Rating HT	LT		
883	Connors Is	1	1	B+SF	1.5 km
884	Barramundi Ck	1	1	B+SF	2.5 km
885	Beach 885	1	1	B+SF	700 m
886	Hucks	1	1	B+SF	1.5 km
887	Sheepwash Ck	1	1	B+SF	1.7 km
888	Beach 888	1	1	B+SF	1.6 km
889	Hughes Ck	1	1	B+SF	2.5 km

Between the eastern shore of Haughton River mouth and Hughes Creek, 35 km to the east, is a series of very low energy beaches, interspersed with mangrove-fringed shores and tidal creeks. There is no vehicle access to any of the beaches and all are backed by an extensive tidal plain.

Connors Island beach (883) is a straight, 1.5 km long, north facing strip of low sand, backed by a low beach ridge, then mangrove-covered tidal flats, and fronted by 1 to 3 km wide, sandy tidal flats.

Barramundi Creek beach (884) is an arching, 2.5 km long, low sand spit, with dynamic recurved spits at either end. Consequently it changes shape and length over the years. It is backed by older spits separated by extensive mangrove forests, and fronted by 3 km wide tidal flats off the creek, that narrow to 500 m in the centre of the beach.

Beach **885** is a low energy, 700 m long strip of sand set in lee of 2 km wide tidal flats. The low beach is, for the most part, fronted by a fringe of mangrove.

Hucks beach (886) is a 1.5 km long, north-east facing beach that lies 3 km north of Hucks Landing, a fishing shack settlement on Barratta Creek. The creek flats border the eastern end of the low energy beach, which is fringed by mangroves along much of its length.

Sheepwash Creek beach (887) and Beach **888** are two similar, low, recurved spits 1.7 km and 1.6 km respectively in length. They are low but dynamic spits of sand, recurved toward their western ends. Both are backed by extensive mangrove-covered tidal flats.

Hughes Creek beach (889) is a straight, north facing, 2.5 km long beach that forms the boundary between the low, sandy beach and the mud and mangrove dominated eastern bay shore. It lies between the western Hughes Creek and an unnamed eastern creek, and consists of a low, low energy beach, that is accessible at low tide by a 4WD track that runs out across the swampy coastal plain from Brandon, 20 km to the south.

Swimming: None of these beaches or tidal flats are really suitable for swimming owing to the shallow water, strong tidal currents in the creeks and possibility of crocodiles.

Surfing: None.

Fishing: Excellent fishing in all the creeks along this fertile marine environment.

Summary: An isolated, low energy shore fronting a wide, flat, swampy coastal plain. It is mainly the domain of local fishers.

Burdekin River Delta

The Burdekin River is the largest in Australia when in flood. As a result, the river has built the largest delta in Australia. The deltaic shoreline extends for 60 km along the coast and it has built out 10 km in places. The present mouth occupies 20 km of shore where it breaks into three major branches. In flood, the river delivers large quantities of sand and mud to the coast, with the sand settling in the wide channels and at the river mouth and the mud on the floodplain and offshore. The mouth has one major and two secondary channels and extends from Groper Creek at Big Hill in the south to the main channel at Rita Island and the northern channel at Anabranch. The extensive river mouth shoals average several hundred metres in width and are moved both onshore and northward by the relentless south-east Trade winds and waves. The sediments are slowly organised into long bars and larger spits. The spits move northward at rates between 100 and 200 m per year. There are usually five to six spits between Anabranch and Alva Beach, each several kilometres in lengths and all slowing moving northward. By Alva Beach, the spits merge with the base of Cape Bowling Green and beyond there move up the cape as multiple shore-parallel bars. As a result of the episodic floods, the large quantities of sand deposited, and the migratory bars and spits, the entire deltaic shoreline is an extremely dynamic feature. While it is gradually building seaward, there are many areas of temporary retreat, as well as migration of the bars, spits, island and creek mouths. A good local knowledge is required to keep up with the present shape of the shoreline. The following is based on the delta's shoreline in the 1990's, and this will certainly change with time. The following shoreline features should therefore be read as a typical, rather than a necessarily accurate, description.

890   CAPE BOWLING GREEN

No.	Beach	Unpatrolled		Type	Length
		Rating			
		HT	LT		
890	Cape Bowling Green	2	3	R+double bar	14.5 km

Cape Bowling Green is the longest spit in Australia, extending north for over 14 km as a narrow, low spit, to form the eastern boundary of Bowling Green Bay. The cape represents the northern terminus for sand delivered by the massive Burdekin River, 40 km to the south. The sand is gradually moved by southerly waves, via a series of spits and bars, along the coast and up the cape. There is no vehicle access to the cape and access from the sea is difficult owing to the numerous shifting bars and usually moderate waves. There is, however, a small automatic lighthouse located 5 km south of the tip.

The beach (890) forms the eastern boundary of the cape and varies considerably along the cape. In places it is tens of metres wide and backed by active foredunes, while elsewhere it is nonexistent as waves cut into the backing mangroves. The beach width is largely dependent on the location of the migrating bars and their points of attachment to the shore. The bars and surf zone are equally variable, but usually consist of two bars extending up to 200 m seaward of the beach, with a deep trough in between. Rips are common, but irregularly spaced along the beach and inner bar.

Swimming: A hazardous beach owing to the bars, troughs and rips, coupled with strong longshore and tidal currents.

Surfing: Chance of variable beach breaks across the bars, particularly when the Trade winds are blowing.

Fishing: A top beach fishing spot full of rip holes and the trough along the beach, with the mangroves just behind.

Summary: A dynamic cape and beach environment; difficult to access when waves are running.

891   ALVA BEACH (AYR SLSC)

Patrols:	
Surf Life Saving Club:	September to May
Lifeguard on duty:	Christmas & Easter

No.	Beach	Rating		Type	Length
		HT	LT		
891	Alva	**3**	4	R+ BR	4.5 km

Alva Beach (891) is part of the Burdekin River delta. The delta extends for 60 km from the tip of Cape Bowling Green south to Beach Hill and includes a number of river entrances and long, low beaches and barrier spits, all fronted by wide and shifting sand shoals, most of which are slowly moving northward.

Alva Beach is located on a spit of sand that extends 20 km northward to Cape Bowling Green. The beach lies immediately north of the shifting mouth of Plantation Creek, one of the five Burdekin River entrances, and extends for 4.5 km to where the small Alva Creek crosses the beach, beyond which is Cape Bowling Green. The beach is 15 km north-east of Ayr, and has been the site of the Ayr Surf Life Saving Club since 1926. Besides the surf lifesaving club, there are two rows of holiday houses stretching for about 500 m along the beach, which is also known as Lynchs Beach (Fig.4.38).

The beach varies considerably depending on the condition of the Plantation Creek entrance. At times it is located south of Alva Beach and the beach is exposed to 0.5 to 1 m waves. These produce a 50 m wide high tide beach fronted by an irregular beach and bar, with many holes and rips. Waves tend to arrive from the south-east and run up the north-east facing beach, thereby producing a northward current. Tidal currents in Plantation Creek can also flow close to shore. Under these conditions, extreme care should be taken at all times.

When the Plantation Creek mouth extends across the front of the beach, the outer spit can block off the beach, forming a quiet lagoon, with a long walk or wade out to the spit required to reach the surf. When this occurred in 1990 a causeway was constructed out to the spit, however by 1994 the spit had moved, the beach was again exposed and the causeway was gone (Figs.4.39a & b).

Swimming: It is best to carefully assess the prevailing conditions at Alva Beach because of the potentially highly variable surf and swimming conditions. Check with the surf lifesavers or talk to the locals it you are at all unsure. Only swim between the flags when the patrols are operating.

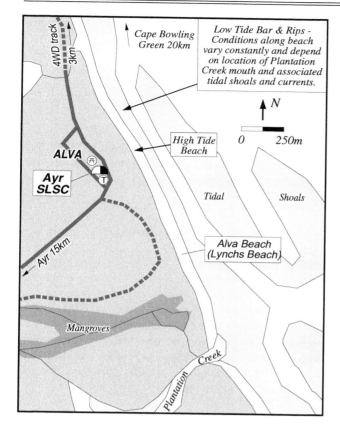

Figure 4.38 *Alva Beach, site of Ayr SLSC, is part of the dynamic shoreline of the large Burdekin Delta. Beach and bar conditions change considerably from year to year, and necessitate additional caution when swimming at the beach.*

Surfing: This is one of the more exposed beaches and commonly has waves averaging up to 1 m, higher in strong south-east winds, which, together with a range of bars, produces some reasonable beach breaks.

Fishing: A very popular spot owing to the large creek and the common rips and gutters along the beach. There is also a boat ramp at the creek.

Summary: A rather lonely, wild and windswept beach, with a potentially hazardous surf and tidal currents. The large surf lifesaving club has a shady picnic area with barbecues and a playground. Worth a drive out through the sugarcane fields to visit, however use care if swimming out here as wave, tide and beach conditions vary constantly.

Figure 4.39 *Two views of Alva Beach, the site of Ayr SLSC. View (a) taken in 1990, shows a sand spit just off the beach, with a temporary causeway providing access to the surf. In (b) taken four years later, the 1990 spit has gone and a new spit is extending north along the beach. Note the numerous wave and tidal shoals and channels off the beach.*

892-897 PLANTATION CREEK
SCOTTIES BEACH

| No. | Beach | Unpatrolled | | Type | Length |
| | | Rating | | | |
		HT	LT		
892	Plantation Creek	2	3	R+BR	4 km
893	Anabranch	2	3	R+BR	8.5 km
894	Rita Island (north)	2	3	R+BR	3 km
895	Rita Island	2	3	B+SF	2 km
896	Rita Island (south)	2	3	B+SF	3 km
897	Scotties Beach	2	3	B+SF	3 km

South of Alva Beach, the real Burdekin delta shoreline dominates the coast. This 30 km of shoreline contains some of the most dynamic beaches and shorelines in Australia. They are all in a continual state of change owing to the influence of the Burdekin and its massive floods, the river and tidal currents and the relentless force of the south-east Trade winds and their waves that push the sand northward along the beaches and bars.

There is vehicle access only to the rear of the Anabranch Beach; the rest are backed by mangroves and/or surrounded by water. There are shack settlements on Anabranch and Scotties. All six beaches are exposed to the south-east Trades and their waves, with waves averaging over 0.5 m in the north, but decreasing slightly to the south. A feature of all the beaches are the extensive bars off Plantation, Anabranch and northern Rita Island, giving way to 1 km-plus wide sand bars off the river mouth in the south. During strong Trades the waves break for several hundreds of metres across these bars, producing an exciting approach to the river and creek mouths.

Plantation Creek beach (**892**) is part of a dynamic, approximately 4.5 km long, low and narrow island-spit that extends north from the mouth of Plantation Creek. The beach is sinuous and fronted by usually two bars, with deep troughs and rips close inshore. The entire beach and island can be awash during cyclones and has no permanent settlement.

Anabranch beach (**893**) has a similar dynamic beach, but is backed by a 150 m wide, more stable, low dune area which, in front of Plantation Creek, houses about thirty fishing and holiday shacks. These lie directly in front of Plantation Creek, before it turns and runs north.

Rita Island consists of three sections (894, 895 & 896) which in total contain 8 km of low, sandy shoreline backed by a prograding shoreline, manifest in multiple beach ridges up to 300 m wide in the north (**894**) and 1 km wide in the centre (**895**), while the southern section (**896**) is a low island detached from the main island. The entire island represents the true mouth of the Burdekin, and the channels separating the beaches, particularly south of the island, carry the bulk of the Burdekin flood waters and sediments. On the south side of the island, the main Burdekin mouth is 2 km wide. As a consequence, the island shore is skewed 3 km seaward and river mouth sand bars extend another 1 km further seaward. The entire area is a zone of shallow, shifting sand bars.

The beaches of Rita Island generally consist of a sandy high tide beach fronted by shore-parallel bars and rips in the north, grading into wider and more variable bars and channels to the south.

Scotties beach (**897**) is a narrow, 3 km long sliver of sand that is eroding into the backing mangroves, in places leaving the mangroves stranded on the beach. There are a handful of shacks on its northern tip, but otherwise it is a natural system. Extensive sand flats lower waves at the beach to usually less than 0.5 m.

Swimming: Swimming is not recommended on these beaches owing to their remoteness and particularly the presence of rip holes and currents and the shifting nature of the bars and holes. If swimming, stay clear of the rips and stay on the attached portions of the bars.

Surfing: Waves break over the bars during the Trades, providing a sloppy beach break to 1 m.

Fishing: The Burdekin is a very popular fishing spot, both in the many creeks and along the beaches. If going offshore, be very careful owing to the wide, shallow and shifting sand bars.

Summary: An interesting and dynamic part of the Australian coast that sees very few visitors. If you do visit, have a boat and a local guide.

Cape Upstart & Upstart Bay

Upstart Bay lies in lee of 540 m high Cape Upstart, so named by Captain Cook in June 1770, owing to the fact that 'being surrounded by low land it starts or rises up singley'. The low land comprises the extensive coastal deltas and plains of the Burdekin and Elliot Rivers, as well as major creeks like the Molongle, Armstrong and Big Jack. The 100 km² cape, that reaches a central crest at 2 400 m high Station Hill, is also surrounded by low land for several kilometres to the south and west.

In lee of the cape is **Upstart Bay**, 15 km across the mouth of the Burdekin and up to 15 km deep. It has approximately 50 km of shoreline between the Burdekin and the cape, the first 20 km containing four generally long, sandy beaches, then 10 km of mangrove-covered tidal flats deep in the bay, and finally 17 small, sandy pocket beaches along the cape's west facing shore, ending at the cape.

898-901 BIG HILL - MOLONGLE CK

No.	Beach	Unpatrolled		Type	Length
		Rating HT	LT		
898	Big Hill	1	2	B+river channel	1.1 km
899	Wunjunga	1	1	B+SF	13.6 km
900	RM Beach	1	1	B+SF	3.5 km
901	Molongle Ck	1	1	B+SF	700 m

The western shore of Upstart Bay has four beaches between the southern channel of the Burdekin River, Groper Creek and Molongle Creek. All four are accessible by vehicle and have some degree of fishing shack and holiday settlements.

Big Hill beach (**898**) is a 1.1 km long, narrow beach that skirts the base of 150 m high Big Hill. It also faces into the mouth of Yellow Gin Creek, which merges with the larger Groper Creek into a 1 km wide mouth. The Beach Mount Road ends at the base of the hill, where there is a fishing-holiday settlement of about 50 houses and shacks, but with no facilities or store. The beach faces north-east into the creek mouth and has a deep channel and strong tidal currents right off the northern half of the beach. The beach widens a little to the south, where it is fronted by tidal flats and has a small creek that drains across the beach, however the channel still lies just offshore. Beyond the 100 m wide, deep channel are another several hundred metres of river mouth shoals, all part of the Burdekin river sediments. The beach ends at some low bluffs, upon which many of the houses have been built.

Wunjunga beach (**899**) begins on the south side of the low bluff and the southern slopes of Big Hill. It is connected to the highway by a 10 km long road that reaches the beach 4 km south of the settlement, then parallels the back of the beach. The beach is 13.6 km long, faces north-east and is backed by up to 15 older beach ridges in the north, while to the south it is a 7 km long recurved spit. It is at times breached in the middle, and a creek flows across the beach just south of where the road turns north. There is also a 4WD track leading to a small southern settlement on the inside of the southern tip of the beach. The beach consists of a moderately steep and narrow high tide beach, fronted by wide tidal flats off the northern and southern creek mouths, but which narrow in the middle to a 100 m wide bar. Waves are usually less than 0.5 m and spill across the bar at low tide.

RM beach (**900**) lies on the southern side of 15 m high RM Point, which forms the southern boundary of RM Creek. There is a fishing settlement on the point containing a string of about 20 shacks, linked by a gravel road to the highway 8 km away. RM Beach runs from the eastern side of the point for 3.5 km to the mouth of Molongle Creek. It initially faces north-east but swings around to face north by the creek. It is backed by a series of old beach ridges and fronted by 1 km wide tidal flats.

Molongle Creek beach (**901**) is a 700 m long strip of north facing sand formed on the east side of the creek mouth. It consists of a narrow high tide beach fronted by the extensive tidal flats of the creek mouth, that extend 1 km into the bay. There is a boat ramp behind the beach, accessible by a road from the highway, 7 km to the south. The Molongle Boat Club has recently developed a marina in the creek mouth, which includes a dredged channel for boat operations at low tide. The beach remains in a natural state.

Swimming: Most swimming takes place at Big Hill Beach. While waves are usually low to calm, you must take care of the deep channel and tidal currents, stay close inshore, and well clear of the channel. Elsewhere you can only swim at high tide, with extensive tidal flats exposed at low tide.

Surfing: The only surf in the area breaks along Wunjunga Beach, however it is usually small and sloppy.

Fishing: The four settlements along these beaches have been primarily developed for fishing communities, and a lot of fishing is done in the creeks and the bay, as well as off the beaches at high tide. Boats are launched off the beaches and at the Molongle marina.

Summary: Four low energy beaches relatively close to the highway, but until the development of the Molongle marina, remained largely overlooked by passing traffic.

On the western shore of **Cape Upstart** is a 10 km long, low energy, west facing, rocky shore. In amongst the rocks are 20 small, sandy beaches ranging in size from 70 m to 700 m. These are all backed by steeply rising, moderately vegetated slopes of the cape. There is no vehicle access to the shore, however 15 of the beaches have been occupied by holiday and fishing shacks, with access by boat. In all, there are over 200 shacks scattered along the shore, making it one of the densest shack developments on the entire coast. All the beaches are sheltered from the Trades that blow offshore, and calm to low waves usually prevail.

902-905 THE SPIT - WINDY POINT

No.	Beach	Unpatrolled		Type	Length
		Rating HT	LT		
902	The Spit	1	1	R+tidal flats	500 m
903	Beach 903	1	1	R+tidal flats	150 m
904	Windy Point (south)	1	1	R+tidal flats	100 m
905	Windy Point (north)	1	1	R+LTT	200 m

The western Cape Upstart beaches extend from the cape down to an elongated spit, 10 km to the south, beyond which tidal flats and mangroves replace the rocks and sand. **The Spit** (902), as it is called, is a low, sinuous strip of sand, facing west and up to 500 m long, but prone to changes in shape and length. It is attached to the rocky shore for half its length, then is backed by mangrove-fringed tidal flats.

Beach **903** lies immediately north of The Spit. It faces south-west into Upstart Bay and is a 150 m strip of high tide sand, fronted by beachrock and shallow sand flats, while ten shacks occupy the back of the beach and backing slopes. A small creek descending steeply from the backing slopes drains out across the northern end of the beach and has deposited boulders at its mouth, which are covered in mangroves.

Windy Point is a protruding tip of rock with small beaches to either side. Windy Point (south) (904) is a 100 m long pocket of south-west facing sand, with rocky headlands to either side, a very small creek draining the backing 100 m high slopes in the centre, two shacks behind the beach and beachrock outcropping along the beach.

On the north side of **Windy Point** is a 200 m long, northwest facing beach (905) that is occupied along its length by about 20 shacks. There is a small creek draining across the northern end, backed by a small area of mangroves. The beach is bordered by headlands, and beachrock runs the length of the beach.

Swimming: These beaches are usually calm with occasional low waves, however swimming is better at high tide because of the shallow sand flats off the beach.

Surfing: None.

Fishing: The locals fish the bay and southern creeks from boats, as well as off the beaches and many small headlands at high tide.

Summary: An isolated but popular spot for the locals and shack owners.

Cape Upstart National Park

Area: 5463 ha
Beaches 911 to 927

Cape Upstart National Park occupies the top half of Cape Upstart - a 55 km² rocky coast and hinterland, with peaks rising to over 2000 m. There is no vehicle access to the park, apart from one 4WD track to the far south-east corner at Coconut Bay. While most of the park is unoccupied and undeveloped, there are a series of shacks along the western shores between Flagstaff Bay and the cape. These are all accessible by water.

For further information:
Whitsunday Information Centre
Cnr Mandalay and Shute Harbour Roads
PO Box 332
AIRLIE BEACH QLD 4802
Phone: (07) 4946 7022
Fax: (07) 4946 7023

906-911   BEACH 906 - BEACH 911

No.	Beach	Unpatrolled			
		Rating HT	LT	Type	Length
906	Beach 906	1	1	R+LT	200 m
907	Beach 907	1	1	R+LT	150 m
908	Flagstaff Bay (S)	1	1	R+LT	150 m
909	Flagstaff Bay	1	1	R+LT	200 m
910	Beach 910	1	1	R+LT	100 m
911	Beach 911	1	1	R+LT	70 m

Between Windy Point and Moonlight Bay are six small pockets of sand, each occupying a small valley or indentation in the rocky cape shoreline. The first four are backed by shacks, with the only access by boat.

Beach **906** is a 200 m long, west facing beach consisting of two small indentations. Rocks and beachrock border both ends as well as outcropping along the beach. There are several shacks toward the northern end of the beach.

Beach **907** is a sandy and rocky, 150 m long indentation in the shore, with about ten shacks overlooking the northern end. The shacks sit right on top of the 30 m wide beach, with tidal flats in front. The shacks merge with the adjoining shacks in Flagstaff Bay.

Flagstaff Bay (908 & 909) is a 200 m long, north-west facing beach contained between a long, rocky point and a small creek that drains out against the northern headland. The western half of the beach is backed by about ten shacks, with the creek and a few mangroves behind the eastern half. The 30 m wide beach is fronted by a flat low tide bar.

Beach **910** is a 100 m long pocket of sand wedged in between rocky outcrops, with two rock outcrops also cutting across the beach. Beach **911** lies around the northern rocks and is similar in length, with rocks at each end. Neither beach has any shacks.

Swimming: These six beaches all offer usually calm to low wave conditions, with the best conditions occurring at high tide. However, watch for submerged rocks off many of the beaches.

Surfing: None.

Fishing: Best in the bay and off the beaches, particularly off the many rocks at high tide.

Summary: Six small beaches used by the shack holders for holidays and fishing trips.

912-914   MOONLIGHT BAY - BEACH 914

No.	Beach	Unpatrolled			
		Rating HT	LT	Type	Length
912	Moonlight Bay	1	1	R+LT	700 m
913	Beach 913	1	1	R+LT	300 m
914	Beach 914	1	1	R+LT	300 m

Moonlight Bay (912) is the largest bay on the western cape shore. The 700 m long bay contains a protected, south-west facing, northern end, and a straighter, north-west facing main beach. Up to thirty shacks are located in three groups along the back of the beach. A small creek drains out toward the northern end, producing a sandy tidal delta in this corner. The rest of the beach has a moderately steep, 30 m wide beach fronted by deeper water, together with two outcrops of rocks.

Beach **913** lies just around the rocks at the northern end of Moonlight Bay. It is a 300 m long, north-west facing, sandy beach, with rocky outcrops dominating the southern half, and about thirty shacks on the beach crest and backing slopes. A small creek drains out against the northern rocks and has a few mangroves in behind the beach.

Beach **914** lies at the base of 200 m high slopes. It is a crenulate, 300 m long, north-west facing, sandy high tide beach, with rocks and rock reefs dominating the low tide and subtidal areas. The southern half of the beach is unoccupied, while fifteen shacks occupy the northern slopes. The beach ends at one of the larger creeks to drain the cape. It emerges from a rocky channel to deposit a sandy tidal delta.

Swimming: These three beaches tend to have deeper water off the beaches, which makes them more suited for swimming. However, while waves are usually low, watch out for submerged rocks.

Surfing: None.

Fishing: Good fishing off the rocks and beaches, especially at high tide.

Summary: Three popular beaches occupied by about 75 shacks, with generally deeper water off the beaches.

a small rocky outcrop, with a second smaller creek crossing the centre of the beach. About ten shacks lies in amongst the trees behind the beach.

Shark Bay (north) (916) is a 200 m long strip of sand bordered by low, rocky outcrops, with about ten shacks behind the beach. The entire bay is fronted by 100 m wide sand flats, then rocky reef flats that widen to the north.

Beach **917** is an undeveloped, 300 m long, sand beach backed by steep, sparsely vegetated slopes, with low, rocky points to either end and some rocks across the beach. Beach **918** lies immediately to the north and is a 150 m long pocket of sand and rock, with several shacks perched on the backing southern slopes. Beach **919** is a straight, 150 m long beach, fronted by a strip of beachrock, then the sand and reef flats. Several shacks sit on the beach crest. Beach **920** is an unoccupied, 150 m long, narrow strip of sand and beachrock, backed by steep slopes (Fig. 4.40).

Swimming: You can only swim on these beaches at high tide because of the continuous sand and rock flats; even then, watch for numerous rocks.

Surfing: None.

Fishing: The tides dictate both boating and fishing along these six beaches, with all activity occurring around high tide.

Summary: Six low energy beaches protected by the rock reef that parallels the shore, together with a near-continuous sand flat.

915-920   SHARK BAY - BEACH 920

| No. | Beach | Unpatrolled | | | |
		Rating HT	LT	Type	Length
915	Shark Bay (S)	1	1	R+SF+rock	500 m
916	Shark Bay (N)	1	1	R+SF+rock	200 m
917	Beach 917	1	1	R+SF+rock	300 m
918	Beach 918	1	1	R+SF+rock	150 m
919	Beach 919	1	1	R+SF+rock	150 m
920	Beach 920	1	1	R+SF+rock	150 m

From Shark Bay to Beach 920, the beaches are fronted by shallow sand flats and then coral-rock flats, resulting in a low energy shoreline that can only be approached at high tide. Shark Bay is a 700 m long indentation in the rocky shore, that extends from the mouth of the creek separating it from Beach 914, up to a rocky headland. It contains a near-continuous, sandy shore, broken by rock outcrops into two beaches. **Shark Bay (south)** (915) is 500 m long, extending from the southern creek mouth to

Figure 4.40   *View of beaches 917 (right) to 920 (left), part of a series of low energy pocket sand beaches on the western side of Cape Upstart, with numerous shacks located in lee of the beaches.*

921 CAPE UPSTART

No.	Beach	Unpatrolled Rating HT LT		Type	Length
921	Cape Upstart	1	1	R+LT	250 m

Twin-peaked Cape Upstart is a 60 m high headland with steep cliffs and slopes on its north side, and a small valley below its southern slopes. The valley is occupied by one of the larger shack settlements on the cape, with a total of about thirty shacks. The back of the valley hangs over a small, rocky gap in the cape, while the front has a 250 m long, sand beach (921) and beachrock, fronted by gently sloping sand flats.

Swimming: Best off the beach at mid to high tide.

Surfing: None.

Fishing: The rear gap offers some higher energy rock fishing, while the deeper water off the beach permits fishing and boating at more stages of the tide.

Summary: The cape is a very prominent landmark and so named by Cook back in 1770. Today, apart from the few shacks, the cape and shoreline remain much as they were when Cook passed by.

922-927 BEACH 922 - COCONUT BAY

No.	Beach	Unpatrolled Rating HT LT		Type	Length
922	Beach 922	1	2	R (boulder)	100 m
923	Beach 923	1	2	R (boulder)	600 m
924	Beach 924	1	2	R+LT rocks	200 m
925	Kingfisher Bay (N)	1	2	R (boulder)	300 m
926	Kingfisher Bay (S)	1	2	R+LT	200 m
927	Coconut Bay	1	2	R+BR	200 m

The northern and eastern shore of Cape Upstart consists of 18 km of rugged, rocky coastline, with steep cliffs in places and the land generally rising rapidly to between 200 and 400 m. In amongst the rocks are four small embayments containing six beaches totalling 1.6 km in length. Three of the beaches are composed of boulders, with only three sandy beaches totalling 600 m in length. All six beaches lie in the Cape Upstart National Park, with vehicle access only possible to Coconut Bay, lying just inside the southern park boundary. The bay is linked

via a torturous, 7 km long 4WD track to The Cape homestead. The easiest access to all beaches is by boat, with Beach 924 and Kingfisher Bay (south) offering relatively safe anchorage.

Beach **922** lies 3.5 km east of the cape. It is a 100 m long, north-east facing boulder beach lying at the base of a steep stream, that has delivered the rounded boulders. It is a steep, narrow beach with deep water off the base, and rock reefs immediately offshore.

Beach **923** occupies the northern part of an open, 1 km long embayment, located on the north-east corner of the cape. It faces east and consists of a 600 m long, graded boulder beach, with scattered, larger boulders along the beach, and deep water just off the beach.

Beach **924** occupies the southern corner of the same embayment. It is a 200 m long, north facing pocket of sand that is protected from direct wave attack by a 50 m high eastern headland. Steep rocks also form the western boundary. A low, tree-covered beach ridge backs the sandy beach, however at low tide a shallow rock and cobble rock flat fronts the beach.

Kingfisher Bay is a 500 m opening on the south-east side of the cape. Within the shallow bay are two beaches. The northern (**925**) is a 300 m long, east facing boulder beach, with deep water right off the beach. The southern (**926**) is a 200 m long pocket of north-east facing sand, wedged in between 50 m high headlands. The beach consists of a low dune ridge backed by a small, usually closed creek, containing a few mangroves, with a 70 m wide, low gradient intertidal beach. Waves are usually less than 0.5 m and decrease to the eastern end of the beach, with only a chance of a rip at the western end, during higher waves.

Coconut Bay (927) is an exposed, headland-bound, 200 m long, south-east facing beach, that is linked by its surf zone to the neighbouring, longer Abbot Bay Beach. A 4WD track reaches the back of the beach, which is also backed by a low spinifex and casuarina-covered foredune, with a small creek draining out along the northern headland. The low gradient beach is over 100 m wide at low tide, with a chance of one or two rips in the low tide surf, particularly against the northern rocks.

Surfing: There is only surf at Coconut Bay, with usually low, spilling breakers produced by the waves that average just over 0.5 m, but can reach over 1 m during strong Trades.

Fishing: This is a popular coast for fishing the many rocky reefs from boats. If ashore, there are numerous vantage points from the predominantly rocky shore and beaches.

Summary: Six relatively isolated and little visited beaches, with Kingfisher Bay (south) offering the best landing and Coconut Bay the best surf.

Abbott Bay

Abbott Bay is a relatively open, north-east facing bay bordered by the eastern point of Cape Upstart and Abbott Point, a distance of 30 km. The shoreline curves in between for a length of 43 km, with most of its sandy shore ranging from relatively high energy beaches immediately south of the cape, to low energy in the more protected southern corner. This corner is also the site of the Elliott River mouth, as well as Saltwater, Splitters, Branch and Mount Stuart Creeks, all with upland sources and substantial flows during the summer wet. Sediments delivered to the bay by the rivers and creeks have built a coastal plain up to 5 km wide, that also links the two capes, together with Moose Hill, Mount Curlewis and Mount Bruce. Fifty metre high Camp Island lies 2.5 km off the Elliott River mouth. While the Bruce Highway comes within 4 km of the bay shoreline, there is essentially no development other than fishing shacks at the Elliott River mouth and on Camp Island. All access to the coast is either by 4WD tracks or boat.

928, 929   BEACH 928, THE CAPE

No.	Beach	Unpatrolled		Type	Length
		Rating			
		HT	LT		
928	Beach 928	1	3	R+BR	600 m
929	The Cape	1	3	R+BR	11.5 km

Between the southern rocky shore of Cape Upstart and the Elliott River mouth is a 12 km long beach, that is cut toward its northern end by a tidal creek, draining a swampy area behind the beach. This beach is, in fact, the major link that ties the cape to the mainland, as it is backed by the swamp, then an older Pleistocene (120 000 year old) barrier and then the extensive salt flats and mangrove-fringed shore of Upstart Bay. The Cape homestead lies on higher ground 2 km in from the northern end of the beach, and there is a vehicle track from the homestead to the main beach as well as Coconut Bay. There is, however, no development on the beaches.

Beach **928** occupies the first 600 m between the jutting 30 m high headland that separates it from Coconut Bay, and the meandering mouth of the tidal creek. The east facing beach is backed for the most part by the rocks of the cape, with waves reaching the rocks at high tide. The southern 200 m is occupied by the dynamic northern spit of the creek mouth. The beach is low gradient, with a 200 m wide low tide bar, that widens at the creek mouth and is fronted by up to four rips in the low tide surf.

The **Cape** beach (**929**) lies on the southern side of the shallow creek mouth. After the tidal shoals, it extends for 11.6 km south, then south-east, to the 500 m wide mouth of the Elliott River. The beach is backed by up to 1 km of low, vegetated dunes, with some areas of wind-blown sand, then a long swamp paralleling the back of the dunes, then the older 300 to 400 m wide Pleistocene barrier. The beach receives waves averaging over 0.5 m high in the north, but these decrease to less than 0.5 m by the Elliott River. For the first 5 km, they produce low tide rips spaced about every 100 m, backed by a 200 m wide, low intertidal beach and then the low dunes covered with casuarina and spinifex.

Swimming: Be careful near the creek mouth as there are strong tidal currents and deep channels, while numerous rips dominate the low tide surf along the northern half of the beaches.

Surfing: Best towards the north where the waves are higher.

Fishing: The creek at high tide and rip holes at low tide offer good fishing.

Summary: A relatively long, undeveloped beach backed by low, mainly vegetated dunes.

CAPE UPSTART to CANNONVALE (Fig. 4.41)

Length of Coast:	234 km
Beaches:	84 (beaches 930-1014)
Surf Life Saving Clubs:	Bowen (Queens Beach)
Lifeguards:	Horseshoe Bay
Major towns:	Bowen, Cannonvale

Regional Map 4: Cape Upstart to Cannonvale

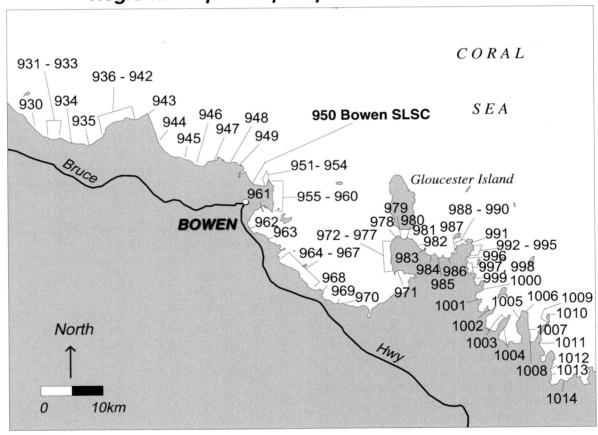

Figure 4.41 *Regional map 4 Cape Upstart to Cannonvale, beaches 930 to 1014. Beaches with Surf Life Saving Clubs indicated in bold.*

930-943 ELLIOTT R, SALTWATER CK & MOUNT CURLEWIS

No.	Beach	Unpatrolled Rating HT	LT	Type	Length
930	Elliott River	1	2	R+LT	5.5 km
931	Saltwater Ck	1	2	B+SF	1.2 km
932	Mount Curlewis (W)	1	2	B+SF	750 m
933	Mount Curlewis	1	2	B+SF	300 m
934	Mount Curlewis (W)	1	2	B+SF	4 km

Mount Curlewis is a twin-peaked, 220 m high hill that dominates the southern shore of the bay. Along its base and to either side are 12 km of low energy beaches, bounded by the Elliott River and Splitters Creek and breached by Saltwater Creek. In addition to the extensive tidal shoals at the stream mouths, there are extensive shallow sand waves extending up to several hundred metres off the beaches.

Elliott River beach (930) is a 5.5 km long spit of sand, backed by a wide tidal channel and mangroves. The spit increases in width toward the river mouth, where it is made up of a series of recurved spits. At the 500 m wide river mouth, tidal shoals extend up to 1 km into the bay. In amongst the spits are approximately ten fishing shacks, comprising the only development on the beach, which is surrounded by water. Just inside the river mouth is a more extensive shack development that is linked by a 7 km long road to the highway at Guthalungra. The beach is narrow and moderately steep at high tide, while at low tide it widens at the river mouth and toward the south, as sandy shoals increase in dominance. Waves average less than 0.5 m.

Saltwater Creek beach (931) is a low, narrow dynamic, 1.2 km strip of sand that is often breached by the backing creek and cyclone waves. It is backed by mangroves, with an older spit lying 300 to 400 m in behind and amongst the mangroves. It is undeveloped and accessible only by water.

Mount Curlewis (west) beach (932) is a 750 m long spit of sand attached to the rocky western shore of the mount, with a natural groyne often forming its western boundary. A small, though mobile, creek mouth separates the beach from Saltwater Creek Beach. The beach is low, narrow and backed by mangroves, however there is a 4WD track along the back of the beach ridge that leads to a solitary shack toward the creek mouth.

Mount Curlewis beach (933) runs for 300 m along the base of the mount. The beach varies in width as sand moves along the shoreline; there are also numerous rocky outcrops on the beach and in the water. The 5 km long vehicle track from the highway skirts the base of the mount, leading to several beachfront shacks on this and the next beach.

Mount Curlewis (east) beach (934) extends in an easterly direction from the base of the mount for 4 km to the broad mouth of Splitters Creek. The beach is attached to the base of the mount for the first 700 m, where there are several beachfront shacks. It then continues as a low, sandy beach, backed by a 1 to 1.5 km wide series of 12 low, vegetated beach ridges and spits, that represent a 6 000 year build out of the shoreline. A 4WD track also reaches a few shacks on the inner spits on the western side of Splitters Creek mouth. The beach is moderately steep and narrow at high tide, with extensive sandy shoals exposed at low tide; these widen and shoal toward the eastern creek mouth.

Swimming: Swimming is best on all these beaches toward high tide and clear of the creek mouth, which has strong tidal currents and deep channels.

Surfing: None, apart from low waves on Elliott River Beach.

Fishing: The shacks along these beaches have been built by fishermen to provide good access to the bay, creeks and beaches. A popular spot with locals.

Summary: Five generally low energy, although in places dynamic, beaches, and the location of the largest shack settlements in the bay.

935, 936 SPLITTERS CK, BRANCH CK

No.	Beach	Unpatrolled Rating HT	LT	Type	Length
935	Splitters Ck	1	1	B+SF	700 m
936	Branch Ck	1	1	B+SF	2.5 km

The southernmost corner of the bay is also the lowest energy, as well as the focus for the summer floods from Splitters and Branch Creeks. The end result is an extensive, 2 to 3 km wide, low-lying mangrove swamp, fronted by two low, low energy barrier islands - namely Splitters Creek and Branch Creek Beaches.

Splitters Creek beach (935) is a low, crenulate, dynamic strip of sand. It is backed by extensive tidal channels and shoals, then mangroves, and faces north across several hundred metres of ridged tidal flats. It changes shape and length during floods and cyclones and is prone to overwashing.

Branch Creek beach (936) is a similar low strip of sand up to 2.5 km long, backed by mangroves, and also fronted by the ridged sand flats. In places the beach has been washed away and eroding mangroves form the shoreline.

Swimming: Neither of these beaches is recommended for swimming.

Surfing: None.

Fishing: Very popular in the backing creeks.

Summary: Two low, dynamic strips of sand set in a low energy corner of the bay.

937-941   MOUNT BRUCE

No.	Beach	Unpatrolled Rating HT LT		Type	Length
937	Mt Bruce (spit)	1	1	B+SF	400 m
938	Mt Bruce (W)	1	1	R+LT	1.2 km
939	Mt Bruce	1	1	R+LT	500 m
940	Mt Bruce (E 1)	1	1	R+LT	1 km
941	Mt Bruce (E 2)	1	1	R+LT	1.5 km

Mount Bruce (also known as Mount Stuart) is the 280 m high, conical twin of Mount Curlewis. It dominates the south-eastern corner of the bay and stands out boldly, as it is surrounded by water and low-lying salt flats. Only Dingo Beach to the east links it to the mainland. The mount has five near-continuous beaches around its base. There is 4WD access across the tidal flats and via Dingo Beach to both sides of the mount, but only a solitary shack on the western beach.

Mount Bruce spit (937) is a 400 m long spit of essentially unvegetated sand that forms occasionally at the mouth of the tidal Mount Stuart Creek. The spit deflects the creek mouth to the west and is tied to the western base of the mount. It has a narrow, moderately steep high tide beach, fronted by extensive tidal shoals.

Mount Bruce (west) beach (938) is a 1.2 km long, crenulate, west to north-west facing strip of high tide sand paralleled by two strips of intertidal beachrock, with indentations in between, then a low tide bar and sand flats. A casuarina-covered beach ridge backs the beach, while the beachrock represents a lithified former beach. The 4WD track leads to this beach, where the only shack is located.

Mount Bruce beach (939) is a 500 m, narrow high tide beach running along the northern side of Mount Bruce. It is bordered by beachrock in the west and the steep slopes of the mount in the east, that rise directly to the crest just 600 m behind the beach. Numerous rocks outcrop on and just off the beach.

Mount Bruce (east) (940 & 941) is a 2.5 km long, north facing beach that runs in an arc between the eastern base of the mount and the 50 m high hill that separates it from Dingo Beach. The central portion is backed by densely vegetated, low dunes, while steep slopes back either end, with rocks along their base. The beach is moderately steep and narrow at high tide, with a 500 m long strip of beachrock along the western half of the beach. The central to eastern half of the beach is fronted by a 50 to 100 m wide low tide bar.

Swimming: All these beaches are best at high tide, however care must be taken with the beachrock and rocky outcrops. Waves usually average about 0.3 m.

Surfing: None.

Fishing: Best off the beaches at high tide.

Summary: Five generally narrow beaches surrounding the base and side of Mount Bruce, together with the beachrock remnants of former beaches.

942   DINGO BEACH

No.	Beach	Unpatrolled Rating HT LT		Type	Length
942	Dingo	1	2	R+LT	2 km

Dingo Beach (941) is a 2 km long, north facing beach that occupies a shallow, 3.5 km wide embayment between two 50 m high, conical headlands; the eastern one is named Bald Hill and is part of Abbott Point. There is 4WD access to the beach across the tidal flats from Caley Valley Homestead and across the retaining wall from Abbott Point; otherwise it is backed by tidal flats.

This is an interesting beach in that it contains a continuous strip of beachrock containing rich evidence of a former beach. The rock is backed by high tide sand, then a low, vegetated, 200 to 300 m wide dune flat, then the high tide salt flats. Waves average less than 0.5 m along the beach and crash on the beachrock at mid to high tide (Fig. 2.18).

Swimming: There is deep water off the beach at high tide, however the beachrock can be a problem.

Surfing: None.

Fishing: Good fishing off the beachrock toward high tide.

Summary: This beach lies 9 km off the highway along periodically flooded 4WD tracks, and consequently has no development and is rarely visited.

943, 944 ABBOTT POINT

No.	Beach	Unpatrolled Rating HT	LT	Type	Length
943	Abbott Pt (groyne)	2	3	R+rocks	50 m
944	Abbott Pt	2	3	R+beachrock	8.5 km

Abbott Point is the site of a large coal loading facility constructed in the 1980's, that consists of a long jetty and shore based coal storage facilities. The coal is railed from the Collinsville mine, located 80 km south-south-west of the point. There is a private road to the port that provides access tracks to both beaches. The point itself is a round, 20 m high hill surrounded by basalt boulders. On its south side is a long beach running down to Euri Creek. A small groyne has been built on the northern tip of the beach to provide a small boat harbour, subsequently forming a small beach in its lee.

The **Abbott Point (groyne)** beach (943) lies on the north side of the hooked groyne that extends about 100 m out from the beach. Inside its moderately protected waters is a 50 m long pocket of sand that grades into basalt boulders and then the rocky point, which in turn is crossed by the coal jetty. The southern corner of the beach provides a moderately protected boat shelter, while the waves quickly pick up toward the point.

The main **Abbott Point** beach (944) has built out against the groyne, widening at its northern end. The beach is 8.5 km long and runs relatively straight south-east toward Euri Creek. It consists of a relatively steep high tide beach, fronted for the first 4 km by a parallel band of beachrock. Beyond the beachrock is a low tide bar that runs for 3 km, before the bar widens into tidal flats toward the more protected reaches in lee of the sand shoals off Euri Creek. Waves average over 0.5 km at the point but decrease toward the south. The beach is backed by low, active dunes up to 200 m wide, and then a series of low, vegetated dune ridges. The port road runs along the crest of these ridges.

Swimming: The groyne beach is the safest location when clear of boats. The main beach is hazardous where fronted by beachrock, while it is safer toward the south where the bar is present.

Surfing: The waves tend to close out on the beachrock in the north, and break over the bar toward low tide in the south.

Fishing: Best off the groyne or beachrock at low tide.

Summary: These two beaches are on company land and, while serviced by road, are generally off limits to the public.

945-949 ERUI CK & DON RIVER (WEST)

No.	Beach	Unpatrolled Rating HT	LT	Type	Length
945	Euri Ck	1	2	B+SF	2 km
946	Don R (W 4)	1	2	B+SF	2.5 km
947	Don R (W 3)	1	2	B+SF	2.5 km
948	Don R (W 2)	1	2	B+SF	2.5 km
949	Don R (W 1)	1	2	B+SF	2.5 km

Euri Creek and the Don River deliver large quantities of sand to their mouths during floods and form a 12 km long deltaic shoreline. The sand is reworked westward by the prevailing south-east winds and waves to form a series of five migrating recurved spits and their attendant multiple bars and shoals, as well as their associated river and creek mouths (Fig. 1.3). In all this is one of the most dynamic and unstable sections of the Australian coastline, and a smaller version of the equally dynamic Burdekin River delta, 75 km to the north-west. All five "beaches" vary considerably in length, width, location and shape in response to floods, waves and tides. They are all low, largely unvegetated and prone to overwash at spring tides and during cyclones. Essentially, they represent a slow movement and rearrangement of the Don and Euri sands westward toward, and finally around, Abbott Point. There is no vehicle access and no development on any of the spits.

Euri Creek beach (945) is approximately 2 km long, varying in length and form across the mouth of Euri Creek, a substantial upland creek. It has a narrow, moderately steep high tide beach fronted by several hundred metres of intertidal sand flats. It is backed by over 1 km of abandoned beaches and spits.

Don River (west 4, 3, 2 and 1) (becahes **946, 947, 948, 949**) are four equally dynamic and similar spits that extend westward for 10 km from the 300 m wide sandy mouth of the Don River at Bowen. All four spits are slowly moving westward, rearranging their length, shape and location. All four are fronted by 300 to 500 m of intertidal sand flats and bars, and backed by bare and mangrove-covered sand flats, and then a series of older beach-spit ridges.

Swimming: These are dynamic and changeable beaches, with strong tidal currents at all the creek mouths, and are not recommended for swimming. Waves are usually less than 0.5 m high.

Surfing: The waves break over a wide range of bars and shoals off the spits and produce a highly variable, low, spilling surf, that changes with the tides.

Fishing: The backing creeks and mangrove areas are popular at high tide.

Summary: A very dynamic section of coast; best left to its own devices.

950   QUEENS BEACH (BOWEN SLSC)

Patrols:
Surf Life Saving Club:
Lifeguard   September to May
Stinger net: November to May

No.	Beach	Rating HT	LT	Type	Length
950	**Queens**	1	1	R+LT/SF	4.5 km

Queens Beach (950) is Bowen's main northern beach and occupies 4.5 km of Queens Bay, a relatively open, north to north-east facing bay located between the western mouth of the Don River and rocky Cape Edgecumbe. The beach lies just 2 km north-west of Bowen town centre and the suburbs of Queens Beach and Edgecumbe Heights back its west and eastern shores respectively. In addition to the beachfront houses, there are caravan parks in the centre and at the western end of the beach, with golf links in between. The houses are fronted by a wide foreshore reserve, a narrow high tide beach and then a low tide bar. The bar is 100 m wide in the north, widening to become ridged sand flats several hundred metres in width, in the more protected east (Fig. 2.16a).

Swimming: This is a popular swimming beach, with best conditions toward high tide. The safest swimming, particularly during stinger season, is in the stinger enclosure. Waves average less than 0.5 m and decrease to the east.

Surfing: Chance of a low beach break at the western end of Queens Beach.

Fishing: Best at high tide over the bar and sand flats.

Summary: Bowen's northern beach suburb and holiday accommodation back this attractive, north-facing beach.

951-954   GREYS BAY to CAPE EDGECUMBE

No.	Beach	Unpatrolled Rating HT	LT	Type	Length
951	Greys Bay	1	1	R+boulders	250 m
952	Beach 952	1	1	R+boulders	150 m
953	Beach 953	1	1	R+boulders	150 m
954	Cape Edgcumbe	1	2	R+boulders	150 m

The eastern end of Queens Bay is bounded by hummocky, 80 m high Cape Edgecumbe that protrudes just over 1 km to the north, with reefs and North Rock extending another 500 m off the point. Both sides of the point have small embayments containing a total of nine beaches (Fig. 4.42). Four are located on the western shore and are protected by the point and their orientation. The road to Horseshoe Bay runs behind the first two, while the outer two require a short walk to reach.

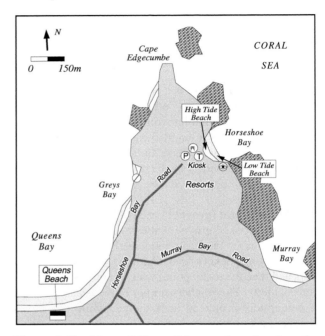

Figure 4.42 *Map of Cape Edgecumbe, showing the southern end of Queens Beach and the smaller Greys, Horseshoe and Murray Bays.*

All four beaches are similar in nature. They are composed of coarse quartz sand derived from the surrounding granite, they have low waves averaging less then 0.3 m and calm conditions are common. They all have relatively steep beaches fronted by a shallow bar, with boulders located on and off the beaches, as well as bordering each end. There is also a small coral reef lying just off the beaches.

Greys Bay (951) is located just around some granite rocks from the eastern end of Queens Beach, with the road running right behind half of the 250 m long, west facing beach. There is a boat ramp at its southern end. Beaches **952** and **953** have a few houses behind each, while Cape Edgecumbe beach (**954**) is backed by vegetation and large boulders, with only foot access.

Swimming: These are four relatively safe beaches with usually calm to low wave conditions. The only hazard is the number of boulders on and off the beaches.

Surfing: None.

Fishing: A popular spot for rock fishing, as well as for launching boats to fish the outside rocks and reefs.

Summary: Four accessible, quiet and picturesque little beaches, all framed by large granite boulders.

Edgecumbe Bay

Edgecumbe Bay is a 30 km wide, 20 km deep, north facing bay lying between Cape Edgecumbe and Cape Gloucester on Gloucester Island. Because of its orientation and protection from the Trades and bay islands, it has a low energy shoreline, with waves only reaching the beaches on the exposed eastern shore of Cape Edgecumbe. Numerous small creeks and the Gregory River drain into the bay, producing a shallow shoreline dominated by intertidal sand flats up to 1.5 km wide, and extensive mangroves, particularly in the south-eastern corner centred on the Gregory River mouth. In addition, apart from the scattered residential and holiday development on Cape Edgecumbe and the town and port of Bowen, the majority of the 65 km shoreline is backed by gently rising grazing land, with only a few shack settlements along the shore. The Bruce Highway only touches the shore at the Mount Gordon rest area 3 km south of Bowen.

---

955   HORSESHOE BAY

No.	Beach	Rating HT	LT	Type	Length
955	**Horseshoe Bay**	1	2	R+LT	150 m

Horseshoe Bay is the most popular of Bowen's twelve beaches. It is a picturesque little beach, lying out on the north-east tip of Cape Edgecumbe, 4 km from the town centre. The whole cape is composed of rounded granite rocks and boulders (Fig. 4.43), including the famous landmark Mother Beddock, a large, round boulder perched seemingly precariously 100 m above Kings Beach.

Horseshoe Bay beach (**955**) is just 150 m in length, faces north-east and is bordered by large, rounded granite boulder-headlands. The road runs right to the back of the beach where there is a car park, kiosk, an amenities block and a shady picnic area, with two small resorts tucked in behind.

The beach receives low waves, usually less than 0.5 m, which combine with the medium to coarse granitic sand to build a steep beach at high tide, with a narrow, shallow bar exposed at low tide. Surf, currents and rips are rare, requiring waves greater than 0.5 m to form.

Swimming: A relatively safe beach under normal low wave conditions; just use care if swimming or snorkelling around the rocks.

Surfing: None.

Fishing: A popular spot to fish off the rocks or beach, particularly at high tide. There is a boat ramp in adjoining Greys Bay.

Summary: One of the nicest beaches on the coast, very popular with locals and tourists.

Figure 4.43: *Horseshore bay contains a protected 150 m long beach bordered by the prominent granite headlands and boulders.*

956-959 MURRAYS & ROSE BAYS, KINGS BEACH

No.	Beach	**Unpatrolled** Rating HT	LT	Type	Length
956	Murrays Bay	1	2	R+LT+ boulders	200 m
957	Rose Bay	1	2	R+LT	100 m
958	Kings (N)	1	2	R+LT+inlet	600 m
959	Kings	1	2	R+LT+inlet	1.2 km

The eastern side of Cape Edgecumbe faces into Edgecumbe Bay. The bay was sighted and named by Captain Cook in June 1770. It is an open, north facing bay, 18 km wide between the cape and the eastern boundary at prominent, 600 m high Gloucester Head. The granite boulders and hillocks of the cape continue to dominate its eastern shore, producing the two small embayments of Murray and Rose Bays, and the longer Kings Beach, linking the cape with Flagstaff Hill. The eastern cape beaches are all accessible by roads radiating out from the eastern end of Queens Beach, and all lie within 4 km of Bowen.

Murrays Bay (956) is a 200 m long, north facing pocket of sand wedged in between and amongst granite headlands and boulders, with rocks dominating the low tide beach. The little bay houses a small resort, fronted by the small high tide beach, a boat ramp and the rock-strewn low tide bar and beach.

Rose Bay (957) is a 100 m long, east facing bay, bordered by granite headlands and backed by four houses, but with a predominantly sandy beach, including a 100 m wide low tide bar. A few granite rocks and boulders are spread across the beach and bar.

Kings Beach is divided in two by a tidal creek draining across the centre. Kings Beach (north) (**958**) is a 600 m long sand beach extending from a low, northern granite headland to the shallow, narrow, meandering creek mouth. It is backed by a road and small residential development consisting of three rows of houses, then the creek. The beach faces north-east and receives waves averaging 0.5 m that break across a 100 m wide low tide bar, and up the low gradient beach face at high tide. The bar widens considerably toward the inlet where it spreads into a series of tidal shoals and channels (Fig. 4.44).

Kings Beach (**959**) runs from the inlet for 1.2 km down to the rocks of Flagstaff Hill; there is an excellent view of the beach from its crest. The beach has a low gradient high tide beach and wider low tide bar, that widen toward the inlet mouth. It is backed by a low dune and the Flagstaff Hill Road that runs the length of the undeveloped beach, providing excellent access.

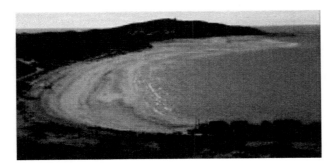

Figure 4.44   *View north along Kings Beach at low tide. Note the wider tidal shoals in front of the northern inlet.*

Swimming: These are relatively safe beaches under the usual low 0.5 m waves. However higher waves, the Kings Beach inlet and the rocks at Murrays Bay all provide additional hazards.

Surfing: Chance of a low beach break at Kings Beach, north of the inlet and at Rose Bay.
Fishing: Best off the rocks and in the inlet and creek.

Summary: Four beaches used by both tourists and locals and representing the main "surfing" beaches for Bowen.

960   FLAGSTAFF HILL

No.	Beach	Unpatrolled		Type	Length
		Rating			
		HT	LT		
960	Flagstaff Hill	1	1	B+SF	500 m
961	Quay Street	1	1	B+SF	1.2 km

Flagstaff Hill is a 60 m high, grassy hillock that signals the entrance to Port Denison, the port of Bowen. A gravel road leads to the crest of the hill and to a small settlement on the tip of the point. Around the base of the hill is a low, rocky shore together with a narrow, sandy beach (**960**) on its north-east side. The beach consists of a sandy high tide strip, however at low tide it is essentially a rock flat.

The port of Bowen lies inside the point and consists of a large boat marina and a 600 m long jetty. Backing the jetty is the town of Bowen, with a narrow high tide beach (**961**) at the foot of **Quay Street**. This beach has been somewhat modified by the marina, jetty and foreshore development. Today the beach is approximately 1.3 km long, faces east and is dissected by the marina, the jetty and a boat ramp, as well as two outcrops of rock, together with a backing seawall and a small creek at its southern end. At low tide, 400 m wide ridged sand flats are exposed (Fig. 4.45).

Figure 4.45   *Bowen Jetty has a narrow high tide beach to either side, with 400 m ridged sand flats exposed at low tide.*

Swimming: Only at high tide.

Surfing: None.

Fishing: The jetty and marina breakwater are very popular with the locals.

Summary: Two very accessible beaches, but little used for swimming owing to the rocks at Flagstaff and the development and shallow water at Quay Street.

962-970   EDGECUMBE BAY (west)

No.	Beach	Unpatrolled			
		Rating		Type	Length
		HT	LT		
962	Mt Gordon 1		1	TSF	3 km
963	Mt Bramston	1	1	TSF	3 km
964	Duck Ck	1	1	TSF	1.5 km
965	Brisk Bay	1	1	TSF	3.5 km
966	Miowera	1	1	TSF	2.8 km
967	Emu Creek (W)	1	1	TSF	2 km
968	Emu Creek (E)	1	1	TSF	1.5 km
969	White Cliffs (W)	1	1	TSF	1.8 km
970	White Cliffs (W)	1	1	TSF	500 m

The nine beaches that occupy the western shores of **Edgecumbe Bay** are essentially very low energy strips of narrow high tide sand, fronted by varying widths of intertidal sand flats, with many also fringed by mangroves. This is a low energy, generally north-east facing shore that is well protected by the large Gloucester Island, as well as the smaller Stone, Thomas and Poole Islands, and shallow rock reefs off Mount Bramston. There is little development other than a few fishing shacks, and limited access apart from the highway rest area at Mount Gordon beach. The rest area is the only point between Cardwell in the north and Clairview in the south where the coast can be viewed from the highway.

Mount Gordon beach (962) is a crenulate, 3 km long, low energy high tide beach fronted by up to 1 km of ridged sand flats (Fig. 2.18). There is good access and parking at the Mount Gordon rest area right on the highway, but otherwise no development. Rocks and rock flats are scattered along the beach, with mangroves increasing to the east.

Mount Bramston beach (963) is a narrow strip of crenulate high tide sand, fronted by a 50 to 100 m fringe of rocks and mangroves, then 1 km wide sand flats. It runs for 3 km and ends at the mouth of Hay Gully Creek.

Duck Creek beach (964) is a 1.5 km, east facing, low energy beach ridge fronted by some mangroves and 1 km wide tidal flats, and backed by a small salt flat. A small creek forms the eastern boundary and the larger Duck Creek mouth the western boundary.

Brisk Bay beach (965) is a 3.5 km long, north-east

facing, rocky shore with a strip of high tide sand, and a 100 m wide, fringing intertidal rock flat. Numerous shacks back the beach and are serviced by a gravel road from the highway. Poole Island lies 1 km offshore and is connected to the mainland by tidal flats at low tide. **Miowera** beach (966) is a 2.8 km continuation of the low rocky shore and high tide sand beach.

Emu Creek (west) (967) is a 2 km long, north-east facing beach, backed by a series of four vegetated beach ridges and fronted by tidal flats up to 800 m wide, with mangrove-fringed Emu Creek mouth forming the eastern boundary. There is no development on the beach or ridges.

Emu Creek (east) (968) extends from the eastern side of Emu Creek mouth for 1.5 km around a low, square, rock-fringed head. It has a narrow high tide beach, then the rock flats grading into deeper sand flats.

White Cliffs (969 & 970) is a 2.3 km long, crenulate, north facing shore consisting of a low, narrow strip of high tide sand, fronted by ridged tidal flats up to 1.5 km wide, with numerous mangroves fringing the beach. A gravel road runs behind the western beach servicing numerous shacks.

Swimming: All these beaches are usually calm except when strong winds are blowing across the bay; even then waves rarely exceed 0.3 m. The best time to swim is at high tide. However, stay clear of the tidal creeks and watch for submerged rocks on many of the beaches.

Surfing: None.

Fishing: This bay is a very popular location for fishing in the creeks and bay fishing over the rock reefs.

Summary: A low energy shoreline, relatively close to the highway, but primarily utilised by locals for fishing and holidays.

971, 972   SINCLAIRE BAY

No.	Beach	Unpatrolled			
		Rating		Type	Length
		HT	LT		
971	Sinclaire Bay	1	1	B+SF	1.2 km
972	Sinclaire Bay (W)	1	2	R+LT	1.2 km

Sinclaire Bay is a south facing, low energy bay located at the eastern end of Edgecumbe Bay. It is bounded by 1 km long Sinclaire spit on the west and a mangrove-fringed, rocky shore to the east, with open Miralda Creek entering the eastern end of the bay. A 7 km long gravel road from Dingo Beach runs along the eastern shore of the bay and across the back of the beach, continuing up along the coast to Cape Gloucester, 6 km to the north. There are houses located on both the main and western bay beaches.

Sinclaire Bay beach (**971**) averages 1.2 km in length, but can be extended through growth of the western spit. The spit is the southernmost and bayward part of a series of several low beach ridges that have built out 200 m into the bay over the past few thousand years. A number of houses are located toward the eastern side of the beach near where it links with the backing hill slopes. The beach is narrow at high tide, widening to a tidal flat at low tide, with waves restricted to wind chop across the 3 km wide bay.

Sinclaire Bay (west) beach (**972**) forms the western boundary of the prograding bay shoreline. It extends from the western tip of the spit in a north-east, then north-west, direction for a little over 1 km, ending at a small, mangrove-filled inlet. It consists of a low energy beach, partially fronted by intertidal beachrock. There are a few houses along the beach front.

Swimming: Both beaches are usually calm or with low waves. Care is needed around the beachrock on the western beach.

Surfing: None.

Fishing: Best in the bay and mangrove-fringed creeks, and off the beachrock at high tide.

Summary: An increasingly popular location with the local holiday-makers and occasional tourists who wander off the beaten track.

Cape Gloucester & Gloucester Head

At daylight on 4 June 1770, Captain Cook sighted what he called a 'lofty promontory', which he called Cape Gloucester. However, the 530 m high cape is known today as Gloucester Head, with Cape Gloucester being the name of a low, sandy foreland located on the mainland immediately south of the island and 10 km south of the head.

973-981 CAPE GLOUCESTER beaches

No.	Beach	Unpatrolled		Type	Length
		Rating HT	LT		
973	Beach 973	1	2	R+beachrock	800 m
974	Beach 974	1	2	R+beachrock	500 m
975	Beach 975	1	2	R+beachrock	600 m
976	Beach 976	1	2	R+beachrock	1.6 km
977	Passage Islet (S)	1	2	R+beachrock	1 km
978	Cape Gloucester (W)	1	2	R +beachrock	1.2 km
979	Cape Gloucester (E 1)	1	2	B+SF	1.1 km
980	Cape Gloucester (E 2)	1	2	B+SF	700 m
981	Cape Gloucester (E 3)	1	2	B+SF	1 km

North of the Sinclaire Bay beach ridges, the coast trends north for 5 km to Cape Gloucester then turns and heads east. This shoreline surrounds a 15 km², 400 m high series of hills that link to the mainland at Dingo Beach. Around the base of the hills are the two Sinclaire Bay beaches, then a series of 11 beaches between the bay and Dingo Beach. All are low energy, generally face west to north and are well protected by their orientation and Gloucester Island, which lies 1 km immediately north of Cape Gloucester. A vehicle track from Dingo Beach runs to Sinclaire Bay then up the coast to Cape Gloucester.

Beach **973** is an 800 m long, curving, generally west facing strip of high tide sand fronted by near-continuous intertidal beachrock. It extends from the northern side of the small inlet that separates it from the Sinclaire Bay beaches, to a low sand and beachrock point. The track to Cape Gloucester runs along the back of the beach, and there is also a landing strip paralleling the beach.

Beach **974** is a more sandy, 500 m long, curving, west facing beach that runs north of the beachrock point to a second low point, fronted by low tide rocks. Water is deep off the southern sandy half of the beach, while sandy-rock flats lie off the northern half.

Beach **975** extends from the northern side of the rocks for 600 m to another rock-fringed point. It has a sandy high tide beach, while the rock flats dominate at low tide.

Beach **976** is a crenulate, west facing, 1.6 km long, continuous strip of high tide sand, that weaves its way along the coast amongst an assortment of rocks lying both on the beach and exposed as low tide rock flats, with a few mangroves as well. The vehicle track to Cape Gloucester winds its way along behind the beach.

Passage Islet is a small, 20 m high islet lying 500 m off the shore. It has a long, sandy tail extending toward the shore and a sandy, cuspate foreland in its lee. On the southern side of the foreland is a 1 km long beach (**977**), consisting of a high tide beach with low tide rock flats and beachrock.

On the eastern side of the foreland is the main **Cape Gloucester** beach (**978**) a 1.2 km long, north-west facing, sandy beach, the eastern half of which is backed by the scattered Cape Gloucester settlement consisting of houses, fishing shacks, a small resort called Monty's and usually several boats moored off the beach. This beach ends at the cape, which is a sandy foreland that protrudes 800 m toward Gloucester Island, located 1 km across Gloucester Pass.

On the eastern side of the cape is an open, 1.5 km wide, north facing bay containing two curving beaches separated by a low beachrock point. The first (**979**) is 1.1 km long, the second (**980**) is 700 m. Both are fronted by 400 m wide intertidal sand and rock flats. The first beach forms the eastern side of the Cape Gloucester foreland and is backed by low beach ridges, while bedrock hill slopes back the shore to the east. The Cape Gloucester vehicle track continues on past the two beaches.

Beach **981** lies beyond a low, rocky point and consists of a crenulate, 1 km long, north-east facing, low energy beach, with a continuous high tide sand beach fronted by both rocks and a few mangroves, then 400 to 500 m wide sand-rock flats.

Swimming: All these beaches usually have low waves to calm conditions and are relatively safe for swimming, They are best at high tide, as rocks and tidal flats are exposed at low tide. Take care of the numerous submerged rocks along many of the beaches.

Surfing: None.

Fishing: This whole area is used by locals for fishing weekends and holidays, with most fishing in the bay and over the tidal flats at high tide.

Summary: An area that is being increasingly opened up by the Whitsunday locals for their shacks and holiday houses, and increasingly attracting a few wayward tourists.

982-990 SHOAL BAY, DINGO BEACH, NELLY, JONAH, LITTLE JONAH & GEORGE bays

No.	Beach	Unpatrolled Rating HT LT		Type	Length
982	Shoal (Hideaway) Bay	1	1	B+SF	2.4 km
983	Black Currant Island	1	1	B+SF	500 m
984	Dingo Beach	1	1	B+SF	2 km
985	Nelly Bay	1	2	B+rock flats	1.2 km
986	Jonah Bay	1	2	B+rock flats	1.3 km
987	Little Jonah Bay	1	1	B+SF	500 m
988	Beach 988	1	2	B+rock flats	200 m
989	Beach 989	1	2	B+rock flats	400 m
990	George Bay	1	1	B+SF	500 m

Between Shoal Bay and George Point is an open, 8 km wide, north facing bay, bordered by high land behind Shoal Bay, with the low ground of Dingo Beach in the centre and hilly terrain rising to over 300 m toward George Point. The shoreline consists of a series of small bedrock-controlled embayments containing nine beaches, and three small islands. All the beaches face essentially north and are low energy, with calm conditions or low waves dominating. The Dingo Beach Road reaches the coast at Dingo Beach; from there a western road runs out to Sinclaire Bay and an eastern road to Nelly and Jonah Bays. The eastern four beaches have no vehicle access.

Shoal Bay (982), more recently renamed Hideaway Bay as part of a land subdivision, is a 2.4 km long, north-east facing, low energy bay, backed by wooded hills rising to 400 m. The lower slopes overlooking the bay were subdivided in the early 1990's. The beach consists of a narrow high tide strip of sand fronted by 200 to 300 m wide, rocky sand flats. In the south-east corner of the bay is a second 500 m long, curving, west facing beach (**983**) that ties small **Black Currant Island** to the mainland via a narrow strip of sand and mangroves.

On the eastern side of Black Currant Island is **Dingo Beach** (984) - the main centre for the area, containing a store and caravan park. The 2 km long, curving, north facing beach is bordered by low, rocky headlands and consists of a steep, 50 m wide high tide beach, fronted by 500 to 1 000 m wide sandy-reef flats (Fig. 2.16b). There is road access to the back of the beach, which also has two rows of houses, together with a boat ramp and a tidal swimming (stinger) enclosure. The eastern end of the beach protrudes northward in lee of a small island 1 km off the beach, with tidal flats connecting the island to the point at low tide.

Nelly Bay (985) lies on the eastern side of the point and is a 1.2 km long, north-east facing, curving beach, bordered on the east by a 60 m high wooded headland. The western end of the beach has been developed for housing and small tinnies are stored on the beach. The high tide beach is relatively narrow and steep, with beachrock outcropping in places. Camping is permitted in the casuarina trees behind the beach.

Jonah Bay (986) lies over the headland east of Nelly Bay and is connected by a gravel road, with a few houses backing the 1.3 km long, north facing beach. The beach consists of a high tide beach fronted by inner rock flats and deeper sand flats, with a low, casuarina-covered dune behind the beach, then wooded slopes rising to 150 m.

Little Jonah Bay (987), beaches 988 & 989 and Georges Bay (990) all occupy small indentations in the hilly shoreline between beach 986 and George Point. There is no vehicle access to these beaches. Little Jonah Bay is contained in a small, 500 m wide, north facing valley. It consists of a low, 200 m wide dune flat covered in casuarinas, a steep high tide beach, and 200 m wide sand and rock flats, with mangroves fringing the eastern headland. Beach **988** is a 200 m long strip of high tide sand fringed by low tide rocks and a few mangroves. Steep wooden slopes back the beach. Beach **989** is another 400 m long strip of sand and rocks at the base of wooden slopes. It consists of two parts separated by 50 m of rocks, with a few mangroves also fringing the eastern end of the beach. The beach is fronted by 100 m wide rock-reef flats. **George Bay** (990) lies immediately east of 100 m high George Point. The bay is 1 km deep and contains a protected, 500 m wide, west facing beach. It is a sand beach fronted by tidal flats that runs the length of the bay, with outcrops of beachrock, and stands of mangroves at either end and against the headlands.

Swimming: The main swimming area is in the tidal enclosure at Dingo Beach. If swimming at other beaches it is always better toward high tide, owing to shallow flats off all beaches, together with numerous rocks.

Surfing: None.

Fishing: There are numerous rock and reef flats off the beaches that can be fished at high tide. Rock fishing is also possible from the many headlands and rocks.

Summary: An area that has seen considerable development since the 1970's and which will continue to grow, owing to its close proximity to the booming Whitsundays.

Dryander National Park

Shoreline length: 48 km
Beaches: 987-997, 1002-1007, 1009-1011 (20 beaches)

Dryander National Park covers hilly, wooded country between George and Grimston points. The backing hills rise to over 800 m and their slopes and valleys produce a highly indented coastline. While the two points are located only 17 km apart, there is 60 km of shoreline made up of seven major headlands with deep bays in between. There are twenty beaches between the two points, sixteen of which are in the national park, together with four on the northern side of Georges Point. The only road access into the park is along the Earlando Road, with the road to Beach 986 ending at a northern park boundary.

The park preserves extensive areas of wooden, hilly land, together with the predominantly low energy, rocky coastline that hosts the generally small, low energy pocket beaches. Most of the beaches are dominated by rocky headlands and intertidal rocky flats, together with several cobble-boulder beaches. Mangroves occur at the heads of most of the deeper bays and fringe many of the beaches, while coral reefs are also found off several of the beaches.

For further information:
Rangers Office	Marine Parks
Whitsundays National Park	Mackay District Office
cnr Shute Harbour Rd	Cnr Wood & River Sts
PO Box 332	PO Box 623
Airlie Beach, Qld 4802	Mackay, Qld 4740
phone: (07) 49 467 022	phone: (07) 47 518 788

991-1000   **GEORGE PT & DRYLANDER N.P.**

No.	Beach	Unpatrolled		Type	Length
		Rating			
		HT	LT		
991	George Pt (E)	1	2	B+rock flats	250 m
992	Beach 992	1	2	Cobbles	50 m
993	Beach 993	1	2	Cobbles	50 m
994	Beach 994	1	2	R+cobbles	100 m
995	Beach 995	1	2	R+cobbles	100 m
996	Beach 996	1	2	B+S-rock flats	250 m
997	Beach 997	1	2	B+S-rock flats	200 m
998	Beach 998	1	2	B+S-rock flats	650 m
999	Beach 999	1	2	Cobbles	120 m
1000	Beach 1000	1	2	Boulders	100 m

George Point is the northern tip of a section of rugged, irregular coast that extends down to Cape Conway, 65 km to the south-east, and contains 170 km of predominantly rocky shoreline, together with numerous high islands of the Whitsundays paralleling the coast. The first 9 km south of the point contains ten small beaches along 12 km of coast. All are low energy, all are located in indentations and small bays along the predominantly rocky coast and most are influenced, if not dominated, by rocks. They are all relatively isolated, with no vehicle access to any of the beaches.

Immediately east of George Point is beach **991**; a 250 m long, east facing strip of high tide sand fronted by a cobble to boulder flat, and bordered by 100 m high, bare to shrubby headlands. Beach **992** is a 50 m long pocket boulder beach backed by a narrow, steep, wooded valley. Beaches **993** and **994** are two nearly adjoining pockets at the mouths of small valleys, composed largely of cobbles and boulders that have been arranged into moderately steep beaches.

Beach **995** has a veneer of high tide sand underlain and fronted by a cobble to boulder field. This small, 100 m long beach lies immediately east of the 250 m long, sandy beach **996**. This beach occupies a slightly larger valley and contains the relatively straight beach, fringed by rocky headlands and a few mangroves, with sand to rock flats extending off the beach.

Beach **997** occupies an adjoining valley of similar dimensions and is a 200 m long, low sand spit, backed by a mangrove-fringed creek and salt flats. It has a sandy high tide beach that is fronted by sand and rock flats.

Beach **998** is one of the longer local beaches at 650 m in length. It faces north-east and occupies a broader valley that has been partially infilled with sediment, resulting in a couple of hundred metres of flat, wooded sand ridges behind the beach, together with the mangrove-lined valley creek that flows through a small inlet across the middle of the beach. The sand beach is fronted by 500 m wide, shallow sand flats and patches of coral reef.

Beach **999** is actually a series of three small, near-continuous cobble to boulder beaches (50, 20 and 50 m in length) backed by steeply rising slopes, while beach **1000** consists of two adjoining 50 m long boulder beaches backed by steep slopes.

Swimming: All these beaches are best at high tide when the rocks and intertidal flats are covered. Be careful toward low tide when shallow water, rocks and some reefs predominate.

Surfing: None.

Fishing: Excellent fishing amongst the many rocks and reefs, together with two small accessible creeks at beaches 997 and 998.

Summary: A relatively isolated and natural part of Dryander National Park; best visited by boat at high tide.

1001 EARLANDO

No.	Beach	Unpatrolled		Type	Length
		Rating			
		HT	LT		
1001	Earlando	1	1	B+rock flats	150 m

Earlando is the site of the sole public vehicle access and the only resort in this part of the Whitsundays. The resort sits between 500 m high hills and a 4 km deep bay, with much of the adjacent shore occupied by mangroves, and only a small, 150 m long beach (**1001**) providing easy access to the bay waters. It has a moderately steep, sandy beach that is usually home to an assortment of recreational craft (Fig. 4.46). There is parking and public access at the top of the beach, with the resort occupying the area immediately south of the beach. The resort caters for camping and caravans as well as having beachfront holiday units, a store and a restaurant. For information phone (07) 4945 7133.

Swimming: Best at high tide when there is deeper water right off the beach.

Surfing: None.

Fishing: Many guests use the boats to fish the surrounding bays and creeks, or wander down to the small rock jetty to fish at high tide.

Summary: The best place to see a part of this spectacular coast and a nice spot to stop, away from the more congested parts of the Whitsunday mainland.

Figure 4.46 *The low energy beach at Earlando is used for swimming as well as launching small boats.*

1002-1011 DOUBLE & WOODWARK bays

No.	Beach	Rating HT LT		Type	Length
		Unpatrolled			
1002	Beach 1002	1	2	R+rock flats	800 m
1003	Double Bay (W)	1	2	R+rock flats	200 m
1004	Double Bay (E)	1	2	R+rock flats	200 m
1005	Datum Rock	1	2	R+rock flats	200 m
1006	Beach 1006	1	2	boulders	600 m
1007	Woodwark Bay (1)	1	2	sand-rock flats	150 m
1008	Woodwark Bay (2)	1	2	sand-rock flats	1.1 km
1009	Woodwark Bay (3)	1	2	R+cobbles	300 m
1010	Woodwark Bay (4)	1	2	R+cobbles	50 m
1011	Beach 1011	1	2	R cobbles+flats	400 m

East of Earlando the shoreline is dominated by four protruding wooded ridges that extend up to 4 km toward the north-north-east, with equally deep **Double and Woodwark bays** in between. The 50 km of shoreline contain ten beaches totalling only 4 km; all are dominated by the presence of rocks on the beach or in the fronting tidal flats, as well as being bordered by rocky shores and headlands.

Beach **1002** is an 800 m long, north facing beach located between two narrow headlands that extend up to 1 km seaward of the beach. The beach is moderately steep, consists of sand to cobbles, is fronted by intertidal rock flats, with mangroves fringing the ends and a small fringing reef off the western end.

Beach **1003** is a 200 m long strip of high tide sand fronted by a continuous, rocky intertidal zone. It faces north toward the 2.5 km wide mouth of Double Bay. Beach **1004** is contained in the second of the "double" bays and lies 3 km inside the bay as a 200 m long, sandy high tide beach, fronted by an intertidal rock flat, with rocky spurs to either end. **Datum Rock** beach (**1005**) lies on the opposite shore of the bay and is similar in length and character, except it faces the other direction and contains a single shack. It is also fronted by a small fringing coral reef, which provides a more protected anchorage off the beach. Beach **1006** lies at the tip of the eastern entrance to Double Bay. It is a 600 m long, north facing, steep cobble to boulder beach.

Woodwark Bay lies immediately east of Double Bay and is a 3 km wide, 5 km deep bay that contains four low energy beaches. Beach **1007** is a 150 m long sand and rock flat, with a few mangroves on the rocks, and coral reef paralleling the beach. Beach **1008** lies at the head of

the bay and is a 1.1 km long, north facing high tide sand beach, backed by a low sand ridge and shack. This beach is outside the park and has 4WD access. Beaches **1009** and **1010** are two sand/cobble/rock flat beaches, 300 and 50 m long respectively, located on the north-eastern side of the bay. The longer beach hosts two old tin shacks.

Beach **1011** is a 400 m long, high tide sand and cobble beach fronted by intertidal sand flats. It is more exposed than many of the neighbouring beaches, with occasional waves up to 0.3 m, and a near-continuous fringing coral reef below low water.

Swimming: All these beaches are fronted by rocks and rock flats and are only suitable for swimming at high tide.

Surfing: None.

Fishing: There are numerous rocks and reefs along this rugged and irregular coast that can be fished by boat, or from the shore in places.

Summary: An interesting part of Dryander National Park, most frequented by boats and yachts, with well-protected anchorages in the bays.

1012, 1013   BEACHES 1012, 1013

No.	Beach	Rating HT LT		Type	Length
		Unpatrolled			
1012	Beach 1012	1	2	R+rock flats	100 m
1013	Beach 1013	1	2	R+rock flats	100 m

At the western end of Pioneer Bay, 3 km north-west of Cannonvale, are two small high tide beaches both located at the heads of small valleys. The valley behind beach **1012** has been cleared and there are a couple of houses just behind the beach, with mangroves fringing the small creek where it flows out at the northern end of the beach. Beach **1013** is similar but fronts an uncleared valley. Both beaches face east, are approximately 100 m long and consist of a high tide sandy-cobble beach fronted by low tide rock flats, with fringing mangroves. Waves are usually very low to calm. Apart from the houses, there is no public land access and no facilities.

AIRLIE BEACH TO NEWRY ISLAND
including WHITSNDAY ISLANDS
(Fig. 4.47)

Length of Coast:	210 km (mainland only)
Beaches:	72 (beaches 1014-1086), plus 17 island beaches
Surf Life Saving Clubs:	none
Lifeguards:	none
Major towns:	Airlie Beach

Regional Map 5: Airlie Beach to Newry Island

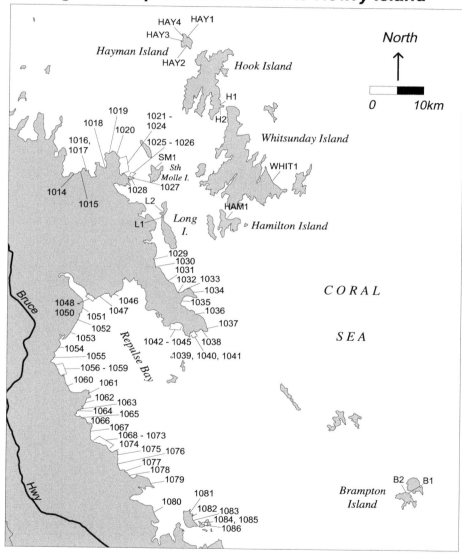

Figure 4.47   *Regional map 5 Airlie Beach to Newry Island, mainland beaches 1014 to 1086, and Whitsunday Islands (17 beaches).*

1014, 1015 CANNONVALE, SHINGLEY

No. Beach	**Unpatrolled**		Type	Length
	Rating			
	HT	LT		
1014 Cannonvale	1	1	B+SF	400 m
1015 Shingley	1	2	B+rock flats	300 m

Cannonvale is the commercial centre of the Whitsundays and has a long history of white settlement. **Cannonvale Beach** (1014) is the traditional recreational beach for the area, although it has been overtaken more recently by Airlie Beach. The main Shute Harbour Road runs past the eastern end of the beach, with Beach Road backing most of the 400 m long, north facing beach. A large, traditional, wooden tidal pool is located in the centre of the beach, with a small, grassy park and an amenities block between the road and the beach. The beach consists of a 50 m wide, moderately steep high tide beach and 200 m wide low tide sand flats, fronted by an equally wide rock flat exposed at the lowest tides. A small creek drains out at the eastern end of the beach, beyond which mangroves fringe the shore.

Shingley Beach (1015) lies immediately east of Cannonvale. It can be accessed via the turn-off to Able Point Marina, with a road running the length of the 300 m long, north facing beach. A boat ramp is located at the western end and parking is available immediately behind the beach. As the name suggests, the beach is composed of coarse sand, cobbles (shingles) and occasional rock, and is fronted by a low tide rock flat. The large Able Point Marina lies immediately east of the beach.

Swimming: The tidal pool at Cannonvale is best at high tide, as it is bone dry at low tide.

Surfing: None.

Fishing: Best off the small jetty next to the Cannonvale pool at high tide.

Summary: Two very accessible beaches, with only Cannonvale used substantially for swimming, while Shingley is more a site for boat launching and fishing.

1016, 1017 AIRLIE BEACH

AIRLIE BEACH				
No. Beach	Rating		Type	Length
	HT	LT		
1016 Airlie (1)	1	1	B+SF	150 m
1017 Airlie (2)	1	1	B+SF	250 m

Airlie Beach fronts the main street and shops of the town of the same name, now the centre of the thriving Whitsunday tourist industry. Consequently this is one of Queensland's better known and more popular beaches. The beach that sits in Airlie Bay lies just off the Proserpine-Shute Harbour Road. It is backed by parking areas, a walkway, the Intercity bus terminal, the shops, a hotel and a park.

The beach faces almost due north and is split in two by the small Airlie Creek. The western side of the beach (**1016**) is just 150 m in length, bordered by two car parks and linked by a footbridge across the creek to the eastern half. The eastern side (**1017**) is 250 m in length and is backed by a shady park and a walkway that ends at a small boat harbour (Fig. 4.48). Both beaches are composed of medium to coarse sand and shells. The bay is usually calm, only receiving low waves during northerly winds. A few mangroves fringe the eastern ends of both beaches. At high tide the beach has a moderate slope and water deep enough to swim in. However at low tide, a rocky tidal flat is exposed and swimming is not possible (Fig. 1.1b).

Swimming: A very popular spot to swim at high tide, however you can only sunbake at low tide. Watch the rocks underfoot on the tidal flats.

Surfing: None.

Fishing: Best off the headlands and breakwater at high tide.

Summary: A very accessible and popular tourist destination, better for sunbathing than swimming. If you do want to get wet, make sure it's near high tide.

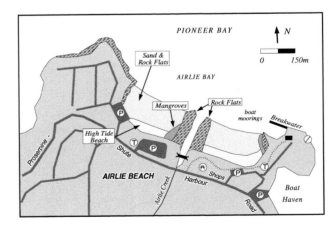

Figure 4.48 *Airlie Beach is backed by the town of the same name and bordered by prominent points. While the beach is sandy at high tide, at low tide it is fronted by wide sand and rock flats.*

1018 FUNNEL BAY

No. Beach	Unpatrolled		
	Rating HT LT	Type	Length
1018 Funnel Bay	1 1	B+LT-SF	400 m

Funnel Bay occupies the northern half of the valley that is used to access Shute Harbour. The main Shute Harbour Road runs through the valley, with a 1.5 km long access road leading to Funnel Bay and its beach. The valley is also the site of a rather confined landing strip surrounded by valley sides rising to over 300 m. Funnel Bay beach (1018) is 400 m long, faces north and is backed by a low coastal plain through which Flame Tree Creek meanders, to reach the shore at the eastern end of the beach. Mangroves fringe the creek and its mouth. A gravel road runs right to the back of the beach, where there is a picnic area on the low dune. The beach consists of a low gradient, 50 m wide high tide beach fronted by extensive intertidal sand flats up to 500 m wide. Waves are usually calm.

Swimming: Only at mid to high tide, owing to the tidal flats.

Surfing: None.

Fishing: Best in the creek and creek mouth when the tide is in.

Summary: An accessible but mainly bypassed beach, used more by the locals than the thousands of passing tourists.

1019-1014 PIONEER PT-SWAMP BAY

No. Beach	Unpatrolled		
	Rating HT LT	Type	Length
1019 Pioneer Pt	1 2	R+boulder flats	1 km
1020 Notch Hill (N)	1 2	R+boulder flats	300 m
1021 Notch Hill (S)	1 2	Cobble becah	300 m
1022 Green Pt (N)	1 2	Cobble beach	200 m
1023 Green Pt (S)	1 2	Cobble beach	250 m
1024 Swamp Bay	1 1	Rock flats	300 m

Pioneer Point forms the eastern boundary of Pioneer Bay, the site of Cannonvale and Airlie Beach. The point is part of an 8 km long, 200 to 300 m high, densely vegetated spur that runs from Shute Harbour to the point. The 12 km shoreline between the point and harbour is dominated by the rock base of the spur, with the few beaches restricted to small valley mouths and mostly composed of rocks, cobbles and boulders. There is no road access to any of the beaches, with a walking trail from the Shute Harbour Road leading across a low point in the spur to Swamp Bay providing the only formal land access.

Pioneer Point beach (**1019**) is a 1 km long, north facing strip of coarse sand, cobbles and boulders running along the northern side of the spur. It is backed by densely wooded slopes, including boab trees, and is cut in two by a small rock outcrop.

Notch Hill is the 350 m high northern crest of the spur, that has two 300 m long, north-east facing beaches on its eastern side, both backed by steep, vegetated slopes. The first (**1020**) has a narrow strip of high tide, white coral rubble, fronted by a cobble to boulder intertidal beach, with large rocks increasing toward the northern end of the beach. The second (**1021**) is similar, but with rocks increasing toward the southern point and a small creek emerging in the centre of the beach.

Green Point forms the eastern extremity of the spur and is backed by 370 m high Mount Mekarra. The steep, forested slopes of the mount descend to the rocky shore, with beaches either side of the point. The northern beach (**1022**) is 200 m long, faces east and consists of a coarse cobble to boulder beach, bordered by increasing rocks toward the point. The southern (**1023**) is 250 m long and faces east but is more irregular, owing to two small creeks depositing cobble and boulders on the beach.

Swamp Bay (1024) is, as the name suggests, a low energy embayment at the junction of the Pioneer Point spur and another point called The Beach. The north facing bay has a narrow high tide beach composed of coral rubble and cobbles, fronted by a 500 m wide rock flat that is studded with mangroves. A 3 km long walking track from the Shute Harbour Road leads to the beach, where there is a water tank and bush camping is allowed.

Swimming: None of these beaches are recommended for swimming owing to the dominance of rocks, especially at low tide.

Surfing: None.

Fishing: Excellent rock fishing from all beaches and their adjoining points, especially toward high tide.

Summary: Six rarely visited beaches, owing to the dominance of rocks and generally unsuitable landings.

1025-1028   CORAL to CANECUTTERS beaches

No.	Beach	Unpatrolled Rating HT	LT	Type	Length
1025	Coral Beach	1	2	Cobble-boulder	400 m
1026	The Beak (1)	1	2	Cobble-boulder	60 m
1027	The Beak (2)	1	2	Cobble-boulder	250 m
1028	Canecutters	1	1	B+SF	250 m

The Beak is a 1.5 km long, 70 m high, densely wooded headland. Its predominantly rocky shore houses three beaches along its more exposed northern shore, as well as a more sheltered southern beach. There is no vehicle or formal land access to any of the beaches.

The three northern beaches are all composed of steep, white coral rubble at high tide, grading into darker cobble and boulders toward low tide. **Coral Beach** (1025) is 400 m long, with rock points to either end and rock flats at low tide. There is a fishing shack located at its eastern end. The beach can be accessed via a 2 km long walking track from the Shute Harbour residential area. Beach **1026** is only 60 m long and is wedged in between two rocky points. Beach **1027** is 250 m long, has a very low sandy crest and is fringed by a small coral reef.

Canecutters Beach (1028) is the traditional site where the canecutters and now farmers spend their holidays, setting up camps in the casuarina trees that back the 250 m long beach. Otherwise it is usually empty and only accessible by boat. The south-east facing beach lies only 1 km from busy Shute Harbour, but is largely bypassed by tourists. It is one of the few sandy mainland beaches and has a narrow, sandy high tide beach, a 50 m wide, sandy intertidal flat, then 500 m wide sand and rock flats that are exposed at low tide.

Swimming: Best at high tide on all four beaches, with Canecutters by far the most popular beach, especially during local holiday periods.

Surfing: None.

Fishing: Best along the northern three beaches and adjoining rocky points.

Summary: Three boulder beaches and one sandy beach just across the bay from Shute Harbour, but only accessible by water.

Cumberland Islands
Whitsunday, Lindeman, Sir James Smith & Cumberland Groups Selected Islands and Beaches

The Cumberland Islands are a group of high islands lying from 1 to 40 km off the Whitsunday and Repulse Bay coast, extending in a south-south-easterly direction for 100 km from Hayman Island in the north to Brampton Island in the south. The islands include three groups - the Whitsunday Group, Lindeman Group and Sir James Smith group. There are several large islands including Whitsunday (10 930 ha) and Hook (5180 ha), together with more than 75 smaller islands. All are high continental islands and formed when sea level rose, flooding this part of the coast, about 6 000 years ago. Hook and Whitsunday have peaks rising to 460 m and 435 m respectively. The islands were sighted by Cook when he sailed though the group in June 1770, naming Whitsunday Passage and the Cumberland Isles.

Most of the islands are part of the Whitsunday National Park and are surrounded by part of the Great Barrier Reef Marine Park. The relatively natural and densely vegetated islands also contain numerous sand and coral beaches, and extensive areas of fringing coral reefs. Resorts have been developed on Hayman, Hook, South Molle, Daydream, Long, Hamilton, Lindeman and Brampton Islands, while camping is permitted on several others. The islands are serviced by ferries from Able Marina at Airlie Beach and Shute Harbour, together with airports on Hamilton, Lindeman and Brampton Islands.

For further information:

Rangers Office	Marine Parks
Whitsundays National Park	Mackay District Office
Shute Harbour Road	Cnr Wood & River Sts
PO Box 332	PO Box 623
Airlie Beach, Qld 4802	Mackay, Qld 4740
phone: (07) 49 467 022	phone: (07) 47 518 788

Selected beaches on the following islands (listed north to south) are described in this section. See Fig. 4.49 for location of islands adjacent to Shute Harbour.

	National Park	Marine Park
Hayman Island		x
Daydream Island	x	x
South Molle Island	x	x
Hook Island	x	x
Whitsunday Island	x	x
Hamilton Island		x
Long Island	x	x
Lindeman Island	x	x
Brampton Island	x	x

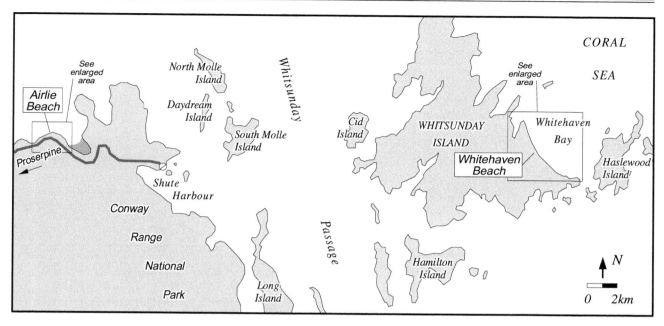

Figure 4.49   *A map showing some of the main islands in the Whitsunday Group.*

HAYMAN ISLAND beaches

No.	Beach	Rating HT LT		Type	Length
HAY1	Rescue Point	1	2	B+reef flat	400 m
HAY2	Barbecue Bay	1	1	B+SF	950 m
HAY3	Blue Pearl Bay (S)	1	2	B+reef flat	250 m
HAY4	Blue Pearl Bay (N)	1	2	B+reef flat	320 m

Hayman Island is located 28 km north-east of Shute Harbour and is the northernmost of the Whitsunday Island group. The island is surrounded by marine park and, while not part of the national park, the island remains in a relatively natural state, with the Hayman Island Resort located behind one of the four island beaches. The island has an area of 300 ha, with 8.5 km of shoreline. The centre of the island rises 250 m to Mount Carousel and its slopes are surrounded by predominantly rocky shore, with the four beaches occupying only 2 km of the shoreline.

Rescue Point beach (HAY1) is the northernmost beach on the island and is located on the north-eastern shore, between the northern Tower Point and Rescue Point, 1.5 km north-east of the resort. It is a 400 m long, north-east facing beach that curves gently between the two points. It consists of a moderately steep, sandy high tide beach, fronted by coral reef flats extending up to 100 m off the beach at low tide. The beach is bordered by jagged rocks on the points and backed by moderately steep,

sparsely wooded slopes. The only access is on foot from the resort or by boat at high tide.

Barbecue Bay beach (HAY2) is the island's main and longest beach. It is 950 m in length and faces south as it curves around in two parts between Groper Point and the western point, which is site of the boat harbour and marina. The resort is located in the small, flat valley immediately behind the beach, with part of the resort, including its famous swimming pool, literally hanging over the beach. A second smaller valley backs the eastern end of the beach and the island's only road runs from this valley to the resort and on to the marina. The beach consists of a 30 m wide, sandy high tide beach, fronted by sandy tidal flats that extend up to 1.5 km south of the beach. Waves are usually low, owing to the protection afforded by neighbouring Hook Island.

Blue Pearl Bay is a 1 km long, west facing bay on the north-west side of the island. It is a popular anchorage owing to its protection from the Trades. At the foot of its steep, vegetated slopes are two small sandy beaches, both fronted by coral reefs. The southern beach (**HAY3**) is 250 m long, faces north to north-west and consists of a narrow, sandy high tide beach, then 200 m wide reef flats. The northern beach (**HAY4**) faces more westerly and is similar in nature, with reef flats extending about 100 m off the beach. While it is possible to walk to these beaches, they are easiest to access by boat at high tide.

Swimming: Barbecue beach is by far the most popular and safest beach on the island, however it is only useable toward high tide. The other three beaches are all relatively safe but only suitable for swimming at high tide, while at low tide it is possible to wade out over the reef flats.

Surfing: None.

Fishing: Best off the marina breakwaters and the rocky shores adjoining the beaches.

Summary: An essentially natural island, containing four beautiful tropical beaches and an international resort.

DAYDREAM ISLAND beaches

No.	Beach	Rating HT	LT	Type	Length
DAY1	North	1	2	B+rock flat	380 m
DAY2	Beach Club	1	2	B+SF	400 m
DAY3	Sunlovers	1	2	B+rock flat	50 m

Daydream Island is one of the closest islands to the mainland, lying just 5 km north-east of Shute Harbour. It is a 1 km long, 100 to 200 m wide island, reaching a height of 50 m. The Daydream Island Resort dominates the island, with accommodation toward the northern end and the Beach Club on the southern point, with its three beaches and all facilities linked by tree-lined walking tracks. The island is protected from the Trades by the larger South Molle Island 2 km to the south-east (Daydream is officially named West Molle Island). As a result, waves are usually low to calm. The island has 2.5 km of shoreline, containing three beaches; two at each end and a small pocket beach on the north-west side.

The northern beach (**DAY1**) occupies the north-eastern tip of the island and is essentially a small, cuspate feature that faces both north and east. A rock groyne has been built at the cusp, cutting the beach into northern and eastern sections, together with a small rock reef capped by three large boulders just off the beach. Both sections are approximately 200 m long and consist of a steep, coarse coral sand and rubble high tide beach, fronted by a rocky low tide flat. A grassy park backs the northern section, with resort apartments behind the eastern section.

Beach Club beach (DAY2) lies at the southern end of the small island and extends from the southern tip for approximately 400 m up the western side. The beach is backed by a seawall, volleyball court and other facilities and is crossed by a helipad at its southern end, a concrete boat ramp in the centre and a small jetty that services the tour boats toward the northern end. The beach is composed of coarse sand and coral rubble and is moderately steep, with rocky flats exposed at low tide.

Sunlovers beach (DAY3) is the main swimming beach for the island, as it faces west and lies protected from the Trades. It is a 50 m pocket of steep coarse sand and rubble, over which a layer of finer sand is maintained by

the resort. Numerous large rocks and boulders dot the beach, however it is usually calm and good for both swimming and snorkelling.

Swimming: Swimming is relatively safe at all three beaches and best at high tide; use the pool at low tide.

Surfing: None.

Fishing: Best off the jetty and northern groyne.

Summary: A small, compact island offering a choice of three beaches and pools, as well as the usual resort facilities.

SOUTH MOLLE ISLAND beach

No.	Beach	Rating HT	LT	Type	Length
SM1	South Molle	1	1	B+reef flats	800 m

South Molle Island was originally cleared for grazing and today houses one of the first resorts built in the Whitsunday Islands. The beachfront resort occupies the flat land behind the main and largest beach on the 400 ha island. The beach and resort face north and are bordered by Lamond Hill and Mount Horn, that rise to over 150 m, and are backed by 200 m high Mount Jeffreys. The island has 12 km of shoreline containing nine beaches of varying size and orientation, together with rocky coast and one mangrove-filled bay. Some of these are linked by graded walking tracks.

The **Resort Beach** (SM1) is 800 m long and faces north, with protection at either end from Lamond Hill and Mount Horn, which protrude 500 m and 100 m respectively to the north. The sheltered beach consists of a moderately steep, 10 m wide high tide beach, that widens to 50 m at low tide, when it is fronted by a 200 m wide rock and reef flat. The island jetty crosses the middle of the beach and extends 300 m into the bay. The beach is backed by a seawall, with a smaller rock groyne west of the jetty. The main swimming beach is to the west of the jetty, as the beach narrows considerably to the east and rocks increase in prominence.

Swimming: Best at high tide west of the jetty, in front of the beachfront units.

Surfing: None.

Fishing: The long jetty is very popular, together with the groyne at high tide.

Summary: A sundrenched, well-protected beach offering excellent swimming and snorkelling at high tide.

HOOK ISLAND beaches

No.	Beach	Rating HT	LT	Type	Length
H1	Resort (E)	1	1	B+rock flats	150 m
H2	Resort (W)	1	1	B+rock flats	200 m

Hook Island is the second largest island in the Cumberland Group and is located immediately north of Whitsunday Island and 22 km north-east of Shute Harbour. The entire island is a national park and its densely vegetated peaks are the highest in the group, reaching 460 m at Hook Peak. Camping is permitted at three locations on the island including the sole island resort in the south-east of the island. The resort faces across a 500 m wide channel to the northern tip of Whitsunday Island. The resort has two beaches separated by a round hill, on the southern side of which is the underwater observatory.

The eastern resort beach (**H1**) is 150 m long and faces south-east toward Whitsunday Island. It is relatively steep and narrow at high tide, with sand and reef flats exposed at low tide. The resort straddles the 200 m of flatter land that links the two beaches (Fig. 4.50). The western beach (**H2**) is 200 m long, faces south-west and, being more protected, is fronted by more extensive sand and reef flats extending up to 500 m off the beach.

Swimming: Best toward high tide on both beaches.

Surfing: None.

Fishing: Best off the rocks to either side of both beaches and off the observatory point.

Summary: A more laid-back resort catering to campers as well as providing accommodation, with two pocket beaches, a headland and reefs surrounding the resort.

Figure 4.50 *Hook Island's two beaches lie either side of a small tombolo which links the small wooden hill to the island, with the resort located between the beaches.*

Whitsunday Island

WHITEHAVEN BEACH

No.	Beach	Rating HT	LT	Type	Length
WHIT1	Whitehaven	1	2	R+LT	5.8 km

Whitsunday Island is the largest in the Cumberland Group with an area of 10 930 ha. It is part of the national park and uninhabited. The island is covered in dense vegetation, much of it above 200 m, with peaks reaching 390 m. The island shore is predominantly rocky and indented with several well-protected bays and inlets. Its eastern shore is more exposed and the south-eastern side houses Whitehaven Beach, one of Australia's most famous and popular beaches. The beach (**WHIT1**) is 6 km long and faces the north-east, as it curves gently from Hill Inlet at the north end to a 100 m high, rocky point at the south end. The southern headland affords some shelter from the south-easterlies and it is here that the tourists are daily discharged onto the sand. There are usually a few yachts at anchor off the end of the beach (Fig. 4.51).

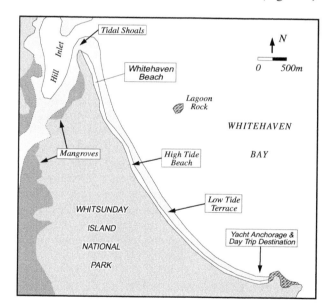

Figure 4.51 *Popular and picturesque Whitehaven Beach is located between Hill Inlet and the southern headland, where most tourist boats let off the daytrippers.*

Whitehaven Beach (WHIT1) is totally natural with no facilities - they are all back on the boats. For the most part it is backed by low, densely vegetated sand dunes, with a few sand blows toward Hill Inlet. The sand is crystal white and very uniform in size. The high tide beach averages about 50 m in width, with a moderate slope down to a narrow low tide bar, with seagrass meadows further out in deeper water. Haslewood Island protects the beach from easterly waves, so waves are usually low except during strong northerly winds.

Swimming: This is a relatively safe beach as long as you can swim. The waves are usually low and the water calm, however it is deep right off the beach at high tide and off the bar at low tide. Many tourists attempt to swim to and from the boats; make sure you are a strong swimmer if doing so. There are strong tidal currents and shifting sand shoals at the Hill Inlet end of the beach.

Surfing: Usually none, except during strong northerly winds and then only at mid to low tide.

Fishing: Best at high tide, in a boat at the inlet or off the southern rocks.

Summary: A long, pristine, beautiful white beach. A must for all visitors to the Whitsundays and well worth the visit.

HAMILTON ISLAND beach

No.	Beach	Rating HT	LT	Type	Length
HAM1	Catseye Bay	1	1	B+reef flats	800 m

Hamilton Island is irregular in shape with an area of 600 ha and 23 km of predominantly rocky shoreline, including several well-protected, mangrove-filled bays. It is one of the more recently and most developed Whitsunday islands with its own airport and marina, as well as the islands' only high-rise resort. It has become a very popular destination and provides a wide range of facilities. The main resort and the high-rise overlook Catseye Bay - site of the island's main beach. The runway has been constructed across Crab Bay, while a small bay has been expanded and transformed into the marina and surrounding facilities.

Catseye Bay beach (HAM1) is 800 m long, faces northeast and is backed by the resort and all its facilities, including a number of beach and water-based activities (Fig. 4.52). The beach is moderately steep and sandy at high to mid tide. However, low tide reveals a several hundred metre wide sand-rock and reef flat. A groyne has been built across the eastern end of the beach and there is a pontoon and small island breakwater off the beach. Waves are usually very low to calm.

Swimming: Best toward high tide when there is relatively deep water off the beach.

Surfing: None.

Fishing: The rocky shore at either end of the beach, the marina breakwaters and beach groyne are all popular locations for shore-based fishing.

Summary: A safe, north facing sandy beach (at high tide), backed by a world class resort and facilities.

Figure 4.52 *Catseye Bay on Hamilton Island is the site of the island's major accommodation and beach recreational activity.*

LONG ISLAND beaches

No.	Beach	Rating HT	LT	Type	Length
L1	Palm Bay	1	1	B+S-rock flats	200 m
L2	Happy Bay	1	1	B+S-rock flats	300 m

Long Island is a 9 km long, 200 m to 2 km wide island that parallels the mainland south of Shute Harbour, with a deep channel called The Narrows separating it from the coast by just 500 m. The entire island is a national park with three areas of settlement, all on the protected western shore of the island, in Paradise, Palm and Happy Bays. Small resorts are located in Palm and Happy Bays. The bays are connected by graded walking tracks maintained by national parks.

Palm Bay (L1) is a small, semi-circular, west facing bay housing a 200 m long high tide beach. It is located at the narrowest point in the island and the beach essentially ties the two parts of the island together (Fig. 4.53). On the flat land behind the beach is a small resort that runs through the palm trees to the eastern side of the island. The beach is backed by palm trees and consists of a narrow strip of high tide sand on a moderately steep beach, that grades into rock and rubble toward low tide. A boat anchorage has been dredged out of the sand and rock flats, and boats can anchor at the base of the beach with access to The Narrows at all tides.

Figure 4.53 *Happy Valley resort is located on the western, protected side of Long Island, with good access also to the eastern side of the long, narrow island.*

Happy Bay (L2) lies 1 km north of Palm Bay and is a larger version, occupying a 1 km long, west facing, semi-circular bay, with a 300 m long sandy beach in the southern corner, backed by a resort known as Club Crocodile. It too is in a narrow section of the island and the flat land runs for 400 m to the eastern shore and, like Palm Bay, ties these parts of the island together. Access to the resort is via a jetty out on the southern Humpy Point and an elevated 300 m long walkway. Both enable access to the resort at all tides. The beach faces north-west and is well protected from the Trades. It is moderately steep and narrow at high tide, with low tide revealing a 200 to 300 m wide sand to rock flat. Boats must anchor in deeper water beyond the flats.

Swimming: Both beaches are best toward high tide when the flats are covered.

Surfing: None.

Fishing: Best in the channel at Palm Bay and off the jetty at Happy Bay, or around the rocks at high tide.

Summary: Two smaller resorts offering well-protected beaches and a long island to explore.

LINDEMAN ISLAND beach

No.	Beach	Rating HT LT	Type	Length
LIN5	Home	1 1	B+S/reef flats	100 m

Lindeman Island is a 600 ha island located 30 km south-east of Shute Harbour in the Lindeman Group of islands. It has 15 km of shoreline including nine sandy beaches; all but one are located in the Lindeman Island National

Park. The park covers most of the island, apart from a southern section housing an airstrip and the Lindeman Island Resort. While the island reaches a height of 212 m at Mount Oldfield, it is generally undulating. A number of walking tracks link the resort with the beaches.

Home Beach (LIN5) is the main beach and lies at the foot of the resort. It is a 100 m long pocket of sand, bordered by rock headlands, together with two terminal groynes, and the island jetty and boat ramp at the western end. In addition a seawall, pool and the resort back much of the beach (Fig. 4.54). The beach is relatively narrow and steep at high tide, widening to 50 m at low tide when it grades into a sand then reef flat up to 300 m wide.

Figure 4.54 *Lindeman Island resort is located on slopes overlooking Home beach.*

Swimming: The island is sheltered from the Trades by larger Shaw Island, located 3 km to the south-east across Kennedy Sound; hence waves are usually low to calm. Best swimming is toward high tide when the flats are covered.

Surfing: None.

Fishing: The rocks and jetty are the most popular locations.

Summary: A largely national park island including several protected beaches, with Home beach sitting at the foot of the resort.

BRAMPTON ISLAND beaches

No.	Beach	Rating HT LT	Type	Length
B1	Sandy Pt	1 2	B+SF	100 m
B2	Sandy Pt (W)	1 2	B+SF	600 m

Brampton Island is one of the original resort islands, with a lodge catering for tourists established in 1933. At that time the island was a station running sheep, cattle and goats. It is now a national park and home to the international standard Brampton Island Resort, located on its north-western tip. A runway provides air service to the island and resort. The island is one of the more southern in the Cumberland Group and is located closer to Mackay (37 km to the south-west) than Shute Harbour.

The resort beach is called **Sandy Point** and consists of two parts - a 100 m long, north facing beach (**B1**) and a 600 m long, north-west facing beach (**B2**). Both beaches consist of strips of high tide sand fronted by 100 to 200 m wide sand flats. The western beach is the main swimming beach as it is more sheltered from the Trades and has a tidal pool on the sand flats, as well as a small jetty for boats. The north beach faces toward adjoining Carlisle Island. Between the two islands is an inlet that narrows to 350 m off the point and through which strong tidal currents flow. At the eastern end of the beach is a seawall for the end of the runway. The resort lies immediately behind the beach.

Swimming: The western beach is the main swimming beach and includes the tidal pool. Be careful on the northern beach as there is deep water and strong tidal currents off the tidal flat.

Surfing: None.

Fishing: Best in a boat in the inlet and over the many reefs. The best shore fishing is at high tide from the jetty and into the inlet.

Summary: A relatively remote resort island offering usually calm beaches and a world class resort.

Conway Range National Park

Area:	23 800 ha
Shoreline length:	3710-3812 km (102 km)
Beaches:	1019-1046 (28 beaches)

The Conway Range extends in a south-south-east direction for 36 km from Pioneer Point near Airlie Beach to Cape Conway, which forms the northern boundary of Repulse Bay. Both the cape and bay were named by Captain Cook in June 1770. The rugged range reaches heights of 430 m at Mount Conway, 555 m at High Mountain and 482 m near the cape itself. The entire range is a formidable barrier consisting of steep relief, densely vegetated hills and valleys, with a predominantly steep and rocky shoreline. As a consequence, apart from timber cutting, there has been little development, apart from the road across to Shute Harbour and the accompanying development. Most of the range and the 23 800 ha national park is in a relatively pristine condition, with little access to

most of the park even today, other than by boat or on foot.

For further information:
The Ranger
Conway Range National Park
Cnr Shute Harbour Rd & Mandalay Rd
PO Box 332
Airlie Beach Qld 4802
phone: 079 467 022
         079 467 022

1029-1034 COW IS, PURITAN BAY, ROUND HEAD

No. Beach	Unpatrolled Rating HT	LT	Type	Length
1029 Cow Is (N)	2	3	cobble+LT	250 m
1030 Cow Is	1	2	boulder+TSF	1.2 km
1031 Boulder 1031	2	4	boulder	300 m
1032 Puritan Bay	2	4	boulder+ S/rock flats	1.5 km
1033 Boulder lagoon	2	4	boulder	500 m
1034 Round Head	2	4	boulder	1.2 km

The coast south of Shute Harbour is both rugged and protected; the protection is afforded by Long Island which parallels the shore for 10 km. Only south of Long Island does sufficient wave energy reach the rocky coast to develop a series of predominantly boulder beaches.

Cow Island (north) beach (1029) is a 250 m long, 30 m wide, high tide cobble beach, fronted by a low gradient sand bar (Fig. 4.55). The beach faces east into the Trades and receives waves over 0.5 m during strong Trades. These surge across the cobble at high tide and spill across the sand bar at low tide. Rocky headlands and dense vegetation border and back the beach.

Figure 4.55 *A typical cobble high tide beach fronted by a fine sand low gradient low tide beach, Cow Island.*

Cow Island is a 40 m high, 1 km long island located 300 m offshore. In its lee is a protected section of coast containing a high tide sand, cobble and boulder beach (**1030**) fronted by irregular sandy and rocky tidal flats up to 1 km wide, together with mangroves in lee of the island. The beach is 1.2 km long and faces east to north-east.

The shore south of Cow Island trends south-east and 2 km south of the island it contains a 300 m long high tide boulder beach (**1031**), fronted by a low tide rock flat. One kilometre further on is the western shore of **Puritan Bay** - a 1.5 km wide, 1 km deep, north-east facing bay. The entire 1.2 km long shoreline curves around inside the bay and consists of a steep high tide cobble beach (**1032**) fronted by a mixture of cobbles, boulders and sand flats up to 300 m wide, together with a patch of coral reef toward the eastern end. Two small creeks drain out across the beach depositing small cobble deltas on the shore.

Round Head extends 2 km east of Puritan Bay and contains two boulder beaches. The first (**1033**) is on its northern side and consists of a 500 m long, north facing boulder beach, which impounds a small, 2 ha lagoon. The second (**1034**) consists of a north-east facing boulder beach that forms the shoreline for 1.2 km north of Round Head. Both are fronted by low tide rock flats.

Swimming: Only the beach north of Cow Island (1029) is suitable for swimming, as all the others are dominated by rocks and boulders, particularly at low tide.

Surfing: Only low spilling waves on the northern Cow Island beach.

Fishing: Excellent fishing off the many kilometres of rocky shore and reef, with most fishing done from boats, as landing can be difficult and dangerous.

Summary: A natural, rocky section of coast backed by densely vegetated slopes rising rapidly to 300 to 400 m.

1035 GENESTA BAY

No. Beach	Unpatrolled			
	Rating		Type	Length
	HT	LT		
1035 Genesta Bay	1	1	UD	1.8 km

Genesta Bay lies on the southern side of Round Head. It contains the only sand beach (**1035**) on the Cape Conway section of the range. It is a 1.8 km long, east-south-east facing beach that receives most waves generated by the prevailing south-east Trade winds. The bay has trapped a substantial amount of fine sand to build the beach, low backing dunes and 400 to 500 m wide, gently sloping low

tide bar and sand flats. The bay has three parts - a small, low area behind the northern end of the beach, a central ridge rising to 400 m and backing the centre of the beach, and a creek draining out of a valley at the southern end. Two small, rocky islets also lie off the southern end of the beach and are attached by the sand flats at low tide. A few mangroves grow amongst the rocks at either end.

Swimming: At both high and low tide there is a wide, shallow surf zone, devoid of rips; however the surf zone is better at high tide when it's closer to shore.

Surfing: Usually a wide, low to moderate, spilling surf.

Fishing: Better toward high tide and off the boundary rocky shore.

Summary: An interesting, natural, flat sandy beach on a coast dominated by steep boulder beaches.

1036-1045 CAPE CONWAY beaches

No. Beach	Unpatrolled		Type	Length
	Rating			
	HT	LT		
1036 Boulder 1036	2	4	boulder	600 m
1037 Cape Conway (N)	2	4	boulder	150 m
1038 Cape Conway (W 1)	2	4	boulder	200 m
1039 Cape Conway (W 2)	2	4	boulder	300 m
1040 Cape Conway (W 3)	2	4	boulder	80 m
1041 Cape Conway (W 4)	2	4	boulder	100 m
1042 Boulder 1042	2	4	boulder+ S-rock flats	600 m
1043 Boulder 1043	2	4	boulder+ S-rock flats	400 m
1044 Boulder 1044	2	4	boulder	250 m
1045 Boulder 1045	2	4	boulder	80 m

Between Genesta Bay and The Inlet, inside Repulse Bay, is over 25 km of rugged and rocky shoreline. In several small indentations in the shore and at the mouths of a few small creeks are ten boulder beaches totalling about 2.8 km in length. Most of the beaches are exposed to Trade wind waves and all are difficult, even dangerous, to land on when the Trades are blowing.

Beach **1036** is a 600 m long boulder beach located 3.5 km north-west of Cape Conway. It faces north-east and its boulders run along the base of slopes that rise to 480 m; the highest in the range.

Beach **1037** consists of two adjoining pockets of boulders located 500 m on the northern side of Cape Conway, totalling 150 m in length. A small creek is dammed behind the longer southern section and its lagoon contains a few mangroves. The dense vegetation behind the beach has been blasted and shaped by the strong Trade winds.

Beaches **1038, 1039, 1040** and **1041** are a series of four boulder beaches located in an open, 2.5 km wide, southeast facing bay that lies immediately west of Cape Conway. They each occupy the mouth of a small valley and are all composed of steep, cobble to boulder high tide beaches, with low tide boulder flats. A few mangroves grow at the eastern end of the first beach.

Beaches **1042, 1043, 1044** and **1045** lie in the second bay west of Cape Conway; a semi-circular 2.5 km wide, south to south-west facing bay. They are all high tide boulder beaches, with beach 1042 fronted by a low tide sand and cobble flat. A creek drains out across beach 1043 and is building a small, rocky, mangrove-covered delta.

Swimming: None of these beaches are suitable for swimming owing to the dominance of rocks and their more exposed locations. Be very careful if swimming, diving or boating around the cape, as very strong tidal currents run in and out of Repulse Bay via the cape.

Surfing: None.

Fishing: Excellent fishing from boats over the many reefs in the area. Little fishing is done from the shore owing to the difficult access.

Summary: A rugged, densely vegetated cape surrounded by 25 km of equally rugged shore, together with a smattering of rocky beaches.

1046-1050   **REPULSE, CONWAY, WILSON**

No.	Beach	Unpatrolled Rating HT LT	Type	Length
1046	Repulse Beach	1 1	UD	1.3 km
1047	Conway Beach	1 1	UD	1.5 km
1048	Beach 1048	1 2	B+SF	150 m
1049	Beach 1049	1 2	B+SF	300 m
1050	Wilson Beach	1 1	B+SF	300 m

Repulse Beach (1046) is the northernmost beach in Repulse Bay; the bay was named by Captain Cook when he was essentially repulsed from entering the bay by the outgoing tide. Repulse beach would still look now much as it did when Cook passed by. It lies on the south-western boundary of the national park and there is no formal land access to the beach, which is backed by a 180 m high, densely vegetated spur that ends at Rocky Point. The beach is 1.3 km long, faces south-east and is bordered by Rocky Point to the south and a small, mangrove-fringed tidal creek and sand delta to the north. The Inlet, a large, mangrove-filled estuary, extends north-east of the beach. The beach is composed of medium to fine sand and has a

30 m wide high tide beach fronted by a 200 m wide low tide bar and sand flat, with the deeper channel of the The Inlet bayward of the bar. Waves spill across the bar, particularly toward low tide.

Conway Beach (1047) is the site of a small residential and holiday settlement and caravan park, all located on low land backing the 1.5 km long beach. Immediately behind the beach is a casuarina-lined foreshore reserve incorporating parking, a picnic area and amenities. The beach has a low gradient high tide beach flattening to 400 to 500 m wide, low sand flats.

Beaches **1048** and **1049** are located along the base of the headland that runs north from the western end of Conway beach. They both face west across the 1 km wide mouth of the Proserpine River, and are fronted by 100 to 200 m of low tide sand flats, then the deep river channel.

Wilson Beach (1050) is a small beach settlement 2 km north-west of Conway Beach. The 300 m long beach faces due south across the sand flats and channel of the Proserpine River. The beach consists of a steep high tide beach, fronted by a 200 m wide sand flat, then the river channel. There is a tidal pool located in the centre of the beach, backed by a small reserve, and a boat ramp on the rocks at the southern end.

Swimming: Wilson beach is the safest, with the tidal pool, while Conway has a shallow, low gradient and usually a low-wave surf zone. Be careful at low tide on the tidal flats in the Proserpine River mouth as there are crocodiles and strong currents that run out of the river.

Surfing: Only a low, spilling wave on Conway Beach.

Fishing: Best in the river from a boat.

Summary: Conway and Wilson beaches are out-of-the-way little settlements 20 km off the Shute Harbour Road and largely bypassed by the Whitsunday tourist trade on the other side of the range.

1051-1055   **NEW BEACH to O'CONNELL RIVER**

No.	Beach	Unpatrolled Rating HT LT	Type	Length
1051	New Beach (N 2)	1 1	B+SF	2 km
1052	New Beach (N 1)	1 2	UD	1.7 km
1053	New Beach	1 2	UD	3.6 km
1054	Letherbrook	1 2	UD	2.2 km
1055	O'Connell R	1 2	UD	1.2 km

Between the Proserpine and O'Connell River mouths is a 13 km long shoreline fronted by low beaches and backed by 2 to 3 km of mangroves, and in places up to 10 km of freshwater marsh. The entire area is part of the low, swampy delta of the Proserpine and O'Connell Rivers, with the Bruce Highway located several kilometres inland. The beaches that form the shore face east to south-east into the northern end of Repulse Bay, making them reasonably well-exposed to the Trades. Most are, however, fronted by extensive low tide bars to sand flats up to several hundred metres wide. Apart from one 4WD track out through the mangroves to New Beach, there is no vehicle access to the shore.

The northern end of New Beach (**1051**) is an eroding, 2 km long, narrow high tide beach, backed by dense mangroves and fronted by the low energy tidal flats of the Proserpine River, which in places extend over 1 km east of the beach. A tidal creek separates the beach from beach 1052.

Beach **1052** is a more substantial beach, with a foredune up to 200 m wide backing the low high tide beach. The tidal flats transform into an irregular 200 to 300 m wide bar. The 4WD track reaches the southern end of this beach.

New Beach (1053) is the most substantial of the beaches; it is 3.6 km in length and faces directly into the Trades. It receives waves over 0.5 m and has a wide, low gradient beach fronted by a low tide bar up to 400 m wide. It is backed by a wide, casuarina-covered foredune up to 7 m high. The beach ends at the 1 km wide mouth of Thompson Creek.

On the south side of Thompson Creek is the irregular **Letherbrook Beach** (1054). This 2.2 km long beach begins inside the southern side of the river mouth. then runs along the open coast to the mouth of the O'Connell River. The beach is backed by older beach ridges and dunes up to 1 km wide. It is fronted by extensive intertidal sand flats up to 2 km wide and is bordered by the deep tidal channels of Thompson Creek and the O'Connell River.

The O'Connell River mouth is 400 m wide and on its southern side has a 1.2 km long beach (**1055**), that ends at a small tidal creek which flows between the beach and the mainland. Mangroves and older beach ridges back the beach, while it is fronted by 2 km of tidal flats.

Swimming: You can only swim at these beaches toward high tide, unless you want a walk of several hundred metres to the low tide shoreline. Stay clear of the tidal creek and river channels as they have both strong tidal currents and the possibility of crocodiles.

Surfing: The best chance is on the most exposed main New Beach where the waves spill across a wide, low gradient surf zone.

Fishing: Best in the creeks and rivers.

Summary: A relatively isolated shore dominated by the wide tidal flats, mangroves and creek and river mouths, with essentially no development or land access.

1056-1060 COVERING BEACH (LAGUNA QUAY)

No. Beach	Unpatrolled			
	Rating		Type	Length
	HT	LT		
1056 Covering (1)	1	2	TSF+rock flats	500 m
1057 Covering (2) (Laguna Quay)	1	2	TSF+rock flats	500 m
1058 Covering (3)	1	2	TSF+rock flats	1 km
1059 Beach 1059	1	2	TSF+rock flats	700 m
1060 Beach 1060	1	2	TSF+rock flats	300 m

At Covering beach the low, sandy and muddy coast of the Proserpine and O'Connell Rivers gives way to a gradually rising upland coast, with the Connors Hills paralleling the coast 2 to 5 km inland and rising to over 300 m. The shoreline faces north-east between the O'Connell River and Midgeton and is dominated by rocky shores, with low energy rock and tidal flats covered by varying degrees of mangroves. In amongst the rocks and mangroves is a near-continuous, irregular strip of high tide sand that comprises the 'beaches'.

Covering Beach is the name given to a 2 km stretch of low, rocky shoreline fronted by the narrow high tide beach, sand and rock flats and mangroves. The beach is divided into three sections (**1056, 1057, 1059**) by a small, rocky outcrop and by Covering Creek. The beach has traditionally been backed by grazing land, however in the mid 1990's the **Laguna Quay** resort, including shorefront golf course and marina, was developed. Now the beach is the centre of the major development.

Past Covering beach the shoreline continues on much the same toward Midgeton. Beach **1059** is 700 m long. It is backed by wooded grazing land with 4WD tracks running behind the shore. A new caravan park has been developed on the Midgeton Road toward the southern end of the beach. A more substantial, low, rocky protrusion separates these beaches from Beach **1060**, which is a 3 km long, more easterly facing beach that becomes increasingly dominated by mangroves toward Midgeton. It ends in a 3 km wide, north facing, mangrove-filled bay located immediately north of Midgeton.

Swimming: While the waters are usually calm along these 'beaches', their low gradient and the presence of rock flats and mangroves makes them unsuitable for swimming.

Surfing: None.

Summary: A 12 km section of low energy shore, now the focus of Laguna Quay.

1061-1064 MIDGETON

No.	Beach	Unpatrolled		Type	Length
		Rating			
		HT	LT		
1061	Beach 1061	1	2	TSF+rock flats	700 m
1061	Midgeton	1	1	UD	1.8 km
1063	Dempster Ck (1)	1	1	B+SF	300 m
1064	Dempster Ck (2)	1	1	B+SF	900 m

Midgeton is the only settlement on the western shores of Repulse Bay. There is nothing north (except the Laguna Quay Resort) until Conway beach, 20 km away, and nothing south until Seaforth, 27 km away. The lack of development of the bay reflects its remoteness from any major towns, the hilly interior (which diverts the highway up to 20 km inland), the dead-end and poor quality of many of the coastal access tracks and the low energy, shallow nature of the bay shore, that has precluded port development.

Midgeton is a small holiday settlement with a store, caravan park and grassy foreshore reserve. At the northern end of the settlement is a low headland, on the eastern end of which is 700 m long beach **1061**. The beach faces east and curves around between two low headlands. It has extensive rock flats off the headlands and 1.5 km wide sand flats off the beach, which at low tide link with a small, mangrove-fringed islet 300 m off the beach. There is vehicle access to the northern end of the beach from the Midgeton Road.

Midgeton Beach (1062) is a 1.8 km long, south-east facing, low gradient, sandy beach. It is backed by a foreshore reserve fringed with coconut trees, then the houses of the settlement. The beach is fronted by a wide, low gradient intertidal beach, with sand flats extending up to 1 km off the southern end in front of Yard Creek mouth, which forms the southern boundary. A smaller, mangrove-fringed creek forms the northern boundary (Fig. 4.56).

Dempster Creek is located 2 km south of Midgeton. It has a 500 m wide creek mouth that flows out between two low energy, rocky shores. On the north side, conical, 60 m high Mount Midge overlooks the creek, while to the south is a 30 m high hill. On the north side of the creek are two headland-bound, low energy beaches. The northern Dempster Creek beach (**1063**) is 300 m long, faces south-east and consists of a sandy high tide beach, fronted by over 1 km of variable sandy tidal flats and the channel of Dempster Creek. The land behind the beach and most of Mount Midge has been cleared and there is

vehicle access to the beach.

The southern Dempster Creek beach (**1064**) lies immediately to the south and is a 900 m long, south-east facing, lower energy beach, with a high tide sandy beach including a few rock outcrops and sandy tidal flats that extend out to the main creek channel. The beach is bordered by 10 m high cleared headlands and there is a 1.5 km long vehicle track off the Midgeton Road to the back of the beach.

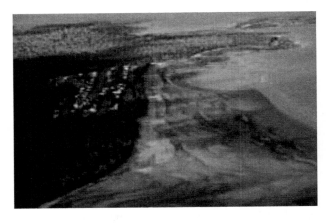

Figure 4.56   *A view north along Midgeton beach and settlement. Note the wide low gradient low tide beach.*

Swimming: Midgeton is the most popular and most suitable beach for swimming, with generally low waves and a low gradient beach and surf zone. The other beaches are either too shallow (beach 1061) or are impacted by strong tidal currents (Dempster Creek beach) and caution should be used.

Surfing: None, apart from occasional low, spilling waves at high tide at Midgeton.

Fishing: The two Midgeton and Dempster Creeks are the favourite spots.

Summary: Midgeton is a locals' settlement off the tourist track. However it does have a nice, accessible beach and basic facilities for the traveller.

1065-1067   TEN MILE BEACH

No.	Beach	Unpatrolled		Type	Length
		Rating			
		HT	LT		
1065	Beach 1065	1	1	B+SF	1 km
1066	Beach 1066	1	1	B+SF	800 m
1067	Ten Mile	1	2	UD	3 km

The 5 km of coast between Dempster and Hervey Creeks consists of a moderately undulating peninsula that has been cleared for grazing, with a farm access track to Ten Mile beach. Around the northern and eastern sides of the peninsula are three sandy beaches, with mangroves fringing the two creeks that form the western and southern boundaries. Apart from the access track there is no development and no facilities.

Beach **1065** is a 1 km long, roughly north facing beach that curves around irregularly between two low points, with rocks outcropping along the beach. It is backed by land that rises to 30 m, and is fronted by a few clumps of mangroves and the tidal flats and channel of Dempster Creek.

Beach **1066** is an 800 m long, north-east facing beach, bordered by a low rocky point in the north and a small, beach-bound tidal creek in the south. There are mangroves off the small creek mouth and a low rocky point beyond it.

On the south side of the point is 3 km long, east facing **Ten Mile** beach (1067), one of the more energetic beaches in the area, with a high tide sandy beach and 100 m wide bar in the north that grades into wider sand flats along the more protected southern end of the beach. Hervey Creek maintains a 100 m wide inlet and tidal channel at the southern end. A vehicle track from the Bloomsbury Road winds for 10 km out to the beach.

Swimming: Ten Mile beach is the most suitable for swimming, particularly toward the northern, more energetic end, although waves are still usually less than 0.5 m. Beach 1065 and 1066 are both fronted by wide, shallow tidal flats, with the strong tidal currents of Dempster Creek further out.

Surfing: Only an occasional spilling wave toward the northern end of Ten Mile beach.

Fishing: Best in Dempster and Hervey Creeks.

Summary: Three beaches fringing a grazing property and essentially inaccessible to the public.

1068-1073 ROSELLA CK

No. Beach	Unpatrolled			
	Rating HT LT		Type	Length
1068 Beach 1068	1	1	B+SF	1.2 km
1069 Beach 1069	1	1	B+SF	400 m
1070 Beach 1070	1	1	B+SF	250 m
1071 Beach 1071	1	1	B+SF	600 m
1072 Beach 1072	1	1	B+SF	300 m
1073 Rosella Ck (W)	1	1	B+SF	1.3 km
1074 Rosella Ck (E)	1	1	B+SF	700 m

To the east of Hervey Creek is 6 km of coast that trends east-south-east, has numerous rocky outcrops, a few small creeks and seven generally curving to irregular low energy beaches, fronted by extensive sandy tidal flats averaging 300 to 500 m. The beaches are backed by gently rising slopes that have been cleared for grazing. There are farm tracks to the coast and beaches, but otherwise no public access and no development or facilities.

Beach **1068** begins on the southern side of Hervey Creek as a low energy, mangrove-fringed sandy shore and extends out into the bay for several hundred metres. It has a sandy high tide beach, with the tidal shoals and channel of Hervey Creek extending up to 1.5 km off the beach. It ends at a small, mangrove-fringed creek, beyond which is beach **1069**, a 400 m pocket of curving low energy beach, the eastern half of which is dominated by sandy, mangrove-covered tidal flats.

Beach **1070** occupies the north side of a low rocky promontory and has a substantial amount of rocky, as well as sandy, tidal flats off the 250 m long beach. It ends at a small, mangrove-fringed tidal creek. On the east side of the creek is beach **1071**, a 600 m long, curving, north-east facing beach, with a mixture of sand and rock flats in the intertidal zone. Beach **1072** is even more dominated by intertidal rock flats. It is 300 m long, occupies a low promontory and faces almost due north.

Rosella Creek (west) is a 1.3 km long, north-east facing beach (**1073**) that extends from a low rocky point to a double creek mouth. Sandy tidal flats off the beach widen from 300 m at the point to over 1 km at the creek mouth. Rosella Creek (east) (**1074**) extends in an irregular fashion from the eastern side of the creek mouth around the northern side of a low rocky point. It is fronted by up to 500 m of sandy flats together with rocky outcrops close to shore.

Swimming: These are seven low energy beaches; usually calm or with low wind waves. The best swimming is toward high tide, while care should be taken near the tidal creeks and their deeper channels.

Surfing: None.

Fishing: Best in the tidal creeks and over the rock flats at high tide.

Summary: As these beaches front private grazing land they are essentially inaccessible to the public, except by boat.

1075-1079   FLAGGY ROCK, MENTMORE, DEWARS

		Unpatrolled		
No.	Beach	Rating HT LT	Type	Length
1075	Flaggy Rock (N)	1 2	UD	1.3 km
1076	Flaggy Rock Ck	1 2	UD	1.75 km
1077	Mentmore (N)	1 1	B+SF	450 m
1078	Mentmore	1 1	B+SF	1.3 km
1079	Dewars	1 2	UD	3 km

Flaggy Rock Creek is a small, mangrove-fringed tidal creek that enters the coast at the southern end of a 3 km long, north-south trending section of coast fronted by two beaches, and backed by gently rising, cleared grazing land. The only vehicle access is via the farm tracks.

The northern beach (**1075**) is 1.3 km long and faces east. It extends from a low, cleared headland to a small, cuspate foreland in lee of an offshore rock reef. It is backed by a near-continuous, linear, land-locked lagoon. On the southern side of the foreland is beach **1076**, which runs another 1.7 km past a small creek mouth to the larger, mangrove-fringed mouth of Flaggy Creek. At low tide the creek drains across the wide intertidal beach. Both beaches receive waves averaging about 0.5 m, with height decreasing in lee of the reef and toward the creek mouth. The low waves and fine sand maintain a wide, low gradient beach, with a 100 m wide bar along the northern beach, while south of the foreland the bar widens to become a 500 m wide, low, ridged sand flat. At the creek mouth the coast turns east and is fringed by mangroves all the way to the northern Mentmore beach headland.

The Mentmore beaches are two headland-bound beaches, 450 m and 1.3 km long, respectively. They both face north-east and are bounded by low, cleared, rocky headlands, with mangroves fringing the southernmost headland. They are backed by cleared grazing land. The northern beach (**1077**) is wedged in between rocky reefs extending off both headlands, together with a small tidal creek exiting across the southern corner of the beach. Two hundred and fifty metre wide, ridged sand flats front the low gradient beach, while a row of casuarinas line the low foredune behind the beach.

Mentmore Beach (1078) is similar but longer, with a slightly larger tidal creek draining across the southern end of the beach, and sand flats that widen from a 100 m wide bar in the north to 600 m wide, ridged sand flats in the south. Waves average less than 0.5 m on both beaches. The only land access to the beaches is via farm tracks, which lead to a solitary shack behind the middle of the beach.

Dewars Beach (1079) is located at the eastern tip of

4 km long, 120 m high Stewart Peninsula. The 3 km long beach runs along the eastern tip of the peninsula. It faces north-east and is sheltered toward its southern end by the Brothers Islands; a series of six small, low islands extending 4 km offshore, with the inner three connected to the mainland by tidal flats. The beach consists of a shelly, moderate gradient high tide beach fronted by a 100 m wide bar that widens in lee of three islands to a 300 m wide, ridged sand flat (Fig. 4.57). Strong tidal currents maintain a channel between the inner island and the beach. A low, rocky headland, surrounded by boulders and a small, perched creek mouth, forms the northern boundary, while rock flats define the southern limit. The backing land is cleared for grazing, and the only access track to the beach must negotiate 100 m of tidal flats behind Mentmore beach to cross from the mainland to the peninsula.

Figure 4.57   *A view south along the wide low gradient Dewars beach, fringed in the north with basalt boulders.*

Swimming: These five beaches all offer wide, low gradient, sandy beaches, usually with waves less than 0.5 m and no rips. Best swimming is toward high tide and toward the northern ends, and clear of the tidal creeks.

Surfing: Only small spilling waves toward the northern ends when the Trades are blowing.

Fishing: Best in the tidal creeks and over the rock reefs at high tide.

Summary: Five low energy, sandy beaches, all backed by grazing land with restricted land access.

1080   ST HELENS BEACH

		Unpatrolled		
No.	Beach	Rating HT LT	Type	Length
1080	St Helens	1 1	B+SF	2.3 km

St Helens Beach houses one of the few settlements along this section of coast. The 2.3 km long, north-east facing beach lies 12 km east of the highway and is backed by two groups of about 100 houses. A low, rocky, mangrove-fringed point forms the northern boundary of the beach (**1080**), while a tidal inlet and 65 m high, conical Skull Knob border the southern end. The beach lies in lee of Rabbit Island, located 5 km to the east, and is well protected from most waves. As a result the beach consists of a narrow, sandy high tide beach, fronted by sandy tidal flats up to 2 km wide, that also contain patches of mangroves and rock flats. In addition, two small rock groynes have been built across the high tide beach.

Swimming: Only at high tide.

Surfing: None.

Fishing: Best in the creek and amongst the rocks flats and mangroves at high tide.

Summary: A quiet local holiday settlement, rarely visited by travellers.

1081-1086 RABBIT & NEWRY ISLAND BEACHES

No. Beach	Unpatrolled			
	Rating HT	LT	Type	Length
1081 Rabbit Is (1)	1	2	R+LT	700 m
1082 Rabbit Is (2)	1	2	R+LT	300 m
1083 Rabbit Is (3)	1	2	R+LT	650 m
1084 Newry Is (1)	1	1	B+SF	250 m
1085 Newry Is (2)	1	1	B+SF	50 m
1086 Newry Is (3)	1	1	B+SF	250 m

Newry Island Group National Park

Area: 485 ka (Rabbt is 348 ha)
Rabbit and Newry Islands are part of Newry Island Group National Park, which also includes nearby Outer Newry, Acacia, Mausoleum and Rocky Islands. Access is by boat from the Port Newry launching ramp located about 4 km south of the islands and accessed via Seaforth. Camping sites are maintained on the three Rabbit Island beaches, and Outer Newry and Rocky Islands. In addition, there is a small resort on Newry Island.

For further information contact:
Mackay District Office
Queensland National Parks & Wildlife Service
Cnr Wood & River Streets
Mackay Qld 4740
Phone: (07) 49 51 8788

The Ranger
Cape Hillsborough National Park
MS 895
Seaforth Qld 4741
Phone: (07) 49 59 0410

Three kilometre long **Rabbit Island** consists of northern and southern rock sections, with maximum elevations of 75 m and 96 m respectively, joined by the southern beach and its low foredune, with a 1 km² mangrove swamp behind the beach. The three beaches (**1081, 1082, 1083**) are similar in character, all facing east-north-east, and are 700 m, 300 m and 650 m in length respectively. Each beach is bounded by densely vegetated, 20 m high headlands that are fringed by rock flats. The beaches are backed by low, narrow, casuarina-covered foredunes and receive waves averaging 0.5 m, which maintain a moderately steep high tide beach fronted by a 100 m wide low tide bar. The southern beach also has intertidal rock flats along its southern half. Apart from four primitive camp sites there are no facilities on the island.

Newry Island lies 200 m off the south-eastern tip of Rabbit Island. It is a smaller island, just under 1 km long, with a maximum elevation of 75 m. The island is protected from the Trades by Outer Newry Island and conditions are usually calm around the island, apart from strong tidal currents. The island has three small beaches. The northern beach (**1084**) is 250 m long and is bounded by densely vegetated, 40 m high headlands. It faces northeast and consists of a 50 m wide high tide beach fronted by 100 m wide sand and rock flats. Beach **1085** is a 50 m pocket of sand immediately east of the northern beach, and is similar in character.

Beach **1086** is the main island beach and site of the **Newry Island Resort**. It is 250 m long, faces east toward Outer Newry Island and consists of a 30 m wide high tide sand beach. The beach is bordered by rocky points and fronted by a 100 m wide sand flat, together with mangroves on the southern rock flats. The resort is located in amongst the trees, runs the length of the beach and is backed by a 95 m high, densely vegetated hill. A walking track circumnavigates the hill, leading to a picnic area on the south-west tip of the island.

Swimming: Because of the low tide rock and sand flats, swimming is best at high tide.

Surfing: None.

Fishing: Best off the rocks at high tide or from a boat in the tidal channels.

Summary: Two natural, tropical islands and part of the six island national park, with essentially no development apart from the small resort on Newry Island.

SEAFORTH-MACKAY-CAPE PALMERSTON (Fig. 4.58)

Length of Coast:	217 km
Beaches:	90 (beaches 1087-1177)
Surf Life Saving Clubs:	Mackay, Sarina
Lifeguards:	Lamberts, Mackay Harbour
Major towns:	Mackay, Sarina

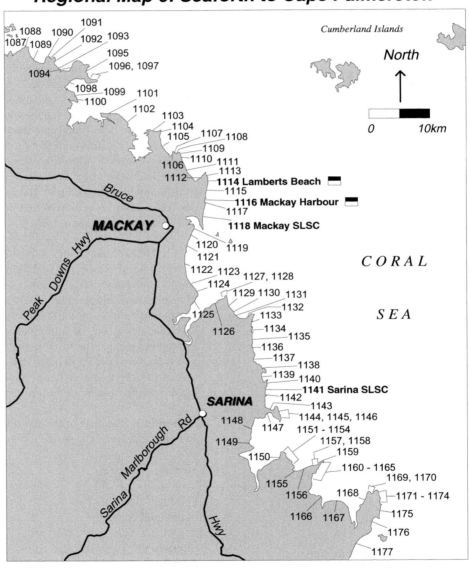

Figure 4.58   *Regional map 6 Seaforth, Mackay region to Cape Palmerston, beaches 1087 to 1177. Beaches with Surf Life Saving Clubs indicated in bold, beaches with lifeguard patrol only indicates by flag.*

1087, 1088 FINLAYSON POINT

| No. Beach | Unpatrolled | | |
	Rating HT LT	Type	Length
1087 Finlayson Pt (W)	1 1	B+SF	300 m
1088 Finlayson Pt	1 1	B+SF	300 m

Finlayson Point forms the northern boundary of Seaforth beach. It is a 1.5 km long, low, rocky, north-east trending point, with two low energy beaches located on its northern side. A vehicle track from Seaforth leads to the point. Beach **1087** is part of a series of vegetated recurved spits located on the very protected northern side of the point and consists of a 300 m gap in the predominantly mangrove-fringed, sandy shoreline. The second beach (**1088**) is out on the point itself and is a curving, north facing, 300 m long, sandy high tide beach fronted by 500 m of sand and rock flats that connect with Redcliff Island.

Swimming: Neither beach is suitable for swimming, owing to the low gradients, with the point beach being the better choice at high tide.

Surfing: None.

Fishing: Only at high tide in the mangroves or over the rock flats on the point.

Summary: Two low energy beaches just north of Seaforth.

1089-1091 SEAFORTH, BLUE & HALLIDAY bays

| No. Beach | Unpatrolled | | |
	Rating HT LT	Type	Length
1089 Seaforth	1 1	UD	5 km
1090 Blue Bay	1 2	R+cobble	150 m
1091 Halliday Bay	1 1	UD	500 m

Seaforth is an older beachfront settlement that has long been popular with people in the Mackay district. It retains much of its early character with many older style beach houses and the old beach tidal pool. It is now growing in popularity and at Seaforth and elsewhere in the area, new residential and holiday developments are occurring. The settlement lies 22 km off the Bruce Highway and consists of a few rows of houses, a large beachfront caravan park and a long, grassy foreshore reserve. There are

also a couple of stores and a service station, all located on the 200 m wide, low, sandy foredune ridges that back the beach.

Seaforth Beach (1089) is 5 km long and runs south from Finlayson Point, then curves around to finally face north at the mouth at Plantation Creek, which forms the southern boundary. The beach receives waves averaging 0.4 m at the point, which decrease in height toward the creek. The beach responds by maintaining a wide, low gradient, 200 m wide bar in the north which, south of the pool, gradually widens to 1 km wide, ridged tidal sand flats. The creek exits via a deep channel across the southern end of the flats.

Blue Bay (1090) is located on the point immediately east of the creek. It is a north facing cobble and boulder beach, bordered by low rocky points. It is used as a boat launching site at high tide. Halliday Bay is located immediately behind the beach.

Halliday Bay (1091) is a small, but growing, residential settlement and larger beachfront resort. It is accessible via a 4 km road off the Cape Hillsborough Road. The resort backs the northern half of this 500 m long, northeast facing beach, with a beachfront reserve running the length of the beach, a tidal pool located against the northern rocks (Fig. 3.7a) and a boat launching area in the southern corner, where there is also an amenities block. The beach receives waves averaging less than 0.5 m that maintain a moderately steep high tide beach, which grades into a 200 m wide, low gradient low tide beach. The beach is bordered by a 20 m high northern headland and 60 m high McBrides Point, that extends 500 m out from the southern end of the beach. Rocks fringe both ends of the beach, with a few mangroves growing on the southern rocks.

Swimming: The Seaforth and Halliday Bay stinger enclosures are the safest location in the summer stinger season, otherwise the beaches are relatively safe with usually low waves. However, stay clear of the deep Plantation Creek channel and tidal currents.

Surfing: Only low, spilling waves.

Fishing: Best off the rocks or in Plantation Creek. There is a good all-tide boat ramp at Port Newry, at the back of Seaforth.

Summary: Two increasingly popular residential and holiday destinations, offering relatively safe beaches in a tropical setting.

low and fine grained, the other steep and coarse grained.

1092-1094 BALL BAY-SMALLEYS BEACH

No.	Beach	Unpatrolled Rating HT	LT	Type	Length
1092	McBrides Pt	1	2	B+S/rock flats	150 m
1093	Ball Bay	1	1	UD	2.2 km
1094	Smalleys	1	1	B+SF	1.3 km

McBrides Point forms the northern headland for 2.5 km wide, north facing Ball Bay. Within the bay are three beaches, the small settlement of Ball Bay and the southern national park camping area at Smalleys beach. Both Ball Bay and Smalleys beach are located on separate roads off the Cape Hillsborough Road.

McBrides Point beach (1092) is a 150 m long, northeast facing pocket of high tide sand fronted by 200 m wide intertidal rock flats. It is hemmed in between two rocky headlands and is only accessible on foot from Ball Bay, 300 m to the west.

Ball Bay beach (1093) is a 2.2 km long, very low gradient sand beach that extends from the southern side of McBrides Point down to the mouth of an unnamed creek that enters the middle of the bay. The northern end of the beach is backed by the small, quiet Ball Bay settlement, which includes a camping reserve and amenities block, but no store. The beach receives waves averaging less than 0.5 m and consists of a very low gradient beach, that is up to 200 m wide at low tide and widens toward the creek mouth into a series of 500 m wide tidal ridges.

Smalleys beach (1094) lies on the eastern side of the creek. It is 1.3 km long and bordered by the creek mouth and a vegetated, rocky knoll in the west, and the prominent 270 m high peak behind Cape Hillsborough in the east. The cape extends 2 km north of the eastern end of the beach. The beach is very protected because of its orientation and the cape, with waves usually very low to calm. As a result the beach consists of a steep, coarse grained, 30 m wide high tide beach and essentially flat, 1 to 2 km wide low tide sand and mud flats, with rocks bordering the cape shoreline. The beach and land east of the creek is part of the Cape Hillsborough National Park.

Swimming: Best at Ball Bay toward high tide.

Surfing: Only low, spilling waves at Ball Bay.

Fishing: The creek and rocks of McBrides Point offers the best locations.

Summary: An interesting bay surrounded by high points and hills, with two adjacent, yet contrasting beaches; one

---

Cape Hillsborough National Park

Area:	728 ha
Length of coast:	12 km
Beaches:	1094-1097

Cape Hillsborough is one of a series of prominent rhyolitic peaks formed as volcanic cores. The peaks dominate this part of the coast, so much so that Captain Cook remarked on the 'pretty high promontory which I have named Cape Hillsborough.' The cape is the highest at 268 m, with Andrews Point reaching 135 m, and backing Pinnacle Rock 229 m. In between is generally low to flat land, including extensive mangrove forests and three beaches. There is car access to Smalleys beach and the main cape beach, and a walking track over the headland to Beachcomber Cove. There is a lovely grassy park behind the cape beach, which is also home to a number of kangaroos, and a caravan park and resort behind the eastern end of the beach.

For further information contact:

Mackay District Office
Queensland National Parks & Wildlife Service
Cnr Wood & River Streets
Mackay Qld 4740
Phone: (07) 49 51 8788

The Ranger
Cape Hillsborough National Park
MS 895
Seaforth Qld 4741
Phone: (07) 49 59 0410

1095-1097 BEACHCOMBER COVE, CAPE HILLSBOROUGH

No.	Beach	Unpatrolled Rating HT	LT	Type	Length
1095	Beachcomber Cove	1	2	UD	380 m
1096	Cape Hillsborough	1	2	UD	2 km
1097	Wedge Island	1	1	B+SF	130 m

Cape Hillsborough is bordered by Smalleys beach to the west, steep rocky shores to the north and small Beachcomber Cove beach in the east. The only formal access to the cove is via a ridge-top walking track from the ranger's office, or at low tide along the beach from Cape Hillsborough beach. **Beachcomber Cove** beach (1095) is 300 m long and is backed by steeply rising, densely vegetated slopes, with Cascade Creek running down the slopes to cross the northern end of the beach. In the summer wet season there is so much water running down onto the beach that you can swim in fresh water in the surf. The beach has a cobble high tide ridge fronted by the 200 m wide intertidal beach composed of fine, grey sand. The actual Cape Hillsborough extends 500 m north of the northern end of the beach, while low rocks separate it at high tide from the adjoining main cape beach (Fig. 4.59).

Cape Hillsborough beach (**1096**) is 2 km long, extending from the northern rocks to the rocks of Andrews Point in a curving north-east to north facing arc. The central 1 km of the beach is backed by a 150 m wide, low foredune that links the cape with Andrews Point. Located on the foredune is the foreshore reserve and resort. The reserve has picnic and barbecue facilities and an amenities block, with boat launching access at its eastern end and a kiosk at the entrance to the resort. A mangrove-filled bay backs the foredune. The beach is composed of fine sand, receives low waves and consists of a wide, low gradient beach that is up to 200 m wide at low tide. It narrows toward Andrews Point, as a few rocks appear off the beach.

Figure 4.59 *Cape Hillsbourough beach (foreground) and Beachcomber Cove, both located in Cape Hillsborough National Park.*

Wedge Island beach (**1097**) lies on the west side of small Wedge Island, which is connected to the main cape beach at low tide by sand and rock flats. The beach is 150 m long, faces west toward the main cape beach and consists of a narrow, sandy high tide beach, a mid tide cobble to boulder beach and low tide sand flats.

Swimming: Swimming is best toward high tide on both Beachcomber and Cape Hillsborough beaches.

Surfing: Usually only low, spilling waves that are highest on Beachcomber Cove beach.

Fishing: Best off the rocks at high tide.

Summary: The cape, with its beaches and facilities, is a very popular location with tourists and touring caravans. It is attractive, has relatively safe beaches, a spectacular setting and several walks around the headlands, plus the resident wildlife.

1098-1103   NOBBIES CK-SHOAL POINT

No.	Beach	Unpatrolled		Type	Length
		Rating			
		HT	LT		
1098	Nobbies Ck	1	1	B+SF	1 km
1099	Belmunda (1)	1	1	B+SF	500 m
1100	Belmunda (2)	1	1	B+SF	1 km
1101	Williamsons	1	1	B+SF	2.5 km
1102	Neils	1	1	B+SF	4 km
1103	Shoal Pt (spit)	1	1	B+SF	1.5 km

Sand Bay is an open, low energy, east facing bay extending south of Andrews Point for 5 km to Williamsons beach. The bay shore and much of the bay is dominated by extensive tidal sand shoals and channels, that give the bay its name. Apart from Andrews Point and two knolls, most of the shore and backing country is low and flat, with several square kilometres of mangroves backing the beaches and comprising the shore. Apart from the Cape Hillsborough Road that clips the northern bay beach and a couple of 4WD tracks across the tidal flats, there is no development on the 30 km of bay shore.

Nobbies Creek exits into the north-western, mangrove dominated corner of the bay. Immediately east of the mangroves and paralleling part of the cape road is a 1 km long, south-east facing, low, narrow, low energy high tide beach (**1098**), fronted by 1.5 km of intertidal sand flats and backed by mangroves.

In the western centre of the bay is a 1 km long, 40 m high bedrock outcrop surrounded by tidal flats and mangroves. Along the eastern side of the bedrock are the two Belmunda beaches. The northern one (**1099**) is 500 m long and backed by a densely vegetated, sandy plain with a southern headland, while the main beach (**1100**) is 1 km long and backed by grassy bedrock slopes, with rocks also outcropping along the beach. Sandy tidal flats extend 1 km off the northern beach, decreasing to 200 m off the southern beach.

South of Belmunda beach is the most extensive area of tidal flats, channels and mangroves, with the next sandy shoreline lying 4 km to the south-east at the western tip of 2 km long **Williamsons** beach (1101). This beach and its companion **Neils** beach (1102) consist of a total of 6.5 km of low energy beach ridges, fronted by up to 2 km of sand flats that even link Green Island, 4 km offshore, to the Neils beach tidal flats on spring low tides. Behind the beaches are older beach ridges and mangroves extending up to several hundred metres west of Neils beach. Both beaches are accessible by 4WD tracks that cross grazing land and tidal flats to reach the beaches.

Shoal Point lies 12 km south-east of Andrews Point and forms the southern end of the low energy shoreline that includes Sand Bay and Williamsons-Neils beaches. Extending for 1.5 km west of the 20 m high bedrock Shoal Point is a low energy sand spit (**1103**), that ends 2 km from the eastern end of Neils beach, with large, mangrove-fringed Reliance Creek in between. Extensive tidal flats extend 2 km north of the western end of the beach, narrowing toward the point where rock flats surround the beach (Fig. 4.60). A sealed road runs out to Shoal Point where there is a small residential settlement, a foreshore reserve and boat ramp, and a seawall backing part of the low energy, though dynamic, shoreline.

Swimming: Apart from Shoal Point, very few people ever reach these low energy beaches, that are dry at low tide and can only provide enough water for a swim toward high tide.

Surfing: None.

Fishing: Best in the many tidal channels and mangroves.

Summary: These are five relatively isolated, little visited beaches that are mainly backed by private grazing land, with accessible Shoal Point forming the southern boundary.

Figure 4.60   *View west along the low energy becah in lee of Shoal Point, with tidal sand flats extending out into Sand Bay.*

1104-1106   SHOAL PT, BUCASIA

No. Beach	Unpatrolled		
	Rating HT LT	Type	Length
1104 Shoal Point	1 3	UD	500 m
1105 Bucasia	1 2	UD	4.5 km
1106 Sunset Bay	1 1	B+SF	300 m
Roving patrols at Bucasia by Mackay SLSC			

Between Shoal Point and Eimeo headland is an open, 4 km wide, east-north-east facing bay that contains three beaches - one at Shoal Point, the long Bucasia beach and the tidal Sunset Bay. Bucasia is one of Mackay's northern beaches and is a popular residential, holiday and traveller destination.

Shoal Point beach (1104) lies below the 20 m high bluff of the point and consists of a 30 m wide, low gradient high tide beach fronted by a 150 m wide intertidal bar, with rocks outcropping on each, particularly toward low tide. A gravel road off the Shoal Point road ends in a foreshore reserve at the northern end of the beach, with a second access track from the houses behind the southern end.

Bucasia beach (1105) begins immediately south of the Shoal Point rocks and runs south for 3 km before swinging around to face north-east and finally north, at the low energy mouth of Eimeo Creek that forms its southern boundary. The beach is backed by vegetated parabolic dunes in the north, with the Kohuna Resort and residential development along the southern 2 km; all fronted by a wide foreshore reserve and in the very south, a caravan park, and finally a narrow sandy spit. Toward the northern end of the reserve is an older style timber tidal pool. There are a few shops and a small shopping centre, together with a boat ramp into the backing southern Eimeo Creek. Several boats are usually moored in the mangrove-fringed creek. The beach itself begins as a low gradient high tide beach and 200 m wide low tide bar, which become more protected in the southern 2 km to form a steep high tide beach fronted by 1 km of intertidal sand flats, widening to 2.5 km off Eimeo Creek.

Directly opposite the southern end of Bucasia spit, 100 m away on the other side of Eimeo Creek mouth, is **Sunset Bay** beach (1106) - a crenulate, low energy, narrow high tide beach, fronted by the wide creek tidal flats, that ends amongst the rocks of Eimeo Point. The beach faces west into the sunset, hence the name. The road to Eimeo runs along the back of the beach, which is also backed by several houses, a boat ramp, park and playground, together with a seawall.

Swimming: Both Shoal Point and Bucasia offer

kilometres of relatively safe swimming, with a tidal pool for summer bathing. Be careful at the creek mouth owing to the deeper, cobble-lined channels and strong tidal currents.

Surfing: Chance of small, spilling breakers at Shoal Point.

Fishing: Eimeo Creek is a favourite spot; tourists also try their luck off Bucasia beach at high tide.

Summary: A lovely embayment with beaches ranging from the more exposed Shoal Point to the very protected Sunset Bay.

1107, 1108 EIMEO, DOLPHIN HEADS

No.	Beach	Rating		Type	Length
		HT	LT		
1107	Eimeo	1	1	B+SF	500 m
	Lifeguard (Christmas & Easter)				
1108	Dolphin Hd	1	2	B+SF	400 m

Eimeo Point is a small, 60 m high, conical headland, capped by a small but famous hotel that offers spectacular views from every room. Sunset Bay lies below to the west, with small Eimeo beach to the east. A road off the Bucasia Road runs out to Eimeo and on to Dolphin Heads and Blacks beach.

Eimeo beach (1107) is a 400 m long, north facing, low energy beach. The beach extends east of the rocks at the base of the point as a single sand spit, backed by a small, mangrove-filled tidal creek. There is one row of houses on the spit, which ends at the creek mouth. The beach consists of a 50 m wide high tide beach fronted by sand flats that widen to 100 m off the creek mouth.

On the eastern side of the creek mouth is **Dolphin Heads** beach (1108) - a 300 m long, north-west facing strip of sand fronted by the creek tidal flats, then rock flats, on the north side of Black Point. The beach is backed by a caravan park and resort on the point, which has its own small lake that has been dredged out of the tidal flats, for swimming and even boating.

Swimming: Both the beach and the lake, for resort guests, offer relatively safe swimming, although the beach can only be used at high tide.

Surfing: Chance of a low, reformed wave wrapping around the point during strong south-easterlies.

Fishing: Best off the Eimeo Point rocks or in the creek.

Summary: A small but popular beach, catering to the locals and also increasingly to tourists.

1109-1112 BLACKS BEACH & SLADE BAY

No.	Beach	Rating		Type	Length
		HT	LT		
1109	Blacks (1)	2	2	Cobble +LT	300 m
1110	Blacks (2)	2	2	Cobble +LT	400 m
1111	Blacks	2	2	UD	4.2 km
	Lifeguard (Christmas & Easter)				
1112	Slade Bay	1	1	B+SF	2 km

Slade Bay is an open, east to north-east facing bay lying between the northern Blacks Point and Slade Point, 5 km to the south-east. Blacks beach forms the western shore, with an extensive tidal creek system draining out in the southern corner, and a low energy sandy shore running from the creek in a north-east direction to rocky Slade Point. There is residential and holiday development along the northern half of Blacks beach and continuous foreshore development backing the Slade Bay beach.

Black Point is a 60 m high, basalt outcrop that has two exposed, east facing beaches (**1109 & 1110**) along its north-eastern tip. These beaches are 300 m and 400 m long respectively, backed by vegetated bluffs rising to 60 m, with houses on top of the bluffs and a 20 m wide cobble to boulder high tide beach at their base. At low tide the boulders give way to a low tide sand bar. Slightly protruding rocky points and rock flats border the ends of each beach.

The main **Blacks Beach** (1111) begins on the southern side of the rocks and extends for 4.2 km down to the tidal creek mouth. The northern 2 km are exposed to waves averaging 0.5 m and have a 50 m wide high tide beach fronted by a low tide bar, that widens from 100 to 200 m to the south. Halfway down the beach the impact of the creek is felt and intertidal sand flats widen to more than 2 km by the creek mouth. Shoreline oscillations caused by the shifting tidal flats have resulted in a seawall being built in front of the beachfront properties along the central section of the beach.

On the eastern side of the 500 m wide creek mouth is the western tip of **Slade Bay** beach (1112). This beach begins out on 35 m high Slade Point and runs back into the bay. It faces north-west and is very protected from waves. It consists of a narrow strip of high tide sand, fronted by 50 m wide rock flats and then up to 2 km of tidal flats. However the shoreline is dynamic, which has resulted in a seawall being built to protect the road and beachfront houses that back the northern half of the beach.

Swimming: Blacks beach is the most popular location and is used by locals and holiday-makers.

Surfing: The highest waves occur up on Black Point, where waves up to 0.5 m spill across the bar toward low tide.

Fishing: The best fishing is in the creek, or off the Black Point rocks at high tide.

Summary: Both Blacks beach and Slade Bay are essentially residential suburbs of Mackay, with some facilities and accommodation for tourists and travellers, particularly at Blacks beach.

1113-1115   SLADE PT, LAMBERTS & NORTH HARBOUR

No.	Beach	Rating		Type	Length
		HT	LT		
1113	Slade Pt	2	5	R+LT rocks	200 m
1114	**Lamberts**	2	3	R+BR	600 m
	Lifeguard Patrols:		Christmas & Easter		
1115	North Harbour	2	3	R+LT	2.4 km

Slade Point is a prominent, 1 km long, 35 m high, rocky headland, capped by a 20 m high water tower and fronted by an extensive, rocky foreshore. Along the north-east facing section of the point is a 200 m long pocket of steep high tide sand and cobbles (**1113**) fronted by a 100 m wide, rocky intertidal zone. A grassy reserve backs the southern end of the beach. While the beach is relatively safe and protected by the rocks at high tide, it is a very hazardous location at low tide. A sand bar runs seaward of the rocks and provides some surf during strong Trade winds.

Lamberts Beach (1114) is the northern of Mackay's two main surfing beaches. It is located just 1 km south of Slade Point, with a lookout on the southern part of the point providing an excellent view of the beach. The Slade Point Road runs right behind the beach. The beach is 500 m long, faces due east and is bordered by a prominent, 40 m high headland and lookout at the northern end, and a lower rocky platform and reef at the southern end. A 100 m wide grassy park, with amenities, backs the beach, with casuarina trees also fringing the back of the beach. The beach is composed of coarse sand and some gravel, which produces a steep high tide beach, while at low tide it is fronted by a 50 m wide bar that is usually cut by three rip currents and channels (Fig. 4.61). Waves average 0.5 to 1 m, making it one of the more exposed and higher energy beaches in the area (Fig. 4.62). Lifeguards patrol the beach during the Christmas and Easter holidays.

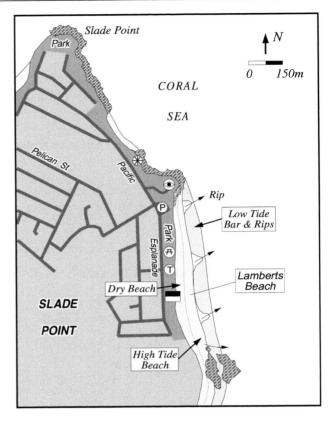

Figure 4.61   *A map of Slade Point and Lamberts beach. There is a foreshore reserve backing the point and beach, and a lookout on the headland overlooking both beaches.*

Figure 4.62   *View north along Lamberts Becah during moderate wave conditions. Note the steep high tide becah and low tide bar, which is cut by rips during moderate to high waves,.*

On the southern side of Lamberts beach is 2.4 km long **North Harbour** beach (1115), which faces due east and runs straight down to the northern harbour wall of Mackay Harbour. The beach is accessible on foot from Lamberts beach, or via a gravel road through the Mackay Harbour area, which leads to the southern end of the beach. Most of the beach is backed by natural, largely vegetated, 10 to 20 m high sand dunes, although there are plans to link the Slade Point Road to the harbour road. The beach has a moderately steep high tide beach fronted by a 150 m wide low tide bar. Waves average 0.6 m and rips occasionally form along the low tide bar.

Swimming: Slade Point beach is only suitable at high tide in low wave conditions. Lamberts is a popular surfing beach as well as being patrolled during the Christmas and Easter holiday period. However swimmers should watch out for the rips toward low tide. North Harbour is less popular owing to the more difficult access and care should be taken as rips do occur along the beach.

Surfing: Lamberts is the main board surfing beach for Mackay and is the place to go when the south-easterlies are blowing. However, waves are short wind waves that are best at mid to low tide.

Fishing: Lamberts has some of the few rip holes in the area that can be fished, along with the rocks and reefs at each end of the beach and on Slade Point.

Summary: Lamberts is a well planned, very accessible and popular beach with locals and the surfers. Slade Point is a nice spot for a picnic, while North Harbour is better for a long beach walk.

1116, 1117 **MACKAY HARBOUR**

No.	Beach	Unpatrolled			
		Rating		Type	Length
		HT	LT		
1116	Mackay Hbr (N)	1	1	R+LT	100 m
1117	Mackay Hbr (boat ramp)	1	1	R+LT	50 m

Mackay Harbour has been constructed by building two 800 m long breakwaters out from the beach, enclosing nearly 1 km² of protected harbour. Tucked in at either end of the harbour are two pockets of sand; the northern (**1116**) is a 100 m long, steep beach wedged up against the northern breakwater. An access track runs to the low dunes backing the beach, with North Harbour Beach also being accessible from this track.

The southern beach (**1117**) is located between the main wharf and the southern breakwater, is the site of the heavily utilised harbour boat ramp and is backed by a seawall. This beach has a steep high tide beach and a sand and rock low tide bar. Its main recreational use is for boat launching at the ramp, with the Mackay Sailing Club utilising the northern end of the beach.

Summary: While these beaches are both calm, only the northern beach is suitable for swimming.

1118 **MACKAY SLSC**

Patrols:	
Surf Lifesaving Club:	September to May
Lifeguard on duty:	September to May

No.	Beach	Rating		Type	Length
		HT	LT		
1118	Mackay (SLSC)	2	3	R+LT	4.3 km

The town of Mackay is located at the mouth of the Pioneer River. On the north side of the river mouth is 9 km of near-continuous beach, now broken in the centre by the breakwaters and seawalls for the Port of Mackay. The main Mackay surfing beach begins on the south side of the harbour walls and runs for 4 km to the rock training wall that lines the north side of the river mouth. This beach houses the Mackay Surf Life Saving Club (founded in 1928). The club is located in the centre of a 500 m long beachfront reserve containing extensive parking, together with shady picnic and plays areas and a kiosk (Fig. 4.63). However, during 1998 the surf club will be relocated 1.5 km to the south, due to the construction of a resort and marina at its present location.

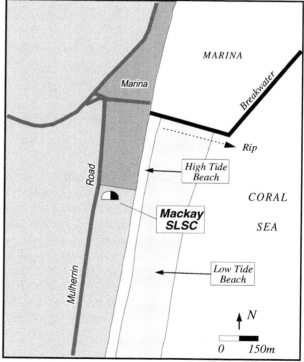

Figure 4.63   *Mackay SLSC is located south of the new Mackay marina and older harbour. It is part of a large foreshore reserve offering a wide range of recreational facilities.*

The beach (**1118**) faces almost due east and is exposed to some of the biggest tides on the east Australian coast (up to 8 m) and biggest waves inside the Great Barrier Reef lagoon, commonly 0.5 to 1 m. As a consequence the beach has a 60 m wide high tide beach which, as the tide falls, is fronted by a continuous low tide beach/bar up to 200 m wide, with the chance of rips cutting across the bar at the low tide level.

Swimming: A relatively safe beach at high tide under normal wave conditions, however be careful at low tide as you will be a long way from the surf lifesaving club and rips will be present when the waves exceed 0.5 m. Also stay clear of the harbour walls, as a permanent rip runs out along the rocks.

Surfing: This is one of Mackay's main surfing beaches. It is best at mid to low tide and when there is a bit of a south-easterly blowing.

Fishing: The harbour walls are the most popular location, with the beach only suitable at high tide.

Summary: A well maintained beach and reserve, providing good access and all facilities for surfers and families for a day at the beach.

1119-1122 SOUTH WALL, TOWN & FAR

No.	Beach	Unpatrolled		Type	Length
		Rating			
		HT	LT		
1119	South Wall	1	1	B+SF	800 m
1120	Town	1	1	B+SF	900 m
1121	Far	1	1	B+SF	2.6 km
1122	Far (S)	1	1	B+SF	2.9 km

The city of **Mackay** is located on the southern side of the Pioneer River mouth. Most of the city is located on river mouth floodplains of the Pioneer. Toward the coast the floodplain is, and in places was, fronted by a few hundred metres of mangroves, then a series of low beaches and dunes that form a 7 km section of east facing shore from the river mouth south to Bakers Creek mouth. Since the 1960's a great deal of the mangrove areas have been infilled for residential development and Town and Far beaches developed for residential, recreational and tourist facilities.

The northern beach (1119) lies in lee of the **South Wall** of the river mouth. It is an 800 m long beach, backed by low beach ridges and mangroves and fronted by up to 3 km of intertidal sand flats. To date there has been no development on the beach, although it is earmarked for tourist development.

Town Beach (1120) was developed in the 1970's as the beachfront suburb of Iluka. However the main beachfront road was located too close to the beach and now the 900 m long beach has been largely replaced with a seawall to protect the road. Extensive tidal flats front the wall and remaining beach.

Far Beach (1121) extends south of Town Beach for 2.9 km. It is the site of a large caravan park and other tourist accommodation and recreational facilities. It is also backed by a seawall in places and consists of a narrow high tide beach fronted by 2 to 3 km of tidal flats. South of the seawall a low grass and casuarina-covered foredune backs the beach. The beach ends at a small creek mouth. Extending for another 900 m on the south side of the creek is the southern extension of Far Beach (1122). At present it is undeveloped, although there is 4WD access to the southern tip that recurves round as a spit into Bakers Creek mouth. This beach is also fronted by 2 to 3 km of tidal flats (Fig. 4.64).

Figure 4.64 *A view of the southern end of Far Beach, with the tidal sand shoals widening to 2 to3 m owing to the imapct of the Pioneer River and protection from Dudgeon Point .*

Swimming: You can only swim at these beaches toward high tide owing to the extensive tidal flats. Due to the tidal flats and offshore Flat Top and Round Top Islands, waves are usually low and conditions relatively safe at high tide, apart from the tidal currents in the creek mouths.

Surfing: None.

Fishing: Best in the creek mouth and off the seawalls at high tide.

Summary: Mackay's eastern beachfront suburbs back this 7 km long, exposed beach. However it is largely protected from waves by the 2 to 3 km wide tidal sand flats and offshore islands.

1123-1125 BAKERS CK, MCEWENS BEACH

No.	Beach	Unpatrolled Rating HT	LT	Type	Length
1123	Bakers Ck (S)	1	1	B+SF	1.5 km
1124	McEwens (N)	1	1	B+SF	3 km
1125	McEwens	1	1	B+SF	1.2 km

Bakers Creek is a 1 km wide tidal creek that essentially forms the southern boundary of Mackay. Dense mangroves fringe the creek shores and 2 to 3 km of tidal flats extend offshore from the mouth. On the south side of the creek mouth is a 1.5 km long, low energy beach and tidal flat (**1123**). The beach consists of a narrow, high tide sandy ridge, backed by dense mangroves and fronted by tidal flats also containing scattered mangroves. There is a gravel road to the western tip of the beach where there is a creek-front shack development.

McEwens beach (north, **1124**) is a 3 km long sand spit that is slowly moving into Baker Creek. Over the past 6000 years similar spits have built the shore out over 1 km, with the low sand ridges separated by extensive mangrove-covered tidal flats. There is a 1.5 km long 4WD track across the tidal flats and ridges to the eastern tip of the spit, but otherwise no development.

McEwens Beach (1125) is a small beachfront settlement strung along the back of the 1.2 km long beach. The beach faces south-east into Sandringham Bay, with the southern shore 2 km away across the bay and tidal flats and channels in between. There is a 6 km long sealed road from the highway to the beach and a playground, park and picnic facilities at the beach, but no store. On the beach there is a small post and wire swimming enclosure.

Swimming: Only at high tide and safest at McEwens beach tidal pool.

Surfing: None.

Fishing: Bakers Creek and Sandringham Bay are both popular fishing locations, with the beaches only fishable at high tide.

Summary: Three very low energy beach-tidal flats mainly used by local fishers.

1126-1128 DUDGEON PT

No.	Beach	Unpatrolled Rating HT	LT	Type	Length
1126	Dudgeon Pt (W 2)	1	2	B+SF	1.3 km
1127	Dudgeon Pt (W 1)	1	2	B+SF	1.1 km
1128	Dudgeon Pt	1	2	B+ rock flats	500 m

Dudgeon Point is a low bedrock point, the first rock south of Slade Point 20 km to the north. The point and its western extension form part of the southern shore of Sandringham Bay, a 2 to 3 km wide, shallow, sand-filled, low energy bay, with mangrove-covered tidal flats dominating its inner reaches. Three kilometres south-west of Dudgeon Point the mangroves give way sufficiently for a sandy beach to form the shoreline, with three beaches between there and the point. The point and backing land are part of a cattle station and the land has largely been cleared for grazing.

Beach **1126** is a 1.3 km long, north-west facing, low energy, sandy beach fronted by 200 m wide sand flats and the deep southern channel of Sandringham Bay. The beach is bordered by a small, mangrove-fringed creek to the west and a smaller tidal creek in the east that exits beside rock flats. A tributary of this creek also runs behind the beach, exiting occasionally across the southern end of the beach. Older recurved beach ridges that have been cleared for grazing back the creek.

Beach **1127** lies on the east side of the low rock flats and is a curving, 1.1 km long, north-west facing, high tide sandy beach fronted by 100 to 200 m wide tidal flats, then the deeper tidal channel. It is bordered by 200 m wide rock flats, with scattered mangroves on the southern rock flats. The station homestead lies on cleared land just behind the beach.

Dudgeon Point beach (**1128**) fringes the low, cleared Dudgeon Point. The beach runs around the perimeter of the point, facing essentially north. It consists of a narrow high tide beach, backed by a few fishing shacks and fronted by Dudgeon Ledge; a 1 km wide, irregular rock flat and reef.

Swimming: These are three low wave energy beaches and are relatively safe at high tide. However, low tide exposes the deep tidal channels of Sandringham Bay off the western beaches and extensive rock flats off the point beach.

Surfing: None.

Fishing: Excellent fishing in the tidal channel and over the rock flats at high tide.

Summary: Three relatively isolated beaches, surrounded by grazing land and essentially off limits to the public.

1129,1130 DUDGEON & HECTOR PTS

No.	Beach	Rating HT	LT	Type	Length
		Unpatrolled			
1129	Dudgeon	1	2	R+LT	2.5 km
1130	Hector Pt	1	3	R+LT +rocks	300 m

Dudgeon Point and its protruding Dudgeon Ledge form the southern entrance to low energy, tide dominated Sandringham Bay, which lies to the west of the point. To the south of the point is a 2.2 km long sandy beach (**1129**) that runs essentially due south to Mount Hector. The beach receives waves averaging 0.5 m and consists of a moderately steep high tide beach fronted by a 100 m wide, continuous low tide bar. There are rock flats at either end of the beach as well as some in the centre, with a small reef just offshore. The beach is backed by a densely vegetated, 50 m wide foredune and a small swampy lake, then cleared grazing land.

At the rock-dominated southern end of the beach is a 300 m pocket of sand in amongst the colourful sandstone rocks at the base of 60 m high Mount Hector. This beach (**1130**) faces east-north-east and receives waves averaging 0.4 m that break across the rocks and 10 m wide low tide bar. The point and backing land are part of a nature reserve.

Swimming: The main beach is relatively safe during low waves, however watch out for the rocks at low tide and rips when waves exceed 0.5 m.

Surfing: Chance of low, spilling waves across the bar toward low tide.

Fishing: Best over the rocks at high tide.

Summary: These beaches are accessible on foot around Mt Hector from Hector beach and it is well worth the walk to see the colourful rocks of the headland.

1131, 1132 HECTOR

No.	Beach	Rating HT	LT	Type	Length
		Unpatrolled			
1131	Hector	1	2	B+SF	1.5 km
1132	Hector (E)	1	1	B+SF	100 m

Hector beach lies at the mouth of Louisia Creek and in lee of the Hay Point coal loading terminal and its 3 km long jetty. The beach and its beachfront houses are accessible via a 1 km road off the Hay Point Road. There is a good boat launching ramp and amenities block at the Louisia Creek mouth which forms the western end of the beach, but no other facilities. The main beach (**1131**) is 1.5 km long and curves around to face north-east. It consists of a 100 to 200 m wide foredune, backed by the mangroves of the creek and fronted by a 100 m wide low tide bar then sand flats extending another 500 m into the bay.

The eastern beach (**1132**) is a remnant of its former self as the northern end of the coal storage facilities, including a 500 m long seawall, lie on top of most of this beach. What is left is a 100 m segment that abuts the western side of the seawall and an even smaller portion on the eastern side. It is separated from the main beach by a low tide rock reef and backed by access tracks leading to the coal facility.

Swimming: Hector beach is protected from the south-easterlies and is a lovely spot for a walk or swim. Be careful at the creek mouth as it contains strong tidal currents.
Surfing: There is a wave over the creek mouth tidal shoals during very strong Trades.

Fishing: The creek is the most popular location.

Summary: A quiet little settlement and beach, with the massive coal loader and jetties to one side and mangrove-filled Louisia Creek to the other.

1133-1136 HAY POINT, HALF TIDE & SALONIKA

No.	Beach	Rating HT	LT	Type	Length
		Unpatrolled			
1133	Hay Pt	1	2	R+LT	2.2 km
1134	Half Tide	1	2	R+LT	250 m
1135	Salonika	1	2	R+LT	2.4 km
1136	Salonika (S)	1	2	B+S rock flats	500 m

Hay Point is a prominent headland, now dominated by a large coal loading facility and two long jetties extending north-east from the point. The point is serviced by the small settlement of Hay Point, that includes a beachfront hotel and caravan park, but no shops. The Hay Point Road runs right out to the point and provides good access to the beach.

Hay Point beach (**1133**) extends for 2.2 km south of the point to a small creek and the 800 m long harbour breakwater. The breakwater was constructed between a natural rock reef and the small Highwater Islet; it now protects the southern end of the beach which is slowly infilling with sand, with the shoreline having built out over 100 m. A road runs 100 m in along the back of the beach to the northern point, with the hotel and caravan park at the southern end. There is also a good launching ramp at the breakwater. The beach receives waves averaging 0.5 m and consists of a moderately steep high tide beach fronted by a 100 m wide, continuous low tide bar. A small creek drains across the southern end where there is also a patch of rock flats.

Half Tide beach (1134) lies on the south side of the breakwater and consists of a 250 m long, cobble and sand pocket beach, wedged in between the rocks and breakwater and southern low rocky point. A tidal creek used to drain across the southern half of the beach until the backing property owner blocked it, forming a freshwater swamp in the former mangrove-filled creek. The beach receives waves averaging less than 0.5 m and has a steep sand to cobble beach which, toward the north, also incorporates cobble and boulders eroded from the breakwater. It is fronted by a 100 m wide low tide bar and backed by a small reserve and amenities block. The Hay Point Air-Sea Rescue Base also has a small building behind the beach.

On the southern side of the rocky point is 2.4 km long **Salonika** beach (1135). This longer sandy beach picks up waves averaging 0.5 m and has a small rip running out against the northern rocks. The beach itself is moderately steep and 50 m wide at high tide, with a 100 m wide low tide bar that increases to 150 m in width at the southern Breens Creek mouth. The beach is backed by a series of foredunes that were cleared and developed for housing in the mid 1990's. There is good access for the length of the beach from backing residential roads.

On the southern side of shallow Breens Creek mouth is a 500 m continuation of Salonika beach (**1136**). This section of beach is backed by the rocks of Point Victor, with small but reef-bound Victor Islet lying 2 km offshore. The beach is fronted by wide sand flats near the creek mouth, with 50 m wide rock flats increasing in dominance toward the point.

Swimming: Hay Point, Half Tide and Salonika all offer reasonably exposed sandy beaches with relatively safe swimming under normal wave conditions. Watch for rips when waves exceed 0.5 m, and tidal currents at Salonika.

Surfing: There is sloppy surf along these beaches during strong south-easterlies.

Fishing: The breakwater is the most popular location, as well as Breens Creek.

Summary: An out-of-the-way though reasonably popular spot with locals, whose numbers are increasing as development spreads along the coast.

1137, 1138 MICK READY

No.	Beach	Unpatrolled		Type	Length
		Rating			
		HT	LT		
1137	Mick Ready	1	1	B+SF	600 m
1138	Beach 1138	1	2	R+LT	120 m

South of Salonika the next 5 km of coast is dominated by the rocks of 20 m high Point Victor and the slopes of 160 m high Mount Hayden. Two small creeks drain to the coast and each has a small sandy beach at its mouth. Both beaches are backed by densely vegetated valleys and slopes and access is restricted to private vehicle tracks to Mick Ready beach. The second beach is accessible around the rocks from the southern headland, where there is street parking at Grasstree beach, located about 200 m east.

Mick Ready beach (1137) is 600 m long, faces east and is bordered by Point Victor and a southern rocky headland. A small creek drains out in the centre of the beach and flows across 300 m wide sand flats that front the narrow high tide beach (Fig. 2.1a).

Beach **1138** is a smaller version of Mick Ready; just 120 m long and bordered by rocky headlands. It has a narrow high tide cobble beach, then a sandy low tide bar with rock flats to either side.

Swimming: These are two difficult to access beaches that are relatively safe under low waves, but prone to rips during higher waves.

Surfing: Chance of a wave over the rock flats at mid to high tide.

Fishing: Good fishing off the rocks at high tide and in the creek at Mick Ready.

Summary: Two little beaches surrounded by private property, but close enough to Grasstree to walk around the rocks at low tide.

1139 GRASSTREE

No.	Beach	Rating		Type	Length
		HT	LT		
1139	Grasstree	1	2	R+LT	1.75 km
Roving IRB patrol from Sarina SLSC					

Grasstree beach is one of the three older beach settlements for the nearby town of Sarina. It was traditionally a site of holiday shacks, that in recent years are being replaced by more substantial residences. Two rows of houses extend along the back of most of the 1.7 km long beach, with a grassy foreshore reserve between the houses and the beach. There is a store, and an amenities block and picnic facilities in the reserve. The beach (**1139**) faces east and is moderately protected by the low northern and southern Coral Point headlands. Cabbage Tree Creek maintains a permanent channel, named Castrades Inlet, at the southern end of the beach. The beach consists of a steep high tide beach fronted by a 150 m wide low tide bar (Fig. 2.14c), with waves averaging less than 0.5 m in the north but decreasing toward the inlet.

Swimming: A relatively safe beach under usually low waves. Watch for rips when waves exceed 0.5 m and the deep channel and tidal currents at the inlet.

Surfing: Low, spilling waves break across the bar during strong Trade winds.

Fishing: The inlet is a very popular spot with locals, as well as rock fishing at high tide off the points.
Summary: A quiet beach settlement mainly used by the Sarina locals.

1140, 1141   CAMPWIN, SARINA (SLSC)

Patrols:				
Surf Life Saving Club:			September to May	
Lifeguard on duty:			September to May	
No. Beach	Rating HT LT		Type	Length
1140 Campwin	1	2	R+LT	350 m
1141 **Sarina**	1	2	R+LT	1.3 km

Campwin and Sarina beaches are the two main surfing beaches for the sugar town of Sarina, located 15 km inland on the Bruce Highway. They share the same 2 km long stretch of beach between Coral and Perpetua Points, with Leeper Reef and adjoining rock flats separating the two at low tide (Fig. 4.65). **Campwin** beach (1140) lies to the north of the rocks and is 350 m long, faces east and receives waves averaging 0.5 m. It has a relatively steep high tide beach and a 200 m wide low tide bar. The beach is backed by four rows of houses and has a store and small beachfront reserve with basic facilities. A boat ramp into Castrades Inlet is located at the back of the beach.

Sarina Beach (1141) lies between the rock flats and 35 m high Perpetua Point. The beach has a caravan park, two motels, a store and a number of holiday and residen-

tial houses, all contained in two streets backing the southern 500 m of the 1.3 km long beach. The beach faces due east and is bordered in the north by some low intertidal rocks that separate it from Campwin beach, while in the south are the rocks of 40 m high Perpetua Point. The Sarina Surf Life Saving Club, founded in 1968, is located at the far southern end of the beach and is the first building you see when arriving at the beach. A reserve and beachfront caravan park are located north of the club house.

The beach has a high tide range, up to 8 m, and receives waves averaging between 0.5 and 1 m. These have built a narrow high tide beach, fronted by a 200 to 300 m wide, flat low tide beach (Fig. 4.66). A small creek crosses the centre of the beach, with the low rocks at the northern end.

Swimming: A relatively safe, patrolled beach when waves are less than 0.5 m. Be careful when waves exceed 0.5 m as they can produce rips, particularly at low tide. Campwin beach is patrolled occasionally with roving IRB patrols from Sarina SLSC.

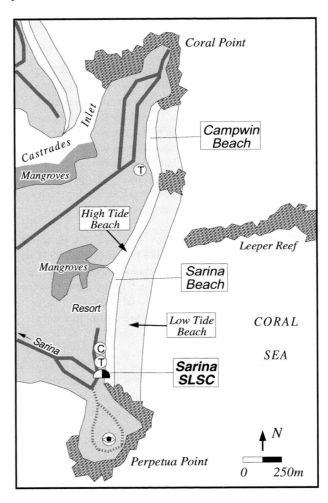

Figure 4.65: *A map of Campwin and Sarina beaches, with Sarina SLSC located at the southern end of the main beach.*

Figure 4.66: *A view of Sarina beach at low tide, showing the 200 to 300 m intertidal beach. The surf club is located at the southern (left) end of the becah.*

Surfing: Usually a low, sloppy beach break, best at mid to low tide.

Fishing: Best off the southern head, as the beach can only be fished at high tide.

Summary: A popular spot with locals and holiday-makers, with all the facilities required for a day trip or a longer stay.

1142-1147   **PT SALISBURY, FRESHWATER PT & DECEPTION INLET**

No.	Beach	Unpatrolled			
		Rating		Type	Length
		HT	LT		
1142	Pt Salisbury	1	2	R+TF	900 m
1143	Sarina Inlet (S)	1	1	R+TF	900 m
1144	Freshwater Pt (N 1)	1	2	R+LT	200 m
1145	Freshwater Pt (N 2)	1	2	R+LT	150 m
1146	Freshwater Pt (N 3)	1	2	R+LT	100 m
1147	Deception Inlet	1	1	R+TF	650 m

Perpetua Point forms the northern entrance to the large, shallow **Sarina Inlet,** that has predominantly mangrove-fringed shores extending up to 10 km inland, where the sizeable Plane Creek flows into the system at Sarina. The inlet entrance is 2 km across, with a 50 m high peninsula ending at Freshwater Point forming the southern entrance. A 500 m long causeway across tidal flats links the peninsula with Armstrong beach. Between Perpetua and Freshwater Points are two low energy beaches in the inlet, three small, exposed sandy beaches, and the low energy Deception Inlet in Llewellyn Bay, located on the southern side of Freshwater Point.

Point Salisbury lies 1 km west of Perpetua Point, with a 900 m long, curving, south-east facing, low energy beach (**1142**) in between. The Sarina beach Road skirts the top end of the beach and there is a 4WD track to some shacks out on the southern end of the beach below Point Salisbury, but otherwise the beach is undeveloped. The beach is protected from direct wave attack by Freshwater Point, which extends 4 km to the east, and the entrance tidal shoals of Sarina Inlet. It consists of a narrow, moderately steep high tide beach fronted by 300 to 400 m wide intertidal sand flats and the deep tidal channel of the inlet (Fig. 2.17a).

Mangroves dominate the shore on the south side of Sarina Inlet until the more exposed rocky shores of the peninsula are reached. Two kilometres due south of Perpetua Point is the beginning of a 900 m long, north facing, crenulate high tide sand beach (**1143**) fronted by some sand, but predominantly intertidal rock flats between 50 and 300 m wide. There is a vehicle track to the northern end of this beach that runs off the end of the main Freshwater Point Road. The land behind this beach was in the process of being subdivided in the mid 1990's and access may have improved by now.

Freshwater Point is a 2.5 km long, east facing point backed by both cleared and wooded slopes that rise to 50 m. Four wheel drive tracks provide access to most of the shoreline, which is dominated by 100 m wide intertidal rock flats. On the exposed eastern shore are three pockets of sand. The three beaches (**1144, 1145, 1146**) all face east, are 200 m, 150 m and 100 m long respectively and consist of a narrow high tide sandy beach, fronted by a 100 m wide low tide bar, with wider rock flats to either side. There is no development on the point or at the beaches.

Deception Inlet lies 2 km north-west of Freshwater Point and faces south-south-east into Llewellyn Bay. The apparent inlet lies between two protruding, 50 m high, wooded headlands and is in fact a very low energy, 650 m long beach (**1147**) consisting of a narrow high tide beach, with fine sand to mud flats extending 400 m into the bay and mangroves fringing the base of the beach. The Freshwater Point Road runs a couple of hundred metres behind the beach and houses are being built between the road and the beach.

Swimming: All six beaches are rarely used for swimming, either because they are too low and often consist of exposed tidal flats, or in the case of the Freshwater Point beaches, are rarely visited. Use care if swimming on Freshwater Point as rips will flow out from the small beaches when there is surf across the bar.

Surfing: Only chance of a wave is on the Freshwater Point beaches.

Fishing: Sarina Inlet is one of the main fishing sites for the region. There is also rock fishing on Freshwater Point and off the Deception Inlet points.

Summary: Six little used beaches that are, however, coming under increasing pressure from development, particularly on Freshwater Point.

1148, 1149 ARMSTRONG BEACH, FIGTREE PT

No. Beach	Unpatrolled		
	Rating HT LT	Type	Length
1148 Armstrong	1 1	B+SF	3.6 km
1149 Figtree Pt	1 1	B+SF	800 m

Armstrong Beach (1148) is the only settlement on the coast between Sarina and Green Hill, 40 km to the south. Even so, this is only a small beachfront community, located 7 km east of Sarina, with about fifty holiday houses, a store and a caravan park. The 3.6 km long beach lies 4 km inside Llewellyn Bay and faces east toward the 5 km wide bay mouth, that extends from Freshwater Point to Allom Point. The beach is bordered by the rocks and rising land of the Freshwater Point peninsula in the north, with a small creek crossing the beach where it meets the rocks, and a small rock outcrop in the south. The main settlement is along the southern kilometre of the beach, although the entire beachfront is beginning to develop. The beach receives low waves and consists of a low gradient high tide beach, fronted by fine sand to muddy tidal flats that extend up to 2 km off the northern end of the beach.

Figtree Point beach (1149) lies immediately south of Armstrong beach, with a small creek and rock outcrop separating the two. This is an even lower energy, 800 m long beach, with the narrow high tide sand beach fronted by 2 km wide sand to mud flats, together with fringing rock flats and mangroves. There is no development on this beach, with the only access being across the small tidal creek.

Swimming: These are relatively safe and usually calm beaches that can only be used at high tide when the tidal flats are covered.

Surfing: None.

Fishing: Most residents either launch their boats from the creek at high tide to fish the creek and bay, or fish the creek from the shore.

Summary: A quiet, though accessible little settlement, that is gradually increasing in population and popularity.

1150-1152 ALLOM PT

No. Beach	Unpatrolled		
	Rating HT LT	Type	Length
1150 Allom Pt (W)	1 1	B+SF	600 m
1151 Allom Pt	1 2	R+rock flats	200 m
1152 Allom Pt (E 1)	1 2	R+rock flats	350 m

Allom Point is a 30 m high, wooded headland that forms the southern entrance to Llewellyn Bay. There is no vehicle access to, and no development on, the point or its adjoining beaches. To the west of the point is the low energy, largely sand-filled, tide dominated bay. Just inside the western side of the point is a 600 m long sand spit (**1150**). The spit is backed by mangroves and high tide salt flats and fronted by sand to mud flats up to 1 km wide, with mangroves increasing toward the point. A small creek drains the backing flats and exits via the southern end of the spit.

On the tip of the point is a 200 m long, north-east facing high tide sand beach (**1151**), fronted by 100 m wide, jagged rock flats with a few mangroves growing on the rocks, and dense shrub behind. On the southern side of the point is a second rockbound beach (**1152**); it is 350 m long, faces more easterly and is also dominated by jagged bordering and low tide rock flats.

Swimming: None of these beaches are suitable for swimming, except at high tide during calm conditions. Be very careful on the point beach due to the dominance of rocks.

Surfing: None.

Fishing: Best in the spit beach creek, or off the point rocks at high tide.

Summary: Three isolated beaches that are only accessible by boat, and even then with some difficulty on the rocky point beaches.

1153-1156 ALLOM PT, DAWSON CK

No. Beach	Unpatrolled		
	Rating HT LT	Type	Length
1153 Allom Pt (E 2)	1 1	UD	1.25 km
1154 Allom Pt (E 3)	1 1	B+SF	800 m
1155 Dawson Ck	1 1	B+SF	750 m
1156 Dawson	1 1	B_S rock flats	2.4 km

Between Allom and Glendower Points is an open, east facing, 6 km wide, low energy bay, with Dawson Creek located in the centre. Apart from one 4WD track that crosses the high tide flats to the southern Allom Point beach, there is no vehicle access or development on any of the beaches.

Immediately south of the Allom Point rocks are two sandy beaches that receive waves averaging less than 0.5 m. The first (**1153**) is 1.25 km long, faces north-east and consists of a narrow high tide beach backed by a 500 m wide, melaleuca-covered beach ridge plain and fronted by low gradient, 200 m wide sand flats that widen to the south. The point borders the northern end and a small area of rock flats the southern end. On the other side of the rock flats is beach **1154**, an 800 m long, east-north-east facing, low energy beach fronted by up to 1.5 km wide tidal flats and backed by a similar low beach ridge plain. This beach ends at a low rocky point that forms the northern side of Dawson Creek. The 4WD track leads to the southern end of this beach.

Four hundred metres to the south, across the Dawson Creek mouth, is beach **1155**, a 750 m long, low energy high tide beach fronted by 2 km of tidal flats and fringed by occasional mangroves. The beach faces due north and is backed by a low, 300 m wide beach ridge plain and bordered to the east by mangrove-covered rock flats.

Dawson Beach (**1156**) makes up most of the eastern shore of the 'bay'. It is 2.4 km long, faces north-north-east and is essentially a highly crenulate, rocky shore, with a narrow high tide sand beach fronted by rock and sand flats. The flats decrease in width from 2 km at the creek to 100 m at the north end of the beach.

Swimming: The Allom Point beaches offer the more attractive sandy beach sites for a swim, although only at high tide, as mangroves and rocks dominate the Dawson beaches.

Surfing: None.

Fishing: Most shore fishers head for Dawson Creek, with the beaches and rocky shore only suitable at high tide.

Summary: Four relatively isolated and difficult to access beaches with no development.

1157, 1158　GLENDOWER PT

No. Beach	Unpatrolled			
	Rating HT	LT	Type	Length
1157 Glendower Pt (W 2)	1	2	R+LT	100 m
1158 Glendower Pt (W 1)	1	2	R+LT	350 m

Glendower Point is a 60 m high, densely vegetated headland that forms the boundary between the Dawson beaches and the larger Ince Bay. There is an 8 km long 4WD track out to the point, where there are a handful of shacks right on the mainland part of the point, with another 1 km of rocks, reefs and islands extending north-east to the tip of the point. On the western side of the point are two small beaches; the first (**1157**) is located 1 km west of the point and is a 100 m long pocket of sand wedged in between rock flats and backed by overhanging trees. It consists of a narrow high tide beach and 200 m wide low tide bar, which is wider than it is long. There is no vehicle access to this beach.

The second beach (**1158**) is right on the point immediately west of the shacks. It is 350 m long and has a 30 m wide, moderately steep high tide beach fronted by 200 m wide sand flats, that are bordered by rocks and rock flats.

Swimming: Both beaches receive low waves and are relatively safe under normal conditions, though rips will form when waves exceed 0.5 m.
Surfing: None.

Fishing: The shack dwellers fish the rocks at high tide and use boats to fish the many nearby reefs and creeks.

Summary: An isolated although accessible point, rarely visited apart from the shack owners.

1159-1164　GLENDOWER PT (S)

No. Beach	Unpatrolled			
	Rating HT	LT	Type	Length
1159 Glendower Pt (S 1)	1	1	R+LT	50 m
1160 Glendower Pt (S 2)	1	1	R+LT	100 m
1161 Glendower Pt (S 3)	1	1	R+LT	100 m
1162 Glendower Pt (S 4)	1	1	B+S rock flats	400 m
1163 Glendower Pt (S 5)	1	1	B+S rock flats	400 m
1164 Glendower Pt (S 6)	1	1	B+SF	300 m

Glendower Point and the backing peninsula protrude 4 km to the north-east, tapering down into two beach-connected islets, ending at the 35 m high point. The point also forms the northern headland for Ince Bay; a low energy, 8 km wide, box-shaped bay that is bordered in the south by Cape Palmerston. The coast for 3 km south of the point is dominated by 10 to 20 m high rocky bluffs of the point, together with a series of five beaches lying between and below the bluffs, with some extensive areas of intertidal rock flats. Waves, which average just under 0.5 m at the point, decrease into the bay as extensive tidal flats begin to front the high tide beaches.

Beaches 1159 and 1160 both lie out on the narrow end of the point and connect higher grassy islets to either side. Beach **1159** is just 50 m long, faces south-east and consists of a steep, narrow high tide beach, bordered by small, irregular rocky points and low tide rock flats and backed by grassy slopes. At low tide, 50 m wide sand flats are exposed. Beach **1160** is similar, only it serves as a 100 m long link between the tip of the point and the mainland and has a steep, sandy eastern front and 100 m wide low tide sand flats, while it is backed by rocks and boulders and a few mangroves on the sheltered western side of the point. Beach **1161** lies on the eastern side of the first small islet off the point and consists of a 100 m long strip of high tide sand fronted by both rock and sand flats, with a low, rocky southern point.

Beach **1162** is firmly on the mainland and extends south for 400 m from a point it shares with the western beach 1158. The only development on the point, five fishing shacks, are located on the 10 m high northern tip of this point, overlooking the beach. The beach consists of a 20 m wide strip of high tide sand fronted by 200 m wide sand and rock flats at low tide. Beach **1163** is similar in length and orientation only with slightly wider sand and rock flats. The 4WD track to the point reaches the coast at the northern end of this beach.

Beach **1164** is the last true beach before the tidal flats. It is a 300 m long, curving, east facing strip of high tide sand below the 10 m high, densely vegetated bluffs. A group of rocks divide the beach into 200 m and 50 m high tide sections with continuous low tide sand flats extending over 500 m off the beach.

Swimming: All these beaches are best at high tide owing to the extensive sand and rock flats exposed at low tide. Waves are usually low to calm and rips absent.

Surfing: None.

Fishing: The locals fish the rocks at high tide, as well as using boats to fish the surrounding reefs and nearby tidal creeks.

Summary: This is a relatively isolated and little visited point, mainly used by the shack owners.

Ince Bay

Ince Bay has an 8 km wide entrance between Glendower Point and Cape Palmerston, and extends 4 to 5 km inland. The bay faces north-east and is well protected from the Trade wind waves by Cape Palmerston. The 35 km of bay shore is dominated by extensive low-lying sand flats and backed firstly by low beach ridges, then mangrove-fringed tidal creeks and flats extending up to 5 km inland. Only the bordering points and backing 200 to 300 m high Mount Funnel Range provide any relief. Apart from

4WD tracks to the two points and the few shacks at Glendower Point, there is no development in the bay, with the southern half lying in Cape Palmerston National Park.

Cape Palmerston National Park

Area: 7160 ha
Coast length: 28 km
Beaches: 1167-1177 (11 beaches)

Cape Palmerston National Park covers an area of 7160 ha, including 28 km of coast roughly divided equally either side of Cape Palmerston. It extends inland for up to 10 km and includes 344 m high Mount Funnel and the eastern slopes of the Mount Funnel Ranges. The only land access is via the rough, 20 km long 4WD track that runs off the Green Hill Road.

Mackay District Office
Queensland National Parks & Wildlife Service
Cnr Wood & River Streets
Mackay Qld 4740
Phone: (07) 49 51 8788

Cape Palmerston was sighted and named (after the Lord of the Admiralty) by Captain Cook in June 1770.

1165-1170 INCE BAY-CAPE PALMERSTON

No. Beach	Unpatrolled		
	Rating HT	Type LT	Length
1165 Ince Bay (W)	1 1	B+SF	1.5 km
1166 Cutlack Is	1 1	B+SF	1.5 km
1167 Hogan's Camp Is	1 1	B+SF	4.5 km
1168 Cape Palmerston (W 3)	1 1	B+SF	1.7 km
1169 Cape Palmerston (W 2)	1 1	B+SF	400 m
1170 Cape Palmerston (W 1)	1 1	B+SF	750 m

These five Ince Bay beaches are all relatively isolated, low energy and with no development, other than partly cleared grazing land behind beach 1165. Much of the bay's 40 km² is dominated by intertidal sand flats that in places extend 3 km out from the shore.

Beach **1165** lies at the north-western end of the bay. It is an irregular, 1.5 km long, east facing beach composed of a low, narrow high tide beach interrupted by rocks and mangroves and fronted by 1 to 2 km wide intertidal sand flats. A small tidal creek drains out across the northern end of the beach against the rocks of Glendower Point.

Beach **1166** occupies the irregular northern side of

Cutlack Island, a low, almost 1 km² bedrock island, backed by high tide flats and fronted by 2 to 3 km tidal flats. The beach is low, narrow and crenulate, with numerous fringing mangroves. A tidal creek flows out along the eastern side of the island.

Hogan's Camp Island beach (1167) runs along the northern side of the 4.5 km long island. The island is actually a series of low, 100 to 300 m wide beach ridges and recurved spits, backed by high tide flats and fronted by 2 to 3 km wide tidal flats, together with some mangroves along the beach. Both ends of the island curve around as spits into wide, mangrove-fringed tidal creek mouths.

Cape Creek drains the eastern end of Ince Bay and has a 2 km wide, funnel-shaped entrance between the eastern side of Hogan's Camp Island and beach **1168**. This beach is a 1.7 km long, north-west facing strip of partially cleared beach ridges and spits, backed by up to 1.5 km of mangroves and fronted by 1 km wide tidal flats. A number of mangroves also fringe the northern portion of the low energy beach. It is bordered by a small, mangrove-fringed creek to the north and a 100 m wide tributary of Cape Creek to the south. There is 4WD access to the beach across the tidal flats and a track running down the spine of the beach ridges.

Beaches 1169 and 1170 lie out on the northern tip of **Cape Palmerston** and are backed by the densely vegetated, 10 to 20 m high slopes of the cape. Beach **1169** faces north-west and consists of a narrow, 400 m long strip of high tide sand fronted by linear, 100 m wide rock flats, then 500 m wide sand flats. Beach **1170** lies 500 m west of the tip of the cape and is a headland-bound, 750 m long, north facing, sandy beach. Rock flats border the western end and dense vegetation backs and overhangs the beach, together with a small freshwater lake-marsh behind the eastern end. The beach consists of a moderately steep, 30 m wide high tide beach fronted by 500 m wide sand flats, together with a few mangroves along the base of the eastern headland. The 4WD track to the point ends on the low headland overlooking this beach.

Swimming: All these beaches have low to calm wave conditions and are relatively safe, with the best swimming toward high tide. However they are rarely used apart from fishers.

Surfing: None.

Fishing: Excellent fishing in the tidal creek, while a few also venture out to fish the rocks of the cape.

Summary: Six natural and isolated beaches, with difficult access to the cape beaches only.

1171-1176   CAPE PALMERSTON (S)

No.	Beach	Unpatrolled Rating HT LT		Type	Length
1171	Cape Palmerston (S 1)	1	2	R+rock flats	200 m
1172	Cape Palmerston (S 2)	1	2	R+rock flats	500 m
1173	Cape Palmerston (S 3)	1	2	UD	400 m
1174	Cape Palmerston (S 4)	1	2	UD	350 m
1175	Cape Palmerston (S 5)	1	2	R+rock flats	1.85 km

From **Cape Palmerston** the bedrock coast runs essentially due south for 4.5 km. The shoreline is dominated by vegetated, 10 to 20 m high bedrock bluffs, together with several small rocky points and extensive intertidal rock flats. Along the base of the bluffs and between some of the rock flats are five sandy beaches. The 20 km long vehicle track from the Green Hill Road runs along the partially cleared land behind the beaches with vehicle access to all beaches, but only two shacks on the cape.

Beach **1171** is a 200 m long, steep, high tide cobble and boulder beach lying on the north-eastern tip of the cape and backed by 30 to 40 m high, grassy bluffs. At low tide it is fronted by 50 m wide irregular rock flats.

Beach **1172** is a similar boulder beach, only 500 m long, with bluffs that decrease in height to the south where one of the two cape shacks is located. The rock flats also widen to 200 m off the southern rocks.

Beach **1173** is a 400 m long, sandy beach wedged in between two small, grassy points with rock flats extending over 200 m seaward. The beach is backed by a low foredune, with a 30 m wide high tide beach fronted by a low gradient, 200 m wide low tide bar. A low, cleared headland separates it from beach **1174**, which is similar but 350 m long, with a shorter bar section owing to rock flats that occupy half the beach.

Beach **1175** is 1.85 km long, faces due east and has an irregular shoreline owing to the rocky bluffs and rock flats. The beach consists of a narrow high tide beach fronted predominantly by 200 m wide rock flats, with two patches of low gradient sand toward the south. The second of the cape shacks is located on the southern point.

Swimming: All five beaches face into the Trades and receive waves averaging 0.5 m. These surge up the beaches at high tide and break across the low tide rocks and bar at low tide. Best swimming is on the sandy beaches and well clear of the rocky points and rock flats that dominate much of the cape beaches.

Surfing: Chance of a low, spilling surf on the sandy beaches when the Trades are blowing.

Fishing: The best shore fishing is from the points at high tide.

Summary: The cape is part of the national park and this section offers 4WD access to a string of attractive bluff- and rock-bound beaches.

1176, 1177   COCONUT PT

No. Beach	Unpatrolled		
	Rating HT LT	Type	Length
1176 Coconut Pt (N)	1 2	UD	2.5 km
1177 Coconut Pt (S)	1 2	UD	6.5 km

Coconut Point is a protruding cuspate foreland formed in lee of Temple Island, which lies 1.5 km offshore, and an extensive rock flat and reef that separates these two beaches. The Cape Palmerston access track runs usually a few hundred metres in behind the beaches, with the best access in the south along the beaches at low tide. There are no facilities and no development on the beaches, much of which lie within the national park.

Beach **1176** extends from the southern rocks of the cape to Coconut Point, a distance of 2.5 km. It initially faces south-east, swinging around to face east in lee of the island. Extensive rock flats lie off the northern point and

Coconut Point, together with a rock reef 500 m off the southern part of the beach. The beach is composed of fine sand which, with the 0.5 m waves, maintains a low, narrow high tide beach and a 200 m wide, low gradient intertidal bar. Behind the beach are older, densely vegetated beach and foredune deposits extending over 1 km inland, including Holocene and possibly Pleistocene blowouts and one larger parabolic dune.

Beach **1177** is one of the longer beaches on this part of the coast and runs from the southern side of Coconut Point for 6.5 km to the small mouth of Daintry Creek. The beach initially faces south-east, swinging to the south by the creek. The national park boundary occurs 4 km down the beach. The beach is similar to its northern extension with a wide, low gradient beach, that becomes increasingly protected to the south by offshore rock reefs and the tidal delta of the creek. The beach is backed by low, casuarina-covered foredunes that are up to 1 km wide in the north, narrowing to the south. The track from Green Hills reaches the shore a few hundred metres north of the creek mouth, with the best access along the firm beach at low tide.

Swimming: Both beaches offer long stretches of low gradient beaches, with low waves and usually free of rips.

Surfing: Chance of low, spilling breakers across the low gradient bar.

Fishing: Best in the creeks and over the rock flats at high tide.

Summary: Two long, natural beaches that can be driven at low tide.

BROAD SOUND TO SHOALWATER BAY (Fig. 4.67)

Length of Coast:	455 km
Beaches:	121 (beaches 1178-1299)
Surf Life Saving Clubs:	none
Lifeguards:	none
Major towns:	none

Regional Map 7: Broad Sound to Shoalwater Bay

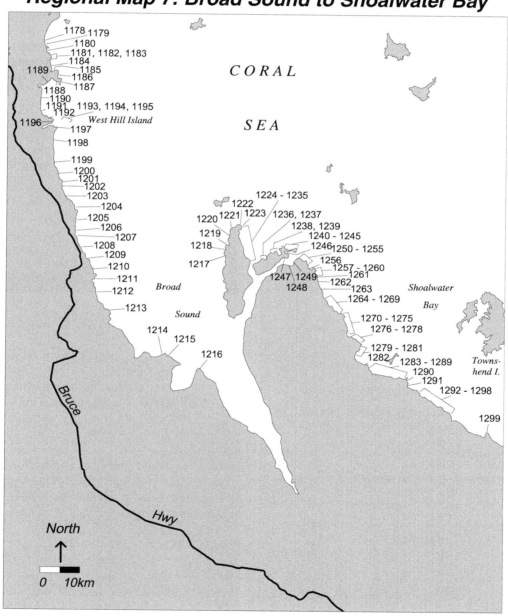

Figure 4.67   *Regional map 7 Broad Sound to Shoalwater Bay, beaches 1178 to 1299.*

1178, 1179 DAINTRY CK

No.	Beach	Unpatrolled			
		Rating HT	LT	Type	Length
1178	Daintry Ck	1	1	B+SF	750 m
1179	Little Daintry Ck	1	1	B+SF	1 km

The two Daintry creeks drain the southern slopes of the Mount Funnel Range, with prominent Mount Funnel, a volcanic neck, lying 5 km west of the northern creek mouth. There is an access track running a few hundred metres behind the two beaches, with tracks to both beaches and to about ten shacks on the southern headland.

Daintry Creek beach (**1178**) runs for 750 m between the mouths of Daintry and Little Daintry creeks. It is a sinuous, east facing beach with a narrow high tide strip of sand fronted by 500 m wide sand and rock flats and more rock reef further out. Little Daintry Creek beach (**1179**) is similar, lying between the creek mouth and the southern headland. It is 1 km long with sand and rock flats, plus extensive rock flats along the northern side of the headland that protrudes 500 m east of the beach. The access track leads to the fishing shacks strung along the eastern and southern sides of the partially cleared headland.

Swimming: Only at high tide as there are extensive tidal flats at low tide.

Surfing: None.

Fishing: Best in the two creeks and out on the rocky headland at high tide.

Summary: Two exposed beaches that are protected by the extensive tidal and rock flats.

1180-1185 KNOBLER CK, GREEN HILL, HALL CK

No.	Beach	Unpatrolled			
		Rating HT	LT	Type	Length
1180	Knobler Ck	1	1	B+SF	1.5 km
1181	Green Hill (1)	1	2	B+S/rock flats	300 m
1182	Green Hill (2)	1	2	B+S/rock flats	400 m
1183	Green Hill (3)	1	2	UD	750 m
1184	Mosquito Hill	1	2	UD	2.5 km
1185	Hall Ck	1	2	B+SF	1.2 km

Green Hill is a 50 m high hill that is surrounded by a number of beaches. Between the hill and the coast is a relatively new residential development containing about 100 housing sites. There is an 11 km long, sealed (then gravel) road from Ibibie on the highway to the settlement. This road provides access to five of the six beaches in this area.

Beach **1180** occupies a 1 km wide, east facing bay, into which Knobler Creek drains against the northern headland. The beach extends for 1.5 km south of the creek mouth to the southern headland (Fig. 4.68). The protected nature of the bay, the beach and fringing rock flats all serve to lower waves in the bay, which has permitted it to largely fill with sand. The narrow high tide beach is backed by up to ten older beach ridges, extending 500 m inland, and fronted by 500 m wide intertidal sand flats and fringing rock flats. A vehicle track off the Green Hill Road runs out to the beach.

Figure 4.68   *Knobler Creek and beach.*

Beaches 1181, 1182 and 1183 are three near-continuous beaches that fringe the rocky shore at the base of Green Hill. All three are accessible by vehicle or on foot from the Green Hill settlement. Beach **1181** is 300 m long, faces north-east and is bordered by sand-covered rocks to the north and a prominent rock flat to the south. Beach **1182** lies on the south side of these rocks and is 400 m long, faces more easterly and is also bordered by extensive intertidal rock flats. Beach **1183** lies at the foot of the settlement and runs from the northern rock flats to the mouth of a tidal creek that flows across the beach.

Beach **1184** begins on the southern side of the 50 m wide creek and runs for 2.5 km south to the mouth of the larger Hall Creek, that runs out along the side of a 2 km long headland. This beach is backed by a 500 to 600 m wide series of ten densely vegetated foredune ridges, with the two creeks linking to surround the beach with creeks and high tide flats. However, there is vehicle access across the high tide salt flats to the beach. Sixty metre high Mosquito Hill lies 2 km west of the beach. Beach **1185** is located along the north facing shore of the headland and is paralleled by the tidal channels and flats of Hall

Creek. It is located on private land accessible from the Yarrawonga track.

Swimming: All these sandy beaches offer relatively safe swimming toward high tide, with conditions usually too shallow toward low tide. Be careful near the creek mouths as they have deep channels and strong tidal currents.

Surfing: During strong Trades there are usually low, spilling breakers across the bars on the more exposed beaches.

Fishing: This is a popular location owing to the number of tidal creeks and rock reefs.

Summary: A small but growing settlement and an area likely to see increasing development and access.

1186, 1187 NOTCH & YARRAWONGA PTS

	Beach	**Unpatrolled** Rating		Type	Length
No.		HT	LT		
1186	Notch Pt	1	2	R+LT	1.8 km
1187	Yarrawonga Pt	1	2	R+LT	1.2 km

Notch and Yarrawonga Points are two prominent 20 to 30 m high headlands that protrude about 5 km to the east. They are accessible via a 12 km long gravel then 4WD track off the Green Hill Road. There are approximately fifty shacks around Yarrawonga Point, while Notch Point is backed by private land.

The **Notch Point** beach (1186) is 1.8 km long, faces due east and is bordered by low, cleared headlands, together with a creek running out against the southern headland. The beach consists of a moderately steep high tide beach fronted by a low tide bar that widens to 100 m at low tide. It is backed by low foredunes extending up to 300 m west of the beach, which in turn are backed by a now drained lake and the tidal creek. There is a private track to a dwelling on the northern headland.

Yarrawonga Point beach (1187) is similar. It is 1.2 km long, faces east and is slightly more exposed, with a moderately steep, 50 m wide high tide beach and 50 to 100 m wide low tide bar. The beach is backed by 10 to 30 m high, vegetated blowouts. There are about thirty shacks on the foredune behind the southern end of the beach and up on Yarrawonga Point (Fig. 4.69).

Swimming: These are two sandy beaches with deep water off the high tide beach and a low, shallow bar at low tide. Waves are usually less than 0.5 m and rips are rare.

Surfing: Only a chance of low, spilling breakers toward low tide.

Fishing: The nearby creeks and rocks at high tide offer the best locations.

Summary: Two attractive, headland-bound beaches, with reasonable access to Yarrawonga.

Figure 4.69   *Yarrawonga Point.*

1188-1192 MARION CK, FOUR MILE

	Beach	**Unpatrolled** Rating		Type	Length
No.		HT	LT		
1188	Yarrawonga Pt (west)	1	1	B+SF	1.5 km
1189	Marion Ck	1	1	B+SF	2.8 km
1190	Four Mile	1	1	B+SF	5.3 km
1191	Four Mile (S)	1	1	B+SF	1.3 km
1192	West Hill (N)	1	1	B+SF	2.3 km

Between Yarrawonga Point and West Hill Island is an unnamed bay, 8 km wide at the entrance between the two points. It extends 5 km inland and contains 20 km of low energy shoreline, dominated by narrow, sandy beaches, extensive intertidal sand flats, and backing tidal creeks and mangrove-covered tidal flats, with the creek channels meandering across the tidal flats at low tide. There are vehicle tracks to the backing five beaches, but apart from some patches of cleared grazing land, there is no development.

The beach on the western side of **Yarrawonga Point** (1188) is a very low energy, highly crenulate strip of sand that snakes off the west of the point and faces generally south-west into the bay. The beach consists of a narrow, relatively steep high tide beach fronted by patches of mangroves and intertidal sand flats 200 to 300 m wide, which end at the meandering channels of Marion Creek. The beach is backed by some low, casuarina-covered beach ridges and patches of cleared sloping land, with the Yarrawonga Point track running roughly parallel to the back of the beach. A number of shacks are located on Yarrawonga Point and overlook the eastern end of the beach, together with a few on the low beach ridges.

Marion Creek beach (**1189**) is a 2.8 km long, straight, south-east facing beach bordered by 100 m wide Marion Creek to the north and 300 m wide Gillinbin Creek to the south. Both creeks drain the coastal range as well as having extensive mangrove-covered tidal systems. The high tide beach is relatively narrow and steep and fronted by 1 to 2 km of intertidal sand flats. It is backed by 200 to 300 m of densely vegetated beach ridges and recurved spits that have built northward along the beach over the past few thousand years. These in turn are backed by extensive mangrove-covered tidal flats in the north, with slightly higher, partially cleared land in the south, on which there is a vehicle track out to the beach.

Four Mile Beach (1190) lies on the southern side of Gillinbin Creek. The meandering, mangrove-fringed creek runs parallel to the 5.3 km long beach for 3 km, with a channel extending over 1 km out across the tidal flats east of the creek mouth. The beach is relatively straight, faces due east and has a 50 m wide, moderate gradient high tide beach fronted by the 1 to 2 km of tidal flats. A small creek forms the southern boundary. The beach is part of the massive sand spit that has been moving sand northward along the beach for the past few thousand years, in the process building out the beach 200 to 300 m and leaving behind a series of recurved spits that today form the eastern shore of the backing creek. The entire spit is densely vegetated, with a straight vehicle track leading from some cleared paddocks to the southern portion of the beach.

Immediately south of Four Mile beach is a smaller, 1.3 km long, east-north-east facing strip of high tide sand (beach **1191**), lying between the small unnamed creek and slightly larger southern unnamed creek. It is fronted by wide tidal flats and backed by a 50 to 100 m wide vegetated beach ridge and 200 m wide high tide salt flats. A vehicle track runs out over the salt flats to the beach.

West Hill (north) beach (**1192**) borders the northern side of some higher ground that is part of the spine of bedrock running out toward the large West Hill Island. The beach is 2.3 km long, faces east-north-east and consists of a crenulate, low energy beach. It is attached to the bedrock in the east, while to the west it becomes a low spit extending into the bay. There is a shack settlement just around the eastern end, from which the beach is accessible on foot.

Swimming: All five beaches are essentially high tide strips of sand fronted by wide tidal flats and, as a consequence, can only be used for swimming at high tide. While waves are low to often calm, be careful near the creeks as they have deep channels and strong tidal currents.

Surfing: None.

Fishing: Excellent fishing in the creeks.

Summary: Five relatively isolated, undeveloped beaches, but accessible to those with a good map and a 4WD.

West Hill Island and West Hill National Parks

West Hill Island National Park
Area: 398 ha
Shoreline length: 6 km
Beaches: 1193-1195 (3 beaches)

West Hill National Park
Area: 685 ha
Shoreline length: 6 km
Beaches: 1197-1198 (2 beaches)

For further information:
Mackay District Office
Queensland National Parks & Wildlife Service
Cnr Wood & River Streets
Mackay Qld 4740
Phone: (07) 49 51 8788

1193-1195 WEST HILL ISLAND

No.	Beach	Rating HT	Rating LT	Type	Length
		Unpatrolled			
1193	West Hill Is (1)	1	2	B+S rock flats	1.5 km
1194	West Hill Is (2)	1	2	B+rock flats	800 m
1195	West Hill Is (3)	1	2	B+rock flats	400 m

West Head Island is a cone shaped, 4 km² island peaking at 282 m high West Hill. The entire island is part of West Hill National Park and is covered in densely vegetated slopes that include rain forest, eucalypt and melaleuca forests. While the hill is an island at high tide, it can be reached at low tide across 300 m wide tidal flats. Most of the 6 km of island shore is rocky, with three narrow beaches located on the more protected northern and north-east shores.

Beach **1193** lies on the north-west tip of the island. It is backed by steeply rising slopes that decrease in height to the west where they end at a small sand spit. The beach is fronted by 500 m wide tidal flats that narrow to the east, while 50 m wide rock flats begin to run along the back of the beach. Beach **1194** extends for 800 m along the northern face of the island and is a narrow high tide strip of sand, fronted by 50 to 100 m wide intertidal rock flats. Beach **1195** lies on the north-eastern, east facing side and is similar with a narrow, 400 m long high tide beach fronted by rock flats. The lower slopes behind this beach have been cleared in the past.

Swimming: Only the more protected beach 1193 is suitable for swimming as the others have extensive rock flats.

Surfing: None.

Fishing: Best off the eastern beaches and rocky shore at high tide, as well as in the channel between the island and the mainland.

Summary: A small island national park containing luxuriant vegetation and a variable sandy and rocky shore.

1196-1199 WEST HILL-CARMILLA

No.	Beach	Unpatrolled			
		Rating HT	LT	Type	Length
1196	West Hill Pt	1	1	B+SF	1 km
1197	West Hill Ck	1	2	B+SF	800 m
1198	Blind Ck	1	2	UD	6.2 km
1199	Carmila	1	1	B+SF	2 km

In lee of West Hill Island lies the 400 m wide mouth of West Hill Creek, with bedrock shore to the north and the northern end of a 6 km long barrier system to the south, 7 km^2 of which lies within the West Hill National Park. Two low energy beaches lie either side of the creek mouth, while the barrier is the last relatively high energy beach on this section of coast before the more protected beaches of Broad Sound.

Beach **1196** runs along the south-east facing, eastern shore of the bedrock bluffs in lee of the island. It consists of low, densely vegetated bedrock country, cut off from the mainland by high tide flats. However, there is a 4WD track across the flats and along the northern side of the creek to a string of about 30 shacks located along the crenulate, 1 km long beach. The beach consists of an irregular, steep, narrow high tide beach fronted by rocks and sand flats that are 200 m wide in the south, widening to link with the island in the north.

Beach **1197** lies on the south side of the creek mouth and is a curving, north facing, 800 m long, low energy beach fronted by sand flats up to 300 m wide, then the deep channel of West Hill Creek. Mangroves dominate both ends of the beach. A small, 20 m high headland forms the western boundary.

Beach **1198** is a 6.2 km long, slightly curving, east facing beach, the northern half of which makes up the bulk of **West Hill National Park**. The beach is anchored by a small, northern headland and runs to the south of the mouth of Blind Creek, which drains part of the backing tidal areas. The beach is part of a beach-to-foredune ridge barrier system that has prograded up to 1 km seaward in the north, while the southern 1.5 km consist of an elongate spit backed by mangroves and high tide flats. The beach has a moderate gradient high tide beach that widens to 100 m at low tide, with some rock patches toward the southern end; remnants of the old Blind Creek mouth. Just south of the national park boundary is freehold land and private road access to a beachfront property.

Carmila beach (1199) provides the first good vehicle access to the coast south of Green Hill, and the best boat ramp. There is a 6 km long road out through the sugar cane farms from the small highway stop of Carmila to the creek mouth of the same name. There is a concrete boat ramp and small shack settlement on the southern side of the creek mouth and a camping reserve with limited facilities immediately to the south. A track runs the length of the 2 km long beach down to the mouth of Feather Creek, that forms the southern boundary. Carmila Creek runs out in lee of some shelly beachrock that has anchored the creek mouth. The beach faces due east, however it is relatively protected by the offshore Flat Islands and extensive tidal flats. It has a relatively steep, 30 to 50 m wide high tide beach, with low tide exposing up to 1 km of intertidal sand flats (Figs 1.26a & 4.70).

Swimming: Carmila is the most popular spot, particularly at high tide, with the long Blind Creek beach also offering relatively safe swimming. However be very careful of the deep channel and strong tidal currents of both West Hill and Carmila Creeks.

Surfing: Only low waves on Blind Creek beach.

Fishing: Both creeks are a haven for fishers, with boat access to the outside rocks and reefs at high tide as well.

Summary: A range of beaches and largely undeveloped West Hill National Park. The beaches include mangrove-covered tidal flats and the last of the wave-dominated beaches before the tide-dominated Broad Sound is entered.

Figure 4.70 *A view from Carmilla high tide beach out across the 1 km wide tidal sand flats.*

1200-1204 THREE MILE-OAKY CKS

No.	Beach	Unpatrolled		Type	Length
		HT	LT		
1200	Three Mile	1	1	B+SF	3.4 km
1201	Flaggy Rock Ck	1	1	B+SF	800 m
1202	Stockyard Ck	1	1	B+SF	700 m
1203	Lantana Ck	1	1	B+SF	4 km
1204	Oaky Ck	1	1	B+SF	4 km

South of Carmila beach the coast continues on as a series of increasingly low energy beaches, bordered by numerous creeks and fronted by extensive intertidal sand flats, with mangroves increasing toward Clairview. While the highway runs just a few kilometres inland, there is only 4WD access to the coast just north of Clairview. Between Feather and Oaky Creeks is 15 km of coast consisting of five beaches, all bordered by creeks draining the backing Connors Range. The land between the highway and the coast increasingly gives way from sugar farms to cattle grazing areas.

Three Mile Beach (1200) is 3.4 km long, faces east and is bordered by Feather Creek in the north and Flaggy Rock Creek to the south. It consists of a narrow high tide beach backed by 200 m of low dunes, then scrubby coastal plain. The beach is fronted by 1 to 2 km wide, rock and sand tidal flats that are widest and rockiest off the creek mouth. A few mangroves also grow at the mouth of Flaggy Rock Creek. Two 4WD tracks provide access to the creek mouths at either end of the beach.

Flaggy Rock Creek beach (1201) runs from the southern side of Flaggy Rock Creek mouth for just 800 m to the mouth of smaller Stockyard Creek. Due to the presence of the creeks, the beach consists of two adjoining sets of beach ridge-recurved spits, with the whole beach system protruding 200 m into the sound. The entire barrier system is densely vegetated, with no development other than an access track from the coastal plain across some high tide salt flats to the southern side of Flaggy Rock Creek mouth, where a few shacks are located. The beach is fronted by 1 km wide tidal flats.

Stockyard Creek beach (1202) is also wedged between two creek mouths, Stockyard and Lantana, located only 700 m apart. The beach runs essentially straight between the creeks and consists of a low beach ridge fronted by 1 km wide tidal flats and backed by a 500 m wide series of eight beach ridges, then some high tide salt flats. There is an access track to the beach across the flats and ridges, but no development.

Lantana Creek beach (1203) represents the end of the 22 km of prograding beach and foredune coast between

West Hill and the bedrock that backs the southern end of the 4 km long beach. The beach begins at the joint mouth of Lantana and Bluewater Creeks, where it forms a 300 m long spit, and runs to the south-east as a slightly curving beach to where the gently rising coastal plain reaches the coast. The narrow high tide beach is fronted by 1 km wide tidal flats and backed by a series of older recurved spits along the northern 1.5 km, which represent northward migration of Bluewater Creek. To the south it is initially backed by a 200 to 300 m wide beach ridge plain, then the coastal plain for the southern 1 km. The beach ends at a small unnamed creek. A 4WD track parallels the back of the southern half of the beach.

Oaky Creek beach (1204) extends for another 4 km from the southern side of the unnamed creek to the prominent mouth of Oaky Creek. The entire beach is backed by the low coastal plain and consequently has a crenulate, east-north-east facing shoreline, with rock flats forming the points in the crenulations, and 1 km wide tidal flats paralleling the beach. The beach is backed by the scrubby coastal plain with a 4WD track leading to the northern end of the beach.

Swimming: All these beaches usually receive low waves and are calm when there is no wind. They are only suitable for swimming at high tide because of the tidal flats, however stay clear of the many creek mouths, which have deep channels and strong tidal currents.

Surfing: None.

Fishing: Best in the many creeks.

Summary: Five undeveloped, natural, low energy beaches facing into the high tides of Broad Sound and all fronted by extensive intertidal sand and rock flats, together with an increasing scattering of mangroves.

Broad Sound

Shoreline length: 225 km
Beaches: 1205-1221 (16 beaches)

Broad Sound is just that: a broad, 35 km wide, 70 km long, funnel-shaped sound that is bordered in the west by the coast south of Oaky Creek and a broad peninsula, and by Long Island to the east. The sound has 250 km of shoreline that receives essentially no ocean waves, but has the highest tide range in eastern Australia, reaching as much as 8 m deep in the sound. A number of small tidal creeks draining off the Connors Range enter the western side of the sound, with the larger funnel-shaped St Lawrence, Waverley, Styx and Herbert Creeks further south, but no major rivers. The shoreline is predominantly wide, low energy tidal flats and much is covered by mangroves, particularly the southern and eastern shores.

Surfing: None.

Fishing: Best off the rocks into Middle Passage.

Summary: Five low energy beaches/tidal flats.

Quail Island

Quail Island is a generally low, undulating, 12 km long island that is separated from the mainland by 1 km wide Thirsty Sound. The entire island is a grazing property, with a single homestead and a few buildings behind beach 1238 and a permanent population of one. Most of the island is covered by woodlands, with the land generally below 40 m in height, apart from two eastern peaks, an unnamed 98 m high hill and 104 m high Pier Head. The island landing lies just 3 km due west of the Plum Tree landing. The island has an area of approximately 20 km^2 with about 20 km of coastline. Mangroves fringe the southern half along the Thirsty Sound and Middle Passage shores, while there are 12 beaches bordered by rocky headlands along the northern and eastern shores.

1236-1239   QUAIL ISLAND

No.	Beach	Unpatrolled Rating HT LT	Type	Length
1236	Quail Is (1)	1 2	B+rock flats	600 m
1237	Quail Is (2)	1 2	B+rock flats	800 m
1238	Quail Is (3)	1 2	B+rock flats	200 m
1239	Quail Is (4)	1 2	B+rock flats	800 m

The north shore of Quail Island is a mixture of three protruding rocky points and mangrove-filled bays. Along the western and central points are four low energy, narrow high tide beaches fronted by varying degrees of rock and sand flats. All are backed by moderately steep wooded slopes, with no formal vehicle access to any of the beaches.

Beaches **1236** and **1237** lie side by side on the western point, facing north into Sand Bank Bay. They are 600 m and 700 m long respectively and consist of narrow high tide beaches fronted by 50 to 100 m of rock flats, then more extensive sand flats, with a few mangroves scattered amongst the rocks.

Beaches **1238** and **1239** occupy the central point and consist of a 200 m pocket of sand and a longer 800 m strip of high tide sand, both backed by wooded slopes that rise to about 50 m. Both beaches are fronted by irregular, 100 m to 200 m wide rock flats.

Swimming: While waves are usually low along these beaches, they are only suitable for swimming at high tide owing to the prevalence of intertidal rocks and tide flats.

Surfing: None.

Fishing: Best off the rocks and points at high tide.

Summary: Four beaches best accessed by boat at high tide.

1240-1245   LUCY RAVEL PT

No.	Beach	Unpatrolled Rating HT LT	Type	Length
1240	Lucy Ravel Pt (W 2)	1 1	R +LT	200 m
1241	Lucy Ravel Pt (W 1)	1 1	Cobble +LT	150 m
1242	Lucy Ravel Pt	1 1	UD	150 m
1243	Lucy Ravel Pt (E 1)	1 1	UD	200 m
1244	Lucy Ravel Pt (E 2)	1 1	UD	400 m
1245	Pier Head	2 2	Boulder	200 m

Between Lucy Ravel Point and Pier Head is 4 km of rock-dominated shore, containing five small and relatively exposed sandy beaches around the point, and a boulder beach on the head. There are patches of cleared land around the point and a holiday shack overlooking beach 1243.

Beaches 1240, 1241 and 1242 are three small pockets of sand immediately west of 20 m high **Lucy Ravel Point**. They face north-north-west and receive waves averaging less than 0.5 m. Beach **1240** is 200 m long and is almost encircled by low rocky points and rock flats, and it is encircled by a low, narrow dune then mangroves. Beach **1241** is a 150 m long high tide cobble beach with a 50 m wide bar, and is backed by partially cleared land. Beach **1242** faces due north and consists of a low gradient, 150 m wide beach with a scattering of rocks and two rocky points.

Beaches **1243** and **1244** face north-east, exposing them to slightly higher waves. They are 200 m and 400 m long respectively, separated by a 20 m high, cleared headland with a solitary shack. Both beaches have 200 m wide intertidal beaches, with longer rocky points and rock flats bordering each beach.

Pier Head beach (**1245**) lies 2 km further east on the north side of 105 m high Pier Head. The head is connected to the rest of the island by a narrow, 400 m long strip of boulders that is almost awash at high tide. The beach is a 200 m long, north facing, cobble-boulder beach wedged in between two rocky points.

Swimming: The best beaches are beaches 1243 and 1244 which offer the most sand and a chance of low waves. Be careful of the intertidal and boundary rocks on all beaches.

Surfing: None.

Fishing: Best off the many rocky points at high tide.

Summary: This is Quail Island's north coast and more attractive shore, with a scattering of slightly more exposed pocket sandy beaches.

1246, 1247   PIER HEAD

No.	Beach	Unpatrolled		Type	Length
		Rating HT	LT		
1246	Pier Head (W)	1	1	B+SF	700 m
1247	Quail Is landing	1	1	B+SF	1 km

The south-east side of the island faces into Thirsty Sound and across to the mainland settlement of Plum Tree. Despite its south-east orientation it is largely protected from waves by the mainland and Arthur Point. **Beach 1246** lies in a 2 km deep bay that extends west of Pier Head. It is a very low energy, 700 m long beach, with the high tide beach fronted by patches of mangroves and a 1.5 km wide tidal flat that fills the bay.

Beach **1247** is the site of the island landing and homestead. It is 1 km long and fronted by a mixture of sand and rock flats and mangroves, with the deep water and strong currents of Thirsty Sound lying off the 200 m to 500 m wide tidal flats.

Swimming: It is only possible to swim at high tide owing to the tidal flats. Beware of the strong currents in Thirsty Sound.

Surfing: None.

Fishing: Best off the rocks at the southern end of the landing beach, which provides shore access to the deeper sound waters.

Summary: Two low energy beach-tidal flats, with the landing being the focus of island activity and transport.

1248, 1249   PLUM TREE

No.	Beach	Unpatrolled		Type	Length
		Rating HT	LT		
1248	Parker Ck	1	1	B+SF	2.5 km
1249	Plum Tree	1	2	R+beachrock	700 m

On the southern side of Thirsty Sound, opposite Quail Island, is the small but growing settlement of Plum Tree - the only settlement on the entire peninsula and exactly 100 km by gravel road from the highway at Kootandra. It has a store, fuel, caravan park and a good boat ramp. Most of the permanent residents who live in the area are fishermen who fish Broad Sound and Shoalwater Bay. While it is a long drive out to this area, it does have a beautiful coastline, with clean, sandy beaches and numerous headlands and creeks, as well as the biggest tides in eastern Australia.

Beach **1248** lies just west of Plum Tree and consists of a 2.5 km long, north facing high tide beach with tidal flats up to 800 m wide. The beach is backed by densely vegetated, rising slopes that are drained by four creeks. The main mangrove-fringed Parker Creek forms the eastern boundary of the beach, with three smaller creeks crossing the beach to the west.

Plum Tree beach (1249) is more beachrock than beach and consists of a narrow, north-west facing high tide strip of sand fronted by a wider band of crenulate, coarse, boulder beachrock, that curves around toward Parker Creek mouth. The boat ramp is located toward the eastern end of the beach, with the store and settlement close behind.

Swimming: Neither beach is really suitable for swimming, because of the tidal flats in the west and beachrock at Plum Tree. In addition there are very strong tidal currents through the sound off both beaches.

Surfing: None.

Fishing: This is the base for most fishing in the surrounding sounds and bays, and Plum Tree has a very busy boat ramp used by both professional and recreational fishers.

Summary: Plum Tree is the focus of the peninsula and a good spot to obtain information on the many surrounding beaches, fishing spots and points of interest.

1250-1253 ARTHUR PT

No.	Beach	Unpatrolled Rating HT	LT	Type	Length
1250	Arthur Pt (W)	1	2	R+rock flats	200 m
1251	Arthur Pt	1	2	R+rock flats	300 m
1252	Arthur Pt (E 1)	1	2	R+LT+rocks	1.3 km
1253	Arthur Pt (E 2)	1	2	R+rock flats	200 m

Arthur Point is the northernmost point of the peninsula. It is a 30 m high conical hill, backed by a second larger, 100 m high hill. Two small beaches lie to either side of the point, with the coast to the east containing a series of sandy beaches well down into Shoalwater Bay.

Beaches **1250** and **1251** lie to either side of the point; they are 200 m and 300 m long respectively, and both consist of high tide sandy beaches fronted by rocky intertidal flats, with prominent rocky boundary headlands. Waves average less than 0.5 m and surge up the beaches at high tide, but break amongst the rocks at low tide.

Beach **1252** is the main Arthur Point beach, with vehicle access to the southern corner where the steep beach is used to launch boats. The beach is 1.3 km long and faces north-east, swinging around in the more protected southern corner to face north, where there are even a few mangroves against the point. The beach is composed of relatively coarse sand which, with the low waves, maintains a steep beach face fronted at low tide by a 200 m wide continuous bar. Toward the northern end is a rocky islet just off the beach, with rock reefs extending to the beach. The beach is backed by a low foredune that contains about 30 shacks, while the backing land, surrounding a small creek, is also being developed for housing.

Beach **1253** lies out on the tip of the eastern headland and consists of a 150 m long high tide beach fronted by 100 m wide rock flats at low tide. The beach is backed by a 30 m high conical hill, with a smaller grass-topped rockstack located on the rock platform and cut off from the shore at high tide.

Swimming: Beach 1252 is one of the main recreational beaches in the area and a popular swimming location. It is relatively safe in the protected southern corner, apart from the boat traffic. The other three beaches all have substantial areas of low tide rocks.

Surfing: Chance of a low, spilling breaker at beach 1252 during strong easterly winds and waves.

Fishing: The many rocky point and platforms are the most suitable shore-based locations.

Summary: Most visitors head for the main Arthur Point beach and it is well worth a visit, offering a moderately low, relatively safe sandy beach, as well as a place to launch a small boat.

1254-1256 ALLIGATOR & STANAGE BAYS

No.	Beach	Unpatrolled Rating HT	LT	Type	Length
1254	Alligator Bay (north)	1	2	R+rock flats	400 m
1255	Alligator Bay	1	2	R+LT+rocks	600 m
1256	Stanage Bay	1	2	UD	1.7 km

Alligator and Stanage Bays are the next two bays south of the eastern Arthur Point headland. The two bays face north-east and are backed by mainly cleared, gently rising land, with the main Stanage Bay Road running less than 1 km west of the two bays, providing road access to each.

Beach **1254** lies at the northern end of **Alligator Bay** and consists of a 400 m long, discontinuous, relatively steep high tide beach, backed and surrounded by jagged rocks and fronted by irregular intertidal rock flats and reefs. There are a few houses on the backing bluffs that overlook the adjoining main Arthur Point beach.

Beach **1255** is the main Alligator Bay beach and is a continuous sandy beach bordered by 20 m high headlands. The beach is moderately steep at high tide, with a 50 m wide low tide bar, then more rock offshore.

Stanage Bay (1256) is the second main bay on the northern peninsula and contains one of the nicer beaches. It is a 1.7 km long, curving, low gradient beach, broken in the centre by a 20 m high islet connected to the low tide bar. Stanage Point, capped by 70 m high Bald Hill, forms the southern boundary, with a 20 m high cleared headland separating it from Alligator Bay. The beach receives waves averaging less than 0.5 m that come right up the beach at high tide, while low tide reveals a 200 m wide bar. There are two creeks draining across the beach, one north of the islet and a second in the southern corner. The beach is backed by cleared land of the Stanage Bay homestead, which is located 400 m in from the beach.

Swimming: Stanage Bay is one of the safer beaches in the area, with usually low waves spilling across the shallow bar.

Surfing: Low, rideable waves break in Stanage Bay only during strong easterly winds and waves.

Fishing: Best off the rocky point and platforms, and in the two Stanage Bay creeks.

Summary: Three headland-bound beaches with Alligator and Stanage Bays offering the best spots for swimming and recreation.

1257-1259   STANAGE PT

No. Beach	Unpatrolled			
	Rating HT	LT	Type	Length
1257 Stanage Pt (1)	1	3	cobble-boulder	100 m
1258 Stanage Pt (2)	1	2	R+LT	250 m
1259 Corisande Hills (N)	1	2	R+LT	800 m

Stanage Point is a substantial 700 m wide headland beyond which the coast trends south-east for 23 km to Macdonald Point. In between are a series of sandy beaches which gradually decrease in wave energy, owing to the increasing protection of Shoalwater Bay. The first three beaches lie between Stanage Point and a 20 m high headland 2 km to the south. The headland and backing land have been cleared for grazing, and farm tracks of the Stanage Bay property lead to each of the beaches.

Beach **1257** is a 100 m long pocket of south-east facing, cobble high tide beach fronted by intertidal rock flats. The beach is backed by an amphitheatre of cleared slopes rising to 70 m on Bald Hill.

Beach **1258** occupies the same small bay and is a 250 m long, east-north-east facing sand beach, backed by a low, vegetated foredune then cleared slopes, and fronted by a sloping, 80 m wide low tide bar. Cleared, rocky headlands protrude 300 m to either side with rocks and reefs rimming their shores.

Beach **1259** is embayed by easterly extensions of the Corisande Hills, which rise to 140 m within 2 km of the coast. The 800 m long beach lies between two protruding, 20 m high cleared headlands. The beach faces northeast, is relatively straight and backed by a low, casuarina-covered foredune, with a small creek draining out across the southern end, and then cleared grazing land. The beach has a moderate gradient and is 80 m wide at low tide.

Swimming: All three beaches receive waves averaging less than 0.5 m and the two sandy beaches are relatively safe with no rips.

Surfing: Usually none.

Fishing: All three beaches are bordered by rocky shore which provides good access to deeper water, particularly at high tide.

Summary: Three embayed beaches, backed by sloping grazing land.

1260-1263   CORISANDE HILLS-HOLLIS CK

No. Beach	Unpatrolled			
	Rating HT	LT	Type	Length
1260 Corisande Hills (S 1)	1	2	R+LT	50 m
1261 Corisande Hills (S 2)	1	2	R+LT	500 m
1262 Hollis Ck (N)	1	2	R+LT	1 km
1263 Hollis Ck	1	2	R+LT	3.2 km

Beaches 1260 and 1261 both occupy a small, 700 m wide, east facing bay, bordered by rocky platforms and reefs extending out 400 m at either end of the bay. Beach **1260** is a 50 m long pocket of sand wedged in below low bluffs and bordered by rocks and rock reefs. Its neighbour beach **1261** is a 500 m long, irregular sandy beach that begins below bluffs, while to the south it becomes increasingly dominated by rocks and rock flats. Both beaches receive waves averaging less than 0.5 m and have a moderate gradient, 80 m wide intertidal beach.

Beach **1262** is linked by a sand tombolo to beach 1261 and they also share a small tidal creek mouth, that has meandering channels draining across either beach at times. This beach is 1 km long and runs to a southern cuspate foreland, formed in lee of a rock reef just off the low tide beach. The beach has a low gradient and is 100 m wide at low tide. It is backed by the creek and a low, 100 m wide, active, although casuarina-covered, foredune.

On the southern side of the foreland is one of the longer beaches in the bay - 3.2 km long beach **1263** that ends at the mouth of Hollis Creek. This beach, which lies in lee of extensive rock reefs 1 km offshore, has built out over the last few thousand years and is now backed by up to 1 km of foredune ridges, then the wetland of Hollis Creek. Part of the dune system has been subdivided and is being developed for housing, with a service road across the northern end of the creek. The beach is moderately steep, with a 100 m wide intertidal bar that merges with 500 m wide tidal shoals at the meandering creek mouth.

Swimming: All four beaches are relatively safe under the usually low wave conditions. Watch for tidal currents at the creek mouths and rocks and reefs near the headlands.

Surfing: Only low, spilling waves during strong easterly winds.

Fishing: Best in the two creeks or from the many rocks at high tide.

Summary: Hollis Creek beach is the first beach readily accessible to the public south of Arthur Point, while all four beaches offer clear sand and relatively safe swimming.

1264-1267 THE SHACK

No. Beach	Unpatrolled			
	Rating HT	LT	Type	Length
1264 The Shack	1	1	B+SF	700 m
1265 The Shack (S)	1	1	B+SF	600 m
1266 Beach 1266	1	2	B+S/rock	400 m
1267 Beach 1267	1	2	R+LT	1.5 km

South of Hollis Creek mouth the coast is dominated by the lower slopes of the Roger Hills, an 8 km line of hills that rise to 140 m and parallel the coast 2 to 3 km inland. The coast continues to be dominantly sandy, with the beaches wedged in between rocky bluffs.

On the southern side of Hollis Creek is a 700 m long strip of sand (beach **1264**) that lies in lee of the extensive tidal shoals and a few rocks off the creek mouth. This is a low energy beach backed by a low, vegetated foredune and gently rising wooded slope, with a small creek exiting across the southern end of the beach. There are also a few shacks beside Hollis Creek mouth.

Beaches **1265** and **1266** are two crenulate, north-east facing beaches 600 m and 400 m in length respectively. They are dominated by the coastal bluffs and consist of 30 m wide high tide beaches with 50 to 80 m wide intertidal bars, backed by 10 m high bluffs and an occasional low foredune, and interrupted in places by rocks and rock reefs. A small protruding bluff separates the two beaches, with a few shacks located on the bluff overlooking the southern beach. Beach 1266 ends at a cuspate foreland in lee of a small offshore reef.

Beach **1267** is a more embayed beach lying between the sandy foreland and a low headland, with a small creek occasionally breaking out across the centre of the beach. It is 1.5 km long, faces north-east and has a sandy high tide beach fronted by a 100 m wide low gradient bar, which is broken in the centre by extensive rocks on the beach and an inshore reef.

Swimming: These are four generally low wave beaches best suited for swimming toward high tide. Be careful of the numerous rocks and reefs on the southern beaches.

Surfing: Usually none.

Fishing: Best off the many rocks around high tide.

Summary: Four natural, rock-dominated beaches only used by the shack owners.

1268-1271 YENYARDINDLE HUTS

No. Beach	Unpatrolled			
	Rating HT	LT	Type	Length
1268 Beach 1268	1	2	R+LT	1.4 km
1269 Yenyardindle Huts (W)	1	2	UD	1 km
1270 Yenyardindle Huts	1	2	R+LT	400 m
1271 Yenyardindle Huts (E)	1	2	R+LT	450 m

The **Yenyardindle Huts** are a collection of four shacks located on 20 m high grassy bluffs overlooking beaches 1269 and 1270. There is 4WD access to the huts, with tracks leading to the adjoining beaches.

Beach **1268** lies 2 km north-west of the huts and is a 1.4 km long, north-east facing beach located between two low, wooded headlands and backed by a 50 to 100 m wide, casuarina-covered foredune, with a small creek exiting against the southern rocks. There are rocky platforms and reefs extending off each headland, with the sandy beach curving in between. The beach receives waves averaging less than 0.5 m and is composed of medium sand, which produces a moderately steep, 30 m wide high tide beach fronted by a continuous, 50 to 100 m wide low tide bar. The vehicle track to the huts runs just south of the southern end of the beach and there is a 4WD track along the back of the foredune.

Beach **1269** lies immediately to the west of the huts and 500 m around the rocks from beach 1268. It is a 1 km long, north-east facing beach, very similar to its northern neighbour, with two low headlands and a slightly larger tidal creek against its southern rocks. However the beach, while receiving similar waves, is composed of finer sand and as a consequence has a low gradient high tide beach that extends for 200 m as a low intertidal bar, which at low tide reaches the end of the rocky points (Fig. 4.71). In addition the beach has built out 400 to 500 m, leaving it backed by a series of low ridges, with a 10 ha stand of mangroves in lee of the creek mouth. The huts overlook the southern end of the beach, while there is vehicle access to the northern end.

Yenyardindle Huts sit atop a 20 m high, largely cleared bluff, with extensive clearing also on the low backing slopes. To the south of the huts are two smaller beaches at the base of the bluffs that extend 1.5 km to Broome Head. The beaches (**1270** and **1271**) are 400 and 450 m long respectively, face north-east, are backed by low, casuarina-covered foredune then the wooded slopes of the bluffs, and are bordered by protruding wooded slopes, rocky points and rock reefs. Both beaches consist of 30 m wide high tide beaches fronted by a moderately steep slope ending in a 50 m wide, continuous low tide bar, which in turn is fronted by a sandy/rock reef sea floor.

Swimming: These are all relatively safe, usually low wave beaches, with only the creek mouths and rocks posing any additional risk.

Figure 4.71 *View south along the beach immediately west of the Yaryardindle huts.*

Surfing: Usually none.

Fishing: Best in the creeks and off the rocks.

Summary: Four attractive, sandy beaches all fringed by low, casuarina-covered foredunes and wooded bluffs, with rocky points and reefs bordering each end, together with the two small creeks.

1272-1274 BROOME HEAD

No. Beach	Unpatrolled Rating HT LT	Type	Length
1272 Broome Hd	1 2	R+LT	2 km
1273 Broome Hd (S 1)	1 2	R+LT	1 km
1274 Broome Hd (S 2)	1 1	B+SF	500 m

Broome Head is a 30 m high, wooded head with a small creek draining out along its southern side and a few mangroves growing in amongst the rocks. The head is backed by land that is being cleared for housing development. A road has been constructed to the land behind the beach and there is a landing strip 300 m in from the northern end of the beach. The beach (**1272**) runs from the shallow creek mouth for 2 km to the south, where it ends at a low headland surrounded by low, rocky outcrops. The beach curves between the two headlands, with an overall east-north-east orientation. It consists of a moderate gradient high tide beach, fronted by a 100 m wide, continuous low tide bar. The beach is backed by low, 100 m wide,

casuarina-covered dunes, the tidal creek in the north and low, wooded slopes in the south.

Beach **1273** continues on the southern side of the low, jutting, 200 m long point and reef. The beach is 1 km long and runs from the shallow mouth of a small creek by the point to a low, sandy headland fronted by rocks and reef. A low dune backs the beach, while a 50 to 100 m wide low tide bar fronts the sandy high tide beach.

Beach **1274** extends from the southern side of the sandy headland for 500 m to a low, sandy point in lee of a 500 m long reef. The reefs at either end reduce wave height along the beach and the sandy beach is fronted by shallow, 400 to 500 m wide sand flats.

Swimming: These are three low energy, relatively safe beaches, usually with low waves and no rips.

Surfing: Usually none.

Fishing: Best in the two creeks and off the rocky points at high tide.

Summary: Three beaches with backing land in the process of being developed, which will bring considerably more attention to the beaches in the coming years.

1275-1278 BROOME HD-MACDONALD PT

No. Beach	Unpatrolled Rating HT LT	Type	Length
1275 Broome Hd (S 3)	1 1	B+SF	2 km
1276 Broome Hd (S 4)	1 1	B+SF	600 m
1277 Macdonald Pt (W 2)	1 2	B+S reef flats	1.2 km
1278 Macdonald Pt (W 1)	1 2	B+S reef flats	2 km

Four kilometres south of **Broome Head** is a 1.5 km wide, 500 m deep bay. The bay is bordered by a long, low reef and a few mangroves on the northern side, with a rock and sand dominated southern boundary. In between is a 2 km long, curving, east-north-east facing, low energy beach (**1275**), bordered at each end by small tidal creeks. The beach consists of a low, narrow high tide beach fronted by ridged sand flats up to 1 km wide. The protected location and abundance of sand has caused the beach to build out over 500 m, with a series of twelve low, casuarina-covered beach ridges backing the beach. There is a vehicle track to the northern headland and a new access road running behind the ridges, but at present no development on the beach or ridges.

Beach **1276** lies on the southern side of the southern creek and runs out in a rectangular fashion for 600 m along the rock-controlled southern side of the bay. It is highly crenulate, fringed by rocks and mangroves and roughly faces north and east into the sand flats of the main beach.

Beaches 1277 and 1278 run from the southern boundary of the bay for 1.2 km and 2 km respectively, to MacDonald Point; a low, wooded headland. The coast in between is dominated by the presence of bedrock, manifest as low, wooded bluffs behind the two beaches, as bordering low, rocky points, and as extensive rock reefs off both beaches. Beach **1277** faces north-east and has a small creek draining out amongst the southern rocks. There is a shack on the bluffs that divides the two beaches. Beach **1278** runs down to Macdonald Point and is backed by the bluffs, apart from a small creek that spills across the middle of the beach, depositing a lobe of sand on the reef flats.

Swimming: The best swimming is on the two MacDonald Point beaches, with snorkelling also possible over the rock reefs at high tide. All beaches receive low waves and are often calm.

Surfing: None.

Fishing: Best in the creeks and at high tide, with bare sand and rocks exposed at low tide.

Summary: Four low energy beaches, that are accessible by boat at high tide and 4WD.

Shoalwater Bay

Shoalwater Bay was named by Captain Cook on 28 May 1770. He rounded and named Cape Townsend and noted the several islands in the vicinity. He soon ran into 'Shoal Water' and noted what appeared to be a large bay extending to the south-east. He had hoped to wait for spring tide to 'lay the Ship a Shore', however a search of the coast revealed a lack of fresh water and he cancelled this idea. He then continued on via Thirsty Sound to Broad Sound, both of which he named. It was during this time he also recorded tides of 16 to 18 feet (4.8 to 5.5 m); the highest he encountered on the Australian coast. The tides can in fact reach 8 m in Broad Sound.

1279   MACDONALD PT

No.	Beach	Unpatrolled Rating HT LT	Type	Length
1279	Macdonald Pt (S)	1 2	B+SF	1.3 km

Macdonald Point is a low, rocky point that forms the northern boundary of West Bight, the last sandy indentation before mangroves begin to dominate the Shoalwater Bay shoreline. The tidal flats are relatively narrow north of the point, but widen considerably immediately south of the point in response to the lowering waves; a factor that also allows mangroves to dominate the southern end of the 5 km long bight. The point and adjacent beach can be reached via the Stanage Bay Road that runs 8 km to the west and is linked by a 4WD track.

Macdonald Point (south) beach (**1279**) is 1.3 km long and faces south-east. The beach's orientation has permitted strong Trade winds to build a 100 to 200 m wide, 10 to 15 km high foredune in lee of the 50 m wide, soft high tide beach. The beach is bounded by the low, rocky headlands of Macdonald Point and an unnamed southern rocky protrusion. Both headlands have extensive rock flats extending a few hundred metres into the bay. The beach itself has a rock flat toward the centre and extensive intertidal sand flats that widen to 1 km by the southern end of the beach.

Swimming: You can only swim at high tide owing to the high tide range and wide tidal flats.

Surfing: None.

Fishing: Only at high tide when the tidal flats are covered.

Summary: A low energy beach backed by the last of the foredune on this side of the bay.

1280   KREUTZER

No.	Beach	Unpatrolled Rating HT LT	Type	Length
1280	Kreutzer	1 2	B+SF	600 m

Kreutzer beach (1280) is named after a hut of the same name located behind the centre of the beach. The beach is a narrow, crenulate strip of high tide sand, 600 m long and bounded by two low, rocky points, and fronted by a mixture of low gradient rock and sand flats that extend up to 2 km off the beach. The southern rocks reach a low islet 400 m offshore. The vehicle track from the Stanage Bay Road reaches the hut, before going on to Macdonald Point.

Swimming: Only at high tide when the rocks and sand flats are covered.

Surfing: None.

Fishing: Only at high tide from the beach and rocks.

Summary: A low energy high tide beach fronted by shallow rock and sand flats.

Shoalwater Military Reserve

The Shoalwater Bay Military Reserve covers the southern half of Shoalwater Bay, all of Townsend Island, the Peninsula Range, Port Clinton and the east coast down to Five Rocks beach. The reserve is used periodically for military training and as a consequence has entry restrictions. No one is permitted above the high water mark, due to the possibility of encountering live munitions. When there are no exercises, restricted entry is permitted for boats and shore landings below the high water mark.

For further information contact:
   Controlling Authority
   District Support Unit
   Australian Army
   68 Western St
   Rockhampton QLD 4700
   Phone: (079) 275 088

1281, 1282   WEST BIGHT

| No. | Beach | Unpatrolled | | | |
		Rating HT	LT	Type	Length
1281	West Bight	1	2	B+SF	1.8 km
1282	West Bight (S)	1	2	TSF	1.2 km

Deep in **West Bight** is a 1.8 km long, narrow, low energy beach (**1281**) fronted by sandy to muddy tidal flats up to 3 km wide. There are scattered mangroves on the inner tidal flats. The southernmost shore of the bight gives way to a wide fringe of mangroves. The beach is accessible at either end via tracks from the Stanage Bay Road, however there is no development and the southern half of the beach lies in the Shoalwater Bay Military Reserve. The beach is backed by 300 to 400 m of low beach ridges, which are crossed by a small creek that drains out at the southern end of the beach.

On the southern shores of the low headland that forms West Bight is a second low energy beach-tidal flat (**1282**), that begins amongst the mangroves of the southern bight and winds its way for 1.2 km to the point. It has a narrow high tide beach fronted by shallow tidal flats with a scattering of mangroves.

Swimming: Only possible at high tide owing to the shallow tidal flats.

Surfing: None.

Fishing: Best at high tide and at the southern creek mouth on the main beach.

Summary: Two very low energy high tide beaches fronted by kilometres of tidal flats at low tide.

1283-1289   CHARCOAL CK-SABINA PT

| No. | Beach | Unpatrolled | | | |
		Rating HT	LT	Type	Length
1283	Charcoal Ck (N 2)	1	2	B+SF	2.2 km
1284	Charcoal Ck (N 1)	1	2	B+SF	800 m
1285	Charcoal Ck (E)	1	2	TSF	3 km
1286	Akens (W)	1	2	B+SF	700 m
1287	Akens (E)	1	2	B+SF	1.2 km
1288	Sabina Pt (W 2)	1	2	B+SF	1.5 km
1289	Sabina Pt (W 1)	1	2	B+SF	2 km

The coast between the southern headland of West Bight and Sabina Point, 10 km to the south-east, consists of a series of low headlands and continuous tidal flats backed by a series of 11 irregular high tide beaches. The whole section is fronted by Akens Island, which lies 3 km offshore and is connected at low tide by tidal flats. When approaching this area from any direction, the view is dominated by conical, 230 m high Pine Mountain, located 5 km in from the coast. There is gravel road access to Pine Mountain and a vehicle track from there along the back of the coast to some shacks on Sabina Point. Otherwise there is no development and no facilities. The entire section faces roughly north-north-east and receives very low wind waves, except at high tide during strong Trades when waves might reach 0.5 m high.

Charcoal Creek drains the southern side of Pine Mountain and converges with Ross Creek and then Pine Creek, to flow into a mangrove-fringed embayment. Two beaches lie to the north of the creek mouth. Beach **1283** is a 2.3 km long, curving beach that faces east and is bounded by a low headland to the north and a low, rocky outcrop to the south. Beach **1284** extends from the south side of these rocks due south for 800 m to the mangroves and creek mouth. Both beaches have low, narrow high tide beaches and 2 to 3 km of tidal flats. They are backed by low, scrub-covered land that rises slowly to Pine Mountain. On the east side of Charcoal Creek mouth is a 3 km long, north facing beach, **1285**, fronted by 200 to 300 m of mangroves and wide tidal flats, and backed by a low, grassy plain.

At the eastern end of beach 1285 the low slopes of the Normanby Range again reach the coast, forming four open embayments that end at the more prominent 10 m high Sabina Point. The first two beaches, **1286** and **1287,** lie directly south of Akens Island. They are 700 m and 1.2 km long respectively and face north. The first beach is a low, vegetated sand spit that ends at a small tidal creek, while the second is backed by a low, casuarina-covered dune.

Beach **1288** occupies a 1.5 km long, north facing bay bordered by low headlands. The beach is part of a sand spit that has developed across the bay, impounding a backing tidal flat that is drained by a tidal creek toward the western end of the beach. Mangroves occur at the creek mouth and are scattered along the tidal flats that front the beach. There is a vehicle track to the eastern headlands and an old shack.

Beach **1289** runs from the headland for 2 km to Sabina Point. The sandy, north facing beach is crenulate alongshore, protruding when low slopes reach the shore and retreating into sand-filled embayments in between. Two small creeks break out across the beach following heavy rain.

Swimming: All these beaches are only suitable for swimming at high tide, with kilometres of low, sandy tidal flats exposed at low tide.

Surfing: None.

Fishing: Only at high tide, with the tidal creeks offering the best spots.

Summary: Seven isolated beaches all located in the military reserve and little used except during training exercises.

1290-1291 SABINA PT

No.	Beach	Unpatrolled		Type	Length
		Rating			
		HT	LT		
1290	Sabina Pt	1	2	B+SF	1.4 km
1291	Rocky Ck	1	2	TSF	700 m

Sabina Point is the most prominent headland in the southern half of Shoalwater Bay. It protrudes 700 m northeast into the bay, but is only 10 m high. Extensive intertidal rock flats lie bayward of the point, while two sandy beaches (1290, 1291) and fronting tidal flats lie to either side. A vehicle track reaches the point where there are a couple of old fishing shacks; otherwise there are no facilities.

Sabina Point beach (**1290**) is 1.4 km long and faces east

into the Trades. It is bordered by the point in the north and the protruding spit and mangroves of Rocky Creek in the south. Due to its orientation a low, casuarina-covered foredune has developed behind the 50 m wide high tide beach. Tidal flats that are 300 m wide at the point widen to over 1 km off the creek mouth.

Rocky Creek beach (**1291**) runs for 700 m south of the creek mouth to the mouth of a larger creek and more dense mangrove-covered tidal flats. It faces east but is a more protected, lower energy beach. Mangroves are scattered over the tidal flats and a low, vegetated beach ridge backs the beach.

Swimming: Only at high tide.

Surfing: None.

Fishing: Best off the point or in the creek at high tide.

Summary: Two accessible although isolated and undeveloped beaches, both in the military reserve.

1292-1298 MOOLY-LITTLE CKS

No.	Beach	Unpatrolled		Type	Length
		Rating			
		HT	LT		
1292	Mooly Ck	1	2	TSF	250 m
1293	Mooly Ck (E 1)	1	2	T+SF	1.2 km
1294	Mooly Ck (E 2)	1	2	B+SF	1.8 km
1295	Little Ck (W)	1	2	B+SF	1.3 km
1296	Little Ck (E)	1	2	B+SF	1.8 km
1297	Oyster fence (W)	1	2	B+SF	2 km
1298	Oyster fence	1	2	B+SF	500 m

Between Rocky Creek beach and the oyster fence beach is a 12 km long section of low energy shoreline, that is backed by slowly rising slopes of the Normanby Range and fronted by tidal flats averaging 1 to 2 km wide. Where the slopes reach the shore they form a series of shallow embayments and low headlands that, together with three tidal creeks, divide the shoreline into seven beaches all linked by continuous tidal flats. A vehicle track runs behind the shore, 2 to 3 km inland, reaching the coast at Mooly Creek and east of Little Creek. Otherwise there is no development, apart from the abandoned Huttonvale outstation, located 2 km south of Mooly Creek mouth.

Mooly Creek beach (**1292**) is just 250 m long, extending from the east side of the creek mouth to a low point. The low beach impounds a small high tide salt flat. There is a vehicle track to the rear of the eastern end of the beach.

Beaches 1293, 1294 and 1295 are three low, arcuate,

north-east facing beaches that extend for 5 km from the eastern side of Mooly Creek beach to the broad Little Creek mouth. Beach **1293** is 1.2 km long, with a small tidal creek at its western end and a low headland to the east. It is backed by some low beach ridges and a small tidal flat. Beach **1294** is 1.8 km long, lies between two low headlands and is backed by slowly rising ground, apart from a small creek and backing salt flat in the middle. Beach **1295** is 1.3 km long and runs from the headland to the shallow, mangrove-fringed mouth of Little Creek. It consists largely of casuarina-covered sand spits.

Beaches 1296, 1297 and 1298 are the southernmost beaches on the western shore of Shoalwater Bay. Beach **1296** is a 1.8 km long beach and sand spit that forms the western side of Little Creek. It is attached to a low eastern headland and is backed by high tide salt flats. A vehicle track runs out to the headland. Beach **1297** runs for 2 km from the eastern side of the headland to a mangrove-fringed tidal creek. It consists of three sections of low beaches and backing casuarina-covered ridges, that are in turn backed by high tide salt flats. Beach **1298** is a 500 m long, east facing strip of high tide sand, with an old fish (oyster) fence running for a few hundred metres out across the 1.5 km wide tidal flats. The fence forms the southern boundary. South of this point mangroves and tidal creeks dominate the bay shore.

Swimming: Only at high tide.

Surfing: None.

Fishing: Best at high tide in the creeks.

Summary: Seven isolated beaches with vehicle access to Mooly Creek beach and the headland east of Little Creek. All are contained in the military reserve.

1299 SHOALWATER CK

No.	Beach	Unpatrolled		Type	Length
		Rating HT	LT		
1299	Shoalwater Ck	2	3	TSF	500 m

Shoalwater Creek is the largest creek draining into the southern end of Shoalwater Bay. It meanders for 8 km though the mangroves before reaching the open bay. Here the shoreline is dominated by mangroves and tidal flats, except for a 500 m section of sand immediately east of the 1 km wide creek mouth. This beach (**1299**) is only suitable for landing boats and fishing at high tide. It is not recommended for swimming owing to the strong tidal currents and presence of crocodiles.

REEF POINT TO CAPE CAPRICORN (Fig. 4.72)
including Great Keppel Island

Length of Coast:	495 km
Beaches:	109 (beaches 1300 to 1409)
Surf Life Saving Clubs:	Yeppoon, Emu Park
Lifeguards:	Yeppon, Emy Park
Major towns:	Yeppoon, Emu Park, Rockhampton

Regional Map 8: Reef Point to Cape Capricorn

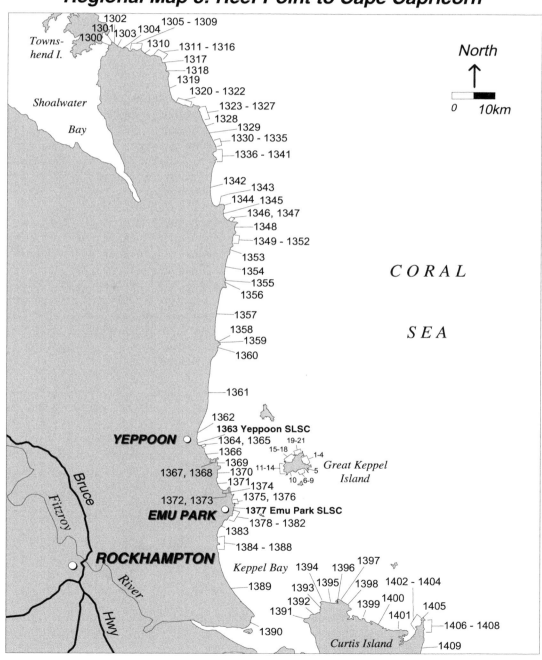

Figure 4.72 *Regional map 8 Reef Point to Cape Capricorn. Beaches with Surf Life Saving Clubs indicated in bold.*

Peninsula Range, Mount Gibraltar and Townsend Island

The **Peninsula Range**, together with Townsend Island, forms the eastern boundary of Shoalwater Bay. The range was formed by intrusive volcanic activity during the Cretaceous period more than 65 million years ago and it rises sharply to over 1000 m, reaching 1800 m at Mount Westall. It has been heavily dissected and forms a rugged backdrop to the wide, mangrove-fringed eastern shore of Shoalwater Bay. The range extends roughly north-south for 25 km and is between 5 and 10 km wide.

An eastern extension of the range is the **Mount Gibraltar** area, which forms the coast for 20 km between Island Head and Port Clinton. This area is largely covered by Quaternary transgressive sand dunes that reach heights of 200 to 300 m. Only the higher bedrock peaks such as Mt Gibraltar and the coastal headlands are uncovered. The Mount Gibraltar area is almost an island, being connected to the range by a low, 500 m wide saddle. Likewise the range is connected to the southern Colcarra Range by a similar low, 500 m wide saddle just south of the old Keiver homestead.

Townsend Island lies immediately north of the range and consists of a series of older Devonian (350 million years) sedimentary (sandstone & siltstone) and metasedimentary (slates, schist, chert, limestone & phyllite) rocks. It forms the northern boundary of Shoalwater Bay and is surrounded by strong tidal passages, including Strong Tide Passage in the south, Canoe Passage, which separates it from smaller Leister Island, and North Passage.

1300-1303   REEF PT

No. Beach	Unpatrolled			
	Rating		Type	Length
	HT	LT		
1300 Reef Pt (W 2)	1	2	TTSF	900 m
1301 Reef Pt (W 1)	1	2	B+SF	300 m
1302 Reef Pt	1	2	R+LT	750 m
1303 Reef Pt (S 1)	1	2	R+rock flats	1 km

Reef Point forms the northern tip of the Peninsula Range. It also forms the southern shoreline of Strong Tide Passage, a 1.5 km wide bedrock channel that separates it from Townsend Island. The point is named after rock reefs that lie just off the point and make negotiation of the passage just that much more difficult. The point itself consists of low, steeply dipping metamorphic rocks. The rocks are strewn about the point and the beaches to either side, as well as forming the passage reefs.

Despite the rocky nature of the point, the shoreline consists of beaches to either side. To the west inside Strong Tide Passage are two low wave beaches, while to the south are two more exposed beaches.

Beach **1300** faces north-west into Shoalwater Bay. It is 900 m long and consists of a narrow high tide beach and sand and rocky tidal flats that extend 200 to 300 m out to the edge of the deep water in the passage. A few mangroves are scattered along its eastern shore. There is an old ruin behind the western end and the beach is used as a campsite by fishers. A military track runs behind the beach.

Beach **1301** sits right in the passage, is just 300 m long, faces north toward Townsend Island and is bounded by low, rocky flats at either end, the easternmost being part of Reef Point.

Reef Point Beach (**1302**) faces north-east and receives moderate waves averaging under 0.5 m. It has a narrow, soft high tide beach, with 200 m wide low tide sand flats/ bar. A few low rocks cross the middle of the beach, while the southern boundary has more extensive rock flats. A military track crosses behind Reef Point to reach the centre of this beach.

Reef Point South (**1303**) is a 1 km long, north-east facing, low gradient sandy beach crossed by four low, rocky ridges and backed in the south by low bedrock.

Swimming: Only swim at high tide close to the beach, as at low tide there are very strong tidal currents off the beaches, particularly those in Strong Tide Passage, as well as numerous exposed and submerged rocks.

Surfing: None.

Fishing: The passage is a very popular spot with strong tidal flow and rock reefs. Be wary of the many submerged rocks along the shore.

Summary: Four attractive, isolated beaches used by boat fishers, except when the military are using the reserve.

1304-1306   NOTCH

No. Beach	Unpatrolled			
	Rating		Type	Length
	HT	LT		
1304 Reef Pt (S 2)	1	2	R+LT	350 m
1305 Notch	1	2	R+LT	650 m
1306 Notch (boulder)	1	2	cobble+LT	50 m

Two kilometres south of Reef Point, high spurs of the Peninsula Range reach the coast providing a dramatic backdrop to the shore as well as providing a series of narrow valleys and gullies, some of which have been occupied by beach deposits. The first three lie either side of a 400 m high spur that runs north for 3 km from Notch Mountain. There are steep military tracks to the two larger beaches, but otherwise boat access is by far the easiest.

Beach **1304** faces north-north-west toward Townsend Island. It is a 350 m long pocket beach surrounded by densely vegetated, 80 to 100 m high bedrock slopes, which also provide two prominent headlands. The beach is moderately steep and narrow at high tide, but over 100 m wide at low tide. A vehicle track reaches the slopes above the western end of the beach.

Notch beach (**1305**) is 650 m long, faces almost due north, is bounded by two prominent vegetated headlands that rise to 150 and 320 m respectively, but is backed by a low, sand-filled valley that extends 1 km inland. A creek drains out along the eastern side of the valley and usually flows across the beach, forming a protruding sand bar. The beach has a low to moderate slope and is up to 150 m wide at low tide. A vehicle track winds across the densely vegetated sand ridges to reach the creek mouth.

Beach **1306** is a 50 m long, steep high tide cobble and boulder beach, fronted by lower gradient sand flats. It is located in a gully on the eastern headland of Notch Beach, 200 m from the main beach.

Swimming: The two main beaches offer good swimming at mid to high tide and a relatively shallow bar at low tide.

Surfing: None.

Fishing: All beaches offer deep water off the headland and the beaches at high tide.

Summary: Three relatively small, sandy beaches surrounded by densely vegetated headlands.

1307-1309 PINETREES

No. Beach	Unpatrolled		
	Rating HT LT	Type	Length
1307 Beach 1307	1 2	boulders+LT	80 m
1308 Pinetrees (W)	1 2	R+LT	650 m
1309 Pinetrees (camp)	1 2	R+LT	100 m

Pinetrees Point is a prominent, pine-covered, rocky point. There are three beaches to the west of the point, each dominated by the rocky coast.

Beach **1307** is an 80 m long pocket of high tide boulders occupying a gully on the eastern side of the 300 m high headland. It faces north and exposes a sandy bar at low tide.

Beach **1308** occupies a small, wide valley that has been partially filled with a series of low sand ridges up to 500 m wide. The beach is 650 m long and bounded by the high headland of beach 1309 and the inner reaches of 1 km long Pinetrees Point. Two small creeks drain across the centre and eastern end of the beach. The beach has a moderate slope and is up to 100 m wide at low tide. A vehicle track reaches the centre of the beach.

Beach **1309** is a 100 m long, north facing pocket of sand backed by a low dune and a small creek. It is protected by headlands that extend 200 to 300 m seaward. It has a moderate gradient high tide beach and a lower gradient, 100 m wide low tide beach.

Swimming: The two sandy beaches offer good swimming in usually calm to low wave conditions at mid to high tide.

Surfing: None.

Fishing: All three beaches have deep water off the beach at high tide, while the rocks can be fished at most tides.

Summary: Three protected, north facing beaches each bordered by high headlands.

1310 PINETREES PT

No. Beach	Unpatrolled		
	Rating HT LT	Type	Length
1310 Pinetrees Pt	1 2	R+LT	2.6 km

On the south side of **Pinetrees Point** is a curving, 2.6 km long, north-east facing sandy beach (**1310**) that runs down to the 1.5 km wide mouth of Island Head Creek. Two kilometres off the southern sandy tip is Island Head, which is almost tied to the beach at low tide by extensive sand shoals. The beach has built out 500 to 1000 m from the backing 500 m high hills, forming a low, densely vegetated sand ridge plain. A vehicle track crosses the plain to reach the centre of the beach.

The beach is composed of fine sand and has a low gradient. The high tide beach is low and about 50 m wide, while at low tide a 200 m wide sand bar is exposed. Waves average just under 0.5 m and provide a small surf (Fig. 4.73).

Swimming: Better at mid to high tide and well clear of the creek mouth, which has strong tidal currents.

Surfing: Usually a very low break across the bar.

Fishing: Best at high tide off the beach and in the creek.

Summary: A long, natural beach backed by an extensive sand plain and bordered by Island Head Inlet.

Figure 4.73   *View along Pinetrees Beach at low tide.*

1311, 1312   ISLAND HEAD INLET

No. Beach	Unpatrolled			
	Rating		Type	Length
	HT	LT		
1311 Island Hd Inlet (1)	1	3	R+LT+inlet	1 km
1312 Island Hd Inlet (2)	1	3	R+LT+inlet	800 m

Island Head Creek is a wide, 25 km long tidal creek draining a large macrotidal estuary that enters the sea in lee of Island Head Island. The inlet is bedrock-controlled and is 1 to 2 km wide at its mouth. On the north side are two beaches that face east-south-east into the inlet channel.

The first beach (**1311**) begins at the sandy junction with Pinetrees Point Beach. The shoreline swings 90° while tidal sand shoals extend 2 km off the point toward Island Head Island. This beach is 1 km long with a narrow high tide beach cutting into higher foredunes, a 50 to 100 m wide low tide beach, then the deep inlet channel. It ends at a small, rocky headland on the other side of which is an 800 m long second beach (**1312**). This beach is similar is shape and extends to a second small, 40 m high headland that is tied to the beach by a small tombolo or spit. Both beaches are backed by 200 to 500 m of densely vegetated foredunes then the high ground rising to 500 m within 1 km of the inlet.

Swimming: While waves are usually low to calm at these beaches, there are strong tidal currents, particularly off the bar. Only swim at mid to high tide and stay close to the shore and well clear of deep water and tidal currents.

Surfing: None.

Fishing: The inlet is a favourite spot and can be fished from the shore or more commonly from a boat.

Summary: Two natural, low energy sandy beaches fronted by the dynamic Island Head Inlet and tidal channels.

1313-1316   ISLAND HEAD

No. Beach	Unpatrolled			
	Rating		Type	Length
	HT	LT		
1313 Island Hd (E 1)	1	2	B+SF	250 m
1314 Island Hd (E 2)	1	2	B+SF	100 m
1315 Island Hd (E 3)	1	2	R+LT	200 m
1316 Island Hd (E 4)	1	2	R+LT	300 m

On the south side of Island Head Creek inlet the coast is dominated by rocky shore that rises to 450 m within 500 m of the shore. Around the northern base of the hills, facing the inlet, are four small beaches. All four are bordered by low, rocky headlands and backed by steeply rising slopes.

The first two beaches (**1313 & 1314**) are located inside the inlet, face north-east and usually receive no waves. They consist of a 250 m and 100 m long (respectively), narrow high tide beach fronted by 100 m wide tidal flats.

The second two beaches (**1315 & 1316**) face north and receive waves averaging 0.3 m. The waves combine with the medium to coarse sand to produce a relatively steep, 50 m wide high tide beach fronted by a 50 m wide low tide beach. The rocks at either end consist of steeply dipping metasedimentary rocks.

Swimming: The second two beaches (1315 & 1316) offer the better swimming, particularly at mid to high tide when the water is relatively deep off the beach.

Surfing: Usually none.

Fishing: Good off the beach at high tide and off the rocks all day.

Summary: Four isolated beaches that can only be reached by boat. Their usually quiet waters provide good anchorages.

1317, 1318 BROWN ROCK, PEARL BAY

No.	Beach	Unpatrolled Rating HT	LT	Type	Length
1317	Brown Rock	1	2	R+LT	400 m
1318	Pearl Bay (N)	2	3	R+LT	400 m

South of Island Head the coast trends south-east, exposing the beaches to the Trades. High terrain rising to 600 m backs the coast, producing a predominantly steep, rocky coast for the first 5 km. There are just two 400 m long beaches in this section, both wedged into small valleys and bordered and backed by rocky slopes.

The first beach (**1317**) lies 1.5 km west of Brown Rock reef. It faces north-east and consists of a sandy high tide beach fronted by a 100 m wide low tide bar. Waves average only 0.3 m. The second beach (**1318**) lies 2 km north of Pearl Bay Beach. It faces east and receives waves averaging 0.4 m. When the waves are larger they produce three to four rips along the beach, including strong rips against the rocks.

Swimming: Both beaches offer relatively safe swimming under normal low wave conditions, however beware of rips on the second beach when waves exceed 0.5 m and particularly at low tide and near the rocks.

Surfing: Only on the southern beach during high waves and occasional southerly swell.

Fishing: Good off the beaches and rocks at high tide.

Summary: Two small, attractive sandy beaches surrounded by rocky valleys and headlands.

1319-1322 PEARL BAY

No.	Beach	Unpatrolled Rating HT	LT	Type	Length
1319	Pearl Bay	2	3	R+BR	5.1 km
1320	Pearl Bay (S 1)	1	2	R+LT	1 km
1321	Pearl Bay (S 2)	1	2	R+LT	1.2 km
1322	Pearl Bay (S 3)	1	2	R+LT	1.7 km

Pearl Bay is an open, 9 km wide, north-east facing bay. Its shoreline is dominated by sandy beaches, separated by small, rocky heads in the south, while extending inland for up to 3 km and rising to 100 m are massive, now vegetated, nested, parabolic dune systems. These dunes were probably emplaced during the Holocene sea level rise and stillstand between 8000 and 5000 years ago. While the dunes extend inland from the main beach, dunes from Port Clinton Beach have in the past cascaded down onto the three southern beaches, and their scarped, 100 m high, densely vegetated slopes now back the beaches. The three beaches are separated by small, rocky headlands. The dune groundwater drains out across the beaches in places, providing a source of clean fresh water.

The main Pearl Bay Beach (**1319**) is 5.1 km long, faces east-south-east and receives waves averaging 0.5 m. These produce a low to moderate gradient high tide beach and a 150 m wide intertidal to low tide bar. The waves can commonly exceed 1 m, up to 2 m, which results in rips cutting across the low tide bar every 200 m, with up to 12 rips commonly present along the northern half of the beach. Wave height lowers along the southern half and rips are usually absent.

At the southern end of the main beach is a small, rocky headland, beyond which the coast begins to trend more south-east and the sand is broken into three sandy beaches, each bordered by low, rocky headlands. The first beach (**1320**) is 1 km long and is backed by densely vegetated sand slopes that rise quickly to over 100 m. The beach consists of a 30 m wide high tide beach and 100 m wide low tide bar. The next two beaches (**1321** & **1322**) are very similar, being just a little longer (1.2 km & 1.7 km respectively) and facing a little more north. Wave energy decreases to the east along these beaches, with the easternmost corner hooked around to face west and provide a safe and popular yacht anchorage. As the waves decrease in height, the high tide beach steepens and narrows, as does the low tide bar.

Swimming: All four beaches provide long, clean, sandy beaches and usually relatively safe conditions under normal low waves. However rips are present when waves exceed 0.5 m, particularly on the northern half of the main beach, cutting across the low tide bar approximately every 200 m.

Surfing: Best on the northern half of the main beach, where there are usually good bars and rips present at mid to low tide.

Fishing: There is good fishing off all the beaches at high tide, as well as the rocks off the small headlands, with a chance of rip holes along the main beach.

Summary: Four beautiful sandy beaches, backed by massive, vegetated dunes and with good anchorage in the south.

1323-1327 DELCOMYN IS - PERFORATED HD

No. Beach	Unpatrolled Rating HT LT	Type	Length
1323 Delcomyn (N)	1 2	R+LT	300 m
1324 Delcomyn (N)	1 2	R+LT	250 m
1325 Boulder	2 3	R+LT	50 m
1326 Perforated Pt (N)	1 2	R+LT	100 m
1327 Perforated Pt (S)	1 2	R+LT	500 m

The easternmost extension of the Peninsula Range-Mt Gibraltar area is a series of rocky headlands and islands that dominate the coast from the southern end of Pearl Bay for 6 km to Perforated Point. Set in amongst the prominent headlands and islands are five small, east facing beaches.

The first two beaches lie in a 1 km wide bay in lee of Delcomyn Island, which lies 1 km offshore. The northern beach (**1323**) is 300 m long and faces south-east, with prominent, 20 to 40 m high headlands extending a few hundred metres from either end. The beach receives waves averaging 0.5 m, which maintain a narrow high tide beach and a 100 m wide low tide bar that is often cut by two rips. It is backed by a small foredune and cleared slopes. The southern beach (**1324**) is 250 m long, faces east and is also protected by a southern headland and small island just off the head. It receives waves averaging 0.4 m that break across a 100 m wide low tide bar. Rips are usually absent. A small creek drains out at the northern end, with a low foredune and rocks backing the rest of the beach.

Beach **1325** is a 50 m wide high tide boulder beach, which at low tide is fronted by a 50 m wide sand bar. The beach is wedged into a 500 m long, narrow, south-east facing, headland-bound bay.

Beach **1326** lies in a small bay between Northeast Point and Perforated Point, with a small island also lying immediately off Perforated Point. The beach is just 50 m long at high tide, widening to over 100 m at low tide. It faces south-east and is backed by slopes that rise quickly to 100 m hills. The beach has a narrow high tide beach and a 50 m wide low tide bar, that receives waves averaging 0.5 m. A small creek flows down the narrow valley behind the beach and saturates the beach. The creek was used by Matthew Flinders in 1802 to replenish his water supplies.

Beach **1327** is a 500 m long, south-east facing beach lying between Perforated Point (and the island) and an 80 m high southern headland. The beach receives waves averaging 0.5 m and is occasionally exposed to higher swell. The waves produce a moderate gradient high tide beach and a 50 to 100 m wide low tide bar, that is cut by a few rips during higher waves. A small foredune and moderate slopes back the beach.

Swimming: All the beaches, apart from the boulder beach, offer relatively safe swimming at mid to high tide under normal low wave conditions. However when waves exceed 0.5 m, the surf breaks across the bars (particularly at mid to low tide) and rips begin to form, intensifying with increasing wave height.

Surfing: Usually a low break across the bars at mid to low tide, and a chance of some waves on the more exposed beaches during higher wind and swell waves.

Fishing: Numerous spots off the beaches at high tide and off the many rocks.

Summary: Five small beaches all surrounded by prominent headlands and pine-covered slopes.

1328, 1329 PORT CLINTON

No. Beach	Unpatrolled Rating HT LT	Type	Length
1328 Port Clinton	2 3	R+BR	6 km
1329 Port Clinton (inlet)	1 3	R+LT +inlet	2.5 km

Port Clinton is a large macrotidal estuary that enters the sea between Perforated Point and the southern Mount Flinders area. The bedrock has been surrounded and covered by massive Quaternary marine and aeolian (dune) sands, and today the 6 km long beach runs south from Perforated Point to partially fill the estuary and form the northern boundary of the 3 km wide inlet. Inside the estuary are well-preserved older shorelines deposited during past high sea levels, at least 120 000 years ago. Clinton Beach (**1328**) runs south then south-west for 6 km from Perforated Point to West Point. It receives a little protection from a small inlet near the northern end and tidal shoals to the south, however for the most part it is fully exposed to south-easterly waves and winds. The waves maintain a narrow high tide beach, with a wider 100 to 200 m wide low tide bar that is cut by rips every 200 m, particularly along the central section of the beach. Tidal shoals and channels extend off the southern West Point end for up to 2 km. These dynamic shoals impact the shape and width of the point and cause the beach to sweep around into the inlet for another 2.5 km.

The inlet beach (**1329**) has a highly crenulate, irregular and variable sandy shoreline, that usually consists of a series of low, sandy spits. These are joined, to varying degrees, to the dune-scarped shoreline that often impounds small lagoons. The spits are produced by the waves and tidal currents moving sand into the inlet and are very dynamic. The beach faces roughly south and receives only low waves at high tide. These dissipate by the western tip, where mangroves may be encountered.

A continuous foredune backs the main beach, behind which are massive parabolic sand dunes that extend up to 10 km inland, reaching heights of a few hundred metres. These dunes have been deposited during past as well as present sea level. They are largely vegetated, though there has been some more recent activity just behind the beach, with bare sand extending up to 500 m inland.

Swimming: The main beach usually receives waves averaging 0.5 to 1 m, which produce a narrow surf zone at high tide that widens at low tide and commonly forms rips across the low tide bar. Safest swimming is at high tide close to the beach. Be careful on the inlet beach because of the deep tidal channels and strong currents off the beach.

Surfing: There is often a low beach break that is better at mid to low tide. During high waves a wide and more energetic break dominates the beach.

Fishing: The northern headland and southern inlet offer the deepest water, with a chance of some rip holes along the beach at low tide.

Summary: A moderately long beach backed by massive sand dunes. It is bordered in the north by pine-covered headlands and in the south by the massive tides and currents of Port Clinton.

1330-1332 HOLTNESS PT-INNER HD

No.	Beach	Unpatrolled Rating HT	LT	Type	Length
1330	Holtness Pt	1	2	B+SF	1.1 km
1331	Bullock Pt	1	2	R+LT	300 m
1332	Inner Hd	1	2	R+LT	50 m

On the south side of Port Clinton, 170 m high Mount Flinders and its surrounding 15 km of rocky shore dominate the coast. To the west the shore faces into the estuarine waters of South Arm, to the north it faces Port Clinton Bay, while on the east it faces the Coral Sea. A total of twelve beaches surround the base of the mount, ranging in size from 50 m to over 1 km. All are dominated by and bounded by prominent headlands and rocks.

On the west side of Port Clinton are two beaches. The first (**1330**) faces west into the estuary and is bordered by Holtness Point in the south and Bullock Point to the

north. It is a low wave, tide-dominated beach, with a narrow strip of high tide sand fronted by sandy tidal flats up to 400 m wide. A few mangroves fringe the southern half of the beach, while rock increasingly dominates the shore of the northern half.

Immediately to the north is beach **1331**, a 300 m long, north-west facing beach bordered by the rocky headlands of Inner Head and 20 m high Bullock Point. The beach receives sufficient wave energy to maintain a 50 m wide low tide bar, rather than tidal flats.

On Inner Head is the third beach (**1332**), a 50 m long, north facing pocket of high tide sand fronted by 100 m wide sand and rock flats, bordered by the rocks of the head.

Swimming: Due to the high tide range, swimming is best at high tide when the sand flats and bars are covered.

Surfing: None.

Fishing: Best at high tide and off the rocks.

Summary: Three beaches that represent a transition from the tide- and mangrove-dominated estuary shoreline to the wave-dominated shores of the open coast.

1333-1336 OBSERVATION ROCK-CAPE CLINTON

No.	Beach	Unpatrolled Rating HT	LT	Type	Length
1333	Observation Rock	1	2	R+LT	1.1 km
1334	Round Is	1	2	R+LT	350 m
1335	Launch Rocks	2	2	R+LT	50 m
1336	Cape Clinton (N)	1	3	R+LT boulders	100 m

The north facing shore of **Mount Flinders** contains four headland-bound beaches. The first, Observation Rock Beach (**1333**), provided anchorage for Matthew Flinders when he made observations from the rocks at the eastern end. The beach is 1.1 km long, faces north and receives waves averaging less than 0.5 m. These combine with the fine to medium white sand to produce a moderately sloping high tide beach fronted by a 100 m wide low tide bar. Dense vegetation runs right to the back of the beach.

Round Island Beach (**1334**) lies 1.5 km in lee of Round Island. It is a 350 m long, north facing beach that receives slightly higher waves and has a more moderately sloping high tide beach and 50 m wide low tide bar. It is surrounded by steeply rising slopes on three sides, including prominent, 60 m high headlands.

Launch Rocks beach (**1335**) is a 50 m long pocket of sand and cobbles, located deep in a narrow, 200 m long gap in the rocks. Likewise, beach **1336** is similar only a little longer and composed entirely of cobbles and boulders arranged in a steep high tide beach, with a boulder flat at low tide. Both beaches make good landing spots at high tide.

Swimming: The best swimming is at high tide at all four beaches, as low tide reveals either shallow sand bars or the rocky sea floor of the two smaller beaches.

Surfing: None.

Fishing: Best off the sloping rocky platforms that border each beach.

Summary: Four relatively protected, natural beaches dominated by their rocky surrounds.

1337-1339   CAPE CLINTON

No. Beach	Unpatrolled		
	Rating HT LT	Type	Length
1337 Cape Clinton (S 1)	1 3	R+BR	250 m
1338 Cape Clinton (S 2)	1 3	R+BR	150 m
1339 Cape Clinton (S 3)	1 2	R+LT	50 m

Cape Clinton is a grassy, sloping headland that rises to 100 m. On its southern side the rocky coast continues as an open, 1.5 km wide bay. Within the bay are three small sand and boulder beaches.

Beach **1337** is the largest and most exposed beach. It is 250 m long, faces south-east and is bordered by sloping, rocky headlands, as well as being backed by a high tide boulder beach that forms a narrow boundary between the low backing grassy slopes and the beach. The high tide boulder and sand beach is narrow and almost awash at spring tide, while the sandy low tide bar is 150 m wide and often cut by two rips against the rocks at each end. Waves average just over 0.5 m and produce a 50 m wide surf zone. Behind the beach a smear of dune sand extends a few hundred metres inland over the backing slopes.

Beach **1338** lies 200 m south of beach 1337, just over a small, 20 m high, grassy headland. The beach is 150 m long, faces east, is rimmed by 10 to 20 m high grassy headlands and backed by a continuous high tide boulder beach. The low tide beach is 100 m wide and usually has one rip exiting against the southern rocks. Waves average 0.5 m and break across the shallow bar at mid to low tide. Just 200 m behind the beach are the inner mangroves of South Arm, Port Clinton.

Beach **1339** is a 50 m long high tide boulder beach that slopes steeply into a small, north facing gap in the rocky coast. A creek drains the backing valley and flows across the beach.

Swimming: The two main beaches offer a chance of small surf, however be careful if waves exceed 0.5 m. Best at mid to high tide close inshore, as rips are usually present toward low tide.

Surfing: Chance of a low beach break across the bars of the two main beaches, particularly if the swell exceeds 1 m.

Fishing: Good fishing off the rocky headlands that border all three beaches.

Summary: Three rock-bound beaches, with the two main beaches the most exposed on Cape Clinton.

1340-1342   QUOIN IS-FRESHWATER CAMP

No. Beach	Unpatrolled		
	Rating HT LT	Type	Length
1340 Quoin Is (1)	2 3	R+BR	100 m
1341 Quoin Is (2)	2 3	R+BR	50 m
1342 Freshwater Camp	2 3	R+BR	12 km

Freshwater Bay is named after the fresh water that drains out of the backing dunes onto the beach, particularly toward the southern corner where there is also a safe anchorage. The bay is 10 km wide between southern Cliff Point and northern Quoin Island and contains 12 km of beaches, for the most part facing east but curving around in the protected southern corner to face north. A 4WD track also reaches the southern corner, which has long been a favourite camping site and anchorage.

At the northern end of the beach in lee of Quoin Island, which lies 1 km offshore, the beach protrudes seaward owing to the lower waves behind the island. As it does so it also merges with the southern slopes of Mount Flinders. The slopes divide the high tide beach into two small pockets of high tide sand (beaches **1340** & **1341**) which are linked together, and with the main beach, at mid to low tide.

The main beach (**1342**) runs continuously from the southern slopes for 12 km to **Freshwater Camp**. Apart from one rocky outcrop in the surf 1 km down the beach and small Single Rock 1 km off the southern end, it is a continuous sandy beach. It receives waves averaging 0.6 m along much of the beach; higher during strong winds and occasional swell. The high wave conditions at times produce a double bar system. The beach has a 30 m wide high tide beach, then a 150 m wide low tide bar that is usually cut by rips every 150 to 200 m, producing up to 50 rips along the beach, then a shore parallel trough off this bar and finally an outer bar that is also cut by more widely spaced rips. Toward the southern end, the wave height drops in lee of Cliff Point and the outer bar merges with the beach, making Freshwater Camp Beach only 50 m wide and free of rips.

Backing the entire beach is a massive, nested, longwalled parabolic dune system, now largely vegetated. This extends on average 2 km inland to the mangrove-fringed eastern shore of South Arm. Five kilometres across South Arm is an older system of beach ridges extending another 3 km inland. These represent an ancient high sea level shoreline and are at least 120 000 years old. At that time Mount Flinders was an island and the waves moved between the island and Cliff Point to reach the earlier beaches. In the southern 2 km the dunes give way to a swamp that is drained by a small creek across the southern end of the beach.

Swimming: There is good surf swimming along the entire beach. However the two Quoin Island beaches and much of the central-northern end is prone to higher waves and rips, particularly toward low tide and when waves exceed 0.5 m. The safest swimming is at the quieter Freshwater Camp.

Surfing: This is one of the better surfing beaches north of Yeppoon, however you need a boat to get there.

Fishing: Freshwater Camp has been used for many years as a fishing camp, with fishers either heading up the beach in 4WDs to fish the rips, or out in boats to fish the rock reefs.

Summary: A long, natural beach offering a good southern camp site and anchorage, and many kilometres of energetic beach. Be warned that it is prohibited to move inland of the beach without permission from the Army Controlling Officer in Rockhampton (see page 222).

1343, 1344   FRESHWATER CAMP-CLIFF PT

No.	Beach	Unpatrolled			
		Rating		Type	Length
		HT	LT		
1343	Freshwater Camp (E)	1	2	R+LT	50 m
1344	Cliff Pt (W)	1	2	R+LT	50 m

Between Freshwater Camp and Cliff Point is 1.5 km of north facing rocky coast backed by steep slopes that rise to 150 m. Tucked into two gullies in this cliff are two 50 m long beaches. The first beach (**1343**) is located just 100 m past the southern end of Freshwater Bay and can be reached by walking around the rock platform. It consists of a high tide boulder beach, fronted by a 50 m wide sand and rock low tide bar, with rugged rock platforms to either side.

The second beach (**1344**) lies in a 100 m wide gully immediately west of Cliff Point. It has a high tide boulder beach and a more sandy, sloping low tide bar, with rock platforms to either side.

Swimming: Both beaches offer relatively calm conditions, particularly at high tide, however watch out for the many rocks.

Surfing: None.

Fishing: Good fishing from the rock platforms that fringe much of the rocky shore along Cliff Point.

Summary: Two pockets of sand offering a relatively safe temporary anchorage.

1345   CLIFF PT

No.	Beach	Unpatrolled			
		Rating		Type	Length
		HT	LT		
1345	Cliff Pt	2	3	R+BR	1.7 km

Cliff Point is a rectangular, rocky section of the coast, where it turns 90° and heads due south. One and a half kilometres south of the point, the 100 m high cliffs give way and the first in a series of sand-filled valleys occur. The first beach (**1345**) is a straight, east facing, 1.7 km long sandy section. It receives waves averaging 0.6 m; higher during strong Trades and occasional south swell. These produce a narrow high tide beach fronted by a 150 m wide low tide beach and bar. The bar is usually cut by eight to twelve rips, including permanent rips against the rocks at each end. Following higher seas a second bar occurs off the beach and is usually cut by six to eight rips, including permanent rips against the headlands.

A foredune and densely vegetated, longwalled parabolic dunes back the beach, extending 2 km inland to the back of Freshwater Camp Beach.

Swimming: There is moderately safe surf swimming under normal waves, which is safer at high tide as rips are more likely at mid to low tide. Be careful near the rocks, as permanent rips are present.

Surfing: There is usually low surf across the bar, however as the beach can only be reached by boat, you will need to anchor offshore to check out the waves as there is no safe anchorage when waves are breaking.

Fishing: The rock platforms at each end and the rips at low tide offer the best locations.

Summary: A straight, energetic beach with surf and massive backing vegetated dunes.

Cliff Point to Corio Bay Dune Field

The 28 km of relatively straight coast between **Cliff Point** and **Water Park Point** on the north side of Corio Bay consists of a series of beaches and intervening headlands. Backing the entire coast is a massive, now stable and vegetated, nested, longwalled parabolic dune field that extends up to 14 km inland and reaches heights of 220 m. In all, the dune field covers approximately 25 km² and represent a massive transfer of sand from the sea floor to the dunes, via the beaches and, in places, cliffs. These dunes were deposited as sea level was rising 10 000 to 6 000 years ago. They blanketed all the present beaches and most of the headlands. Today the beaches are backed by the dunes, while clifftop dunes remain as remnants on many of the headlands, the former sand ramps eroded thousands of years ago. This type of massive dune formation was common on high energy, exposed sections of the southern and eastern Australian coast.

1346, 1347   CAPE MANIFOLD (W)

No. Beach	Unpatrolled		
	Rating HT LT	Type	Length
1346 Cape Manifold (W 2)	1 2	R+LT	200 m
1347 Cape Manifold (W 1)	1 2	R+LT	500 m

At the southern end of Cliff Point Beach the coast trends east again and is dominated by rocky coast for 4 km to Cape Manifold. Wedged in on either side of a 500 m long, protruding headland are two relatively small boulder and sand beaches.

The first beach (**1346**) is 200 m long at low tide. It faces north, receives waves averaging about 0.3 m and consists of a steep high tide boulder beach and an 80 m wide intertidal to low tide beach. Narrow rock platforms fringe the 100 m high cliffs to either side.

The second beach (**1347**) is 500 m long, receives slightly higher waves and is also backed by a steep high tide boulder beach, the boulders extending to low ridges at either end. A 100 m wide, low gradient sand bar is exposed at low tide. The entire beach is rimmed by vegetated slopes rising steeply to 100 m, with steeply dipping cliffs bordering each end.

Swimming: At high tide the waves wash over the boulders, so any swimming should be done at mid to low tide. Waves are usually low and rips absent, unless waves exceed 0.5 m.

Surfing: Only a low beach break likely on the second beach.

Fishing: Most fishers arrive in boats to fish the reefs off the cliffs. The best shore fishing is from the rock platforms on either side of the first beach.

Summary: Two rarely visited pockets of boulders and sand, only accessible by boat when waves are low.

1348   CAPE MANIFOLD

No. Beach	Unpatrolled		
	Rating HT LT	Type	Length
1348 Cape Manifold (S 1)	2 3	R+BR	1.6 km

Cape Manifold was named by Captain Cook on 28 May 1770. He noted numerous fires and smoke from aboriginal camps on the mainland. Today the area is uninhabited except during military exercises.

The cape is a narrow, 20 to 60 m high, finger-like headland protruding 500 m eastward. An 80 m high island lies 400 m east of the cape. Immediately south of the cape is a 1.6 km long, slightly curving, east facing beach (**1348**). It receives waves averaging 0.6 m as well as any higher wind waves and swell. These maintain a narrow high tide beach fronted by a 150 m wide, low gradient intertidal to low tide beach, with up to ten rips present at low tide. A second bar with more widely spaced rips is often present further seaward, as well as permanent rips against the headlands.

Behind the beach is a well vegetated foredune and then extensive vegetated, long walled parabolic dunes that extend up to 10 km inland.

Swimming: A moderately safe swimming beach under normal low wave conditions. However beware of rips toward low tide, against the rocks and when waves exceed 0.5 m.

Surfing: Usually a moderate beach break that picks up with the Trades and occasional swell.

Fishing: Best off the extensive rock platforms on either headland and in the rip holes at low tide.

Summary: An isolated beach only accessible by boat when waves are low. There is a nice watered camp spot in the southern corner under a melaleuca grove.

1349-1353 CAPE MANIFOLD (S)

No. Beach	Unpatrolled			
	Rating		Type	Length
	HT	LT		
1349 Cape Manifold (S 2)	2	3	R+LT	400 m
1350 Cape Manifold (S 3)	2	3	R+BR	650 m
1351 Cape Manifold (S 4)	2	3	R+BR	200 m
1352 Cape Manifold (S 5)	2	3	R+BR	200 m
1353 Cape Manifold (S 6)	2	3	R+BR	1.1 km

South of Cape Manifold Beach the coast continues roughly due south and consists of prominent, 80 m high headlands separating a series of five boulder and sand beaches. All beaches and headlands are backed by the massive, now vegetated parabolic dune field.

The first beach (**1349**) is a 400 m long high tide boulder beach and low tide sand bar that is bordered by a rock platform and small boulder beach in the north, while a small rock outcrop separates it from the next beach at high tide. Beach **1350** is a sandy extension of the boulder beach; both share the same low tide beach and surf zone, with the 650 m long neighbour having a sandy high tide beach. It also has two small creeks, at either end, that drain fresh water across the beach. The southern end also has a 4WD access track, making it one of the few beaches in this area accessible by vehicle. The continuous low tide bar is usually cut by five to six rips, including permanent rips against the rocks.

Beaches **1351** and **1352** lie 500 m further south, around a small headland. These two adjoining beaches are both 200 m long, face east and are separated at high tide by a low, rocky outcrop. At low tide they form one continuous beach and bar. Beach 1351 has a low, narrow, sandy high tide beach backed by bluffs, while 1352 has a steep, cobble-boulder high tide beach and the same low gradient, 150 m wide intertidal to low tide bar (Fg. 4.74). The waves average 0.6 m and there are usually three rips, one at each headland and one in the centre.

Beach **1353** is a kilometre further south of 1352. It faces east, is bordered by 60 to 80 m high headlands and backed by steep, rising, vegetated slopes, together with some rock outcrops on the upper beach. A creek drains across the southern half and a 4WD track descends down the slopes on the southern side of the creek. The entire beach is almost awash at high tide, while a low gradient inter-

tidal to low tide bar extends 200 m across the beach at low tide. Up to eight rips cut across the end of the bar, with a second bar often located just offshore.

Figure 4.74 *A cobble high tide becah fronted by a sandy low tide beach, just south of Cape Manifold.*

Swimming: All these beaches are moderately safe under normal low wave conditions, however they also have a number of potential hazards including high tide boulder beaches, low tide rips, permanent rips against all the headlands and strong rips when waves exceed 0.5 m. So use care if swimming at these beaches.

Surfing: The five beaches all pick up any sea or swell in the area and usually have a low surf across the bar, which is better at mid to low tide. Higher seas accompanying strong Trades and occasional swell will result in more sizeable waves.

Fishing: All beaches have varying degrees of rock platforms at the base of their headlands, and all usually have rip holes that can be fished at mid to low tide.

Summary: Five pockets of sandy beach and prominent headlands, often together with backing boulder beaches. Two of the beaches can be reached by vehicle, with a permit, however most visitors come by boat, seeing but not landing on the beaches.

1354 FIVE ROCKS

No. Beach	Unpatrolled			
	Rating		Type	Length
	HT	LT		
1354 Five Rocks	2	3	R+BR	5.3 km

Five Rocks beach (1354) is a straight, 5.3 km long, east facing beach that lies between The Three Rivers headland and Five Rocks Point. Three Rivers is named after three small creeks that converge on the northern corner of the beach, while Five Rocks is due to the presence of five distinctive sea stacks that sit atop the narrow, 600 m long point. The entire beach is backed by stable, vegetated parabolic dunes that extend up to 8 km inland and reach heights of over 200 m.

The beach is accessible via 4WD tracks at each end, as well as along the low gradient beach at low tide. The beach itself has a 50 m wide, moderate gradient high tide beach, fronted by a 150 m wide low tide bar. Wave average 0.6 m and occasional higher waves cut rips every 200 m. A trough and outer bar lie just seaward, with more widely spaced rips cutting the outer bar.

Swimming: This is a moderately safe beach when waves are less than 0.5 m; above that the rips are active at mid to low tide.

Surfing: Five Rocks is the northernmost of the more accessible beaches (from Yeppoon) and is occasionally visited by local surfers looking for a beach break, when the Trades are blowing or during occasional swell conditions.

Fishing: The southern Five Rocks Point has extensive rock platforms and is a popular spot to fish.

Summary: A relatively accessible beach, just 10 km by sandy 4WD track from the Byfields Road. It is also partly located in the national park, while a sign on the beach marks the beginning of the Shoalwater Bay Military Reserve and this area is often off limits.

1355, 1356 FIVE ROCKS-STOCKYARD PT

No. Beach	**Unpatrolled**		
	Rating HT LT	Type	Length
1355 Five Rocks (S)	2 3	R+BR	300 m
1356 Stockyard Pt (N)	2 3	R+BR	300 m

Five Rocks Point and **Stockyard Point** are two prominent headlands, 40 m and 70 m high respectively and both protruding a few hundred metres east. The two points are located 2 km apart, forming a shallow bay within which are two 300 m long, exposed sandy beaches. The Stockyard Point 4WD track leads down to the back of the southern beach, where there is also a shack. Several more shacks are located on freehold land up amongst the trees on the crest of Stockyard Point.

The northern beach (**1355**) is 300 m long, faces east and is wedged in between Five Rocks Point and a small spur and rocks that partially separate it from the southern beach at high tide (Fig. 4.75). At low tide the beaches are linked by a continuous low tide bar and surf. The southern beach

(**1356**) extends from the spur to the inner reaches of Stockyard Point. It is also 300 m long and receives the same level of wave energy.

The waves, which average 0.6 m, almost reach the backing steep, rocky slopes on the north beach and the low foredune on the south beach at high tide. At low tide a 200 m wide bar is exposed, usually cut by three rips along each beach, including permanent rips against the end rocks.

Figure 4.75 *View of Five Rocks Point from Stockyard Point and the two intervening beaches.*

Swimming: The southern beach is the more accessible and best for swimming at high tide, as the northern beach may be completely awash. Watch out for the rips at low tide and when waves exceed 0.5 m.

Surfing: Chance of a reasonable break across the bar at mid to low tide.

Fishing: A popular spot for locals and visitors, very accessible for 4WD vehicles and easy access to the rock platforms at either end and the rip holes at low tide.

Summary: Two attractive and accessible beaches, by foot and vehicle. An excellent view of both is provided from Stockyard Point.

Byfield National Park

Length of coast: 12 km
Number of beaches: 4 (Beaches 1357-1360)

Byfield National Park extends along the coast from the midway alomng Nine Mile Beach to the souther headland, Water Park Point, and the northern shores of Corio Bay. This section of coast is largely backed by the massive parabolic dune systems that extend up to 10 km inland. Access to the park is via the Byfield Road and then sandy 4WD tracks. Its features include Nine Mile Beach and dune systems. Sweeping views are provided from Stockyard Point and Water Park Point. Nestled on

Stockyard Point is a small freehold area containing a number of holiday shacks.

The park provides no facilities. For information contact:

Queensland Parks & Wildlife Service
Yeppoon and Norman Roads
PO Box 3130
Rockhampton Shopping Fair, Qld 4701
Phone: (07) 4936 0511
Fax: (07) 4936 2212

Swimming: The best swimming is at mid to high tide, as rips are often present at low tide.

Surfing: This is the closest big wave beach outside of Yeppoon and is often visited by locals when they think the surf is up. The breaks are better at mid to low tide.

Fishing: This beach is popular with the locals and visitors who usually fish the rips or drive down the beach to fish and catch crabs in Corio Bay.

Summary: A long and relatively energetic beach with freshwater soaks and massive dunes.

1357 NINE MILE

No.	Beach	Unpatrolled Rating		Type	Length
		HT	LT		
1357	Nine Mile	2	3	R+BR	13 km

Nine Mile Beach (1357) is a 13 km long, east facing beach that runs essentially due south from the southern flanks of Stockyard Point to Water Park Point. The beach and massive backing dune fields lie within Byfield National Park and are accessible by 4WD along a 7 km sandy track off the Byfields Road. The track reaches the beach 2 km south of Stockyard Point. There is no development on the beach, however fresh water can be obtained from the many soaks that drain across the beach, including a small waterfall toward the northern end. A track also leads up to the Stockyard shack settlement and the entire beach can be driven at low tide when firm sand is exposed (Fig. 4.76).

The beach receives waves averaging 0.6 m and consists of a 50 m wide high tide beach, fronted by a 150 m wide low tide bar which, during high waves, is usually cut by rips every 200 m, resulting in up to 50 rips along the beach (Fig. 2.13a & b). A shore parallel trough and outer bar, cut by more widely spaced rips, lie just offshore.

Figure 4.76 *The northern end of Nine Mile Becah showing the wide low tide beach and surf zone.*

1358-1360 WATER PARK PT-LITTLE CORIO BAY

No.	Beach	Unpatrolled Rating		Type	Length
		HT	LT		
1358	Water Park Pt (1)	2	3	R+LT	100 m
1359	Water Park Pt (2)	2	3	R+LT	50 m
1360	Little Corio Bay	2	3	R+BR	750 m

Water Park Point is an irregular, 80 m high headland that forms both the southern boundary of Nine Mile Beach and the northern head of expansive Corio Bay. Much of the 4 km of headland shore is rock, however set in the northern flanks are two small beaches, with longer Little Corio Bay just north of the southern tip. The base of the point is accessible via Nine Mile Beach, however there is no formal vehicle access to the point or the beach and no development.

The two Water Park Point beaches face north-east and are wedged into two small adjoining valleys. The first beach **(1358)** has an irregular, rocky backshore, then a 100 m wide intertidal to low tide sandy beach. The second beach **(1359)** is backed by a 50 m long boulder beach, flanked by wide rock platforms and fronted by a similar sandy beach. The two beaches link at low tide to form a continuous strip of sand 200 m long. Waves average about 0.4 m and spill across the low gradient bar.

On the south side of Water Park Point is the 750 m long, east facing **Little Corio Bay**. It is a relatively straight sand beach **(1360)**, backed by a vegetated dune field (that at one time spilled into backing Corio Bay), a stable foredune, then a 50 m wide high tide beach and a 150 m wide low tide beach and bar. Usually three to four rips cut across the base of the bar, including a permanent rip against the southern headland. Some rocks also lie off the southern end of the beach, with rock platforms bordering the headlands at each end.

Swimming: Little Corio Bay is the more attractive beach, with less rock. However it receives higher waves and is prone to rips at mid to low tide, so use caution.

Surfing: Little Corio Bay usually has a beach break and picks up most swell.

Fishing: There are excellent rocks to fish from at all three beaches, as well as the rip holes at Little Corio Bay.

Summary: It is well worth a slow walk around Water Park Point to see the views of the beaches, headlands and the massive tidal flats of Corio Bay, as well as stopping for a swim, surf or some fishing at one of the beaches.

1361, 1362 FARNBOROUGH, BARWELL CK

No.	Beach	Unpatrolled			
		Rating		Type	Length
		HT	LT		
1361	Farnborough	1	2	UD	17 km
1362	Barwell Ck	1	2	UD	2.6 km

Corio Bay is a dynamic, sand-filled tidal inlet. It has a 2 km wide mouth between Water Park Point and the low Sandy Point. Sandy Point is the northern spit of the 17 km long **Farnborough Beach (1361)** and barrier system that has built out up to 3 km seaward and accumulated over 20 km² of marine beach and dune sands over the past 6000 years. The beach is actually a slightly protruding foreland; a response to wave refraction around North Keppel Island, which lies 11 km offshore. The apex of the foreland is located at the Capricorn International Resort (Fig. 4.77). Today this large accumulation of sand and sandy beaches remains in a natural state for the first 10 km, with the southern 7 km housing the resort and the small settlements of Farnborough and Bangalee. The resort has a low profile and is set well back from the beach; only the access tracks leading to the beach suggest there is more behind the foredune. There is good road access to the three settlements and the resort and a gravel road running up the back of the spit past the extensive mangroves of Corio Bay, as well as 4WD access via Bangalee along the firm beach at low tide.

The main beach ends in the south at the small **Barwell Creek**, beyond which the beach **(1362)** continues on for 2.6 km to the low rock of Spring Head. This beach is backed by rising land toward the head and the northern suburbs of Yeppoon. The main road parallels the back of most of this beach, with a foreshore reserve and caravan park located between the road and the beach.

The beaches are composed of fine sand and exposed to waves averaging 0.8 m. These factors maintain a low, narrow high tide beach fronted by a 200 m wide, low gradient low tide beach. Waves spill across the low gradient bar at all stages of the tide, and rips are usually absent.

Figure 4.77 *The wide Farnborough Beach at low tide, backed by the Capricorn International Resort.*

Swimming: These are relatively safe beaches under normal low, spilling wave conditions. However, higher waves will induce rips toward low tide, and possibly longshore currents. Be very careful at Corio Bay as there are strong tidal currents and deep channels off the spit.

Surfing: Chance of low, spilling waves; biggest and usually best up toward the spit.

Fishing: Most fishers head along the beach to Corio Bay, while some fish the beach at high tide.

Summary: One of the longer beaches on the coast with good access and an international standard beachfront resort, as well as many kilometres of natural beach.

1363 YEPPOON BEACH (SLSC)

Patrols:			
Surf Lifesaving Club:		September to May	
Lifeguard on duty:		September to May	
No.	Beach	Rating	Type	Length
		HT	LT	
1363	Yeppoon	2	3	UD	1.4 km

Yeppoon has traditionally been the main surfing beach for Rockhampton and inland mining towns like Mount Magnet. The Yeppoon Surf Life Saving Club was established in 1926, although a reel was placed on the beach as early as 1917. The railway line used to bring the holiday makers, however now they arrive on a four lane highway that makes it a fast 40 km drive out from Rockhampton.

Today Yeppoon is a thriving residential and tourist area, offering all the facilities of a Queensland tourist centre. There are several good beaches in the area, with the main

Yeppoon Beach, site of the surf lifesaving club, located at the end of the main shopping street (Fig. 4.78). The beach is 1.4 km long, extending from the low intertidal rocks at Spring Head almost due south to the mouth of Ross Creek, where there is a small breakwater. Unfortunately erosion has plagued the beach and today it is backed by a continuous seawall, behind which there are parking areas and a grassy reserve, with the main road paralleling the northern half of the beach. The surf club is located in the reserve toward the northern end of the beach and the Keppel Bay Sailing Club toward the southern end.

The seawall is reached by the waves at high tide, resulting in little to no high tide beach and a low, flat, 200 m to 300 m wide low tide beach composed of fine sand (Fig. 4.79). The beach is partially protected from waves by the Keppel Island group, and waves average only 0.5 m.

Swimming: A relatively safe beach; just be careful at low tide when you have to walk a long way to reach the water, and near the southern Ross Creek entrance which has both deeper water and strong tidal currents.

Surfing: Usually only low, sloppy waves.

Fishing: Best in Ross Creek; the beach is usually too shallow.

Summary: A major tourist destination and a popular and very accessible beach, backed by all tourist facilities.

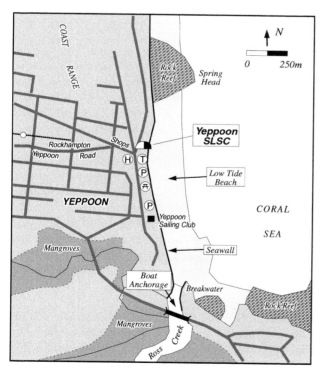

Figure 4.78 *Yeppoon town centre and beach.*

Figure 4.79 *Nippers on Yeppoon Beach at low tide. Note the wide, shallow low tide surf zone.*

1364-1368   WAVE PT, COOEE BAY, LAMMERMOOR BEACH, STATUE BAY, ROSSLYN BOAT HARBOUR

No. Beach	Unpatrolled			
	Rating		Type	Length
	HT	LT		
1364 Wave Pt	1	2	UD	500 m
1365 Cooee Bay	1	2	R+LT	500 m
1366 Lammermoor	1	2	UD	2.5 km
1367 Statue Bay	1	2	R+	
			rock flats	600 m
1368 Rosslyn Boat Hbr	1	1	B+SF	30 m

Bedrock reaches the coast at Ross Creek and dominates much of the coast between here and Cocoanut Point, 17 km to the south. The main Emu Park Road runs close to the coast because of the backing higher ground and provides good access to all the beaches.

Immediately south of Ross Creek is **Wave Point** beach (1364), which begins on the southern side of the Ross Creek groyne and curves around for 500 m, with a roughly north-east orientation. There are rocks on the beach, which ends against the rocky, 20 m high Wave Point. The beach is backed by beachfront houses wedged between the main road and the beach, with houses also located on the point. There is a small car park toward the southern end of the beach. Waves average less then 0.5 m and break across a wide, low gradient low tide bar and beach.

On the southern side of Wave Point is the small **Cooee Bay**, which has a 500 m long, curving beach (**1365**) extending from the point to the southern Wreck Point. The beach is bordered by the 20 m high headlands and their rocky platforms and reefs, with rocks extending along the northern section of the beach. The remainder is a moderately steep high tide sand beach fronted by a continuous, 50 m wide low tide bar. Waves average less than 0.5 m and rips are usually absent.

On the south side of Wreck Point lies 2.5 km long **Lammermoor Beach** (1366). The beach is backed by a casuarina-covered foredune, with houses in the north and the main road running right behind the southern half of the beach. The best access is at a small bluff-top car park in the centre of the beach, below which are a few rocks across the beach. The beach is composed of medium sand and has a relatively steep high tide beach, with a lower gradient, 100 m wide, continuous low tide bar, that widens in the southern end where a small creek drains across the beach. The beach ends at the low energy Statue Rocks.

Just beyond Statue Rocks is the small **Statue Bay** which houses a low energy, 600 m long, north facing beach (**1367**). The beach is backed by a foreshore reserve, containing a car park in the centre of the beach and access to the beach for high tide boat launching. The main road backs the reserve, with a few houses on the other side of the road. The beach consists of a coarse, steep, sand and cobble high tide beach fronted by low tide rock flats, including a few mangroves on the eastern rocks. The breakwater for the Rosslyn Bay boat harbour extends off these rocks.

Between Statue Bay and Double Head lies **Rosslyn Bay** which, prior to development in the 1970's, was a low energy, 500 m long beach. The bay has now been developed as a boat harbour, with breakwaters and boating facilities replacing most of the beach. All that remains is a 30 m long pocket of sand (**1368**) wedged in beside the large boat ramp.

Swimming: The most popular beaches are Cooee Bay and Lammermoor Beach, with rocks dominating Wave Point and Statue Bay, while Rosslyn Bay is given over to boating activities.

Surfing: Chance of low, spilling waves at Cooee and Lammermoor beaches.

Fishing: Most shore fishers use the Rosslyn Bay boat harbour breakwaters, while you can also fish from the several rocky points at high tide.

Summary: These five beaches are a southern extension of Yeppoon and are very accessible either from the backing streets or the main road. They are mainly used by locals and largely bypassed by the tourists enroute to Rosslyn Harbour and the ferries to Great Keppel Island.

Double Head, Bluff Point and Mulambin National Parks

Double Head, Bluff Point and Pinnacle Point are three conical shaped, 50 m, 100 m and 60 m high headlands (respectively), all or part of which are small national parks. The three border neighbouring Kemp and Mulambin Beaches.

For further information:
Queensland Parks & Wildlife Service
Yeppoon and Norman Roads
PO Box 3130
Rockhampton Shopping Fair, Qld 4701
Phone: (07) 4936 0511
Fax: (07) 4936 2212

1369-1373 KEMP, MULAMBIN & KINKA

No.	Beach	Unpatrolled Rating HT	LT	Type	Length
1369	Kemp	1	2	UD	2 km
1370	Mulambin	1	2	UD	2 km
1371	Kinka	1	1	B+SF	3 km
1372	Kinka (S 1)	1	1	B+SF	1.2 km
1373	Kinka (S 2)	1	1	B+SF	500 m

Double Head is a 50 m high double point that partially protects Rosslyn Harbour, and rocks were also quarried from the point to build the harbour breakwaters. The largely cleared, steep sided head is now a national park and forms the northern boundary of 2 km long **Kemp Beach** (1369), while 100 m high Bluff Point, also a small national park, forms the southern boundary. The beach is backed by the main Emu Park Road and there is car parking right along the back of the beach, together with amenities and two shady picnic areas. The beach receives waves averaging less than 0.5 m that break across a 200 m wide, low gradient low tide bar and up a moderate gradient high tide beach. A 10 m high, grass and casuarina-covered foredune backs the beach, with the road running along the back of the foredune to provide good access for the length of the beach.

Mulambin Beach (1370) lies on the south side of Bluff Point and runs due south for 2 km to Pinnacle Point, with the 200 m wide low tide beach extending out to, and at times around, the point (Fig. 2.12b). Like Kemp Beach, Mulambin also receives low waves and has a low gradient, fine sand beach with waves spilling across a wide, shallow surf zone at low tide and a moderately steep beach at high tide. The beach is backed by a 10 m high, casuarina-covered foredune and the main road, with a caravan park and camping reserve located between the road and the beach. There is also access at the northern end for launching boats off the beach at high tide. The northern slopes of Pinnacle Point comprise a small national park and also provide excellent views of both Mulambin and the southern Kinka Beaches.

Kinka Beach (1371) occupies the southern half of Shoal Bay, the name referring to the extensive sand and tidal flats that lie off the beach (Fig. 4.80). The beach is 3 km long, faces east-north-east and consists of a narrow and, in places, eroding high tide beach fronted by several hundred metres of variable low tide sand flats and ridges. The creek that used to drain across the northern end of the beach has been dammed by The Causeway and the tidal flats are still readjusting to the absence of the tidal currents that flowed in and out of the creek. To combat the resulting shoreline dynamics, a seawall and more recently an artificial sand barrier have been built along the northern end of the beach. The beach is backed by the main road, with parking areas and access provided at each end, and a caravan park, motel and houses along the western side of the road.

Figure 4.80 *A view across the wide tidal sand flats fronting Kinka Beach.*

The protected southern end of Kinka Beach ends at a mangrove-fringed tidal creek, on the other side of which is the very low energy beach **1372**, a 1.2 km long, crenulate, curving, north facing, narrow high tide beach fronted by several hundred metres of tidal flats and numerous mangroves. There is a vehicle track off the main road to the rear of the beach. The beach ends at a low, rocky, mangrove-fringed point, beyond which is a 500 m long, west facing high tide beach (**1373**) fronted by extensive sand and rock flats.

Swimming: Kemp and Mulambin are two relatively popular and very accessible beaches. Their usually low waves and very low gradients also provide relatively safe swimming.

Surfing: Only low, spilling waves at Kemp and Mulambin.

Fishing: Best at the Causeway and off the points at high tide.

Summary: Five low energy beaches, three of which lie beside the main road and are bounded by three small headland national parks.

1374, 1375   TANBY, FISHERMANS

No. Beach	Unpatrolled			
	Rating		Type	Length
	HT	LT		
1374 Tanby	1	2	UD	2 km
1375 Fishermans	1	2	UD	2 km

South of Shoal Bay the coast trends more to the south and a series of headlands and rocky shore dominate the coast. The first two beaches, Tanby and Fishermans, are both 2 km long, face east and are bounded by rocky headlands.

Tanby beach (1374) is backed by a 100 m high ridge for most of its length and, while there is a private residence at the northern end of the beach, public access is restricted to the lower southern end of Haven Road. The beach is bounded by the ridge of rock in the north, with rocks also outcropping in the centre to divide the low tide beach in two. In the south, Tanby Point protrudes 500 m to the north-east and is also occupied by a private residence. The beach receives waves averaging about 0.5 m that break across a moderately steep high tide beach, while at low tide there is a 100 m wide, shallow, continuous bar.

Fishermans beach (**1375**) runs due south of Tanby Point for 2 km, curving around in lee of Emu Point to almost face north. The protection afforded by Emu Point has long been used to launch fishing boats, thereby giving the beach its name. Today the southern end is backed by a large, grassy foreshore reserve and caravan park, with a long boat ramp protected by a groyne in the southern corner. Twenty metre high vegetated dunes back the northern half, with access in the north also from Tanby Point. The beach is somewhat protected by the point and receives low waves in the south, rising to less than 0.5 m to the north. The waves break across a wide, low gradient low tide beach, with a moderately steep high tide beach.

Swimming: Fishermans Beach is by far the more accessible, popular and safer beach in the area, servicing the Emu Park township and the influx of holidaymakers.

Surfing: Both beaches offer generally low, spilling waves that are usually higher on Tanby Beach.

Fishing: Fishermans Beach is the focus of boat fishing, however there are plenty of rocks and rock platforms to fish at high tide.

Summary: Fishermans is the traditional beach for Emu Park and has excellent access and backing facilities.

1376, 1377   EMU PARK (SLSC)

Patrols:				
Surf Lifesaving Club:		September to May		
Lifeguard on duty:		September to May		
No. Beach	Rating HT LT		Type	Length
1376 Ladies	2	3	R+LT +rocks	100 m
1377 Emu Park (Mens)	1	2	R+LT	1.2 km

Emu Park is a small holiday town that has long been a popular tourist destination. Traditionally holidaymakers have come from Rockhampton and surrounding country towns and even today this is still much the case, with the addition of travelling tourists and backpackers. The town is located in lee of low Emu Point, with attractive beaches either side. The two main surfing beaches immediately south of Emu Point were called Ladies and Mens Beaches. The small Ladies Beach still goes by that name, while Mens is better known today as Emu Park and is the site of the surf lifesaving club.

Ladies beach (1376) is a 100 m long pocket of sand backed by 20 m high bluffs and bordered by rock platforms. There is a large car park and the Singing Ship Memorial on the bluffs. The beach receives waves averaging 0.5 m that break across a 50 m wide low tide bar; the high tide beach is usually absent, owing to the rocks. While the bluffs and rocks no doubt provided some privacy for the ladies, they also produce one of the more hazardous beaches in the area, particularly when waves are breaking over the rocks.

The main surfing beach is **Emu Park** beach (1377), located on the south side of the point. It is 1.2 km long and faces due east, with low Rocky Point forming the southern boundary and a rock outcrop on the centre of the beach. Although the northern half is called Mens Beach, south of the rocks is also known as Shellys Beach. A lifesaving reel was placed on the beach in 1917 and the Emu Park Surf Life Saving Club was founded in 1925. The modern club house is located at the northern end of the beach and is fronted by a seawall (Fig. 4.81). It is surrounded by an extensive grassy foreshore reserve and parking areas. The reserve runs the full length of the beach, with houses set well back from the shore.

The beach receives waves averaging 0.5 to 1 m, which break across a wide, low gradient beach. At high tide the beach is 50 m wide, extending to over 200 m wide at low tide. There are low, rocky points at either end, together with the low intertidal rocks in the centre of the beach (Fig. 4.82).

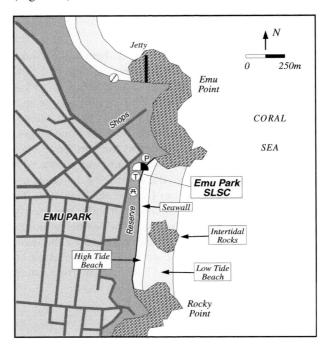

Figure 4.81   *Emu Park Beach is bordered by rocky reefs and headlands and backed by a foreshore reserve, with the surf club located at the northern end.*

Figure 4.82   *A view north from Zilzie Point Beach to Emu Park and Ladies Beaches, with Fishermans Beach at top. This low tide view shows the wide intertidal beach and small low tide surf.*

Swimming: A relatively safe beach, best at mid to high tide. Be careful when waves exceed 0.5 m as rips may occur, at low tide and around the rocks.

Surfing: The surfing beach for Emu Park, offering a usually low beach break.

Fishing: Best off the rocks at each end. There is a boat ramp on the north side of Emu Point.

Summary: Emu Park and its beaches comprise an attractive, relatively quiet, regional Queensland coastal town. The place to head if you do not need all the trappings of Yeppoon, with the added attraction of a nice patrolled beach.

1378-1380 ROCKY & ARTHUR PTS, ZILZIE BEACH

No.	Beach	Unpatrolled		Type	Length
		Rating			
		HT	LT		
1378	Rocky Pt	1	3	R+LT+rocks	150 m
1379	Arthur Pt	1	3	R+LT+rocks	300 m
1380	Zilzie	1	1	UD	600 m

South of Rocky Point are three small, rock dominated beaches - Rocky Point, Arthur Point and Zilzie Beach. The three beaches face east and are all bordered by rocky points and intertidal rock flats.

Rocky Point beach (1378) lies on the southern side of the point and is a 150 m long pocket of high tide sand, narrowing to 50 m at low tide as the rocks expand. The beach is backed by 20 m high bluffs occupied by beachfront houses and the best access is from the northern point. Waves spill across the low tide bar and can generate rips against the rocks.

Arthur Point beach (1379) is similar, only longer at 300 m and backed by higher bluffs also occupied by houses. It has a relatively steep high tide beach and a 50 m wide low tide bar with rocks to either end. Due to the backing houses, access is restricted to around the rocks from Rocky or Zilzie.

Zilzie Beach (1380) is the longest and most accessible of the three beaches. It is 600 m long, extending from Arthur to Zilzie Points. It is also composed of finer sand and has a 200 m wide, low gradient high and low tide beach. There is a small car park at the northern end, an amenities block in the centre and a picnic area toward the south, with houses backing the beachfront road.

Swimming: These are three relatively exposed beaches

dominated by rocky boundaries. Be careful at Rocky and Arthur, as rips can run out against the rocks when waves are breaking. Zilzie is the most accessible and safest of the three.

Surfing: Chance of low, spilling breakers at Arthurs and Zilzie.

Fishing: Best off the many rocks at high tide.

Summary: Three beaches largely surrounded by an expanding residential area, with formal public access to Zilzie only.

Keppel Bay

Keppel Bay is an open, 30 km wide bay, bounded by Zilzie Point in the north and Cape Keppel on Curtis Island to the south. It opens to the north-east and is fed in the south by the large Fitzroy River, the second largest in Queensland. Beaches occupy much of the shore north of the river mouth, while the mouth itself is a massive area of high tide salt flats and fringing mangroves. Along the northern side of Curtis Island there is sufficient wave energy for beaches to again occupy the shoreline. The bay was sighted and named by Captain Cook in May 1770.

1381-1383 COCOANUT BEACH-CASTLE BELLAS

No.	Beach	Unpatrolled		Type	Length
		Rating			
		HT	LT		
1381	Cocoanut	1	2	B+SF	1 km
1382	Cocoanut Pt	1	3	B+SF	300 m
1383	Castle Bellas	1	2	B+SF	3.5 km

At Zilzie Point the coast swings to the south-west, toward Cocoanut Point and Castle Bellas beyond that. At the same time, extensive sand flats associated with the 1.5 km wide mouth of Cawarral Creek lie off the beaches.

Cocoanut Beach (1381) runs for 1 km between Zilzie and Cocoanut Point; a low, rocky point. The beach faces south-east into the Trades, but is protected from high waves by the sand flats. The beach consists of a low gradient high tide beach fronted by low sand flats up to 300 m wide in the centre of the beach. A tidal channel runs along the outside of the flats, coming closer inshore off the two points. The beach is backed by beachfront houses of a relatively new housing estate, with limited access to the beach.

Cocoanut Point beach (1382) lies immediately north of the point and is bounded by the point and a low, rocky outcrop that occupies the high tide beach. It is a 300 m long continuation of the main beach, however the tidal channel lies just off the southern rocks, truncating the sand flats and producing a deep channel and strong tidal currents at low tide. The beachfront houses terminate overlooking the northern end of the beach, with parking and access to the northern end from the street.

Castle Bellas beach (1383) is a 3.5 km long, low energy beach running to the south-west from Cocoanut Point to the deep mouth of Cawarral Creek, where the beach recurves around into the wide creek mouth. The beach is backed by low, swampy beach ridges and swales used for grazing cattle, and two small creeks drain out across the beach. It is fronted by a low gradient high tide beach, a few low sand ridges and sand flats extending out to the tidal channel up to 500 m offshore. There is a vehicle track to the top of Cocoanut Point overlooking the beach, but apart from the farm tracks there is no public access to the beach.

Swimming: These are three low energy beaches best for swimming at high tide. Be careful if out on the tidal flats at low tide, as they are bordered by the deep tidal channel.

Surfing: None.

Fishing: Best at high tide or in a boat.

Summary: The three southernmost Emu Park beaches, including the two most recently developed and one still given over to farming.

Great Keppel Island

Great Keppel Island is a high, 13 km² island lying 14 km due east of Rosslyn Harbour. The island has 30 km of shoreline containing 21 sandy beaches, all backed by wooded slopes that rise to three 100 m high ridges, with the highest, Mount Wingham, peaking at 175 m. The island is surrounded by several smaller islands and areas of patchy coral reef, in addition to some fringing reef on the island. The western tip of the island is the site of a major tourist resort, with visitors arriving by plane and by ferry from Rosslyn Harbour.

Grest Keppel Island - East Coast
GK1-5   **WRECK-RED BEACH**

No.	Beach	Rating		Type	Length
		Unpatrolled			
		HT	LT		
GK1	Wreck	1	3	UD	1.4 km
GK2	Wreck (1)	1	3	R+LT	350 m
GK3	Wreck (2)	1	2	R+LT	150 m
GK4	Wreck (3)	1	2	R+LT	100 m
GK5	Red	1	3	R+BR	250 m

The eastern side of Great Keppel Island extends from the northernmost point, Big Peninsula, down to Red Beach 5 km to the south. In between is an open bay containing the four Wreck Beaches, with Red Beach lying in its own small, south-east facing bay. This is the most exposed and highest energy side of the island and consequently has the two highest energy beaches - Wreck and Red. Access to any of the beaches is by foot and there are no facilities.

Wreck Beach (GK1) is 1.4 km long and curves around within the bay to face generally east. It is bordered by the southern slopes of 115 m high Big Peninsula in the north and a 60 m high headland to the south. It receives waves averaging 0.5 to 1 m which break across a 200 m wide, continuous, low gradient bar and the high tide beach. The northern half of the beach is backed by climbing, largely vegetated dunes that rise to 40 m before slipping over into a beach (GK20) on the north side of the island.

South of Wreck Beach are three smaller beaches, each separated by 40 m high, grassy headlands and which receive increasing protection from the southern Sykes Rock headland. The first (**GK2**) is 350 m long and faces east-north-east, receiving waves averaging 0.5 m. It has a moderately steep high tide beach fronted by a 100 m wide bar. Beside the bordering headland it is backed by a 20 m high, degraded foredune and an older sand blowout that has run up the backing bedrock. Much of the headland and dune degradation is due to the island's goat population, with goats commonly being seen on the headlands.

Beach **GK3** lies around another 40 to 60 m high headland, is 150 m long, faces north-east and receives waves less than 0.5 m. As a result of the lower waves, it has a steep high tide beach fronted by a 50 m wide bar, with a 10 m high foredune backing the beach. The final beach (**GK4**) is the smallest and lowest energy in the bay. It is just 100 m long, faces due north and is often calm. The beach is composed of sand and coral rubble and is relatively steep, with the high tide beach ending amongst the backing bedrock, which then rises steeply to a 170 m high ridge line. There are well developed rock platforms around the bordering headlands.

On the southern side of the ridgeline is the energetic, east facing **Red Beach** (GK5), so named because of the exposed red Pleistocene dunes behind the beach. It is 250 m long and receives waves averaging 0.5 to 1 m that break across an 80 m wide bar, which is backed by a steep high tide beach and a veneer of white sand over the older red dunes, then an amphitheatre of scrubby bedrock rising to a peak of 170 m.

Swimming: Wreck and Red Beaches are the two highest energy on the island and, while rips are usually absent from Wreck Beach, care should be taken in the surf. The southern Wreck beaches are increasingly protected and delightful little, relatively safe beaches.

Surfing: Wreck is the most popular surfing beach on the island, while Red Beach has surf but is more difficult to access.

Fishing: The main headlands, points and rock platforms provide numerous places to fish deeper water at all tides.

Summary: These beaches lie 4 to 5 km from the main resort area, but are well worth the walk for some stunning natural beaches and the chance of usually low to moderate waves.

Great Keppel Island - South Coast
GK6-10 CLAM BAY, MT WINDHAM, LONG BEACH

No.	Beach	Unpatrolled Rating HT	LT	Type	Length
GK6	Clam Bay (E)	1	2	R+reef flat	200 m
GK7	Clam Bay	1	2	R+reef flat	550 m
GK8	Mt Windham (1)	1	2	LT	50 m
GK9	Mt Windham (2)	2	3	boulder+LT	50 m
GK10	Long Beach	1	2	R+LT	1.6 km

The south coast of Great Keppel Island extends for 5 km from the southern Red Beach point to Monkey Point on the south-western tip of the island. It faces south-south-east and contains five beaches of varying nature and size. Two small islands, Halfway Island 1.5 km to the south and Humpy Island 2.5 km south, provide some protection from the southerly winds and waves. Only Long Beach is readily accessible, with the eastern beaches requiring a longer and more difficult walk to get to.

Clam Bay is a 2 km long, open, south facing bay containing four beaches. The first beach (**GK6**) is a 200 m long pocket of south facing high tide sand fronted by a coral reef flat that extends 100 to 150 m off the beach. The reef flat continues on past the main, 550 m long, sandy high tide beach (**GK7**), which is backed by a 10 m high foredune and a small sand flat. The beaches only receive low, attenuated waves at high tide because of the reefs, while at mid to low tide the waves break across the reef. These are delightful beaches at high tide for a swim, or for a wander over the reefs at low tide.

The western end of Clam Bay lies at the base of 175 m high Mount Windham, the highest point on the island. Along its rocky, indented base are two small pocket beaches. **GK8** is a 50 m long low tide bar wedged in between and under surrounding rocks, with no high tide beach. It is relatively exposed and receives waves averaging over 0.5 m, which produce small rips against the rocks. Its immediate neighbour (**GK9**) is of similar

length, but contains a 50 m wide boulder high tide beach, that grades into sand toward low tide.

The western slopes of Mount Windham form the eastern boundary of **Long Beach** (GK10) which, at 1.6 km, is the longest beach on the south side. The long, low gradient beach faces south-south-east and, because of protection from Humpy Island, receives waves less than 0.5 m high. It has a narrow, moderately steep high tide beach backed by a 10 m high foredune and a 100 m wide low tide bar. This beach lies 1.5 km due south of the resort and 500 m south of the runway, and is a popular sunbathing and swimming beach.

Swimming: Long Beach is the most accessible, safest and longest of the south coast beaches.

Surfing: There is no real surf along the south side, apart from the small Mount Windham beaches.

Fishing: There are many excellent rock platforms, plus reef flats spread along the southern shore.

Summary: An exposed and at times windy shore, with Long Beach being a popular location when the wind is not too strong.

Great Keppel Island - West Coast
GK11-14 MONKEY, FISHERMANS, PUTNEY

No.	Beach	Unpatrolled Rating HT	LT	Type	Length
GK11	Monkey	1	1	R+reef flats	550 m
GK12	Monkey (N)	1	1	R+reef flats	150 m
GK13	Fishermans	1	1	R+LT	1.2 km
GK14	Putney	1	1	R+LT	1 km

The western side of Great Keppel Island is the closest to the mainland, lies in lee of the prevailing Trade winds and is the centre of all island activity, containing the resort, runway, local housing and the beach landing for ferries, as well as the two major recreational beaches. The western side is just 3 km long, running from Monkey Point in the south to Lecks Beach in the north. It contains four accessible, generally low energy, west facing beaches.

Monkey Beach (GK11) is a 550 m long, curving, south-west facing beach bordered to the south by 20 m high and 500 m long Monkey Point. In lee of the point is a fringing reef, with 200 to 300 m wide sand flats between the reef and the narrow high tide beach. The beach receives low, lapping waves at high tide and is protected by the reef at low tide. There is a coastal walking track along the beach, that follows the rocks along the northern end and around a steep, wooded point to the northern beach (**GK12**). This is a 150 m long, low energy beach sitting in a small valley, with a sand and rubble high tide beach, including scattered boulders. The beach is fronted by a fringing reef off each headland and a small, central sand flat. It is the first undeveloped beach south of the resort and is used by pedestrians and sunbakers.

Fishermans beach (GK13) is one of the two resort beaches and the centre of most island activity. It is a 1.2 km long, curving, west facing beach and is backed by some freehold housing in the south, with the major resort and the runway backing most of the tree-lined beach. The beach receives waves averaging less than 0.5 m and is often calm. It has a steep, narrow high tide beach, fronted by a continuous, shallow, 100 m wide bar. Boats are moored off the more protected southern end and the entire bay is relatively shallow. Due to its protected nature, the ferries from Rosslyn Harbour nose into the deeper northern end of the beach to land and pick up visitors.

Putney Beach (GK14) is the second resort beach and the site of some of the more active beach activities, including sailing and boating. There is a boat access ramp, with launching from the steep high tide beach. The beach is backed by a low foredune, with fenced access tracks leading to the resort camping ground. Waves average less than 0.5 m and break across a 50 to 100 m wide low tide bar.

Swimming: These are four low energy, relatively safe beaches under the usually low to calm wave conditions.

Surfing: None.

Fishing: Best off the rocks around the Monkey beaches, or off the beaches at high tide.

Summary: These are the four most visited and more popular beaches on the island, with Putney and Fishermans the most accessible and heavily utilised.

Great Keppel Island - North Coast
GK15-21 PARASAIL, LECKS, SWENSDEN

No.	Beach	Unpatrolled Rating HT	LT	Type	Length
GK15	Parasail	1	1	R+LT	500 m
GK16	Lecks	1	1	R+LT	1.8 km
GK17	Beach 17	1	1	R+LT	450 m
GK18	Svensden	1	1	R+LT	750 m
GK19	Svensden (N)	1	1	R+LT	150 m
GK20	Beach 20	1	1	R+LT	400 m
GK21	Beach 21	1	1	R+LT	400 m

The northern side of Great Keppel Island faces northwest and extends for 4.5 km from Parasail Beach to Big Peninsula headland. It contains seven open, sandy beaches

and with the Trades blowing offshore is usually a quiet, out of the way part of the island. Parasail and Lecks Beaches are the easiest to access on foot from the resort.

Parasail beach (GK15), as the name implies, is where the offshore Trades are used to assist people parasailing off the beach. It is a north-north-east facing, 500 m long, low gradient sandy beach backed by a patchy foredune and vegetated bluffs rising to 40 m. Rocks form the western point and outcrop across the eastern end of the beach, separating it from neighbouring Lecks Beach.

Lecks beach (GK16) is the longest beach on the island and site of the island's only tidal creek. The beach is 1.8 km in length and faces initially north, curving around by the eastern headland and creek mouth to face northwest. Backall Creek, the only major creek on the island, drains the only salt flats and mangroves on the island. The creek flows out at the eastern end of the beach, producing a small tidal delta at the mouth. Elsewhere the beach is low gradient and fronted by a 100 m wide, continuous bar. Waves are usually low to calm.

Beach **GK17** is a 450 m long, north-west facing, sandy beach wedged in between two rocky points and backed by steep, sparsely vegetated slopes rising to 50 m. It has a low foredune, with casuarinas draping over the moderate gradient high tide beach, which is in turn fronted by a continuous, 50 m wide low tide bar.

Svensden beach (GK18) is a 750 m long, curving, west facing beach that has a few shacks located on low land behind the northern end. The beach is sheltered and receives low waves, that maintain a low gradient beach fronted by a 50 m wide bar and relatively shallow bay, which is used as a boat anchorage by the shack owners. The beach ends at a northern jagged, 20 m high point, on the eastern side of which is a patchy, 150 m long, steep sandy beach (**GK19**) lying in amongst the rocks. This beach and backing Svensden Beach form a small tombolo that ties the point to the island.

Beach **GK20** lies at the rear of Wreck Beach and the dune of Wreck Beach spills over onto the beach. Consequently it consists of a low, narrow high tide beach and small foredune backed by 20 to 30 m high, partially vegetated dune from the bay. The beach faces north and is sheltered from the Trades, creating low wave to calm conditions and a narrow, continuous low tide bar.

Beach **GK21** lies immediately west of Big Peninsula. It is largely surrounded by an amphitheatre of grassy slopes rising steeply to the 15 m high headland, making foot access steep and difficult. The sheltered beach has a moderately steep, 20 m wide high tide beach fronted by a 30 m wide low tide bar, with deep water off the bar.

Swimming: All seven north side beaches are relatively safe, owing to the usually low wave to calm conditions. Some have deep water off the beach, particularly at high tide.

Surfing: Usually none.

Fishing: There are many excellent rock platforms along the north side, as well as rock reefs and Blackall Creek, the only tidal creek on the island.

Summary: The leeward side of the island and the place to head to get away from the Trades and to find more secluded beaches.

1384-1387 KEPPEL SANDS

No.	Beach	Unpatrolled			
		Rating HT	LT	Type	Length
1384	Keppel Sands (N 2)	1	1	TSF	500 m
1385	Keppel Sands (N 1)	1	1	B+SF	100 m
1386	Keppel Sands	1	1	B+SF	1.3 km
1387	Keppel Sands Pt	1	2	UD	200 m

Keppel Sands is a quiet beach settlement located 15 km off the Emu Park Road and 40 km from Rockhampton. The road ends at the Sands and, apart from the locals, not many tourists make it out to this quaint, older style settlement. The Sands are surrounded by wide Cawarral Creek to the north, extensive tidal flats to the east and the long, swampy beach ridges of Cattle Point to the south, while mangrove-filled tidal creeks make up much of the backing land. It is essentially a land island by the sea.

There are four beaches around the settlement; two are essentially tidal flats and two are main beaches. Beaches 1384 and 1385 both face north across the mouth of 1.5 km wide Cawarral Creek. The first beach (**1384**) is a 500 m long, north-west facing spit of sand that extends south of the northern Keppel Sands headland. The beach ends at a small car park, beyond which it grades into mangroves. West of the car park and behind the mangroves is the Keppel Sands caravan park. The beach is fronted by almost 1 km of tidal flats, then the deep creek channel. There is vehicle access to the beach, which is used at high tide to launch small boats. On the north side of the headland is beach **1385**, a 100 m long, north facing strip of sand fronted by linear rock formations, a 100 to 200 m wide sand flat and the deep channel. It is only accessible on foot from the adjoining beaches.

The main **Keppel Sands** beach (1386) is 1.3 km long, faces east and is bounded by the northern 40 m high, conical headland and a southern protruding, 30 m high headland. The beach is backed by the bulk of the settlement,

two rows of streets, then the backing mangroves and Pumpkin Creek mouth at the southern end. The beach has experienced some erosion and a rock seawall lines much of the beach. Between the seawall and the road is a narrow foreshore reserve which widens a little to the northern end, where there is a car park. Toilet blocks are located at each end of the beach. The beach consists of a narrow high tide beach fronted by several hundred metres of tidal flats. Waves average less than 0.5 m and calm conditions are common.

On the southern point is a 200 m long, low sandy beach (**1387**) wedged in between two rocky points. It is backed by 40 m high grassy bluffs, with an Air-Sea Rescue station located on the crest, affording a commanding view into Keppel Bay. There is a road to the station, but no formal access to the beach.

Swimming: Keppel Sands is the most popular beach but only useable at high tide, while the point beach can be used at high and low tide.

Surfing: None.

Fishing: Keppel Sands is a fishing community, with most fishing taking place in the creeks and the bay. The best boat access is via Pumpkin Creek, where there is a boat ramp and a few jetties. You can only fish from the shore at high tide.

Summary: An out of the way, older style settlement, with a caravan park for visitors.

1388-1390 PUMPKIN & CATTLE PTS, RUNDLES

No.	Beach	Unpatrolled			
		Rating HT	LT	Type	Length
1388	Pumpkin Ck	1	1	B+SF	2.3 km
1389	Cattle Pt	1	1	UD	17 km
1390	Rundles	1	1	B+SF	3 km

South of Keppel Sands is an extensive 20 km long beach ridge plain that has built out 2 to 3 km on the northern side of the Fitzroy River mouth. The entire area is largely undeveloped, apart from land cleared for cattle grazing.

Beach **1388** is a 2.3 km long beach and low beach ridge plain, that is bordered by Pumpkin Creek and Keppel Sands in the north and a wide, shallow, unnamed creek in the south. The beach receives only low waves and consists of a low gradient high tide beach fronted by 300 to 600 m of tidal flats and ridges.

Cattle Point beach (1389) runs for 17 km to the southeast, from the small creek to low, sandy Cattle Point. The entire beach and backing beach ridge plain has built out over the past 6000 years using sediments supplied by the Fitzroy River. The beach receives sufficient wave energy to maintain a wide, low gradient beach that is up to 300 m wide at low tide. There are farm access tracks to parts of the beach, but no formal public access.

At Cattle Point the shoreline turns south and is called **Rundles** beach (1390). It consists of a low energy beach backed by a 3 km truncated series of beach ridges, housing a few fishing shacks at the southern end. The beach is also fronted to varying degrees by sand flats and a dynamic sand spit that at times extends the length of the beach, forming a long, shallow lagoon between the spit and the beach. The spit is usually fronted by a few hundred metres of tidal flats and then the northern channel of the Fitzroy River. However the entire area changes over time, particularly during and following major floods.

Swimming: The main Cattle Point beach offers the best and safest swimming, particularly toward high tide.

Surfing: Only low, spilling waves on the main beach.

Fishing: Most fishing occurs in the creeks or along the beach at high tide.

Summary: Three relatively isolated beaches and backing swampy beach ridge plains.

Fitzroy River Delta & Curtis Island

The **Fitzroy River delta** is one of the largest in Queensland, with an area of approximately 500 km². The river delivers vast quantities of sediment to the relatively quiescent southern Keppel Bay. Here strong tidal currents, rather than waves, have reworked the sediments and the delta largely consists of large tidal channels and an extensive network of tidal creeks and flats, fringed by hundreds of kilometres of mangroves. Some sands have been moved north of the mouth to build the wide Cattle Point beach ridge plain. The only access to and development on the delta proper is the small port of Port Alma.

The eastern boundary of the delta is the large **Curtis Island**, approximately 800 km² in size, extending for 45 km from Cape Keppel in the north to Tail Point in Gladstone Harbour in the south, and reaching a maximum width of 22 km. The island has a series of north-south trending hills, that reach a maximum height of 165 m at Mount Barker toward the north-western corner. The island is separated from the mainland by The Narrows, a 50 km long tidal channel connecting Keppel Bay with Gladstone Harbour. While an island in the true sense, at low tide it is possible to drive across the narrowest part of The Narrows, at Ramsay Crossing. The entire island is a cattle property called Monte Cristo, with very little development other than for cattle grazing and station tracks. The

island has 80 km of more exposed northern and eastern shore, containing 25 sandy beaches. Apart from a few shacks and fishing camps, there is no development on the beaches and a few station tracks comprise the only access, other than by boat.

1391-1396 SEA HILL, WARNER & STATION PTS, CAPE KEPPEL

No. Beach	Unpatrolled Rating HT LT		Type	Length
1391 Sea Hill Pt	1	1	B+SF	2.3 km
1392 Warner Pt	1	1	B+SF	2 km
1393 Station Pt (W)	1	1	B+SF	1 km
1394 Station Pt	1	1	B+SF	200 m
1395 Station Pt (E)	1	1	B+SF	3.5 km
1396 Cape Keppel (spit)	1	1	B+SF	600 m

Sea Hill Point is named after 84 m high Sea Hill, an isolated hill at the north-west tip of the island. Extending west of the hill is a 2 km long series of north facing recurved spits, that curves around to face west into the northern end of The Narrows. There is a lighthouse located on a small hill in lee of the outer spit, with some buildings and a landing also located on the inner spit. The 2.3 km long beach (**1391**) borders the outer spit, consisting of a moderately steep, curving high tide beach fronted by up to 500 m of low tide sand flats, that narrow into The Narrows tidal channel. The spit is attached in the east to Warner Point, a northern projection of the hill.

Warner Point Beach (**1392**) extends for 2 km east of the straight point that protrudes 100 m to the north. The beach bulges slightly to the north and curves around, ending at a small, mangrove-fringed creek. The steep, 20 m wide high tide beach is fronted by 500 m wide intertidal sand flats.

Station Point is a low, sandy point anchored by intertidal rock reefs. On the western side of the point is a low, 1 km long sandy spit extending to the south. The dynamic spit partially encloses a 50 ha area of mangroves. The beach (**1393**) fronts the spit and consists of a moderate gradient high tide beach fronted by 500 m wide sand flats with outcrops of rock reefs also occurring along the beach. There is a fishing camp in amongst the casuarina trees and 4WD access to the camp.

On **Station Point**, between two sets of rock reefs, is beach **1394**. It is a 200 m long, north-west facing strip of low gradient high tide sand, with a continuous, 150 m wide low tide bar. This is the first beach on the island's northern shore to regularly receive low waves. Immediately east of the boundary rocks is the longer beach **1395**, a 3.5 km long, low gradient beach that to the east become increasingly dominated by sand flats associated with Cape Keppel and the mouth of a small tidal creek. In lee of the beach is a 500 m wide series of twelve densely vegetated beach ridges, representing a substantial amount of sedimentation and shoreline progradation of the northern end of the island. There is a row of holiday shacks in amongst the casuarinas running for a few hundred metres east of the point.

The sand spits of Station Point merge with another series of active spits that are tied to the western side of **Cape Keppel**. The cape is capped by grassy, 60 m high Bald Hill. There are station buildings on the slopes of the cape about 500 m south of the point, with a vehicle track leading to the beach. However it can only be used for boating at high tide. The western beach (**1396**) is approximately 600 m in length, however the length, width and shape change with the movement of the sand spits, produced by sand moving around the cape and into the infilling, shallow bay. A series of rocks extending north of the cape form the northern boundary, while the spits merge into sand flats and the mangrove-fringed creek to the south.

Swimming: The best beaches for swimming on are the three Station Point beaches, with deep water off the beaches at high tide.

Surfing: Only a chance of waves on Station Point, with a peeling right-hander down the cape spit during strong easterly winds.

Fishing: Best at high tide off the beaches, over the rock reefs and in the tidal creeks.

Summary: Five remote beaches on the northern end of Curtis Island mainly used by shack owners and fishers, who set up camps on some of the beaches.

1397, 1398   CAPE KEPPEL

No. Beach	Unpatrolled		
	Rating HT LT	Type	Length
1397 Cape Keppel	2 3	UD	500 m
1398 Cape Keppel (E)	1 2	UD	3 km

Cape Keppel forms the eastern boundary of Keppel Bay and is one of two northern points on Curtis Island, the other being Cape Capricorn. Between the two points is an open, 17 km wide bay containing two high energy beaches on the eastern side of the cape, then a series of increasingly lower energy beaches and spits toward the cape.

Cape Keppel beach (1397) lies on the northern tip of the 70 m high, grassy cape and consists of a 500 m long, north facing, low gradient sandy beach bordered by low, partially buried lines of rocks to either end, with additional rocks scattered in the centre. It is backed by a low foredune and the vegetated slopes of the cape, and fronted by a 150 m wide low tide bar. Waves average about 0.5 m.

Immediately south of the southern rocks is a 3 km long beach (**1398**) extending from the cape to a dynamic spit, that forms the northern shore of a highly variable tidal inlet. The beach has a 30 m wide, moderately steep high tide beach fronted by a 150 m wide, continuous low tide bar, with waves averaging 0.5 m. A low foredune parallels the back of the beach, which is in turn backed by an extensive series of casuarina-covered foredune and beach ridges extending up to 1.5 km inland and ending in recurved spits to the south. There is a shack toward the northern end of the beach, but otherwise no development.

Swimming: These are two relatively safe, sandy beaches with low gradient bars, however be careful at the point as rocks dominate parts of the beach and surf.

Surfing: Chance of low waves during strong easterly winds.

Fishing: Best down at the creek mouth, or off the cape rocks at high tide.

Summary: A prominent cape with two energetic beaches on its eastern side.

Cutris Island National Park

Cutris Island National Park is located on Cape Capricorn and includes the beaches to ether side of the cape.

1399-1402   CAMP IS, RED BEACH

No. Beach	Unpatrolled		
	Rating HT LT	Type	Length
1399 Camp Is (N)	1 1	B+SF	2 km
1400 Camp Is	1 1	B+SF	4 km
1401 Camp Is (N)	1 1	B+SF	3.5 km
1402 Red	1 2	R+tidal channel	600 m

The open, north facing bay between Cape Keppel and Cape Capricorn contains a relatively protected though highly dynamic shoreline, consisting of a series of sandy spits and low beaches, backed by a combination of beach ridges and two tidal creeks. The creeks produce tidal currents and shoals that, together with sand feed around Cape

Capricorn, maintain the highly dynamic nature of the bay. During the past 6000 years, a tremendous volume of sediment has been delivered around the cape into this bay, resulting in the shoreline prograding bayward by up to 4 km, leaving behind multiple series of beach ridges and recurved spits in amongst the high tide salt flats.

Beach **1399** is the first of the barrier island-spits. The 2 km long beach lies between the first two tidal inlets east of the cape. The first inlet in particular is very dynamic and the 500 m wide channel, tidal shoals and adjacent spits and beach vary considerably in shape and length from year to year. The southern inlet is smaller but still dynamic. The result is a low beach that is at times building out and at other times eroding back into the mangroves, exposing mangrove stumps and mud on the shore.

Beach **1400** is a longer version of its northern neighbour; it faces north-east and is up to 4 km long. This beach is also eroding back into extensive mangrove woodlands, with the mangrove mud and stumps outcropping along the beach. It is bordered in the south by a small creek mouth, beyond which is beach **1401**, a 3.5 km long, curving, north facing barrier island. This beach consists of a series of recurved spits fronted by up to 2 km of dynamic tidal shoals, including 2 to 3 km long sand spits. As a consequence of the shoals, the beach receives low waves only at high tide, while towards the eastern end it is fringed by mangroves adjacent to the large, 1 km wide creek mouth.

Beach **1402** lies just inside the eastern side of the open, though shallow, creek mouth. The creek drains an area of several square kilometres and its tidal currents maintain the dynamic tidal shoals and spits. The main channel parallels this 600 m long, north-west facing beach composed of red sand. The sand is derived from the backing dunes that spill over from the higher energy eastern Cape Capricorn beaches. The colour is a result of long term weathering and gives **Red Beach** its name. The beach receives no ocean waves, but is paralleled by the tidal channel and there is deep water right off the narrow beach. As a result, this is a favourite anchorage and landing.

Swimming: The Camp Island beaches are little used and relatively safe owing to their usually low wave to calm conditions. While the inlet 'red' beach receives no ocean waves, it does have deep water and strong tidal currents right off the beach, so take care.

Surfing: None.

Fishing: There is excellent fishing in the many creeks, as well as the main channels.

Summary: An interesting section of coastline with low waves, strong tidal currents and an abundance of shallow sand shoals, all combining to produce a highly dynamic shoreline.

1403-1405   CAPE CAPRICORN

No. Beach	Unpatrolled		Type	Length
	Rating			
	HT	LT		
1403 Cape Capricorn (spit)	1	1	B+SF	2 km
1404 Cape Capricorn (W 2)	1	1	R+LT	100 m
1405 Cape Capricorn (W 1)	1	1	R+LT	50 m

Cape Capricorn forms the north-eastern tip of Curtis Island and represents a major change in shoreline direction. Due to its importance for navigation, a lighthouse is maintained on the 90 m high cape. The cape was sighted and named by Captain Cook in May 1770, the name deriving from the fact that it lies almost precisely on the Tropic of Capricorn, at a latitude of 23°30' S.

In lee of the cape are three beaches, two small pocket beaches and a longer dynamic sand spit. The spit beach (**1403**) extends for up to 2 km from a 40 m high headland that lies 500 m south-west of the cape. The sand for the beach has arrived via the dunes from the higher energy eastern beach, as well as moving around the cape. The dune sands form a narrow, low energy beach fronted by 300 to 500 m wide sand flats. The spit periodically builds out from the headland to extend for up to 1 km in front of the backing beach, impounding a shallow lagoon over the tidal flats, that contain a few mangroves. The beach beyond the spit periodically erodes, causing casuarina trees to fall onto the beach.

Between the headland and the cape are the two small pocket beaches. Beach **1404** is 100 m long, faces north-west and is bordered by two 40 m high headlands, while beach **1405** is a 50 m pocket of sand wedged in between the cape and the first headland. Both beaches have moderately steep high tide beaches and a narrow, continuous bar. They are both backed by sand spilling over the cape from the eastern cape beach. Beach 1405 serves as a landing for the lighthouse on the cape, with a trolley system running from the back of the beach up to the lighthouse. Because of their protected nature and relatively deep water off these two beaches, they are also used as boat anchorages.

Swimming: The two cape beaches usually offer calm conditions and deep water off the beach, particularly at high tide.

Surfing: None.

Fishing: There is deep water off the cape beaches and adjoining rocks.

Summary: Two small pocket beaches and a low energy but dynamic sand spit.

1406-1409 CAPE CAPRICORN (S)

No.	Beach	Unpatrolled				
		Rating		Type		
		HT	LT	Inner Bar	Outer Bar	Length
1406	Cape Capricorn	2	3	R+LT	TBR /RBB	60 m
1407	Cape Capricorn (S 1)	2	3	R+LT	TBR /RBB	1.8 km
1408	Cape Capricorn (S 2)	2	3	R+LT	TBR /RBB	1 km
1409	Cape Capricorn (S 3)	2	3	R+LT	TBR /RBB	6 km

Cape Capricorn is the north-eastern tip of Curtis Island and is an exposed headland, bordered to the south by exposed, east facing beaches. A lighthouse sits atop the 90 m high grassy cape, that has shaley sides which produce steep, unstable cliffs. While the western side of the cape offers quiet waters and a safe anchorage, the eastern side is exposed to all easterly winds and waves and has the highest energy beaches between Mackay and Round Head.

On the southern side of the cape is a series of four near-continuous beaches totalling 9 km in length. Halfway along the beaches, the trend of the shoreline bulges slightly eastward as a foreland in lee of the two Rundle Islands, which lie 4 km offshore. The first beach (**1406**) is a 60 m long pocket of sand and cobbles wedged in at the base of the cape, slightly protected by 200 m long, protruding eastern rocks and by facing south down the beach. It has a cobble high tide beach in amongst the lower rocks of the cape and a sandy low tide bar that merges with the main beach at low tide. A trough and second bar lie past the inner bar, with a permanent low tide rip against the rocks.

Once free of the rocks, beach **1407** runs for 1.8 km down to a 40 m high headland that truncates the inner and outer

bars. On the south side of the headland is 1 km long beach **1408**, that ends at a smaller headland, which only truncates the beach and inner bar. The outer bar continues on to beach **1409**, which runs for another 6 km down to where bedrock begins to dominate the coast. The three beaches (1407, 1408 and 1409) all have a straight, low to moderate gradient high tide beach fronted by a low gradient, 150 m wide low tide bar, a shore parallel trough, then a rip-dominated outer bar, with rips spaced approximately every 100 m and permanent rips against the boundary rocks. More than 80 rips can operate along the beaches during moderate to higher waves. Beach 1409 steepens to the south, owing to slightly coarser sand and resulting in a narrower bar. It ends at a low, grassy, rock-bound headland.

The beaches are all backed by low to scarped foredunes and between 1 and 4 km of vegetated parabolic dunes that average 40 m in height, reaching a maximum of 70 m. The rear of the dunes spills over onto the western cape beaches and tidal flats. The southern ends of beaches 1407 and 1408 become increasingly backed by rocks of the point and form, at high tide, a series of pocket beaches.

Swimming: These are more energetic beaches, with waves averaging between 0.5 and 1 m and commonly exceeding 1 m. The higher wave conditions form and maintain rips along the low tide beach. Special care should be taken in the surf, particularly toward low tide.

Surfing: There are usually 0.5 m to 1 m waves over the bars, that are better toward low tide.

Fishing: Both the rocks at high tide, and many rip holes and troughs at low tide, make this an excellent beach fishing location.

Summary: Nine kilometres of dune systems, fronted by a wide and energetic double bar beach system, all in a relatively inaccessible part of the island.

CURTIS ISLAND TO BAFFLE CREEK (Fig. 4.83)

Length of Coast:	276 km
Beaches:	80 (1410 to 1490)
Surf Life Saving Clubs:	Tannum Sands, Agnes Water
Lifeguards:	none
Major towns:	Gladstone

Regional Map 9: Curtis Island to Baffle Creek

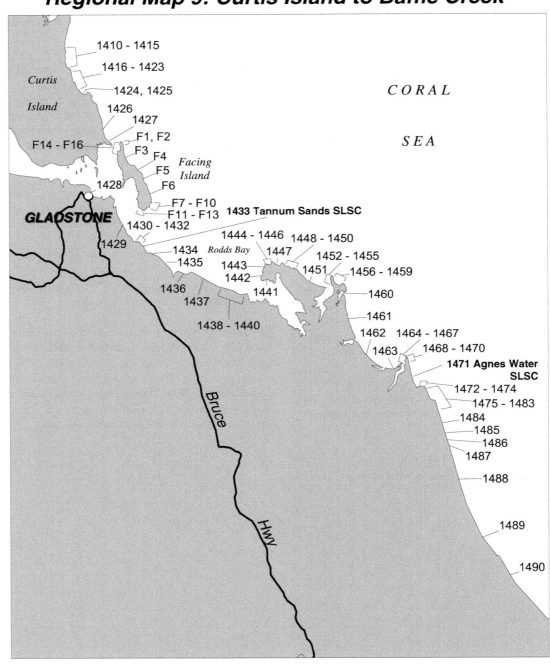

Figure 4.83   *Regional map 9   Curtis Island to Baffle Creek. Beaches with Surf Life Saving Clubs indicated in bold.*

1410-1415   CURTIS IS, SANDHILL

No. Beach	Unpatrolled		Type	Length
	Rating			
	HT	LT		
1410 Curtis Is 1410	1	3	R+rock flats	150 m
1411 Curtis Is 1411	1	2	R+LT	250 m
1412 Curtis Is 1412	1	2	R+LT	400 m
1413 Curtis Is 1413	1	2	R+LT	400 m
1414 Curtis Is 1414	1	2	R+LT	400 m
1415 Sandhill	1	2	R+LT	800 m

Between the end of the long beach 1409 and Sandhill Trig Station, 5.5 km to the south, is a predominantly rocky coast containing six exposed, rock-bound beaches. The beaches are all backed by vegetated, dune covered rocks and there is a landing strip 1 km west of beach 1413. There are 4WD tracks leading to all the other beaches and to shacks located behind beaches 1411, 1413 and 1414, but otherwise no development or facilities.

Beach **1410** lies immediately south of the long beach 1409 and consists of a 150 m long pocket of high tide sand, surrounded by 20 m high, dune-capped, rocky bluffs and fronted by 100 m wide intertidal rock flats. Its neighbour, beach **1411**, is also rock-bound, but longer at 250 m and fronted by a low gradient bar, with rock flats extending off each of the points and a narrow cobble high tide beach. There are a few shacks on the grassy dunes behind the beach.

Beach **1412** is a narrow, 400 m long high tide sand beach fronted by a 150 m wide, low gradient, continuous bar, with rocky points to either end and a patch of rock flats to the south. It is backed by a low, spinifex-covered foredune that grades into the larger vegetated parabolic dunes. A 4WD track skirts the foredune and low headlands.

Beaches 1413 and 1414 are two adjoining 400 m long beaches, divided in two by rock flats. Beach **1413** is backed by low, rocky bluffs, with patches of rock flats on the low tide bar, while beach **1414** has a continuous, 200 m wide low tide bar. Both beaches receive waves averaging 0.5 m that spill across the low gradient bar and beach. Vegetated older dunes cap the bluffs, while a grassy foredune separates the southern half from the backing dunes. There are a few shacks behind both of the beaches.

Beach **1415** lies at the base of 70 m high Sandhill Trig Station; a dune-capped, bedrock hillock. The beach is 800 m long and curves around to face north in lee of a densely vegetated, 30 m high point. It is bounded by low, rocky points and backed by densely vegetated, dune-capped bluffs, with three patches of rocks across the 100 m wide low tide beach. Fresh water drains out of the

dunes and bluffs and saturates the beach, resulting in no high tide beach in the southern corner and melaleuca trees hanging over the beach.

Swimming: These beaches receive waves averaging between 0.5 m and 1 m and occasionally higher. They are relatively safe toward high tide and clear of the rocks, however the falling tide reveals numerous rocks and rock flats, with rips more likely to form against the many rocky points. Use care if swimming in the area.

Surfing: Chance of low beach, point and reef breaks when the waves exceed 0.5 m, with considerable variation from beach to beach.

Fishing: Excellent fishing locations all along this predominantly rocky shoreline.

Summary: These beaches are used by the shack owners, who can boat or even fly in to access this otherwise remote part of the island.

1416-1421   CURTIS IS

No. Beach	Unpatrolled		Type	Length
	Rating			
	HT	LT		
1416 Curtis Is 1416	1	2	R+LT	200 m
1417 Curtis Is 1417	1	2	R+LT	500 m
1418 Curtis Is 1418	1	2	R+LT	500 m
1419 Curtis Is 1419	1	2	UD	100 m
1420 Curtis Is 1420	1	2	R+LT	500 m
1421 Curtis Is 1421	1	3	R+rock flats	50 m

South of Sandhill Trig Station is a relatively straight, 3 km section of rocky shore containing six small, sandy beaches, all bordered by rocky points and backed by dune-capped, rocky bluffs between 20 and 40 m high. There is an access track to beach 1418, but otherwise no development.

Beach **1416** consists of a 50 m long strip of high tide sand that widens at low tide to a 200 m long low tide bar, with bordering rock reefs, including one off the southern end of the beach. Fresh water drains from the backing red bluffs and flows across the beach.

Beaches **1417** and **1418** are two 500 m long, low gradient, sandy beaches backed by narrow strips of dry high tide sand and a low, grassy foredune. Fresh water drains across both beaches at low tide. Both are bounded by rocky points, while 1417 has a central rock bluff behind the high tide beach and 1418 a more protruding bluff and rock flats in the centre of the beach, that divide the beach in two at high tide.

Beach **1419** is a 100 m long pocket of sand completely surrounded by rocks, with backing 20 m high bluffs; rock platforms and reefs to the side; and linear bedrock reefs paralleling the low tide beach. The beach is usually awash at high tide.

Beach **1420** is a 500 m long, east-north-east facing beach, which receives slight protection from a 200 m long, 30 m high southern headland. The beach is backed by low, vegetated bluffs and has a low high tide beach fronted by a 150 m wide, low gradient low tide bar, with waves averaging less than 0.5 m. Just around the southern point is 50 m long beach **1421**; a narrow high tide cobble and sand beach fronted by intertidal rock flats and bordered by 20 m high bluffs to the side and rear.

Swimming: The sandy beaches usually receive waves averaging about 0.5 m and are relatively safe under these conditions. However, periodic higher waves induce rips against the rocks and reef, particularly toward low tide. Beaches 1419 and 1421 are dominated by rocks and are unsuitable for safe swimming.

Surfing: Only low beach breaks during higher waves.

Fishing: Numerous spots on the many rock platforms and over the rocks and reefs.

Summary: Six relatively isolated, natural pockets of sand surrounded by rocky bluffs and reefs.

1422-1425   CURTIS IS, BLACK HD

No. Beach	Unpatrolled		
	Rating HT LT	Type	Length
1422 Curtis Is 1422	1 2	R+LT	200 m
1423 Curtis Is 1423	1 2	R+LT	600 m
1424 Black Hd (W)	1 2	R+LT	2.3 km
1425 Black Hd	1 2	R+LT	200 m

Black Head is a 60 m high, black, basaltic headland that forms the only major inflection in the eastern shore of Curtis Island between Cape Capricorn and Southend. There are two beaches in lee of the head and another two smaller beaches immediately to the north. The beaches are undeveloped, with a 4WD access track running up the eastern side of Black Head and reaching the coast at the western end of the western Black Head beach (1424).

Beaches 1422 and 1423 are two similar sandy beaches, 200 m and 600 m long respectively. They both face east-north-east and are bounded by 20 to 30 m high headlands, with fringing rock platforms and reefs. Beach **1422** has a narrow high tide beach fronted by a 100 m wide bar. It is backed by a low, grassy foredune, with a small, usually

blocked creek and its lagoon behind the centre of the beach. Beach **1423** has the same beach characteristics and also has a small, blocked creek in the centre.

Black Head (west) beach (1424) is 2.3 km long and curves around from the north-east to face north in lee of the head. It is partially protected by the head and receives waves averaging less than 0.5 m, which break across a 100 m wide low tide bar. The high tide beach is about 20 m wide. A small creek, containing a few mangroves and small backing salt flats, exits at the eastern end of the beach and there is an informal camping area in amongst the trees behind this end of the beach. The 4WD track reaches the western end of this beach.

Along the western side of Black Head is a narrow, crenulate sand beach (**1425**) fronted by the 100 m wide tidal shoals of the creek mouth.

Swimming: The three sandy beaches are all relatively safe under the usually low wave conditions, although rips are generated against the rocks during higher waves.

Surfing: Only during strong easterly winds, with beach and occasionally reef breaks.

Fishing: Excellent spots both on the rock platforms and reefs and at the small creek mouth in lee of the head.

Summary: Four relatively remote beaches offering a nice camping spot at Black Head.

1426, 1427   CONNOR BLUFF, SOUTHEND

No. Beach	Unpatrolled		
	Rating HT LT	Type	Length
1426 Connor Bluff	2 3	R+BR	3.5 km
1427 Southend	1 2	R+rock flats	1.5 km

Connor Bluff is a 40 m high, rocky bluff at the southern end of a 6 km long cliffline that extends from Black Head. Beyond the bluff are two beaches totalling 5 km in length, terminating at Southend, the south-eastern point of the island. There is a small settlement at Southend and 4WD trails leading from Southend up behind the beaches to Connor Bluff and on to Black Head.

Connor Bluff beach (1426) is 3.5 km long and runs in a slightly curving southerly direction to the reef flats of Southend. The beach is exposed to waves averaging 0.5 to 1 m and these combine with the coarse sand and gravel of the beach to maintain a steep high tide beach, fronted by a 50 m wide low tide bar, with rips commonly forming along the northern half of the beach. The beach is backed by a grassy foredune and 100 to 400 m of now cleared sand blowouts and parabolic dunes, that average between 15 to 20 m high and have been somewhat degraded by the cattle grazing. A small creek is usually impounded behind the middle of the beach, but periodically breaks out across the beach.

Southend beach (1427) is 1.5 km long and curves around to face north-east. The entire length of the beach is fronted and dominated by rock flats, resulting in a narrow, crenulate high tide strip of sand, with fronting 50 to 100 m wide rock flats. The beach narrows to the south, eventually being almost completely replaced by the rock flats. The Southend settlement backs the southern 800 m of the beach and contains a store and some tourist accommodation. The ferry service from Gladstone lands at a jetty just inside North Entrance.

Swimming: These are two beaches that need some care, owing to the rips on Connor Bluff Beach and the rock flats along Southend.

Surfing: Chance of beach breaks along Connor Bluff Beach, particularly toward low tide.

Fishing: Southend is a popular fishing destination offering beach and rock reefs, and boat fishing in the many creeks and offshore reefs.

Summary: The most accessible part of Curtis Island, serviced by a regular ferry from Gladstone 12 km to the south-west.

Facing Island

Facing Island lies 8 km due east of Gladstone and the 15 km long island forms the eastern shore of Gladstone Harbour. The island is separated from Southend on Curtis Island by 1 km wide North Entrance, while in the south there is a 4 km wide channel between the southern Gatcombe Head and Boyne Island. The island averages 1 to 2 km in width and rarely rises above 20 m. It has an area of approximately 28 km^2 and has traditionally been used for grazing. There are also holiday-fishing shack settlements at Farmers Point and Gatcombe. There is a regular ferry service from Gladstone to these two settlements. The island has 36 km of shoreline, containing 16 beaches and a few kilometres of mangroves along the western shore. Rock reefs border and fringe the beaches on the exposed eastern shore, while the south-east Trades have developed low dunes, between 100 and 500 m in width, in lee of much of the eastern shore.

Facing Island
F16, F1-3   NORTH PT, CASTLE ROCKS, PEARL LEDGE

		Unpatrolled			
No.	Beach	Rating		Type	Length
		HT	LT		
F16	North Pt (W)	1	3	R+rock flats	250 m
F1	North Pt (E)	1	3	R+rock flats	250 m
F2	Castle Rocks	1	3	R+rock flats	500 m
F3	Pearl Ledge	1	3	R+LT&rocks	5.3 km

North Point is the northern tip of Facing Island and consists of 100 m wide intertidal rock flats backed by two sandy beaches and capped by low dunes. On both sides of the point are 250 m long, curving, north-west facing (**F16**) and north-east facing (**F1**) high tide beaches, both fronted by the rock flats. The eastern beach ends at a bend in the shoreline, on the southern side of which is 500 m long Castle Rocks Beach (**F2**). This beach faces more easterly and is also fronted by 100 m wide rock flats, and is backed by now stable dunes that have in the past blown across the 500 m wide North Point. The southern end of the beach has a 100 m long, low, protruding headland called Northcliffe.

On the southern side of the headland is the longest beach on the island; a curving, east to north-east facing, 5.3 km long sandy beach (**F3**) that ends at Pearl Ledge, a 100 to 200 m wide intertidal rock ledge. Waves average 0.5 m and there is usually a low surf along the beach. The beach has a moderately steep high tide beach fronted by a continuous, 100 m wide low tide bar, with the rock flats dominating the southern 1 km of the beach. It is backed by a low foredune and then up to 400 m of now largely stable dunes, that in the past have crossed to the western shore of the island. There are also a few holiday shacks on the dunes, toward the northern end of the beach.

Swimming: These are four relatively safe beaches that are better toward high tide. However watch for rocks on the northern beaches and toward the Ledge, and strong tidal currents in the North Entrance channel, off the point beaches.

Surfing: Chance of low beach and reef breaks along the Pearl Ledge beach.

Fishing: North Point is a favourite spot, particularly with the deeper North Entrance channel off the point, as well as the many rock reefs on all three beaches.

Summary: Three undeveloped beaches, with dunes that have been degraded in the past by overgrazing, but which are now being stabilised.

Facing Island
F4-6 SABLE CHIEF ROCKS, CASUARINA CLUMP

No.	Beach	Unpatrolled		Type	Length
		Rating			
		HT	LT		
F4	Sable Chief Rocks	1	3	R+rock flats	2.3 km
F5	Facing 5	1	3	R+rock flats	2.2 km
F6	Casuarina Clump	1	3	R+rock flats	3.5 km

The easternmost section of the island is an 8 km long section of east-north-east facing shore, consisting of three beaches fronted by continuous rock flats and backed to varying degrees by sand dunes. South of Pearl Ledge is a 2.3 km long, north-east facing beach (**F4**). The continuous high tide beach winds between the rocks flats, with a few more open patches containing a 50 to 100 m wide low tide bar. The beach is backed by a low foredune and 100 to 200 m of grassy, stabilising dunes. The beach ends at Sable Chief Rocks, which extend 500 m east of the beach..

On the southern side of the rocks is a 2.2 km long, more easterly facing beach (**F5**) which, while bordered by prominent rocks, is fronted by deeper reefs and as a consequence has a straighter, more continuous high tide beach and low tide bar, although there are some rocks toward the centre. A low foredune runs the length of the beach, with some minor sand blows and largely stabilised, low dunes up to 300 m wide.

Casuarina Clump is the name given to the casuarinas that used to cover the now cleared dunes. The Clump is fronted by a 3.5 km long, crenulate beach (**F6**) with a shape that is dominated by four major rock reef systems along the shore, producing a crenulate shoreline that protrudes in lee of the reefs, with more open water in between. The northern end is sufficiently open to receive some waves and surf, and commonly has a few rips. Grassy, stabilised, 200 m wide, low dunes back the northern end of the beach, decreasing in width toward the south.

Swimming: The best swimming is on the more open sandy patches away from the reefs, however watch for rips against the rocks and reefs when waves exceed 0.5 m.

Surfing: Chance of a surf in the more open area free of the reef and over some of the reefs at high tide.

Fishing: Excellent shore fishing over the many rock reefs and in the beach gutters.

Summary: The most exposed section of the island, consisting of three longer beaches, numerous reefs and backing degraded dunes.

Facing Island
F7-10 **EAST & SETTLEMENT PTS**

No.	Beach	Unpatrolled		Type	Length
		Rating			
		HT	LT		
F7	East Pt	1	3	R+rock flats	100 m
F8	Settlement Pt (1)	1	3	R+LT	500 m
F9	Settlement Pt (2)	1	3	R+LT	300 m
F10	Settlement Pt (W)	1	2	R+LT	700 m

The south-eastern end of Facing Island consists of 3 km of south-east facing shore between East Point, the eastern tip of the island, and Gatcombe Head, the southern point. In between are two small bays bordered by rocky bluffs and low cliffs.

East Point Beach (F7) is a 100 m long strip of high tide sand fronted by rough rock flats, that narrow to the east, and backed by a grassy, 15 m high bluff. There is 4WD access to the point and beach.

Settlement Point is a protruding, 20 m high, grassy headland that has beaches to either side. To the east, between the point and rocks of East Point, are two sandy beaches. The first (**F8**) is 500 m long, faces east and is bordered by 500 m long East Point in the north and a smaller rocky point to the south, with intertidal rock reefs off the beach, apart from the central 100 m which is open to the sea. The second beach (**F9**) is 300 m long, also faces east and has continuous rock reefs along the beach, becoming more prominent at the small boundary points.

The western beach (**F10**) is a curving, 700 m long, south-east facing sandy beach, with a low gradient high tide beach fronted by a continuous, gently sloping, 80 m wide low tide bar. Rocks only border the ends of the beach. All three beaches are backed by low foredunes and hummocky grassed dunes. There is a solitary house on Settlement Point overlooking the western beach.

Swimming: The western beach is best for swimming owing to the absence of rocks, which dominate the other three point beaches.

Surfing: The three longer beaches are exposed to southerly swell and can produce some waves over the bar at low tide and reefs at high tide.

Fishing: The rock flats off the points and the beach rock flats offer numerous spots to fish the rocks and gutters.

Summary: Four point-bound beaches largely dominated by rocks, apart from the western beach.

Facing Island
F11-13 **GATCOMBE HD**

No.	Beach	Unpatrolled Rating HT	LT	Type	Length
F11	Gatcombe Hd	1	3	R+rock flats	50 m
F12	Gatcombe (1)	1	1	R+LT	100 m
F13	Gatcombe (2)	1	1	R+LT	200 m

Gatcombe Head is the southernmost point of the island. It is a grassy, sloping, 20 m high headland, with beaches to either side and the small settlement of Gatcombe, containing several houses and shacks. On the eastern side of the head is a 50 m pocket of high tide sand (**F11**) which faces almost due south. It is backed by the grassy bluffs of the head and fronted by open water, with rocks and reefs to either side.

On the western side of the head is most of the Gatcombe settlement, lying on the grassy slopes behind two small sandy beaches. The first (**F12**) is 100 m long, faces south-west toward the mainland and is bordered by the head to the east and low rock flats to the west. It has a jetty on the point and a boat launching area on the beach. The second beach (**F13**) is 200 m long, also faces south-west and is the main swimming beach, with toilets located on the backing sloping bluffs. The beach is bordered by the rock flats to the east and a rocky bluff to the west.

Swimming: These three beaches usually receive low waves and calms are common. However, if swimming off the beaches beware of strong tidal currents that run just off the head.

Surfing: Usually none.

Fishing: Excellent fishing from the rocks and jetty, as well as off the beach at high tide.

Summary: A small settlement with three small, accessible, sandy beaches and a ferry service to Gladstone.

Facing Island
F14-15 **FARMERS PT, THE OAKS**

No.	Beach	Unpatrolled Rating HT	LT	Type	Length
F14	Farmers Pt	1	1	R+LT	500 m
F15	The Oaks	1	2	B+SF	2 km

The western shore of the island is 18 km long and largely dominated by low energy rocky shore, with mangrove-fringed Shoal Bay in the south and a sandy strip in the north-west containing two beaches. The first beach is

Farmers Point (F14); site of the island's main settlement, a collection of several houses. The sandy point protrudes into North Entrance and has a continuous, moderate gradient beach to either side, with rock flats bordering both ends. The beach is backed by a low foredune, then the rows of houses and beach shacks in amongst the scattered trees. The ferry from Gladstone lands at the point, using the beach for landing.

One kilometre north of Farmers Point is an open, 1.5 km long, west facing bay that is bordered in the north by North Point. The area is called **The Oaks** and the bay beach (F15) is very protected from waves and more dominated by the strong tidal currents of North Entrance, which parallels the bay. The bay shore contains a 2 km long, curving and crenulate low energy beach, which in the south is fronted by tidal flats, together with one of only three creeks on the island, that crosses the beach and flows across the flats. The northern end however, parallels the North Entrance tidal channels and has deep water and strong tidal currents right off the beach. In the past, sand has blown onto this beach from the eastern beaches but the sand blows have now been stabilised.

Swimming: There is deep water off the beaches at high tide, while at low tide it ranges from tidal flats in the south to deep channels along the northern half, so use care, particularly if swimming off the flats at low tide, as strong currents run past both beaches.

Surfing: None.

Fishing: Best in a boat in the channel, or off the beach at high tide.

Summary: Farmers Point is the main settlement, while The Oaks is an undeveloped beach.

1428 **BARNEY PT**

No.	Beach	Unpatrolled Rating HT	LT	Type	Length
1428	Barney Pt	1	1	B+SF	300 m

Barney Point (**1428**) is the only swimming beach in the township of Gladstone and is located 1 km east of the city centre, on Barney Point. It is a 300 m long, east facing beach that usually only receives low wind waves generated inside the harbour. It is composed of coarse sand and pebbles and consequently has a relatively steep beach face, fronted by 200 m wide tidal flats. It is backed by a reserve that commemorates the founding of Gladstone, together with a picnic, playground and barbecue area. While the access, facilities and easterly views are all attractive, the adjoining northern half of the point is occupied by a large coal storage facility, with the jetty and wharf running off the northern tip of the point, all within 1 km of the beach. A seawall for the facility forms the northern boundary of the beach, with rock flats and a 20 m high grassy bluff to the south.

Swimming: Relatively safe, apart from pollution from the adjoining docks and shipping.

Surfing: None.

Fishing: Best off the seawall and point, and the beach at high tide.

Summary: Gladstone's closest beach; great for a picnic in a well laid-out and attractive park, but check the clarity of the water before swimming.

1429 BOYNE IS

No. Beach	Unpatrolled Rating HT LT	Type	Length
1429 Boyne Is	1 1	B+SF	7.5 km

Boyne Island is a 7.5 km long island consisting of a series of recurved spits that widen to 600 m in the north. The island is bounded by a tributary of South Trees Inlet in the north and the Boyne River to the south. Higher ground behind the southern 2 km of the beach has been developed for the Gladstone aluminium smelter, and the smelter and its extensive siltation fields occupy more than 1 km² of the island's area. There is road access to the smelter and along the back of the island to the bauxite wharf, 3 km north. The land between the road and the beach remains in a natural state, with access to the shore and beach restricted to 4WD tracks that run the length of the beach.

The beach (**1429**) faces east to north-east as it meanders between the two creeks. The high tide beach averages 30 m in width and has a moderate gradient. It is fronted by 200 to 300 m wide, ridged sand flats along the northern 2.5 km, with a 150 m wide bar along the centre and up to 1 km wide tidal shoals of the river mouth off the southern 2.5 km. Waves average less than 0.5 m and rips are usually absent.

Swimming: A relatively safe beach away from the tidal channels.

Surfing: Only during strong south-easterly winds and waves.

Fishing: Best at the creek and river mouth, or off the beach at high tide.

Summary: A long, natural, undeveloped beachfront backed by Australia's largest aluminium smelter.

1430, 1431 CANOE PT

No. Beach	Unpatrolled Rating HT LT	Type	Length
1430 Canoe Pt (W)	1 1	R+tidal shoals	1 km
1431 Canoe Pt (E)	1 2	R+LT	500 m

The Boyne River is a medium size upland river that reaches the coast between rocky Canoe Point and the southern end of Boyne Island. Houses occupy the bluffs along the northern side of the river mouth, while the low Canoe Point has a large picnic area. There is road access and parking at the picnic area and a coastal walking track running right around the point to Tannum Sands, providing good access to both beaches. The point consists of low rock flats, with a series of low, vegetated recurved spits forming the western beach.

Beach **1430**, on the western side of the point, is 1 km long and faces due north across the 200 m wide river mouth. The beach consists of a moderately steep, sandy high tide beach, with tidal shoals that widen from 50 m in the west to 500 m off the point. At the point the rock flats extend 100 to 200 m seaward. Beach **1431** lies immediately south of the rock flats and is exposed to the southerly waves, resulting in a similar high tide beach, with a 100 m wide low tide bar that is bounded by rock flats, as well as having some rocks toward the centre of the 500 m long beach. This beach is accessible from the car park at the northern picnic area, as well as from the road at the southern end.

Swimming: While the western beach, closest to the picnic area, has low waves and is usually calm, it does have deep water and strong tidal currents off the tidal flats, so beware. The eastern beach has higher waves, but is usually relatively safe and free of rips under normal low wave conditions.

Surfing: Only on the eastern beach when southerly winds and waves are prevailing.

Fishing: Best over the rock flats at high tide, or into the Boyne River channel.

Summary: A nice sheltered picnic area and river beach to the west of the point, with a slightly more energetic 'surfing' beach to the east.

1432, 1433 TANNUM SANDS (SLSC)

Patrols:

Surf Lifesaving Club: September to May
Lifeguard on duty: School holidays

No. Beach	Rating HT LT	Type	Length
1432 Tannum Sands (N)	1 2	R+LT	800 m
1433 **Tannum Sands**	1 2	R+LT/SF	600 m

Tannum Sands is the surfing beach for the city of Gladstone. Originally it was established as a holiday settlement, mainly occupied by typical holiday 'shacks'. The beach was then known as Wild Cattle Beach and the Tannum Sands Surf Life Saving Club was established in 1936. Expansion of Gladstone in the 1980's saw the sleepy settlement become a beachfront residential area and an increasingly popular holiday destination. There is road access to and good parking right along the beach, together with a caravan park and beachfront camping reserve.

The beaches are located 30 km south-east of Gladstone and 10 km off the Bruce Highway. The settlement has two caravan parks, including a beachfront camping reserve and a small shopping centre. The beach is divided in two by low tide rock flats. The northern half (**1432**) runs for 800 m between the Canoe Point and Tannum Sands rock flats, with the main southern half (**1433**) extending for 600 m to the mouth of Wild Cattle Creek (Fig. 4.84). The surf lifesaving club is located in the reserve 200 m north of the creek mouth.

The beaches are partially protected from southerly waves by Rodds Peninsula, with waves averaging only 0.5 m, while the tide range can reach 6 m. The waves and tides interact with the sand to form a 50 m wide high tide beach fronted by a 150 m wide low tide bar. South of the camping reserve the bar widens to 300 m as the tidal currents from Wild Cattle Creek maintain a series of ridged sand flats (Fig. 4.85). The surf lifesaving club is located right at the apex of the bar and shoals, resulting in a wide surf zone at high tide south of the club house.

Swimming: A relatively safe beach, particularly at high tide. However, south-east Trades will generate a sweep up the beach and care must be taken near the mouth of Wild Cattle Creek, especially on a falling tide when strong tidal currents flow seaward. Also watch out for the rocks in the centre and northern part of the beach, and for children wandering out across the wide, southern ridged sand flats.

Surfing: Tannum Sands receives waves up to 1 m during a strong south-easterly and can produce rideable beach and reef breaks, which are best at mid tide.

Fishing: The southern creek mouth and northern rocks are the best locations.

Summary: A popular beach with locals and summer holiday-makers, with a good range of facilities, including an excellent beachfront reserve.

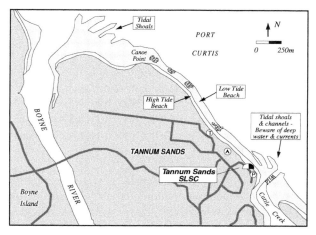

Figure 4.84 *Tannum Sands Beach is located at the southern end of a series of beaches that extend south from the Boyne River mouth. The main beach is also bordered in the south by the tidal channel and shoals of Cattle Creek.*

Figure 4.85 *Tannum Sands Beach and surf club are located immediately north of Wild Cattle Creek (centre), which produces extensive tidal shoals and currents.*

1434, 1435 WILD CATTLE IS

No. Beach	**Unpatrolled** Rating HT LT	Type	Length
1434 Wild Cattle Is (N)	1 2	R+LT	3.2 km
1435 Wild Cattle Is (S)	1 1	R+tidal shoals	3 km

Wild Cattle Island is a low, 6.2 km long sand barrier island, bordered by Wild Cattle Creek and Colosseum Inlet and completely backed by a tributary of the creek and inlet. As a result, there is no vehicle access to the island and the only development is about twenty shacks on the western shore of Colosseum Inlet. The island consists of a series of multiple beach ridges in the centre, with 1 to 2 km long recurved spits to either end. Dense casuarina woodlands cover the ridges.

The first of the island's beaches is a 3.2 km long northern section (**1434**), which faces north-east and ends as a narrow spit on the southern side of the 100 m wide creek mouth, just opposite Tannum Sands Beach. It is composed of medium sand and has a relatively steep high tide beach fronted by a continuous, 150 m wide bar, with wide tidal shoals off the creek mouth. Waves average 0.5 m and spill across the bar at low tide, while they surge up the beach face at high tide.

Beach **1435** commences at an inflection in the island, at the point where the 3 km wide tidal delta shoals of Colosseum Inlet join the beach. These extensive shoals protect the southern beach, with no waves reaching the beach during low tide. However, the high tide beach has a steep, narrow beach face and a 100 m wide bar, before the tidal shoals begin.

Swimming: Both beaches are relatively safe, clear of the creek and inlet channels and currents.

Surfing: The northern beach receives low to moderate waves during southerly winds, which are best across the bar at mid tide.

Fishing: The creek and inlet are the most popular spots, although you can also fish from the beach at high tide.

Summary: A low, natural beach that is only accessible by boat, but offers many kilometres of little used shoreline.

Rodds Bay and Harbour

Rodds Bay is an open, north facing bay bordered by Tiber Point on Hummock Hill Island to the west and Richards Point on Rodds Peninsula to the east, a distance of 15 km. The bay has a 70 km long, low energy shoreline, including two large protected inlets; Boyne and Seven Mile Creeks in the west and Rodds Harbour to the east. However the entire shoreline is undeveloped, apart from the small settlement of Turkey Beach, deep inside Rodds Harbour. The bay contain ten beaches; four on Hummock Hill Island and the remainder on the more exposed section of Middle Head and the western Rodds Peninsula shoreline. The only vehicle access to the entire bay is at Turkey Beach, with 4WD access to Middle Head.

1436-1439 HUMMOCK HILL IS, NOTCH PT

No. Beach	Unpatrolled			
	Rating HT	LT	Type	Length
1436 Hummock Hill Is (W)	1	2	B+SF	3.1 km
1437 Hummock Hill Is (centre)	1	2	B+SF	5 km
1438 Hummock Hill Is (E)	1	2	B+SF	4 km
1439 Norton Pt	1	2	R+tidal creek	1 km

Hummock Hill Island is a 12 km long, bedrock and sand island that is named after 124 m high Hummock Hill in the centre of the densely wooded island. The island is bounded by 600 m wide Colosseum Inlet and sandy Tiber Point to the west and the 2 km wide Boyne-Seven Mile Creek inlet in the east, while it is backed by interlinking tidal creeks. There is however a low tide causeway across Boyne Creek, which connects with a network of 4WD tracks on the bedrock part of the island. There is a house on the central northern tip of the island, but otherwise little development. To either side of the bedrock core are extensive wooded beach ridge plains, that are 2 km wide bordering the Colosseum Inlet and 1 km wide at the eastern Norton Point.

There are four beaches along the island's exposed northern shore. Beach **1436** runs for 3.1 km from the Colosseum Inlet to the only low, rocky point on the island. The beach faces generally north as it curves around between the inlet and the point. Two kilometre wide tidal shoals extend off the inlet end of the beach, while 300 m wide sand flats front the rest of the beach. The high tide beach is relatively steep and narrow and in places eroding back into the casuarina trees. The house on the eastern point overlooks this beach.

Beach **1437** runs from the low, rocky point for 5 km to a small tidal creek that cuts the high tide beach and drains across the low tide bar. Beach **1438** extends for another 4 km on the other side of the creek. Both beaches are relatively straight and face north-north-east, receiving waves averaging less than 0.5 m. Beach 1437 is backed by 100 to 300 m of wooded beach ridges, then the higher bedrock core of the island. A few rocks outcrop toward the centre of the beach, with rock reefs off this area. Beach 1438 runs straight from the eastern side of the creek to Norton Point and is backed entirely by a beach ridge plain up to 1 km wide. Both beaches have moderate gradient high tide beaches that are eroding back into the trees, particularly toward Norton Point. The beaches are fronted by a 150 to 200 m wide, low gradient bar for most of their length, with tidal shoals extending up to 2 km off Norton Point.

Norton Point Beach (1439) is a 1 km long strip of high tide sand that faces east into the 2 km wide inlet of Boyne and Seven Mile Creeks. The back of the beach truncates the multiple beach ridges that have built out over the past 6000 years. It has a narrow, moderately steep high tide beach fronted by the deep inlet channel and its strong tidal currents.

Swimming: The main beaches all provide relatively safe swimming. However, use care at any of the tidal creeks and particularly the two large inlets, which have wide, deep channels and strong tidal currents.

Surfing: Only low, spilling waves over the more exposed barred beaches during easterly winds.

Fishing: Most fishers arrive by boat to fish the two inlets, with the beach and central creek being best at high tide.

Summary: A 12 km long, relatively inaccessible and undeveloped island, offering many kilometres of sandy beaches and backing mangrove-fringed tidal creeks, with two large bordering inlets.

1440 MIDDLE HD

No.	Beach	Rating HT	LT	Type	Length
		Unpatrolled			
1440	Middle Hd	1	1	R+LT	700 m

Middle Head is located towards the middle of the southern shore of Rodds Bay. The head juts into Rodds Bay and is bordered by Seven Mile Creek to the west and Mangrove Bay to the east. Between the 10 m high, wooded, rocky headlands of Middle Head and Innes Head is a 700 m long, north facing beach (**1440**). The high tide beach is relatively steep and ends at a shelly, flat, 50 m wide low tide bar, with rocks and rock flats bordering each end. Some rocks also occur along the beach. Waves average less than 0.5 m during northerly winds, and calms are common. A small, grassy dune backs the beach, with an informal camping area behind that is accessible by boat or along a 4WD track.

Swimming: Relatively safe apart from strong tidal currents off the points.

Surfing: None.

Fishing: Most fishing is done from boats in the adjoining creeks, however you can fish the rocks and beach at high tide.

Summary: An isolated little beach surrounded by wooded slopes and the two points, with a sheltered camping site.

Rodds Peninsula

Rodds Peninsula is a 13 km wide peninsula lying between Rodds Bay in the west and Pancake Creek to the east. The 50 km² peninsula has two high bedrock ridges centred on 89 m high Bray Hills and 140 m high Table Hill, with low, undulating bedrock to the west and low beach ridges and tidal flats to the east. The peninsula has more than 40 km of shoreline; the southern half is fringed by mangroves, while the more exposed northern half has 20 km of predominantly sandy beaches, with a total of ten beaches. At high tide the peninsula is an island, with 4WD access to the peninsula possible only by crossing a 200 m wide high tide salt flat. Apart from the 4WD access, a few fishing camps and cattle grazing, there is no development on the peninsula.

1441, 1442 SPIT END

No.	Beach	Rating HT	LT	Type	Length
		Unpatrolled			
1441	Spit End	1	1	B+SF	1 km
1442	Spit End (N)	1	1	B+SF	1.6 km

Spit End is a series of vegetated, recurved sand spits located at the south-western tip of Rodds Peninsula. There is no vehicle access to the spit, which remains in a natural state. Beach **1441** faces south into Rodds Harbour and consists of a narrow high tide beach fronted by 100 to 200 m wide tidal flats, with deeper water off the flats maintained by the tidal currents. A small tidal creek drains across the middle of the beach.

Beach **1442** is the main spit and is backed by a series of eight spits, that have been built by northerly waves pushing sand down into the entrance of the harbour. They have built the shoreline out up to 300 m and deposited the 1.6 km long, crenulate, west facing beach. The narrow high tide beach is fronted by 200 to 300 m wide, ridged sand flats, that are evidence of continuing sand movement down the spit. A small creek behind the beach is deflected to the south, exiting at the southern end of the beach. The spit and sand flats are attached to the rocks of the peninsula at their northern end.

Swimming: These are two remote beaches that offer relatively safe swimming at high tide and sand flats at low tide. Be careful of strong tidal currents on the flats.

Surfing: None.

Fishing: The mouths of the two small tidal creeks offer the best locations.

Summary: Low energy, though periodically active, beach and sand flats.

1443-1446 FLORA & RICHARDS PTS

No. Beach	Unpatrolled		
	Rating HT LT	Type	Length
1443 Flora Pt	1 1	B+SF	150 m
1444 Richards Pt (W 2)	1 1	R+LT	50 m
1445 Richards Pt (W 1)	1 1	R+LT	100 m
1446 Richards Pt	1 2	R+LT	800 m

The western side of Rodds Peninsula is a 3.5 km long, west facing shore containing Spit End in the south; low, rocky Flora Point in the centre; and Richards Point to the north, the northernmost point of the peninsula. Between Flora Point and Richards Point are four sandy beaches, which gradually become more exposed toward Richards Point.

Beach **1443** lies on the northern side of the low, wooded, rocky slopes of **Flora Point**. It is 150 m long, faces west into Rodds Bay and receives low waves, with calms being common. It has a 30 m wide, low gradient high tide beach fronted by 150 m wide low tide sand flats, with a scattering of the red rocks of the points across the beach and flats. A low, rocky headland forms the northern boundary.

Beach **1444** lies on the northern side of the rocks and headland, and consists of a narrow strip of north-west facing high tide sand welded up against the backing rocky slopes, fronted by a 100 m wide, low gradient low tide bar, with rocks flats to either side.

Beach **1445** lies immediately beyond the rocks and has a straight, sandy high tide beach and a small foredune, fronted by a 150 m wide bar, with red rock flats to either side, as well as some rock reefs about 100 m off the bar.

Richards Point beach (1446) extends west of the point for 800 m. It faces north, curving around to face north-west toward the point. It has a 50 m wide, low gradient high tide beach fronted by a 200 m wide bar in the west, which widens to 400 m wide sand and rock flats at the more protected eastern end. There are a few stunted mangroves over the flats in lee of the point. The beach is backed by a low, grass and casuarina-covered foredune.

There is 4WD access to a shack at the point and the area is used as a fishing and holiday camp, although there are no facilities. Much of the point has been cleared and the low, grassy slopes run down to the rocks and beach.

Swimming: These are four relatively safe beaches that usually receive low waves, with calms being common. Watch out for the many rocks, as well as deeper water and tidal currents off the bars and flats.

Surfing: Chance of a right-hand point break off Richards Point Beach during very strong southerly winds and waves.

Fishing: These beaches are used for fishing camps, and shore fishing off the rocks and beaches is possible at high tide.

Summary: A relatively remote and difficult to access section of the peninsula, offering a series of usually low energy, sandy, rock-fringed beaches.

1447-1450 ETHEL ROCKS

No. Beach	Unpatrolled		
	Rating HT LT	Type	Length
1447 Ethel Rocks	1 2	R+BR	2 km
1448 Ethel Rocks (E 1)	1 2	R+BR	900 m
1449 Ethel Rocks (E 2)	1 2	R+BR	1 km
1450 Ethel Rocks (E 2)	1 2	R+BR	800 m

Ethel Rocks is a rock reef lying 400 m north of Richards Point. The shoreline for 5 km to the east is dominated by three low, rocky points, with four sandy beaches in between. Interestingly, the sand along the three more embayed beaches tends to be finer at their eastern ends and coarser toward their western ends, representing some wave grading of the beach sand, possibly combined with headland bypassing of dune sand. There is 4WD access to the rear of these beaches and some cleared land and station buildings in lee of beach 1447.

Beach **1447** is the longest of the four beaches, at 2 km in length. It faces initially north-east in lee of Richards Point, then curves around to face north against the eastern point. The beach has a 20 to 30 m wide high tide beach, which is interrupted by rocks against Richards Point, and has a layer of black sand. Once free of the point, it is a continuous beach fronted by a low tide bar that widens from 50 m in the west to 150 m in the east. Waves average 0.5 m, occasionally with higher waves producing a few low tide rips toward the centre and western end. A 5 to 10 m high, grassy foredune backs the beach, then the cleared Richards Point and some cleared land behind the beach.

Beach **1448** is a 900 m long, north-east facing, curving beach tied between two low, wooded rock points and rock flats. It is composed of coarse sand and has a relatively steep, 20 m wide high tide beach fronted by a 50 m wide bar, with rips occurring along the low tide line during and following higher than average waves. It is backed by a 5 m high, casuarina-covered foredune, then low, wooded slopes.

Beach **1449** is very similar at 1 km in length and facing north-east, but with finer sand and a low gradient high tide beach fronted by a 100 m wide low tide bar, with occasional high wave-generated rips. Low woodlands and a casuarina-covered foredune back the beach, while low, rocky points form the boundaries.

Beach **1450** runs for 800 m between extensive rocky points and rock reefs. These lower the waves at the beach, leaving it usually free of rips. It has a low gradient high tide beach fronted by a 100 m wide low tide bar. A 10 to 15 m high foredune backs the beach.

Swimming: Under normal low waves, that average about 0.5 m, these are relatively safe beaches. However when waves exceed 1 m, and particularly toward low tide, rips approximately 100 to 150 m apart are generated along the more exposed, western sections of the beaches. Also be careful of the bordering rocks and occasional rock reefs.

Surfing: Chance of beach and point breaks during high southerly waves.

Fishing: Best at high tide off the beaches and rocks.

Summary: Four relatively remote and undeveloped beaches, usually with low waves and occasional surf.

1451 PANCAKE PT

No.	Beach	Unpatrolled			
		Rating HT LT		Type	Length
1451	Pancake Pt	1	2	R+BR	7 km

Pancake Point forms the western point of the 1 km wide Pancake Creek tidal inlet. Extending north-west of the point is a 7 km long sandy beach, which ends amongst the rocks that form the shore, 1.5 km north of Table Hill.

The beach (1451) consists of essentially two parts. The western section is bordered by the low, rocky point, with rocks off the low tide bar. It then runs for 2.5 km to a small creek that drains across the beach, with rock reefs off much of the low tide beach, as well as a few rocks along the beach. Beyond the creek, rocks are absent all

the way to Pancake Point and, as well as the sandy beach and low tide bar, the beach is backed by up to 3.5 km of low, wooded beach and foredune ridges. The approximately twenty ridges represent former shorelines and are evidence of a substantial amount of sediment accumulation. Most of the beach has a moderate gradient high tide beach fronted by a 100 m wide low tide bar, with rips more common toward the western half of the beach. Waves average over 0.5 m in the west, decreasing toward the point, where tidal shoals extend over 1 km north of the inlet.

Swimming: This beach usually receives low seas and swell waves and is relatively safe under these conditions. However, waves greater than 1 m produce rips along the low tide beach and against the rocks and reefs.

Surfing: Chance of usually low beach breaks, increasing during bigger southerly waves.

Fishing: Best along the western end amongst the rocks and reefs, and at the two creek mouths.

Summary: A long beach that offers 7 km of natural beach and foredunes, with some surf and rocks and reefs off the western half.

1452-1455 CLEWS PT

No.	Beach	Unpatrolled			
		Rating HT LT		Type	Length
1452	Clews Pt (S 3)	1	1	B+SF	200 m
1453	Clews Pt (S 2)	1	1	B+SF	200 m
1454	Clews Pt (S 1)	1	2	R+tidal channel	100 m
1455	Clews Pt	1	2	R+LT	150 m

Clews Point is a prominent, 50 m high headland capped by a shipping beacon. There is 4WD access to the beacon via the 40 m high, wooded ridge line that forms the spine of the point. The point forms the northern tip of a 4 km long bedrock ridge, that includes Bustard Head. On the western side of the point is 3 km of south trending, rocky coast that forms the eastern side of Pancake Creek inlet. A 500 m wide tidal channel and sand shoals parallel the point on the other side of the inlet. In amongst the rocky point shoreline are four small, west facing beaches.

Beach **1452** is a crenulate, 200 m long high tide beach fronted by some rocks and 300 m wide, ridged sand flats, then the deep inlet channel. Beach **1453** is very similar, 200 m long, with rocks along the southern half of the beach and ridged sand flats between the beach and the tidal channel.

Beach **1454** is a small, 100 m long pocket of sand fronted by a 50 m sand flat, then the deep inlet channel. Beach **1455** lies immediately west of the point and is a slightly curving, 150 m long, north-west facing beach bounded by wooded, rocky slopes, with rocks outcropping on the beach and in the narrow surf zone. The beach is almost awash at high tide, with a 100 m wide low tide bar fringed by the rocks and boulder beaches on each point.

Swimming: While these are relatively low energy beaches, they are also fronted by the deep inlet channel with its strong tidal currents. There are also rocks along all four beaches, so use caution if swimming here.

Surfing: Usually only a low beach break amongst the sand and rocks on the point beach.

Fishing: Excellent fishing into the inlet channel, as well as off the rocks at high tide.

Summary: Four pockets of sand and rock fronting into the dynamic Pancake Creek inlet and shoals.

1456-1459   CLEWS PT-BUSTARD HD

| No. Beach | Unpatrolled | | |
	Rating HT LT	Type	Length
1456 Clews Pt (E)	2 3	R+BR	150 m
1457 Clews Pt (S)	2 3	R+BR	2 km
1458 Bustard Head (N 1)	1 2	R+LT	50 m
1459 Bustard Head (N 2)	1 2	R+LT	50 m

Between Clews Point and Bustard Head is a 3 km wide, north-east facing embayment containing four beaches. Beach **1456** is a 150 m long, rock-bound beach on the eastern side of Clews Point. It is backed and bordered by rocks of the point, with the beach usually awash at high tide, but with a 50 m wide low tide bar usually fronted by an outer bar cut by two rips, located at each end against the rocks.

Beach **1457** runs from the south side of the Clews Point rocks for 2 km in a curving, north-east facing spiral to the north side of Bustard Head. The beach is composed of fine sand and receives waves between 0.5 and 1 m, which maintain a 30 m wide high tide beach fronted by a 100 m wide low tide bar. Along the more exposed northern half, the low tide bar is fronted by an outer bar that is usually cut by low tide rips every 150 m, with a resulting eight to ten rips along the beach. Both beaches are backed by a largely stable foredune, with some past sand blows extending up onto the backing 40 m high bedrock ridge behind the northern end.

Along the north side of 80 m high Bustard Head are two

small pockets of sand. Beaches **1458** and **1459** are both 50 m long, face north-west and are bordered by bare, sloping rocks. They both have a narrow high tide beach and a 50 m wide low tide bar, with some rock reefs off the beaches. They are backed by the wooded slopes of Bustard Head.

Swimming: The main beach offers the best swimming and a chance of surf, however beware of rips along the northern half, as well as on the Clews Point beaches. The southern beaches generally receive lower waves but are dominated by the surrounding rocks and will have rips against the rocks when surf is breaking.

Surfing: Best chance is toward the northern end of the main beach, over the low tide bar and rips.

Fishing: You can fish the rocks, reefs and rip holes along these four rock-bound beaches.

Summary: Four natural, relatively remote beaches, with access via the 4WD track to Bustard Head.

Bustard Head and Bustard Bay

Bustard Bay was sighted by Captain Cook in May 1770 and named on account of the bustards that, together with ducks, he saw in the bay. He also noted the spring tide to be 8 foot (2.4 m), the first indication he had that he was moving north into a higher tide range coast. This is very close to the official spring tide range of 2.5 m. Today the bay has changed little since Cook sailed by. There is a lighthouse on Bustard Head, a beacon on Clews Point, and a handful of building associated with the lighthouse and a fishing shack on Middle Creek, but otherwise no development.

The shoreline faces north-east and extends for 20 km between Bustard Head and the southern Round Hill Head, where Cook actually landed. It has a predominantly sandy shoreline containing three beaches totalling 21 km in length, together with a small beach on the southern side of Bustard Head and four beaches along the western side of Round Hill. There is also the growing settlement of Round Hill, now called Seventy Seventy, 2 km south of Round Hill Head.

1460-1463   MIDDLE IS, BUSTARD BAY

| No. Beach | Unpatrolled | | |
	Rating HT LT	Type	Length
1460 Jenny Lind Ck	2 3	R+tidal inlet	100 m
1461 Middle Is	2 4	R+BR	9 km
1462 Bustard Bay (S)	2 4	R+BR	7.5 km
1463 Eurimbula Ck	1 2	R+LT	3.5 km

The southern side of Bustard Head looks south along 20 km of sandy beaches, broken by four inlets. Hard against the head and facing into the tidal shoals of the first inlet, **Jenny Lind Creek**, is beach **1460**. The beach is just 100 m long, containing a strip of high tide sand at the base of cleared, 20 m high bluffs. It faces south-east across the dynamic inlet channel and shoals that extend up to 300 m off the beach.

Jenny Lind Creek is named after the schooner "Jenny Lind" that was wrecked at the creek mouth in 1857, while attempting to find safety from a storm.

On the southern side of the inlet is the 9 km long **Middle Island** beach (1461). This east to east-south-east facing beach curves around gently and ends at Middle Creek. Middle Creek connects with Pancake Creek, on the western side of Clews Point, to surround the island at high tide. There is however a low tide crossing, which provides 4WD access to the island and creek mouth, and on to Bustard Head. The island is entirely composed of Pleistocene and Holocene beach-foredune ridges, with up to twenty Holocene ridges prograding up to 3.5 km into the bay. It is, together with the adjoining islands, the most massive foredune ridge plain south of Cowley Beach, 2500 km to the north.

The beach is exposed to all easterly waves, which average 0.5 to 1 m. These and occasional higher waves maintain a 30 m wide high tide beach, a 100 m wide low tide bar and an outer bar that is commonly cut by rips every 150 m, with up to 50 rips occurring along the beach. A 10 m high, casuarina-covered foredune, scarped in places, backs the beach, while off each inlet are 200 to 300 m wide tidal shoals.

Beach **1462** runs for 7.5 km from the southern side of Middle Creek to the smaller Eurimbula Creek inlet. The beach curves around to face north-east and, while receiving increasing protection to the south from Round Hill Head, it receives sufficiently high waves to maintain a rip-dominated low tide bar along most of the beach. The beach is backed by a 3 km wide, casuarina-covered beach to foredune ridge plain, and the modern beach appears to be continuing the progradation of the shoreline. The beach has a relatively low gradient, with a continuous low tide bar, cut by rips every 100 to 150 m, resulting in up to 40 rips along the beach. Towards Eurimbula inlet, the waves decrease in height and rips are less frequent. At the inlet are 100 to 200 m wide tidal shoals. There is 4WD access to the northern creek mouth, where a house is located.

Beach **1463** faces north and extends for 3.5 km from Eurimbula to Round Hill inlet. It is backed by one of the most extensive beach ridge plains on the coast, with 3 km of Holocene ridges, backed by another 3 km of Pleistocene ridges, in total representing a 6 km progradation

of the shoreline. The two tidal creeks border each side, and there is a 4WD track out to the western end of the beach where a few shacks are located. Parts of the backing beach ridges, but not the beach, are contained in the Eurimbula National Park. The beach is moderately protected from south and easterly waves, with waves usually less than 0.5 m. These maintain a high tide beach fronted by a continuous low tide bar, which narrows to the east. Round Hill Creek inlet flows out along the side of the head for 3 km, with tidal shoals paralleling the western side of the channel.

Swimming: These are four relatively isolated, little used beaches, which toward the north are rip-dominated and exposed to occasional higher waves. In addition there are deep channels and strong currents in the four creeks and inlets. Use caution if swimming, particularly if waves are breaking, toward low tide and in the creeks.

Surfing: Best beach breaks are along the more exposed northern beaches and over the tidal inlet shoals.

Fishing: Excellent fishing from the beach into the rip holes, and in the tidal creeks and inlets.

Summary: Bustard Bay's sandy shoreline remains much as it was when Cook sailed by and named the bay in 1770.

Round Hill

Round Hill is famous in Queensland as the first landing site of Captain Cook. Cook and his colleagues landed in May 1770, explored the shore and climbed Round Hill to view the coast and hinterland. Today Round Hill, also known as Seventy Seventy, is a small but growing coastal settlement.

1464-1467 ROUND HILL

No.	Beach	Rating HT LT		Type	Length
		Unpatrolled			
1464	Seventy Seventy	1	1	B+SF	1 km
1465	Monument Pt	1	1	B+SF	150 m
1466	Round Hill (W 2)	1	2	B+SF	100 m
1467	Round Hill (W 1)	1	2	R+LT +rocks	50 m

Round Hill Creek is a 6 km long, funnel-shaped creek that flows out along the western side of Round Hill Head (Fig. 1.26b). On the western side of the creek channel is a 3 km long tidal shoal. This shoal and 40 to 50 m high Round Hill Head provide protection for the anchorages and beaches along the western shore of the hill. A road runs along the shore to the head, where there is a memorial to Cook's visit.

Seventy Seventy Beach (1464) parallels the road and Seventy Seventy, the newer name of the Round Hill settlement. It is 1 km long and curves around to face roughly west. It is backed by a shady beachfront camping and caravan reserve, with houses and a store along the eastern side of the road. The beach is protected from ocean waves by the tidal shoals and is usually calm. It consists of a relatively steep, 20 m wide high tide beach fronted by 50 to 100 m wide sand and rock flats, then the deep creek channel. Boats are usually moored in the channel along the length of the beach, with small boats pulled up on the beach.

Monument Point has a second monument to Cook's landing. There is a car park at the monument and a track down to the beach below the point's northern slopes. The beach (**1465**) is a 150 m long, north-west facing high tide beach fronted by a 50 m wide sand and rock flat, then the deep, 200 m wide tidal channel.

Beach **1466** lies a little further out along the head and is a crenulate, 100 m long, west facing strip of high tide sand fronted by the same rocks and narrow tidal flats, then the deeper channel.

On the north-western tip of the head are two adjoining pockets of sand, totalling 50 m in length (**1467**). They are far enough out on the head to receive low ocean waves and each consists of a narrow, high tide cobble and sand beach fronted by a 50 m wide low tide bar, with rocks to either side. The channel veers to the north and misses this beach. The beach lies to the north and just below the headland car park.

Swimming: These four beaches essentially line the Round Hill Creek channel and, except for the point beaches, receive no ocean waves. While they appear calm, there is deep water in the inlet channel and strong tidal currents, so use caution.

Surfing: None at the beaches, however during big southerly swell, waves break along the extensive tidal shoals on the western side of the inlet channel.

Fishing: Round Hill Creek, the inlet, beaches and rocky points are all very popular locations for shore based fishing.

Summary: Round Hill is one of the last remaining gems on the Queensland coast. However it is now being developed and as a result is losing its isolation and quaint charm.

1468-1470 ROUND HILL (E)

No.	Beach	Unpatrolled		Length
		Rating HT LT	Type	
1468	Round Hill (E 1)	2 4	Cobble +LTT	50 m
1469	Round Hill (E 2)	2 4	Boulders	30 m
1470	Round Hill (E 3)	2 4	R+LTBR	100 m

The eastern side of **Round Hill Head** is exposed to sea and swell waves and, unlike its western side, consists of 1.5 km of predominantly bare rock, with three small beaches tucked into three gaps in the rocks. There is a car park and monument on the head, which lies directly to the west and above the first beach, with access to the other two beaches from residential streets immediately south of the head.

Beach **1468** is a 50 m pocket of high tide cobbles fronted by a 50 to 80 m wide low tide bar, with rocks bordering each side and the 40 m high head behind. There is a track down the bluffs to the beach. Waves average 0.5 to 1 m, and low surf and rips are common.

Beach **1469** is a more protected, 30 m long pocket of cobbles and boulders lying immediately below the small housing estate on the head. It has a rocky beach face and sea floor and is unsuitable for safe swimming, unless calm conditions are prevailing.

Beach **1470** is a 100 m long sand beach backed and bordered by large rocks and boulders, then vegetated bluffs rising to 30 m. The beach has a relatively steep beach face and usually has a deep trough off the beach, as well as a rip running over an outer bar that lies seaward of the boundary rocks. The bar continues south to link with Agnes Waters Beach.

Swimming: These are three relatively hazardous beaches, particularly if waves are breaking. All three are dominated by rocks and can have strong rips.

Surfing: Too much rock for a safe surf.

Fishing: Excellent rock fishing into the deeper holes and gutters.

Summary: The entire head has an interesting shoreline, with the low wave energy, inlet dominated western shore and the wave dominated eastern shore, with its three small, but potentially hazardous, beaches.

1471   AGNES WATER (SLSC)

No.	Beach	Rating HT	LT	Type	Length
1471	Agnes Water	2	3	R+BR	6 km

Patrols:
Surf Lifesaving Club: September to May
Lifeguard on duty: Christmas & Easter

Agnes Water used to be a relatively remote beach located 70 km north of Bundaberg and 60 km off the Bruce Highway. Furthermore, most of the road was only sealed in the mid 1990's. It really only began opening up in the 1990's, many comparing it to Noosa Heads before the boom. The Agnes Water Surf Life Saving Club, reflecting its 'newness', was established in 1989 at the southern end of the beach (Fig. 4.86) and was moved to its present location 2 km further up the beach in 1998. Today the small settlement boasts a small shopping area, new tavern and motel, together with a few caravan parks and a growing number of houses. The name comes from the schooner "Agnes" which, in 1873, disappeared in heavy weather after taking shelter in Pancake Creek.

The main beach is 5.5 km long, running from Round Hill in the north down to Agnes Water. The beach is relatively straight and faces east-north-east. It is famous amongst surfers as being the most northerly beach on the east coast to regularly receive Tasman Sea swell which, when it arrives, provides some excellent surf. Most of the beach is backed by a low dune and natural vegetation. At the southern Agnes Water end there is an extensive foreshore reserve, including a camping reserve.

The beach usually receives waves averaging about 1 m, which combine with the medium sand to build a moderately steep high tide beach (Fig. 2.9a), with a continuous bar exposed at low tide (Fig. 4.87). During and following higher swell, up to 30 rip channels are cut across the lower section of the bar and an outer bar forms along the central and northern sections of the beach. The rip channels will persist for some weeks during lower wave conditions.

Swimming: A relatively safe beach in the southern patrolled area. Care must be taken if the swell exceeds 1 m as rips will be present, particularly at low tide and up the beach.

Surfing: It all depends on the swell. A top spot when it arrives, producing a right off the point, otherwise usually a low beach break.

Fishing: Good off the beach when rip holes are present, otherwise best off the rocks at each end.

Summary: An attractive beach, still relatively unspoiled; however the surrounding region is experiencing rapid development.

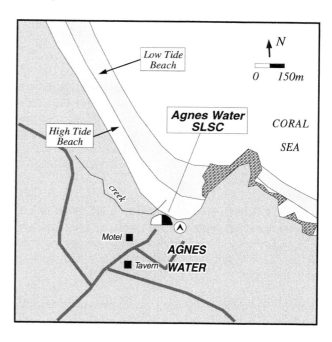

Figure 4.86   *A map of the southern end of Agnes Water Beach showing the location of the original surf club.*

Figure 4.87   *A view of Agnes Water in 1994. The original surf club, located toward the southern end, has since been relocated approximately 2 km up the beach.*

1472-1474   AGNES WATER (S)

No.	Beach	Unpatrolled Rating HT	LT	Type	Length
1472	Agnes Water (S 1)	2	4	R+BR+rocks	60 m
1473	Agnes Water (S 2)	2	4	R+BR+rocks	50 m
1474	Agnes Water (S 3)	2	4	R+BR+rocks	300 m

Immediately east of the southern end of Agnes Water Beach is a 200 m long, rocky headland, on the southern side of which begins a series of generally small, rock-bound beaches that extend for 7 km to Red Rock. The first three beaches lie within 300 m of the point and, although close to Agnes Water, they are only accessible by 4WD or on foot.

Beach **1472** consists of two 20 m wide high tide pockets of sand and cobbles, that link at mid to low tide into a 60 m long, north-east facing beach. The beach is backed by steep, 20 m high, vegetated bluffs and bordered by sloping rocky points and platforms, with a large slab of rock protruding in the centre. Waves average over 0.5 m and rips run out against the rocks.

Beach **1473** lies immediately to the south, is 50 m long and bordered by sloping rocky headlands and platforms. The beach has a moderately steep high tide beach and a 60 m wide low tide bar, usually with one rip running beside the rocks.

Beach **1474** is a 300 m long beach that faces initially north-east, swinging around to face north in lee of the next headland. It is backed by wooded, sloping land, with rocky points to either end, as well as some rocks along the beach. Waves average over 0.5 m at the northern end, but decrease in height in the more rock-dominated southern corner. There is 4WD access to the beach and some clearings in amongst the trees behind the beach.

Swimming: Rips occur on these beaches whenever waves are breaking and particularly toward low tide, so use caution. The southern end of beach 1474 offers the lowest waves and safest swimming, however beware of the rocks.

Surfing: Chance of a surf on the northern end of beach 1474; the other two are too rocky.

Fishing: This entire strip of sand and rock coast offers rock fishing, as well as rock gullies and rip holes.

Summary: Well worth a walk around the head from Agnes Water for three more secluded beaches, that are also accessible by 4WD.

1475-1479   **ROCKY PT**

No.	Beach	Unpatrolled		Type	Length
		Rating			
		HT	LT		
1475	Beach 1475	2	4	R+BR +rocks	20 m
1476	Beach 1476	2	4	R+BR +rocks	350 m
1477	Beach 1477	2	4	R+BR +rocks	1.3 km
1478	Rocky Pt (N 2)	2	4	R+BR +rocks	500 m
1479	Rocky Pt (N 1)	2	4	R+BR +rocks	900 m

Between the second headland south of Agnes Water and Rocky Point is a near-continuous, 3 km strip of sandy shore, broken into five beaches by a series of small, rocky points. All the beaches face east-north-east and generally receive waves averaging 0.5 to 1 m. The waves, together with the medium sand, produce generally moderate to steep high tide beaches fronted by a 50 m wide low tide bar, with rips prominent on all beaches and permanent rips against the rocks.

Beach **1475** lies on the eastern tip of the second headland and is a 20 m pocket of sand with low, rocky points extending 100 m either side of the beach. It is fronted by a bar in line with the points, with a rip always running out of the small bay.

Beach **1476** is a 350 m long, steep, sandy beach with a large rock outcrop toward the southern end. It is backed by wooded slopes and fronted by a 50 m wide low tide bar, usually cut by rips against the bordering and southern rocks.

Beach **1477** is the longest of the five beaches, at 1.3 km. It is bordered by low, rocky points, in addition to some rocks along the centre and at the southern end of the beach. It has a continuous low tide bar that is cut by up to six rips spaced every 150 m, plus rips against the boundary rocks.

Beach **1478** is a 500 m long, rock-bound, sand and gravel beach, with both boundary and central rock outcrops. The beach curves between the rocks, producing a steep, crenulate shoreline with a disjointed low tide bar, that is also cut by rips against the rocks.

Beach **1479** runs south, then south-east, as it curves around in lee of Rocky Point. It is 900 m long and has a relatively steep high tide beach interrupted by rocks in the north and south, with backing wooded slopes and a continuous low tide bar, that is usually cut by two to three rips toward the northern end. Rips are usually absent from the more protected southern corner. There is 4WD access to the southern point and onto the beach, and a few shacks in amongst the trees toward the southern end.

Swimming: All five beaches are dominated by rips when waves exceed 0.5 m and all have permanent rips against the boundary rocks, as well as some larger rocks. The safest spot is in lee of Rocky Point, at the southern end of beach 1479. Otherwise use caution, particularly if waves are breaking, as rips and rip holes will be present.

Surfing: These beaches offer some reef and point breaks as well as beach breaks. The longer beaches offer the best locations, with a right hand break around the lee side of Rocky Point.

Fishing: There are numerous spots to fish from the rocks at high tide, and into the rock gullies and rip holes at low tide.

Summary: A relatively natural string of five rock-bound, sandy beaches, with the best access via Rocky Point.

1480-1483　ROCKY PT-RED ROCK

No.	Beach	Unpatrolled Rating HT	LT	Type	Length
1480	Rocky Pt	2	4	R+BR+rocks	30 m
1481	Rocky Pt (S 1)	2	4	R+BR+rocks	250 m
1482	Rocky Pt (S 2)	2	4	R+BR+rocks	100 m
1483	Red Rock (N)	2	3	R+BR	1.4 km

The coast from Rocky Point to Red Rock is 3 km in length and consists of a rocky shore containing four east facing, steep, sandy beaches. All four are exposed to waves averaging 0.5 to 1 m and all are dominated to varying degrees by low tide rips. There is access to the coast at Rocky Point, where there is a small housing subdivision immediately south of the point, as well as 4WD access to all the beaches.

Rocky Point Beach (1480) is a 30 m wide pocket of sand lying 100 m inside a rock-bound bay. There is road access to houses on the slopes at the rear of the beach. The beach has a steep high tide beach fronted by a 50 m wide, low tide bar with rips forming against the rocks during higher waves.

Beach **1481** lies to the south of the houses and is a 250 m long, rock-bound beach, with rocks also outcropping in the centre to produce a crenulate sand and rock shoreline. The high tide beach is relatively steep and narrow with a 50 m wide low tide bar. Rips form along the rocks during high waves. There is 4WD access to the southern

point, where there is an informal camping site behind the beach.

Beach **1482** consists of three patches of sand and rock, totalling 100 m in length. The rocks form the northern boundary of 1.4 km long beach **1483**, which runs straight down to the northern side of Red Rock headland. Beach 1483 is backed by a 20 m high, climbing foredune. It is relatively steep and commonly cusped, with a low tide bar that is usually cut by five to six rips, including permanent rips against the boundary rocks. In addition, a few rocks outcrop toward the northern end of the beach. There is 4WD access to the southern end, however be careful driving on the beach as it is very soft.

Swimming: These four beaches are all exposed to moderate waves and have rips when waves are breaking, so use caution. Due to their steepness, there is deep water off the high tide beaches, while surf and rips prevail at low tide and against the rocks.

Surfing: Best toward low tide over the low tide bar on the longer beaches.

Fishing: Good access to the points for rock fishing, as well as rip holes off the rocks and beaches at low tide.

Summary: Four reasonably accessible beaches, offering some informal camping, sand and surf.

Deepwater National Park

Shoreline length:　　9 km
Beaches:　　　　　　1484-1488 (5 beaches)

Deepwater National Park is a small park that runs along the coast for 9 km from Red Rock to just south of Wreck Point. The park encompasses four beaches and their backing vegetated, parabolic sand dunes and heath land. In addition there is an elongate, 30 km² freshwater swamp that extends south from Agnes Water and is drained via Deepwater and Broadwater Creeks to enter the sea 20 km to the south.

For further information:
Phone:　(079) 749 224

1484-1488   RED ROCK-WRECK ROCK

| No. Beach | Unpatrolled | | | |
	Rating HT	LT	Type	Length
1484 Red Rock (S)	2	4	R+BR	3.7 km
1485 Wreck Rock (N 2)	2	3	R+BR	150 m
1486 Wreck Rock (N 1)	2	3	R+BR	2.1 km
1487 Wreck Rock	2	3	R+BR	100 m
1488 Broadwater Ck	2	3	R+BR	14 km

Red Rock is a blunt, 30 m high, 500 m long headland fringed by red rocks. South of the headland, the few remaining rocks soon give way to sand and long, straight, sandy beaches (Fig. 3.1a) dominate the coast all the way to the Burnett River mouth. The first five beaches are part of the Deepwater National Park. They all face east-north-east and are well exposed to southerly and easterly waves, that average 0.5 to 1 m in height. The beaches are all composed of relatively coarse sand and have steep high tide beaches fronted by low tide bars, cut by rips every 150 to 200 m on the longer beaches. There is 4WD access from Agnes Water to Red Rock and the beaches to either side, as well as 4WD access from the Deepwater Creek Road to the Wreck Rock picnic and camping area.

Red Rock (south) beach (1484) runs essentially straight for 3.7 km to Wreck Rock. It has a steep high tide beach, with a low tide bar usually cut by twenty rips, in addition to beachrock that parallels much of the intertidal beach. The beach is backed by a 10 to 20 m high foredune and some small, vegetated parabolic dunes, with Wreck Rock forming the southern boundary. There is 4WD access to either end of the beach, with informal camping at the southern end, in amongst the trees in lee of the rocks.

Beach **1485** is bounded by three sets of rocks that loosely impound the 150 m long beach. It is backed by a low foredune, with the access track reaching the shore in lee of the rocks. On the south side of the rocks, beach **1486** runs for another 2.1 km to Wreck Rock. This beach is backed by a 10 m high foredune that fronts some vegetated parabolic dunes up to 500 m in length. It has a steep high tide beach, with the low tide bar cut by an average of ten rips. It is bordered by the low rocks at each end, with 4WD access to either end.

Wreck Rock beach (1487) is another small, rock-bound beach, just 100 m long between two sets of low rocks. The foredune continues on past the rocks while the bar is truncated, resulting in permanent rips against the rocks. The national park 4WD access track runs to the back of the rocks, where there is a picnic and camping area. The name Wreck Rock derives from the wreck of the sailing ship "Countess Russel" in 1873.

South of Wreck Rock, beach **1488** runs straight for 14 km down to the mouth of Deepwater Creek, with the first several kilometres located within Deepwater National Park. This beach has a steep high tide beach, with a low tide bar that is cut by up to 75 rips along its length. There are also the tidal channels, shoals and currents of Deepwater Creek extending 150 m off the beach. The beach is backed by a continuous series of vegetated parabolic dunes that extend up to 500 m inland. Toward the south, the meandering Broadwater Creek backs the beach. There is access to the northern Wreck Rock end of the beach, with some farming behind the dunes in the centre of the beach.

Swimming: These beaches offer steep high tide beaches fronted by deep water, while at mid to low tide a shallow bar cut by rips every 150 to 200 m dominates. Relatively safe during low waves, however be careful when waves are breaking.

Surfing: Best during moderate to high swell conditions over the low tide bar and rips.

Fishing: Numerous rip holes along all beaches, as well as two sets of rocks.

Summary: A relatively difficult to access national park offering beachfront camping and many kilometres of natural beaches.

1489-1491   RULES BEACH, BAFFLE CK, KOLAN R

| No. Beach | Unpatrolled | | | |
	Rating HT	LT	Type	Length
1489 Rules	2	3	R+LT	14 km
1490 Baffle Ck	2	3	R+LT	8.5 km
1491 Kolan R	2	3	R+LT	11 km

South of Deepwater Creek the coast trends south-south-east then south-east for 34 km to the 2m wide, shallow mouth of the Kolan River. In between are three long beaches, each bordered by creek mouths. All three beaches are accessible by vehicle. Wave height gradually decreases down the coast, as the beaches each receive increasing protection from Fraser Island 100 km to the south-east.

Rules beach (1489) is a 14 km long, straight beach lying between the shifting mouth of Deepwater Creek and the larger Baffle Creek mouth. Rocks outcrop halfway down the beach causing a slight inflection in the orientation, otherwise sand dominates the entire beach. Waves average 0.5 m and maintain a relatively steep high tide beach, with a continuous, 50 m wide low tide bar that is usually free of rips. A low foredune backs the beach. The northern 5 km are backed by the meandering Mitchell Creek, with the narrow strip of land between the beach and creek comprising an environmental park. In the centre is a mixture of older foredunes, while to the south are recurved spits of Baffle Creek mouth. The beach is accessible in the south via the Wartburg Road to the small Rules Beach settlement and adjacent macadamia farms.

Baffle Creek beach (1490) begins as a narrow, dynamic spit at the broad creek mouth, which has tidal shoals extending 1 km offshore. It then runs south-east for 8.5 km to the smaller mouth of Littabella Creek, where it ends as a second narrow sand spit. The beach is composed of relatively coarse sand and consequently has a steep high tide beach backed by a low, casuarina-covered foredune in the north, that increases in height to the south. A series of older foredunes extend 2 to 3 km inland of the beach. Waves average 0.5 m along the beach and maintain a continuous low tide bar, with rips usually absent. However, adjacent to both creek mouths are shifting tidal shoals and channels. There is 4WD access to the centre of the beach off the Winfield Point Road and to Noval Park inside the mouth of Littabella Creek. The northern spit of the beach is part of Baffle Creek Environmental Park.

Kolan River beach (1491) is an 11 km long, north-east facing beach. The first 1 km is a bare, narrow recurved spit, followed by 6 km of relatively straight and more stable beach, with the final 5 km comprising a large, dynamic series of spits and tidal shoals associated with the mouth of the Kolan River. The river mouth shoals and channel extend over 2 km offshore. The high tide beach has a moderate slope and is fronted by a continuous low tide bar, that usually receives waves averaging 0.5 m. The entire beach is backed by a beach and foredune ridge plain that is 2.5 km wide in the north, expanding to 5 km wide to the south. This plain is part of the larger Burnett River delta. There is 4WD access to the beach toward the northern end and in the south via Miara, a small settlement located just inside the river mouth.

Swimming: These three beaches offer many kilometres of relatively safe swimming, with deeper water off the high tide beach and usually low waves breaking over the bar at low tide. Be careful near the creek mouths and spits, as there are shifting tidal shoals and channels, with the latter containing strong tidal currents.

Surfing: Lower seas and swell reach these beaches, producing low, spilling waves across the low tide bar and over the tidal shoals.

Fishing: Best at the creek mouths and in the backing channels and mangrove-lined creeks. The best beach holes are near the tidal deltas.

Summary: Three long beaches with limited access and little development.

MOORE PARK TO HERVEY BAY (Fig. 4.88)	
Length of Coast:	166 km
Beaches:	36 (1491-1527)
Surf Live Saving Clubs:	Bundaberg, Elliott Heads, Hervey Bay
Lifeguards:	The Oaks, Bargara, Kellys
Major towns:	Bundaberg, Hervey Bay

Regional Map 10: Moore Park to Hervey Bay

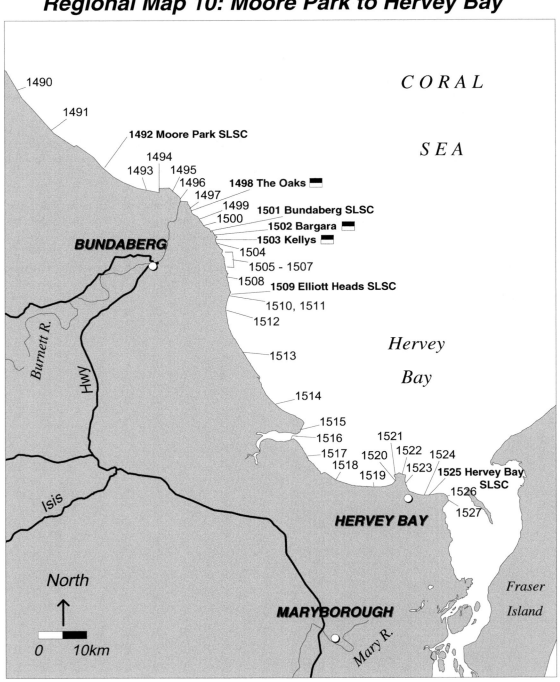

Figure 4.88 *Regional map 10 Moore Park to Hervey Bay. Beaches with Surf Life Saving Clubs indicated in bold, lifeguard patrolled beaches indicated by flag.*

1492 MOORE PARK (SLSC)

Patrols:			
Surf Lifesaving Club:		September to May	
Lifeguard on duty:		School Holidays	

No.	Beach	Rating HT LT	Type	Length
1492	Moore Park	2 3	R+LT	14.5 km

Moore Park is a small holiday settlement 21 km north of Bundaberg. The settlement has a store, caravan park and limited facilities for tourists. It is located on a 20 km long, gently curving beach (1492) that runs from the dynamic mouth of the Kolan River to Barubbra Island at the mouth of the Burnett River. The Moore Park section is 14.5 km long, faces north-east and ends at the mouth of Croome Creek. The entire beach is backed by a 3 to 5 km wide beach to foredune ridge plain, which represents a substantial build-out of the shoreline over the past 6000 years, with much of the sediment coming from the Burnett River.

The Moore Park Surf Life Saving Club was established in 1954. It is located in the foreshore reserve in the centre of the settlement, immediately north of the caravan park (Fig. 4.89). The beach receives waves averaging 0.5 to 1 m. It has a moderately steep high tide beach, with a low, flat bar exposed at low tide (Fig. 4.90). Rips run across the bar when waves exceeds 0.5 m.

Swimming: A relatively safe beach when waves are 0.5 m or less. Higher waves will induce rips, particularly toward low tide, and strong easterly winds and waves will cause a current to run north along the beach.

Surfing: Usually a low beach break that is best at mid to low tide. Low swell occasionally reaches the beach, improving the quality of the waves.

Fishing: Most fishers head down to the Croome Creek mouth, 3 km south of the town. Otherwise the beach is best at mid to high tide.

Summary: A quiet little settlement with adequate facilities for visitors, with a long, natural beach and a patrolled section.

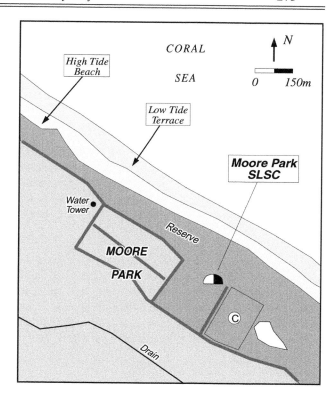

Figure 4.89 *The central section of 15 km long Moore Park Beach showing the town centre and the surf club.*

Figure 4.90 *An aerial view of Moore Park Beach and surf club, shown here at low tide.*

1493-1496 CROOME & WELCOME CKS, BARUBBRA IS, CLARK PT

		Unpatrolled		
No.	Beach	Rating HT LT	Type	Length
1493	Croome Ck	2 3	R+LT	4.2 km
1494	Welcome Ck	1 2	B+SF	1 km
1495	Barubbra Is	2 3	R+tidal shoals	4.5 km
1496	Clark Pt	1 2	B+SF	900 m

Croome Creek forms the southern end of Moore Park Beach. Between the creek and the large Burnett River mouth, 8 km to the east, is a low lying, dynamic section of shore that is variously eroding and building out in response to migration of a large sand spit called Barubbra Island; which in turn is influenced by floodwaters and sand delivered by the Burnett River.

Croome Creek beach (1493) is a north-north-east facing, 4 km long strip of sand that consists of a low beach with no foredune, while a continuous, 50 to 100 m wide bar fronts the beach. The beach is often narrow and eroding, with the backing mangroves visible from the shoreline. It ends at the mouth of Welcome Creek.

Welcome Creek beach (1494) is variously exposed to and protected from waves, depending on the location of the Barubbra Island sand spit. Even when exposed, it is fronted by 200 to 500 m wide tidal flats and shoals. Since the northern Burnett River channel was closed, the spit has been extending and may completely cut off the beach from the sea.

Barubbra Island beach (1495) represents a large store of sand on the north side of the river mouth, that is slowly moving north as a 2 to 3 km long sand spit. Once past Welcome Creek, it merges with Moore Park Beach to continue its northern journey. The masses of sand are also manifest as extensive tidal shoals that extend up to 1 km off the northern end of the spit. The island beach is composed of coarse river sand and is relatively steep, with no foredune but some backing beach ridges and mangroves. There is a shack settlement on the spit, but no vehicle access.

Clark Point is the former northern point of the river mouth, now replaced by a training wall. A 900 m long beach (**1496**) has developed between the low, sandy point and the wall. It is bordered by the 1 km long wall to the south and the sands of Barubbra Island to the north. The beach is a narrow strip of steep sand backed by vegetated overwash flats and mangroves. It is fronted by a shoaling bay between the wall and point.

Swimming: These four beaches are difficult to access and little used, except by the shack owners on the island. While waves are usually low, there are numerous tidal shoals and channels, and strong tidal currents in the creek and channel mouths.

Surfing: Usually only low swell and seas, with variable breaks over the bars and shoals on the more exposed sections.

Fishing: This is a popular fishing area, with the two creeks and the large closed channel, as well as deep water off the high tide beaches.

Summary: A low lying, dynamic section of the Burnett River delta that is adjusting to the closure of the northern channel, as well as still being supplied by river sands.

Hervey Bay

Hervey Bay is a large, 70 km wide, north facing bay that is bordered by Fraser Island to the east and the mainland between Burnett Head and Urangan to the south. The 100 km of mainland shore predominantly comprises long, sandy beaches, with two sections of rock between Burnett and Elliott Heads and at Point Vernon. The back of Fraser Island forms the 70 km of eastern shore which is entirely low energy, sandy beaches. Major settlements along the mainland bay shore include the Bundaberg coast, centred on Bargara, and the Hervey Bay towns between Pialba and Urangan, with smaller settlements at Woodgate, Burrum Heads and Toolgoom.

The Bundaberg Coast

The 22 km of coast between the Burnett and Elliott Rivers lies 10 to 15 km east of the city of Bundaberg and hosts a string of beaches, small towns and residential and tourist facilities for the length of the coast. The entire shoreline is dominated by low, basaltic bluffs and boulder beaches, with the 15 sandy beaches restricted to generally small pockets of sand. The longest beach, Mon Repos, is famous for its turtle rookery. There is good access the length of the coast and numerous facilities for locals, visitors and tourists. There are surf lifesaving clubs at Bargara (patrolling Nielsen Park and Kellys Beach) and at Elliott Heads, with an additional summer lifeguard on The Oaks Beach.

1497-1499 OAKS BEACH

No. Beach	Rating HT LT		Type	Length
1497 South Hd	2	4	R+rocks	100 m
1498 Oaks	2	3	R+LT	200 m
Lifeguard:	School Holidays			
1499 Ricketts	2	3	R+LT	100 m

In contrast to the low, sandy coast extending north of the Burnett River mouth, the coast to the south is dominated by low basalt rocks. The first 5 km south of the river is all basalt, apart from three small pockets of sand.

South Head forms the southern entrance to the river and 1.5 km south of the head is a 100 m long pocket of sand and rocks, bordered by basalt boulders. There is road access to, and a small car park at the back of, the beach (**1497**).

Oaks Beach (1498) is the surfing beach for Burnett Heads. It is located 2 km south of South Head. The beach is a 200 m long, north-east facing pocket of sand wedged in between basalt rocks and boulders to the north, and a groyne that was built by moving basalt boulders off the beach to both trap beach sand and help clear the beach (Fig. 4.91). The beach is backed by a low, narrow dune, with seawalls at each end and a road right behind the dune. There is a small park at the southern end, with parking restricted to the road. The beach receives waves averaging between 0.5 and 1 m which have built a narrow high tide beach fronted by a 50 to 100 m wide low tide bar. When waves exceed 0.5 m, rip currents form against the rocks at each end.

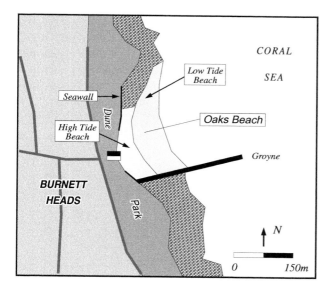

Figure 4.91 *Small Oaks Beach is located between a groyne and basalt boulders. It is patrolled by lifeguards during the summer holidays.*

Ricketts Road runs to another small, 100 m long beach (**1499**) that is partially backed by private property, with access from the road down a small creek drainage easement on the north side of the beach. The beach has a narrow, high tide cobble and boulder beach fronted by an 80 m wide, low gradient sand bar that covers some of the bordering basalt boulders. There are a few tidal pools on the northern boulder flats.

Swimming: These beaches are relatively safe on the bars and clear of the rocks, however if there is any surf care must be taken owing to the rocks and rip currents, usually against the northern boundary rocks.

Surfing: These beaches receive both low wind waves and occasional swell, with the best surf over suitable reefs toward high tide.

Fishing: The rocks and groyne at The Oaks are the best places to reach deep water.

Summary: Three small and very accessible beaches, providing sand access to the sea on an otherwise rocky coast.

1500   MON REPOS

| No. | Beach | Unpatrolled | | | |
		Rating HT LT	Type	Length
1500	Mon Repos	2 3	R+LT	1.8 km

Mon Repos is the longest beach on the Bundaberg coast and one of the least developed. There is road access to either end and the Mon Repos caravan park toward the southern end, but otherwise the beach is backed by a low foredune and farm land. The beach has become a summer tourist attraction, as between November and February up to several hundred turtles land on the beach to lay eggs. The National Parks and Wildlife Service have built a boardwalk and visitor centre to enable people to safely watch the turtles.

The beach (**1500**) is 1.8 km long, faces north-east and is bordered by low basalt rocks, with an outcrop also toward the centre. There is a small tidal creek draining a backing tidal flat and lagoon across the southern end. Waves average just over 0.5 m and maintain a relatively steep, sandy high tide beach, with a continuous low tide bar, usually free of rips. Some past sand blows in the foredune have now been stabilised and most of the dune is now covered with casuarina trees.

This beach is also famous as the site of the first trans-Pacific cable to New Caledonia, laid by the French government in 1890; and where aviator Bert Hinkler first flew his gliders in the then barer sand dunes.

Swimming: Access is restricted to the beach during turtle season, otherwise it is a relatively safe beach under normal low wave conditions. However, waves over 0.5 m will induce rips, particularly against the bordering and central rocks.

Surfing: Chance of a break when the swell is up.

Fishing: Best off the rocks and creek mouth.

Summary: A beach with an interesting history and guaranteed future, owing to the turtles.

1501-1504 NIELSEN PARK, BARGARA, KELLYS, BUNDABERG (SLSC)

Patrols:				
Surf Lifesaving Club:		September to May		
Lifeguard on duty:		School holidays		
No. Beach	Rating HT LT		Type	Length
1501 Nielsen Park				
Bargara SLSC	2	3	R+LT	500 m
1502 Bargara	2	3	R+LT+rocks	300 m
1503 **Kellys**	2	3	R+LT	600 m
Lifeguard:	School Holidays			
1504 Kellys (south)	2	3	R+LT	400 m

Bargara is part of the string of holiday settlements that fringe the Bundaberg coast from Burnett Head south to Elliott Heads. Bargara is fringed by four beaches; Nielsen Park to the north, small Bargara Beach in the centre, which is the home of the Bundaberg Surf Life Saving Club, and Kellys Beach to the south, which consists of two parts and is also patrolled by the lifesavers.

Nielsen Park beach (1501) is located 13 km from Bundaberg, on the north side of Bargara, and consists of the 500 m long beach fringed at either end by extensive basalt rock flats, which at low tide contain large tidal pools (Fig. 4.92). A groyne has been built across the northern end of the beach to retain the sand over the underlying rocks. The beach is backed by a wide, palm-lined foreshore reserve, that also contains the large Bundaberg Surf Life Saving Club, a picnic area and extensive parking. Across the road from the reserve is a large caravan park. The beach has been popular since the 1900's, with an unofficial lifesaving squad patrolling the beach as early as 1914. The Bundaberg Surf Life Saving Club was officially formed here in 1924. Since then the lifesavers have averaged 22 rescues each year, saving well over 1000 people in all.

The beach faces north-east and receives waves averaging 0.5 m, with occasional swell arriving from the east. The beach consists of a moderately steep, 50 m wide high tide beach, with a 100 m wide, low gradient bar exposed at low tide.

Bargara Beach (1502) is located immediately east of the town centre and is backed by a beachfront road, with good beach access and parking. The beach is crossed by three basalt groynes, designed to keep the sand in place and define the boat ramp area. The beach is approximately 300 m long and consists of two parts. There is a 100 m long, high tide sandy section fronted by rock flats north of the main groyne, and a more sandy, 200 m long section on the south side of the groyne, where the groyne-

defined boat launching area is also located.

Kellys Beach (1503) is on the south side of Bargara (Fig 4.92). It is 600 m long, faces due east and is backed by a string of houses on either side of the main road, then the golf course. There is a small lagoon behind the golf course, which is linked to the sea by a small creek crossing the centre of the beach. Like all the beaches along this section of coast, low basalt promontories and rounded boulders fringe either end of the beach. There is a large tidal pool in the northern rocks which is a popular swimming area. The beach has a sandy high tide beach fronted by an 80 m wide and generally rock-free low tide bar.

South of the creek mouth is the southern section of Kellys Beach (**1504**); a 400 m long, high tide sandy beach fronted to varying degrees by sand and rocks and ending at the boulder-strewn southern headland. This beach is backed by a narrow reserve, then a street and houses. There is good access to the length of the beach, with an informal beach boat launching area toward the southern end.

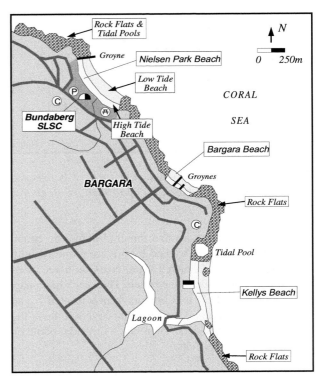

Figure 4.92 *Map of Bargara showing Nielsen Park, site of Bundaberg SLSC, Bargara Beach and Kellys Beach which is patrolled by lifeguards.*

Figure 4.93 *Nielsen Park, site of Bundaberg SLSC, shown here in the centre.*

Swimming: Bargara Beach and Kellys Beach are relatively safe for swimming under normal conditions. However, care must be taken on all beaches near the rocks and especially if swimming in the tidal pools on a rising tide. Higher waves also induce rips against the groynes and rocks on all the beaches. The safest and best locations are at the two patrolled beaches, with best conditions at mid to high tide.

Surfing: Usually low beach breaks, with occasional swell waves.

Fishing: Best over the rock flats at high tide.

Summary: Nielsen Park and Bargara have well laid-out and maintained beaches and reserves, ideally suited to daytrippers and visitors in the caravan park. Kellys Beach is more given over to residential and holiday homes, with a beachfront resort and a caravan park at the northern end.

1505-1508   WOONGARRA, INNES PK, BAROLIN ROCKS

No.	Beach	Unpatrolled			
		Rating		Type	Length
		HT	LT		
1505	Woongarra	2	4	R+LT	200 m
1506	Innes Park (N)	2	4	R+LT	400 m
1507	Innes Park	2	4	R+LT	400 m
1508	Barolin Rocks	2	4	R+LT	50 m

South of Kellys Beach, the coast trends roughly due south for 10 km to Elliott Heads, at the mouth of the Elliott River. Most of the shoreline consists of low basalt bluffs and boulders, with only four small pockets of sand.

Woongarra beach (1505) lies 2 km south of Kellys Beach and immediately south of a new residential development, with good access to the northern side of the 200 m long beach. The beach occupies a small depres-

sion, with a creek draining out across the southern end. It faces east and consists of a narrow high tide beach backed by a grassy foredune, with the low tide bar ending at continuous boulder flats that extend out to low tide. Consequently the beach can only be used for swimming at mid to high tide.

Innes Park is a residential area located on a low, protruding, rocky section of coast, with two small beaches to either side. The northern beach (**1506**) is 400 m long and has a high tide sand beach fronted by a mixture of sand and boulders at low tide. There is good road access at the southern end, with a small foredune behind the beach and a now stable sand blow at the northern end. The southern Innes Park beach (**1507**) straddles a small creek mouth. It is 400 m long and consists of a narrow strip of high tide sand fronted by a continuous, sloping boulder field, with some sand in the small creek mouth. The beach is backed by a casuarina-covered foredune and a small park and car park toward its northern end.

Barolin Rocks is a low, protruding, rocky point, 1 km south of which is a small, 50 m pocket sand beach (**1508**) in amongst a shoreline consisting of continuous basalt boulders. The nearest road ends just south of the point, with foot access to the beach along the low shoreline. The beach consists of a narrow high tide beach and an 80 m wide low tide bar that merges with the rocks both seaward and to the side. A small foredune backs the beach, beyond which is an older, partly active, low dune field extending up to 1 km inland.

Swimming: All four beaches are only suitable for swimming toward high tide, with low tide generally revealing a rocky shoreline.

Surfing: There are various reef breaks along this coast, which need to be checked out with the locals.

Fishing: Good rock fishing the length of the coast, as well as in the small creek at Innes Park.

Summary: An essentially basalt boulder coast, with enough sand to provide some high tide beach access.

1509-1511   ELLIOTT HEADS (SLSC)

Patrols:				
Surf Lifesaving Club:		September to May		
Lifeguard on duty:		School holidays		

No.	Beach	Rating		Type	Length
		HT	LT		
1509	Elliott Heads	2	3	R+LT	150 m
1510	Elliott Hds (groyne)	2	4	R+LT +inlet	50 m
1511	Elliott Hds spit	1	3	R+river channel	200 m

Elliott Heads forms the northern boundary of the shifting mouth of the 1 km wide Elliott River. There is a small islet just off the mouth, while Coonarr sand spit forms the southern shore. The river mouth and settlement of the same name are located 15 km south-east of Bundaberg and contain a number of holiday houses, that surround a large foreshore reserve containing a beachfront caravan park and the Elliott Heads Surf Life Saving Club, founded in 1965. The reserve continues around into the river mouth, with a large kiosk overlooking the mouth. There are three small beaches at the heads: the main beach backed by the surf lifesaving club, a small beach south of the groyne, and a highly variable beach/sand spit just inside the river mouth (Fig. 4.94 & 4.95).

Elliott Heads beach (1509) is 150 m long, with a rocky foreshore and rock groyne forming the northern boundary and a large rock groyne at the southern end. The groyne helps separate the beach from the extensive tidal shoals and currents of the river mouth. At high tide the beach is 50 m wide, however at low tide the combination of low waves and tidal currents has built sand flats that extend 200 to 300 m seaward.

Beach **1510** lies on the south side of the groyne and consists of a 50 m pocket of high tide sand, bordered by rocks and fronted by a widening low tide bar, that grades into tidal shoals extending 300 m off the beach. The main river channel parallels the south side of the shoals.

Beach **1511** is attached to the western side of the head and runs into the river mouth as a dynamic, bare sand spit that can be up to 200 m in length. Although waves are low, the deep river channel and strong tidal currents run along the base of the beach.

Swimming: The main beach is a relatively safe beach close inshore at mid to high tide. However, be very careful at low tide and near the river mouth. Do not swim in the river as the river channel is deep and contains strong tidal currents, plus there are shifting holes and shoals out on the sand flats, even in front of the surf lifesaving club.

Surfing: Usually a low beach break, with the best surf over the river mouth shoals. Waves average 0.5 m, with occasional higher swell coming in around Fraser Island.

Fishing: A very popular spot owing to the deep river entrance. Most people fish from boats or from the river channel or groynes at high tide.

Summary: An interesting little beach equally popular with swimmers and fishers. It has a well-maintained foreshore reserve, together with the added safety of the surf lifesaving club.

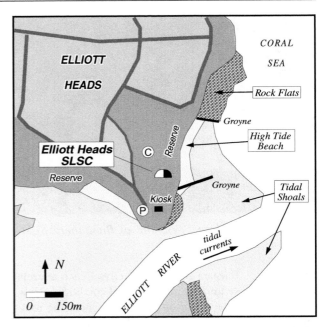

Figure 4.94 *Elliott Heads lies at the mouth of the Elliott River which maintains wide tidal shoals and a deep channel, with the small patrolled beach wedged between the northern headland and a groyne.*

Figure 4.95 *Elliott Heads is located on the north side of the Elliott River. The surf club and patrolled beach (centre) are located between the groyne and headland.*

1512, 1513 **COONARR CK**

No.	Beach	Unpatrolled		
		Rating HT LT	Type	Length
1512	Coonarr	2 3	R+LT	4.2 km
1513	Coonarr Ck	2 3	R+LT	13.7 km

South of Elliott Heads is a 60 km long series of sandy beaches and large inlets sweeping down to the rocks of Point Vernon. Much of the coast is undeveloped, with only three small settlements at Woodgate, Burrum Heads and Toogoom. The first two beaches extend for 18 km south of the heads.

Coonarr beach (1512) is a 4.2 km long, east facing beach that begins amongst the tidal shoals of the Elliott River mouth, then runs straight down to the smaller mouth of Coonarr Creek. There is a gravel road out to the small beachfront settlement of Coonarr, located toward the southern end of the beach. The beach receives waves averaging over 0.5 m, which maintain a relatively steep and narrow high tide beach fronted by a continuous low tide bar, up to 100 m wide.

The small creek maintains a relatively dynamic mouth, which migrates over a 500 m long section of coast. Once established, the southern beach (**1513**) then continues for 13.7 km to the larger Theodolite Creek mouth. There is no development along this beach, with only 4WD access to the shore and a few fishing shacks. The beach consists of a steep high tide beach fronted by a continuous low tide bar, which merges with 1 km wide tidal shoals as it approaches Theodolite Creek. The beach is backed by a low foredune, then a 1 to 2 km wide series of Pleistocene dune ridges.

Swimming: Two relatively safe beaches, usually with low waves and free of rips.

Surfing: Only a low beach break right along the two beaches.

Fishing: Best at high tide off the steep beaches and in the bordering river and creeks.

Summary: Two long, little-developed beaches, with best access at Coonarr.

Woodgate National Park

Area:	5498 ha
Coast length:	10 km
Beaches:	1514 & 1515 (2 beaches)

Woodgate National Park occupies the southern half of Woodgate Beach. It includes the beach and a 3 km wide series of low beach ridges and swamp that extend to the Burrum and Gregory Rivers. There is access from Woodgate via the Walkers Point Road that runs to the small, freehold Walkers Point settlement on the Burrum River. There is a vehicle track off this road to Burrum Point. The start of the 6 km long Melaleuca walking track is located at the point, which takes in beach ridges and swamps in lee of Burrum and North Shore points.

1514-1515 WOODGATE, BURRUM PT

No.	Beach	Rating		Type	Length
		HT	LT		
1514	Woodgate	2	3	R+LT	13.2 km
1515	Burrum Pt	1	2	R+tidal channel	3.2 km

Woodgate is a small, but growing, coastal settlement located 30 km south of Bundaberg. The settlement has basic services and facilities and extends for 2 km along the centre of the 13 km long beach. The beach faces northeast and runs from the mouth of Theodolite Creek in the north to Burrum Point in the south (Fig. 4.96).

Woodgate beach (1514) consists of a low high tide beach fronted by a 100 m wide low tide bar (Fig. 4.97). Towards the creek and Burrum Point, the bar widens to incorporate extensive tidal shoals extending more than 1 km seaward. The beach is backed by a low, casuarina-covered foredune, then a 1 to 3 km wide series of twenty Holocene foredune ridges, that are in turn backed by an equal number of older Pleistocene ridges extending up to 5.5 km inland. The entire beach and barrier system represents a substantial accumulation of Quaternary sands, deposited at present and past sea levels by the low waves of the bay.

Burrum Point beach (1515) forms the northern boundary of the 1 km wide Burrum River mouth. It faces southeast and runs for 3.2 km from its junction with Woodgate Beach at Burrum Point to sandy North Shore Point, the northern head of the river mouth. The extensive river mouth shoals protect the beach, which is calm at low tide. However, the deep river channel and strong tidal currents parallel the beach. The beach is accessible from Woodgate via a road off the Walkers Point Road. The Melaleuca walking track parallels the southern half of the beach.

Swimming: Woodgate is a relatively safe beach with a wide, low gradient beach and surf zone and usually low waves. However when waves exceed 0.5 m, stronger currents, including rips, are generated. There are also strong tidal currents and deep channels at Theodolite Creek and particularly along Burrum Point Beach.

Surfing: Strong southerly winds or a large south swell are required to produce rideable waves at Woodgate.

Fishing: Best off the beach at high tide and in the adjacent creeks.

Summary: An attractive, tree-lined beach and quiet settlement with sufficient facilities for those who want to get well off the main road.

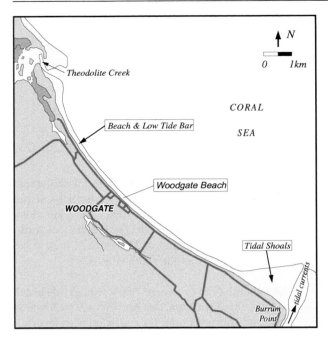

Figure 4.96   *Woodgate Beach is a 13 km long, north-east facing beach bordered by Theodolite Creek in the north and Burrum Point in the south.*

Figure 4.97   *Aerial view of Woodgate beach and settlement.*

1516-1519   BURRUM R, TOOGOOM, DUNDOWRAN

No.	Beach	Rating		Type	Length
		Unpatrolled			
		HT	LT		
1516	Burrum R	1	2	R+river channel	1 km
1517	Burrum Hds	1	2	B+SF	9.5 km
1518	Toogoom	1	1	B+SF	7 km
1519	Dundowran	1	1	B+SF	8 km

Between The Burrum River mouth and Point Vernon is 30 km of sandy shore interrupted by two creeks, resulting in three long beaches. The shore initially faces north-east, swinging around to face north by Point Vernon. At the same time, wave height decreases to less than 0.5 m and the surf zone bar widens to become 1 km wide sand flats.

Burrum River beach is a narrow, 1 km long, north facing beach (1516) paralleling the south side of the river mouth. The beach fronts the small, but growing, Burrum Heads settlement, with the main road running beside the beach. A seawall runs right around the settlement; a result of the houses being located too close to the dynamic river mouth. There is a reserve and boat ramp toward the western end of the beach. The beach is usually calm, but has strong tidal currents just off the beach and in the deep river channel.

The main **Burrum Heads** beach (1517) faces north-east and has a gentle curve between the river mouth and southern Beelbi Creek mouth. Tidal shoals associated with the river and creek extend 2 to 3 km off each end of the beach and influence the majority of the beach. As a result, high tide waves are low, while low tide can reveal many hundreds of metres of sand flats. The land backing the beach consists of both Pleistocene and Holocene beach and foredune ridges up to 2 km wide, as well as tidal creeks and high tide flats, particularly toward the south. Apart from the access and settlement at Burrum Heads, most of the beach is undeveloped.

Toogoom is an old but small settlement on the southern banks of Beelbi Creek mouth. It was originally settled on the creek by timber cutters, while today it is gradually spreading down the open beach. The beach (**1518**) extends from the creek mouth for 7 km toward the east-south-east to the smaller mouth of Oregan Creek. The Beelbi Creek tidal shoals extend up to 2 km off the creek and the northern 3 km of the beach, with the central section having 1.5 km wide sand flats, and more, smaller, 1 km wide tidal shoals off Oregan Creek (Fig. 4.98). As a result, waves are low at high tide and non-existent at low tide, unless you make the long walk out to the shoreline. The actual high tide beach is 30 m wide and relatively steep, contrasting with the wide, flat sand shoals. The beach is backed by a low foredune and foredune ridges that extend up to 1 km inland. A road parallels the back of the beach with a number of access tracks leading to the beach.

Dundowran beach (1519) faces due north and runs for 8 km between Oregan and Eli Creeks, with the latter fringed by mangroves and flowing out in the protected lee of Point Vernon. The two creeks maintain 1 km wide, ridged tidal shoals, with 1.5 km wide sand flats in between. The result is a steep, narrow, low energy high tide beach fronted by the extensive sand flats at low tide. A road runs right behind much of the beach, providing good access, but otherwise there are no facilities.

Figure 4.98 *View west along Toogoom beach and extensive tidal flats, with mangroves fringing the southern Eli Creek mouth.*

Swimming: The main beaches offer relatively safe swimming at high tide, with sand flats at low tide. However be careful at the river and creek mouths, as there are strong tidal currents and deep water, particularly at Burrum Heads.

Surfing: Low wave conditions usually prevail.

Fishing: The river mouth and tidal shoals and channels are the most popular locations.

Summary: Three long, low energy and little-developed beaches, providing relatively safe high tide swimming, or a walk over the tidal flats at low tide.

1520-1524 GATAKERS BAY, PT VERNON, PIALBA

No.	Beach	Unpatrolled Rating HT	LT	Type	Length
1520	Eli Ck	1	1	B+SF	600 m
1521	Gatakers Bay	1	2	R+LT +rock flats	800 m
1522	Pt Vernon	1	2	R+LT +rock flats	1.5 km
1523	Gables Pt	1	2	R+LT +rock flats	3.5 km
1524	Pialba	1	1	B+SF	1 km

Point Vernon is the largest rocky promontory on the coast between Noosa and Bundaberg, a distance of 170 km. Sandy beaches extend to the north, while to the south the last beaches end in the morass of the Great Sandy Strait, before the massive Fraser Island is encountered. The generally low, 5 km² rocky point is surrounded by five low energy beaches.

Eli Creek beach (1520) lies between the small Eli Creek mouth and Eli Point on Point Vernon. It is a low energy, north facing strip of crenulate high tide sand, fronted by scattered mangroves and up to 2 km of ridged tidal flats of the bay and the creek mouth. The beach is backed by sugar cane fields that are now giving way to residential development.

Gatakers Bay beach (1521) occupies 800 m of the shoreline that runs between northern Eli Point and central Point Vernon. It is backed by a shady reserve, then a road and houses. The beach faces east and follows the crenulate nature of the backing rocks. It consists of a narrow, sandy high tide beach fronted by 100 m wide, linear rock flats. There is a boat ramp in the centre backed by a car park and a marina, with a boat ramp and dredged channel across the rock flat in the southern corner. The bay is a popular fishing spot and is also used by windsurfers on windy days.

Point Vernon beach (1522) is a 1.5 km long, east facing, crenulate, sandy high tide beach that runs east from the northern tip of Point Vernon to Gables Point. It is fronted by 100 m wide tidal flats, dominated by linear rock flats, and is backed by a continuous foreshore reserve and The Esplanade.

Gables Point beach (1523) runs for 3.5 km along the south facing side of Gables Point. It is backed by The Esplanade and a continuous foreshore reserve. There is a small boat ramp at the eastern tip. The beach consists of a steep, narrow high tide beach fronted by 50 to 100 m wide, linear rock flats.

Pialba Beach (1524) begins where the point and rock flats end. It fronts the centre of Pialba township and is a 1 km long, east-south-east facing, sandy high tide beach fronted by sand flats up to 500 m wide. The beach ends at small Tooan Tooan creek that drains across the beach and flats. The main road backs the beach, along with a caravan park and foreshore reserve.

Swimming: All five beaches usually receive low waves, less than 0.5 m, and are relatively safe apart from the dominance of rock flats on the point beaches. The best swimming is at high tide.

Surfing: None.

Fishing: You can only fish the beach and rocks at high tide because of the tidal flats.

Summary: A string of low energy, rock and tidal flat dominated beaches, backed by a near-continuous foreshore reserve, with good access to all.

1525 HERVEY BAY (PIALBA, SCARNESS, TORQUAY, URANGAN BEACHES) (HERVEY BAY SLSC)

Patrols:				
Surf Lifesaving Club:	September to May			
Lifeguard on duty:	School holidays			
No. Beach	Rating HT LT		Type	Length
1525 Scarness-Torquay	1	2	B+SF	6 km

Scarness and **Torquay** are at the centre of the thriving Hervey Bay urban area. Beside a growing resident population, the whole area fills with summer and holiday tourists, including increasing numbers of international tourists drawn to see the whales that can be seen over winter in Hervey Bay.

Scarness-Torquay beach (**1525**) runs between Tooan Tooan Creek and Urangan (Fig. 4.99). It is a 6 km long, north facing beach backed by a shady foreshore reserve and The Esplanade, with caravan parks and numerous facilities in the reserve and the two towns. In addition there are boat ramps, the Maryborough and Hervey Bay sailing clubs and one small jetty spread along the beach. At the eastern end is an 870 m long jetty, built in 1917 to reach the deep water of the strait and once used to connect trains to the ships that moored at the end.

The beach is protected from swell by Fraser Island and usually receives low wind waves less than 0.5 m high. The tide range is just over 2 m and the beach ranges from 50 m wide at high tide to 200 m to 300 m wide at low tide. The sand flats widen off Urangan where there are extensive tidal shoals, at the mouth of Tooan Tooan Creek at Pialba, and toward Point Vernon where they merge with rock flats. They also widen off Urangan as the massive tidal shoals (Fig. 4.100) of the Great Sandy Strait run north of Dayman Point.

The beach is patrolled by the Hervey Bay District Surf Life Saving Club. The club was established in 1955 and today patrols the Torquay section of the beach, also known as Shelly Beach.

Swimming: A relatively safe beach, that is best at mid to high tide. Low tide presents wide, exposed or shallow sand flats. Beware of strong tidal currents over the Urangan sand flats.

Surfing: None usually. A cyclone is required to push waves down into the bay.

Fishing: A very popular fishing area, with most fishers heading out in boats to fish the strait, with beach and rock fishing best at high tide. There are a number of boat ramps available.

Summary: One of Queensland's major holiday destinations offering a full range of facilities and accommodation, a usually quiet, safe beach and excellent fishing.

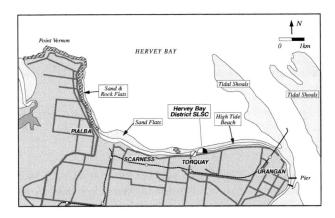

Figure 4.99 *Map of Hervey Bay showing the main centres and the location of the lifeguard patrolled section of beach.*

Figure 4.100 *View of Urangan town and beach, with tidal shoals extending out into the bay.*

1526, 1527 URANGAN

No. Beach	**Unpatrolled**			
	Rating HT LT		Type	Length
1526 Urangan	1	2	B+tidal shoals	400 m
1527 Urangan (S)	1	1	B+SF	500 m

Urangan is the easternmost town in the continuous Hervey Bay urban area and it sits on low Dayman Point, looking north into Hervey Bay and east into the dynamic Great Sandy Strait, with Fraser Island 10 km away on the other side of the strait. There is a small rock outcrop at the point which, together with the change in shoreline direction, defines the boundary with Torquay Beach. **Urangan beach (1526)** faces north-east and runs from the rocks to the Urangan Boat Harbour, where it has built out almost 100 m against the breakwater. It is backed by a shady foreshore reserve in the north and a large car park servicing the boat harbour to the south. The beach is usually calm and has a narrow high tide beach, with tidal flats leading out to the deep strait channels and their extensive tidal shoals.

On the south side of the boat harbour is the 500 m long southern Urangan Beach (**1527**). The beach is bordered in the north by the southern harbour wall and to the south by the start of the strait mangroves, at the small Moolyyir Creek mouth. The beach faces south-east into the strait and receives only low wind waves. These maintain a narrow, steep high tide beach and 200 m wide tidal flats, before the more dynamic tidal shoals and currents of the strait are encountered. The beach is backed by a shady reserve and caravan park, with a small boat ramp at the northern end.

Swimming: Two low energy beaches, come tidal flats, only suitable for swimming at high tide. Beware of the strong strait tidal currents off both beaches.

Surfing: None.

Fishing: The harbour walls are popular at high tide.

Summary: Hervey Bay's southernmost beaches separated by the busy boat harbour.

FRASER ISLAND TO BRIBIE ISLAND (Fig. 4.101)

Length of Mainland Coast:	269 km
Length of Fraser Island Coast:	206 km
Mainland Beaches:	30 (1528-1558)
Fraser Island beaches	17
Surf Life Saving Clubs:	Rainbow Bay, Noosa Heads, Sunshine Beach, Peregian Beach, Coolum Beach, Marcoola, Maroochydore, Alexandera headlands, Mooloolaba, Kawana Waters, North Caloundra, Caloundra, Bribie Island
Lifeguards:	Sunrise, Hyatt Resort, Twin Waters, plus all above surf clubs
Major towns:	Rainbow Bay, Noosa, Sunshine Coast, Woorim

Regional Map 11: Fraser Island to Bribie Island

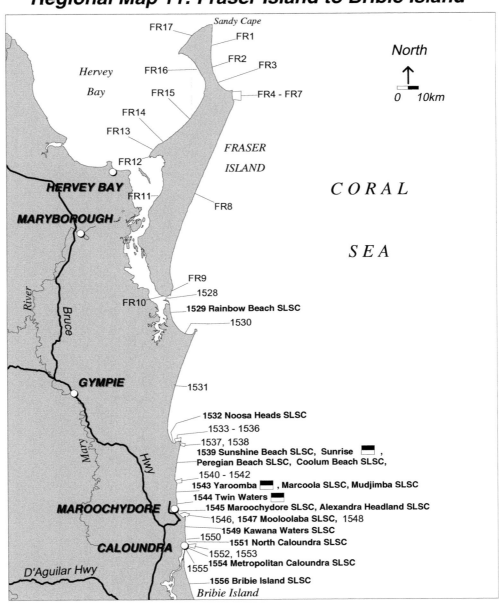

Figure 4.101 *Regional map 11   Fraser Island, the Sunshine Coast and Bribie Island. Beaches with Surf Life Saving Clubs indicated in bold, lifeguard patrolled beaches indicated by flag.*

Fraser Island

Fraser Island forms the eastern boundary of Hervey Bay. The massive, 125 km long island is the largest sand island in the world. It ranges in width from 5 to 25 km, averages 100 m in height, reaching a maximum of 244 m, and contains a massive 1840 km^2 of wind-blown dune sand, totalling about 184 km^3 of sand. The island represents successive layers of wind-blown sand arranged as multiple parabolic dunes, that have been emplaced every time sea level has risen to near its present level, with the oldest and bottom-most dunes in the order of 2 million years old. Each rise in sea level has triggered a massive input of sand to the island that has added another layer, both increasing the width and height of the island. Today most of the dunes are stable and densely vegetated by forests, including some rainforests, with only a small proportion of still active sand blows.

The island offers a wide range of natural attractions including the dunes, forests, perched lakes and 17 beaches totalling 206 km in length, including Queensland's longest beach - 89 km long Seventy Five Mile Beach. In all, 70% of the island shore is sand, the remainder primarily occupied by mangrove-covered sand flats in the Great Sandy Strait. To cater to tourists and visitors, there are three car ferries servicing the island, a number of resorts and a wide range of formal and informal camping sites on the beaches and in the dunes. In the peak holiday season, the main beach resembles a congested road, with hundreds of 4WD vehicles and camps strung along most of Seventy Five Mile Beach. The northern half of the island is contained in Great Sandy National Park, while the southern half is a Sate Forest Reserve.

The island was sighted by Cook in May 1770, who named prominent Indian Head, Sandy Cape and Breaksea Spit.

Great Sandy National park

Area: approx 75 000 ha
Coast length: 150 km
Beaches: 1-8, 13-16 (12 beaches)

Great Sandy National Park occupies the northern half of Fraser Island, with an area of approximately 75 000 ha. It incorporates massive dunes, numerous perched lakes and the only rock headlands on the island, at Waddy Point and Indian Head, as well as the northern tip of Sandy Cape.

For further information:
  Queensland National Parks and Wildlife Service
  Rainbow Beach Road
  Rainbow Beach Qld 4581
  phone:(071) 86 3160

Change in beach character

The 106 Queensland beaches south of Sandy Cape all the way to Coolangatta are distinctly different in character to the 1500 beaches to the north. Unlike the more protected central and north Queensland beaches, those south of Sandy Cape are exposed to persistent higher ocean swell and seas and they have a substantially lower tide range. As a consequence, they are far more energetic and more hazardous for swimmers. They also have the energy to supply massive amounts of sand to build the world's greatest series of sand islands, beginning at Stradbroke and including Moreton, Bribie, Cooloola and the great Fraser Island.

Fraser Island-east coast
FR1-3   MANNAN, NGKALA, ORCHID

No.	Beach	Rating	Unpatrolled Type Inner Outer bar		Length
FR1	Manann	6	TBR	TBR	18 km
FR2	Ngkala	6	TBR	TBR	500 m
FR3	Orchid	5	LTT	RBR	14.5 km

Sandy Cape forms the northern tip of Fraser Island. The island in fact continues on north of here as a 15 km long, shallow sand spit, called Breaksea Spit. The cape separates the energetic eastern shore from the more protected western shores of Hervey Bay. The east side of the island extends south for 130 km to Hook Point. It is divided into two sections north and south of the sole rocky section at Waddy Point-Indian Head. Between the cape and Waddy Point is a 33 km long beach that runs due south, finally curving around to face north in lee of the point. There are three beaches along this section.

Manann Beach (FR1) faces due east and extends for 18 km from Sandy Cape to the coffeerock outcrop at North Ngkala Rocks. This beach, like all of Fraser Island, is composed of fine sand. It has a wide, low gradient swash zone fronted by a double bar system. The inner bar is usually attached to the beach and cut by rips every 200 m, with a deep trough paralleling the bar and a deeper outer bar 100 m off the beach, also cut by more widely spaced rips. Waves average 1 to 1.5 m and break over the two bars, maintaining the up to 90 strong rips right along the beach. A large sand sheet, extending up to 1 km inland, backs the northern 8 km of the beach, with otherwise largely vegetated parabolic dunes averaging 50 m and up to 100 m in height and extending up to 12 km across the island to the Hervey Bay shore.

This beach is little used by the tourist hoards to the south, owing to its considerable distance from the nearest vehicle landing, Moon Point 50 km to the south, as well as the difficulty in negotiating the Ngkala Rocks in a 4WD.

It therefore offers some of the more remote, natural and little used environments on the island.

Ngkala Rocks are two large and several smaller sets of coffeerock outcropping on the beach. The rocks are part of the slow soil forming process that goes on beneath the dunes. They are actually an accumulation of dark, soluble minerals that cement the sand grains to form the dark, coffee-coloured 'rock'. As the overlying dunes are eroded, the rocks are exposed on the beach. The rocks extend right across the high tide beach, requiring vehicles to negotiate a difficult, sandy track over the backing 20 m high dunes. The two main rock outcrops are 500 m apart, with a beach (FR2) identical to Manann in between, apart from the rocks in the swash and inner surf zone. A strong permanent rip runs off the southern rocks.

Orchid Beach (FR3) runs for 14.5 km from the South Ngkala Rocks to rocky Waddy Point. The wreck of the Marloo is located down the beach. The beach undergoes a transformation as it heads south and curves into the lee of the point. The first several kilometres are identical to Manann, with an energetic, rip-dominated double bar system. However, as waves are lowered and refracted around the point, the beach begins to swing around to face north-east, then north, while the lowering waves cause the inner bar to become more continuous with fewer, smaller rips. A deep trough parallels the bar and separates it from the outer bar that runs to within 2 km of the point. Over the last 2 to 3 km of lower energy, north facing beach, another phenomenon contributes to the beach. Large, submerged sand waves periodically move around Waddy Point and manifest themselves as a long sand spit that parallels, then joins, Orchid Beach, often enclosing a backing, elongate lagoon (Fig. 2.1b). The surf zone also widens considerably over the shallow sand wave. As these sand waves move along and merge with the beach, they cause it to vary considerably in width and topography. At times they produce an excellent surfing break along the edge, in lee of the point.

Orchid Beach is also the location of the national park headquarters and the Orchid Beach Resort; a low profile resort set in amongst the casuarina trees, with its own landing strip. The southern end, in lee of the point and sand waves, is also a very popular sheltered camping area, used by fishers and surfers.

Swimming: These are three energetic, rip-dominated beaches. Use extreme care in the surf as rips are always present, together with the deep longshore trough and large rips offshore. The safest swimming is toward the southern end of Orchid Beach, where waves are lower and rips less frequent.

Surfing: The main surfing spot is *Waddy Point*, a long right-hander over the sand waves, as well as 33 km of beach breaks.

Fishing: The entire east side of Fraser Island is a mecca for beach fishers who fish the many rip holes, with rock fishing also at Ngkala Rocks and Waddy Point.

Summary: The north-eastern third of Fraser Island; a little less congested than the southern half and well worth the drive if you have the time and the right vehicle.

Fraser Island-east coast
FR4-7 **WADDY PT, MIDDLE ROCKS, INDIAN HD**

No.	Beach	Rating	Unpatrolled Type		Length
			Inner	Outer bar	
FR4	Waddy Point	6	TBR	RBB	1.4 km
FR5	Middle Rocks (1)	7	TBR	RBB	150 m
FR6	Middle Rocks (2)	7	TBR	RBB	400 m
FR7	Indian Head	6	TBR	RBB	2.2 km

Twenty metre high Waddy Point (Fig. 2.1b) is formed of tertiary basaltic rocks and, together with Middle Rocks and Indian Head 5 km to the south, forms the only bedrock outcrop on the entire island. The point causes a major inflection in the shoreline, as it swings east into Marloo Bay and Orchid Beach, while to the south the great Seventy Five Mile Beach is anchored to Indian Head. Between Waddy Point and Indian Head are four headland-bound beaches, the only ones of their kind on the island.

Waddy Point Beach (FR4) runs due south of the point for 1.4 km to Middle Rocks. It is a straight beach containing a moderate gradient swash zone, usually with eight rips across the inner bar including permanent rips against the headlands, then a longshore trough and outer bar. There is a large, bare sand sheet backing the beach, with the sand blowing up to 1.5 km on to the backing Orchid Beach. To assist revegetation, the sand sheet is closed to vehicles. A well-developed foredune and planted casuarinas are now established behind the beach to prevent sand loss from the beach and, over time, the entire 100 ha sand sheet should revegetate.

Middle Rocks is a sloping, 80 m high, vegetated headland that impounds two beaches. The northern beach (**FR5**) is 150 m long and bordered and backed by bare basalt rocks, with high waves washing right to the rocks. It has permanent rips against the rocks at each end, a deep longshore trough and an outer bar. The southern beach (**FR6**) is similar; only 400 m in length, with a small foredune in the centre and vegetated backing rock slopes. The southern headland has a rock platform containing two large rock pools, called The Aquarium and the Champagne Pools. They were traditionally used by the aborigines as a natural fish trap. The beach itself has a moderate gradient swash zone, with rips against each headland and usually one central rip, then the longshore trough and outer bar. There is a 4WD track running along the slopes behind the two beaches.

Indian Head is a solitary, 50 m high, grass and scrub-covered basalt headland surrounded by bare sand dunes. The head forms the southern boundary of a 2.2 km long, east to north-east facing beach that ends at the northern Middle Rocks. The head protrudes far enough seaward to afford some protection to the southern corner, lowering the waves and causing it to curve around. The protected southern end is a popular 4WD destination for fishers, surfers and campers. The beach (**FR7**) has a moderate gradient and a continuous inner bar usually cut by several rips, which increase in intensity toward Middle Rocks. A low foredune and bare sand dunes extend up to 1 km behind the beach, backed in turn by the vegetated parabolics of the island.

Swimming: These are four more hazardous beaches owing to the strong permanent rips against the boundary rocks, particularly the Middle Rock beaches. The safest swimming is at the southern end of Indian Head Beach, where waves are usually lower and rips less intense.

Surfing: *Indian Head* is a very popular destination for surfers and usually offers a good, long right hand break in the southern corner, as well as camp sites on the beach and head. Elsewhere are beach breaks along all four beaches.

Fishing: This section of coast offers the only concentration of rocks and permanent rip holes, as well as beach rips. Consequently it is a popular destination for fishers, with Indian Head being the prime location.

Summary: Four headland and rock-bound beaches offering some good scenery, surf and fishing, as well as sheltered camp sites in lee of Indian Head.

Fraser Island-east coast
FR8 **SEVENTY FIVE MILE**

No.	Beach	Unpatrolled Rating	Type Inner	Outer bar	Length
FR8	**Seventy Five Mile Beach**				
FR8A	Corroboree	6	TBR	RBB/LBT	14 km
FR8B	Cathedral	6	TBR	RBB/LBT	14 km
FR8C	Maheno	6	TBR	RBB/LBT	9 km
FR8D	Happy Valley	6	TBR	RBB/LBT	7 km
FR8E	Poyungan	6	TBR	RBB/LBT	15 km
FR8F	Eurong	6	TBR	RBB/LBT	8 km
FR8G	Five Mile	5	R/LTT	RBB	22 km
			Total length		89 km

Seventy Five Mile beach (FR8) is the longest beach on the island and in Queensland and amongst the longest in Australia. The beach is a very popular destination of campers, fishers, surfers and tourists who arrive by their thousands in all manner of 4WD vehicles (Fig. 4.102). Some are daytrippers while many stay and camp along the beach. In addition, there are beachfront resorts at Cathedral Beach, Happy Valley and Eurong. The entire beach faces east-south-east, exposing it to the prevailing southerly waves and winds, with only the southern Five Mile Beach swinging around to face more easterly, as waves decrease in lee of the extensive tidal shoals off Wide Bay harbour, the southern boundary of the island. The waves and fine sand combine to maintain a low gradient beach fronted by a double bar system. Rips approximately every 200 m dominate the inner bar, with usually over 400 rips along the beach.

Figure 4.102 *View north along Seventy Five Mile beach, just north of Yidney Rocks.*

Seventy Five Mile Beach, while essentially continuous, can be divided into seven sections based on local names and characteristics. The first is 14 km long **Corroboree** beach (FR8A) that runs south from Indian Head, past Akuna Creek to Wyuna Creek, both small, freshwater creeks draining the backing high dunes. The beach is backed by bare sand sheets for the first several kilometres, however the remainder, like most of Seventy Five Mile Beach, is backed by scarped Pleistocene dunes that range in height, up to 100 m, providing a dramatic backdrop to the beach and surf (Fig. 4.103). The beach is composed of fine sand and has a low gradient, ideal for vehicle traffic. It is fronted by a continuous inner and outer bar system, with rips cutting across the inner bar every 200 m and more widely spaced rips on the outer bar.

Cathedral beach (FR8B) occupies the next 14 km from Wyuna Creek to The Pinnacles, a prominent coffeerock formation. This beach encompasses a park ranger station and southern park boundary and includes the Cathedral Resort and camping area. There is a ranger station 4 km north of the boundary. The Cathedral refers to the coloured sand dunes that are exposed in all the freshly scarped dunes. The colour, layers and coffeerock all derive from concentrations of soluble minerals produced by hundred of thousands of years of soil formation in the older and lower dunes. Cathedral Beach maintains the low gradient swash zone, rip-dominated inner bar, longshore trough and outer bar. It is backed by 60 to 100 m high scarped dunes, rising to 200 m inland. While most are densely vegetated, this section is also backed by the Koorooman, Beemeer and Nulla Kunggur sand blows."

Figure 4.103   *Seventy Five Mile beach is backed for the most part by scarped colourful Pleistocene dunes, capped in places by more recent Holocene sand blows.*

Maheno beach (FR8C) is named after the ship Maheno, wrecked on the beach in 1935 and still a prominent attraction. This section is 9 km long, extending from The Pinnacles past the wreck and freshwater Eli Creek to Chard Rocks, a more prominent coffeerock formation. Beach and surf conditions are the same as the northern sections, with rips dominating the inner bar.

Happy Valley beach (FR8D) runs for 7 km from Chard Rocks to Rainbow Gorge and includes the beach resort and camping area at Happy Valley (actually two small adjoining valleys). It also includes the Enchanted Valley and the small settlements at the Oaks and on protruding Yidney Rocks. The beach ends at the colourful Rainbow Valley. Beach conditions remain the same, with a rip-dominated inner bar, trough and outer bar. The backing 100 m high dune field contains Yidney Lake, Lake Garawongera and the Kirrar sand blow.

Poyungan beach (FR8E) is a 15 km long, straight section from Rainbow Valley to One Tree Rocks. The Lake Garawongera and Cornwells Break 'roads' reach this section of beach. There is a ranger station behind Poyungan Rocks, otherwise no formal establishments. However, beach camping is popular the length of the beach. Wave height still averages 1 to 1.5 m and rips dominate the inner bar. Lake Wabby lies 1.5 km in behind the beach.

Eurong beach (FR8F) runs for 8 km from One Tree Rocks to Eurong, the largest settlement on the island, south to First Creek where the scarped dunes become vegetated as they are fronted by a low, widening foredune plain. Eurong represents the southern end of the continuous, straight, high energy section of the beach, with persistent rips all the way to Indian Head. The settlement includes a beach resort, ranger station and holiday cabins and homes. It is nestled in the lower dunes along a 1 km section of beach. The dunes, that rise to over 100 m inland, contain five perched lakes - Jennings, Birrabeen, Benaroon, Boemingen and Walameboutha.

Five Mile beach (FR8G) represents a transition in the long beach. This 22 km long southern section runs from the mouth of First Creek, which together with Second, Third and Fourth Creeks, drain the 20 km long Jabiru Swamp. This section consists of a low, prograded foredune ridge plain, with the swamp between the ridges and backing high dunes. The plain increases to a maximum width of 3 km at the southern North Spit. The progradation is partly a result of a decrease in wave height and wave refraction around the tidal shoals of Wide Bay Harbour channel, that extend up to 5 km off the spit. The decreasing waves cause a change in the beach and surf, with the rip-dominated inner bar transforming to a continuous bar with no rips (Fig. 2.8b), then a steep beach with no bar. Likewise the outer bar deteriorates and is absent from the southern 5 km, south of Fourth Creek.

Swimming: While this is a very popular beach, it is also a hazardous beach, with up to 450 rips operating across the inner bar alone. Basically there are rips approximately every 200 m for the length of the beach. Use caution if you are not an experienced surf swimmer. Stay inshore on the attached section of bar and well clear of the rips and trough.

Surfing: There are 90 km of beach breaks along this beach to chose from, together with breaks at the *Maheno* wreck and off some of the more prominent coffeerock protrusions at high tide.

Fishing: Most people come here to fish and during the autumn fishing season thousands drive, camp and fish along the beach.

Summary: Most Australians want to go to Fraser Island at some time. They and an increasing number of overseas tourist do go each year. Most head for some section of Seventy Five Mile Beach to drive, tour, camp, fish, relax and surf.

Fraser Island-south end
FR9-10   **NORTH SPIT, HOOK PT**

		Unpatrolled		
No.	Beach	Rating	Type	Length
FR9	North Spit	4	LTT	4.5 km
FR10	Hook Point	3	R	6.5 km

North Spit lies at the northern end of the large Wide Bay Harbour tidal shoals, the point or 'spit' forming where the shoals reach the coast. The shoals both reduce wave height in their lee and change wave direction. The result is a low energy, east facing beach (Fig. 104). This beach (**FR9**) extends for 4.5 km between North Spit and Hook Point, where the channel proper begins. Waves average 0.5 to 1 m, decreasing toward Hook Point. At the same time, the rips decrease and the bar becomes more continuous and narrow to the south (Fig. 3.1b). Because of the generally steeper gradient and at times narrowness of the beach, a gravel road has been constructed paralleling the back of the beach, to provide safe access for vehicles from the Hook Point ferry to the low gradient beach north of Fourth Creek.

Figure 4.104　*View along Hook Point toward the southern end of Seventy Five Mile becah.*

Hook Point is the northern entrance to the 1 km wide Wide Bay Harbour channel. It curves around to face south across the channel to Inskip Point on the mainland. A vehicle ferry crosses the channel from Inskip Point to Hook Point, delivering a regular flow of 4WDs and visitors to the southern tip of the island. As the 6.5 km long beach (**FR10**) runs into the channel, it is increasingly protected from waves. As a result, it has a steep reflective beach face with no bar. This permits the ferry to pull right up to the beach to transfer vehicles to and from the island.

Swimming: These are two lower energy, east side beaches that are little frequented, apart from the in-transit ferry traffic, as the main vehicle track runs behind both beaches. They are relatively safe under normal low waves, however beware of the deep water and strong tidal currents off Hook Point and in the channel.

Surfing: Only on North Point Beach, which usually has a low beach break.

Fishing: Best off Hook Point straight into the deep channel, with a chance of rip holes along North Point during and following higher seas.

Summary: The southernmost beaches on Fraser Island and the first sight of the island for many visitors.

Fraser Island-west coast
FR11 WHITE CLIFFS

No.	Beach	Unpatrolled		Type	Length
		Rating			
		HT	LT		
FR11	White Cliffs	1	1	B+SF	7 km

White Cliffs refers to the scarped or cliffed white sand dunes on this section of the western side of the island. The beach (**FR11**) parallels the Great Sandy Strait and the cliffs are in fact scarped by movement in the deep tidal channel undermining the slopes, rather than waves. The beach is 7 km long and faces west-north-west across to the mainland at Bingham. Calm wave conditions dominate, with the tidal currents providing most of the energy. The beach consists of a narrow strip of high tide sand, fronted by 100 to 200 m wide sand flats and backed by steep cliffed sand dunes up to 60 m high. A few small creeks and valleys breech the cliffs, with the older settlement of Balarragan, a former quarantine station, occupying one, the newer Kingfisher Bay resort another, and Dundonga Creek valley forming the northern boundary.

The beach has been used as a quarantine station, while it now has the large Kingfisher Bay resort residing on the backing slopes.

Swimming: A usually calm and relatively safe beach, apart from the strong tidal currents off the sand flats.

Surfing: None.

Fishing: Good fishing off the beach at high tide and from the Balarragan and Kingfisher jetties.

Summary: A low energy, tide dominated beach on the western side of the island.

Fraser Island-west coast
FR12-16 SANDY PT-PLATYPUS BAY

No.	Beach	Unpatrolled		Type	Length
		Rating			
		HT	LT		
FR12	Sandy Pt	1	2	B+SF	1 km
FR13	Moon Pt	1	2	B+SF	6.5 km
FR14	Coongul Ck	1	2	B+SF	5 km
FR15	Triangle Cliff	1	2	B+SF	24.5 km
FR16	Platypus Bay	1	2	B+SF	23.5 km

The north-western side of Fraser Island runs as a near-continuous, 60 km long, low energy sand beach from Sandy Point to Rooney Point. A number of creeks break the beach into five sections. This is a little used part of the island, with no settlements and only a handful of 4WD tracks reaching the shore, including the busy Moon Point track that is used by 4WD vehicles set down at Moon Point by the Urangan ferry.

Sandy Point (FR12) is a 1 km long recurved spit at the southern tip of the long beach system. It curves around into the northern end of Great Sandy Strait to face south-west toward Urangan, 10 km across the strait. The beach consists of a narrow, steep high tide beach with relatively deep water off the beach. The spit partly encloses a mangrove woodland drained by a creek that exits at the southern tip of the point. On the north side, a small creek separates it from the adjoining Moon Point Beach.

Moon Point beach (FR13) is a 6.5 km long, north-west facing, relatively straight beach that lies between Moon and Coongul Points. It consists of a moderately steep high tide beach backed by a low foredune then an elongate swamp, that usually drains out right at Moon Point.

Coongul Creek beach (FR14) is a relatively dynamic, 5 km long beach that is controlled both by the low waves of Hervey Bay and the location of the creek mouth. The predominantly northerly waves deflect the mouth up to 5 km to the south. Occasionally the mouth re-enters to the north and the southward migration begins again. As a consequence, the steep, narrow beach is backed by a narrow, generally bare to sparsely vegetated sand spit, then the creek and backing mangroves.

Triangle Cliff beach (FR15) is a 24.5 km long, north-west facing, slightly curving beach that runs from the northernmost mouth of Coongul Creek to the larger Wathumba Creek. In addition, Woralie, Bowarrady, Awinga, Bowal and Towoi Creeks cross the beach. This beach is backed by a mixture of coloured, cliffed dunes up to 90 m high and some low swamps. The beach is often narrow and awash at high tide. Waves average up to 0.5 m and maintain a continuous, 50 m wide low tide bar.

Platypus Bay is the name of the waters off the northern end of the beach system that ends at hooked Rooney Point. Between the large Wathumba Creek mouth and the point is a curving, south-west facing, 23.5 km long, low energy beach (**FR16**). The creek mouth maintains a 50 m wide inlet that drains the large Wathumba Swamp. A 6 km long, low, narrow sand spit encloses the swamp, north of which the beach is predominantly backed by scarped dunes up to 60 m high. The beach is relatively flat and usually narrow and often awash at high tide. It is increasingly fronted by sand flats to the north that reach 100 m to 200 m in width. Strong southerly winds do however generate small surf across the flats.

Swimming: These are five generally low energy beaches, only active during strong south-west to northerly winds. Best conditions are toward high tide and clear of the creek mouths.

Surfing: None.

Fishing: Best at high tide and in the creek mouths.

Summary: A 60 km long, little used section of the island, with no facilities and usually difficult vehicle access along the beaches, owing to the many creeks and narrow beaches.

Fraser Island-north end
FR17 SANDY CAPE

No.	Beach	Unpatrolled Rating	Type	Length
FR17	Sandy Cape	3	R/LTT	20.5 km

Sandy Cape is the northern tip of the island. Between the cape and Rooney Point is a slightly convex, 20.5 km long, north-west facing beach (**FR17**). The beach truncates largely vegetated dunes up to 100 m high that originate on the eastern beaches. The highest dune houses the Sandy Cape lighthouse on a pocket of Commonwealth land located 7 km south-west of the cape. The beach receives waves averaging 0.5 m that maintain a moderately steep beach, usually fronted by a low tide bar. Most of the backing land is scarped dune, with only Bool Creek occasionally breaking out across the beach to drain a low, central swampy area.

Swimming: A moderately safe beach under usually low wave conditions. Be careful toward the cape as strong tidal currents flow around the point.

Surfing: Usually a low beach break.

Fishing: Best at high tide off the steep beach.

Summary: The long north-west face of Fraser Island, little visited owing to its remoteness from the main island landing points and tracks.

The Sunshine Coast

The Sunshine Coast contains 140 km of sweeping sandy coast between Fraser and Moreton Islands. It begins in the north at 24 km long Rainbow Beach, with its brilliantly coloured sand dunes. From Double Island Point, the next beach, Cooloola, sweeps for 50 km (much of it national park) to Noosa Heads. On the south side of Noosa Heads, which is also a national park, are a series of generally long beaches that run for 45 km to Caloundra, including Sunshine, Peregian, Coolum, Marcoola, Maroochydore, Alexandra Headland, Mooloolaba, Kawana and Caloundra. Finally, south of Caloundra, Bribie Island sweeps on for another thirty uninterrupted kilometres to Woorim. In all there are over 120 km of brilliant sandy beaches, most receiving moderate to occasionally high swell and possessing energetic surf zones.

Backing the beaches are both the national parks in the north and a series of thriving holiday towns, mostly named after the above beaches. In recognition of both the popularity of the beaches and the hazards of the surf, there are fifteen surf lifesaving clubs between Rainbow Bay and Bribie Island, the oldest having been established in 1915, together with lifeguards on five additional beaches.

1528-1530   RAINBOW BEACH (SLSC)

Patrols:
Surf Lifesaving Club: September to May
Lifeguard on duty: Christmas & Easter

No.	Beach	Rating	Type Inner	Outer Bar	Lengt
1528	Inskip Pt	2	R		2.7 kr
1529	**Rainbow Beach**	6	TBR	LBT	12 km
1530	Rainbow Beach (S)	4	TBR LTT		13 km

Rainbow Beach is a 23 km long, curving beach located in Wide Bay between sandy Inskip Point, opposite Fraser Island, and rocky Double Island Point. The open bay faces north-east and the beaches in the bay face variously north, east and north-east as they swing around between the two points. The growing town of Rainbow Beach is located in the centre of the beaches, with the northern end of the large Cooloola National Park extending south of the southern half of the bay.

Inskip Point beach (1528) is a 2 km long recurved spit that forms the southern shore of the 1 km wide Wide Harbour channel, with Hook Point of Fraser Island forming the northern shore. The beach is relatively narrow and

steep and receives low waves, however strong tidal currents and the wide, deep channels run just off the beach. The beach is used as a landing site for the Fraser Island car ferry. A road from Rainbow Beach runs up behind the main beach to the westernmost tip of the point, where the ferry landing is located.

Rainbow Beach is named after the colourful sand dunes that dominate the southern half of the 24 km long beach. The road from Gympie arrives at the middle of the beach where the town of Rainbow Beach is located. The town offers accommodation, a small shopping area and the Rainbow Beach Surf Life Saving Club, which was founded in 1965. The club house is located on a wide, bluff-top foreshore reserve overlooking the beach (Fig. 4.105, 4.106). There is a ramp from the club to the beach, with additional vehicle access, parking and amenities below the bluffs immediately north of the surf club. Just south of the surf club are Eight Mile Rocks; brown 'coffeerock' outcropping on the beach. The rocks are part of the Pleistocene sand dunes that provide the coloured sands.

The main beach (**1529**) extends from Inskip Point in the north, running due south for 12 km to Eight Mile Rocks. Wave height, which averages nearly 1.5 m in the north, decreases to about 1 m by the town. As the waves decrease, so too the bars change from a rip-dominated double bar in the north, to a single, though still rip dominated, (Fig. 3.2d) bar by the surf club.

Between Eight Mile Rocks and Double Island Point 13 km to the south-east, wave height continues to decrease. The beach (**1530**) faces initially north-east, then north and finally north-west against the point. The rips decrease in size to the south and are soon replaced by a continuous, rip-free low tide bar. The final 2 to 3 km to the point are dominated by elongate sand waves, in the form of migrating sand spits, that periodically form off the point and move westward along the beach, often impounding a backing lagoon. They gradually merge with the beach and the sand continues up the beach toward Inskip Point. The most spectacular part of the beach is, however, the brilliantly 'rainbow' coloured, scarped sand dunes, averaging 100 m in height and up to 135 m high.

There is 4WD access across the point from Cooloola Beach, then up the beach to Rainbow Beach. However the southern beach is often awash at high tide and care need be taken when driving this beach.

Swimming: The central and northern beach is dominated by rips and is potentially hazardous. Be very careful if swimming here, particularly when waves are breaking. Watch out for the deep rip channels, as well as currents in the longshore troughs. Stay close inshore and on the attached parts of the bar. Definitely swim between the flags at Rainbow Beach. Beware of the strong tidal currents at Inskip Point and, while the southern beach is usually the safest with the lowest waves and few rips, be careful of the vehicle traffic on the beach.

Surfing: There are beach breaks the length of this beach, with conditions depending on winds and swell size, as well as bar shape. The southern *Double Island Point* has the long spits of northward moving sand which produce long right-handers during large swell.

Fishing: A very popular spot for beach fishing, with deep holes and gutters common along the beach, particularly north of the surf lifesaving club, and off Inskip Point into the deep tidal channel.

Summary: A long, attractive, natural beach framed by brilliant sand dunes along the southern half. Vehicles are permitted either side of the patrolled section and many people use this as a jumping off point to visit Fraser Island to the north and Cooloola National Park to the south.

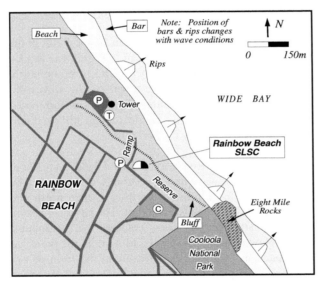

Figure 4.105 *Map of the centre section of 25 km long Rainbow Beach, showing the town centre and surf club.*

Figure 4.106 *A view of Rainbow Bay town centre, surf club (centre) and beach car park.*

Cooloola National park	
Area:	23 150 ha
Coastline length:	56 km
Beaches:	1530 & 1531 (2 beaches)

Cooloola National Park occupies the central and northern part of the massive Cooloola Beach and sand dune system, in all 200 km². It begins at Eight Mile Rocks and incorporates part of Double Island Point and 40 km of beach down to Mount Seewah, a 40 m high dune 1 km north of Teewah, all the backing dunes, and much of the backing swamps, including the upper Noosa River system.

For further information:
Queensland National Parks and Wildlife Service
Rainbow Beach Road
Rainbow Beach Qld 4581
Phone: (071) 86 3160

1531   COOLOOLA BEACH

No.	Beach	Rating	Unpatrolled Type Inner	Outer bar	Length
1531	Cooloola	6	TBR	LBT	53 km

Cooloola Beach and the backing dunes rival Fraser Island in nature and origin. Cooloola is, however, half the size of Fraser and is connected to the mainland by swamps. The first 40 km of the beach and backing dunes are part of Cooloola National Park, and offer a more accessible beach, dune and lake system, compared to Fraser. The northern tip of the beach is anchored by the prominent Double Island Point, named by Cook on account of its twin 90 m high peaks. The point also houses a lighthouse on the eastern peak. The beach begins amongst a field of basalt boulders at the foot of the point, then runs south for 53 km from Double Island Point to the Noosa River mouth. For most of its length it is exposed to energetic waves averaging 1.5 m, which maintain a wide, low gradient beach and double bar system (Fig. 2.5b). Rips dominate the inner bar with up to 250 operating along the beach (Figs. 2.6b & 3.2c). The inner bar and rips are fronted by a wide, deep trough, then the outer bar. Two small creeks, Freshwater and Little Freshwater drain the dunes and flow across the beach. From Teewah village down to the river mouth, the 12 km of beach become increasingly protected by Noosa Head. Wave height slowly decreases and the double bar deteriorates to one rip-dominated bar.

The beach (1531) is totally accessible by 4WD from the ferry across the Noosa River in the south, and from a number of tracks off the Rainbow Beach Road, or Rainbow Beach via Double Island Point in the north. Much of the beach is backed by high, scarped, coloured dunes (Fig. 4.107), while freshwater creeks and some springs cross the beach. Three kilometres south of Double Island Point is the wreck of the 'Cherry Venture', a popular tourist drawcard (Fig. 2.7a). There are protected camping sites at Freshwater Creek and Kings Bore in the national park, as well as the Teewah settlement, which contains about 50 houses all only accessible by 4WD and tucked in behind the lower southern dunes.

Fig. 4.107   *View along Cooloola beach with the backing scarped coloured Pleistocene dunes.*

Swimming: This is an energetic, rip-dominated beach, so use caution if swimming. Stay well clear of the rips and on the attached portions of the inner bar.

Surfing: There are 53 km of beach breaks to choose from.

Fishing: A very popular beach for beach fishing in the many rip holes and gutters.

Summary: A magnificent, long, reasonably accessible beach by 4WD, offering the beach, surf, coloured dunes, massive, largely vegetated sand dunes, a few large sand blows, perched lakes and several protected camping sites.

1532   NOOSA HEADS SLSC

Patrols:				
Surf Lifesaving Club:				September to May
Lifeguards at Noosa Main beach:				all year
	Noosa West beach:			all year

No.	Beach	Rating	Type	Length
1532	**Noosa**	5	TBR-LTT	1.2 km

Noosa Heads is one of Australia's favourite tourist destinations, with large summer and holiday crowds filling the town and its main beach. The town is located at the mouth of the Noosa River and in lee of 2 km long Noosa Head, with much of the head now forming a national park. Immediately north of the river is the more extensive Cooloola National Park. Today Noosa boasts a thriving tourist industry, with major resorts and a wide range of accommodation and facilities. Noosa has long been a popular summer destination, with a surf lifesaving reel placed on the beach in 1915 and the Noosa Heads Surf Life Saving Club founded in 1927.

The main beach (**1532**) runs from the base of the heads to the mouth of the river. The river is now trained with an entrance wall that forms the northern end of the 1.2 km long beach. In addition, to combat beach erosion and maintain some of the sand dumped on the beach, a rock groyne has been built across the middle of the beach and a seawall constructed along the southern half of the beach.

The beach faces almost due north, and receives low waves which have to pass around Noosa Heads. They average between 0.5 and 1 m high at the beach, where they usually form a continuous bar that is cut by rips during and following higher waves. Waves are higher and rips more prevalent at, and north of, the groyne (Fig. 4.108, 4.109).

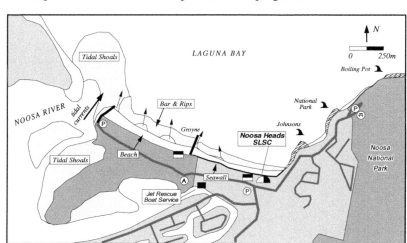

Figure 4.108   *Map of Noosa Beach and river mouth. The Noosa Heads SLSC is located at the main beach, with lifeguards also patrolling the beach west of the groyne.*

Figure 4.109 *A view along Noosa Beach from the river mouth, past the groyne to the main beach, shown here in an eroded state. Note the prominent rips running along the beach on both sides of the groyne.*

Swimming: The safest bathing is in the main patrolled area east of the groyne. However, there is also a lifeguard stationed west of the groyne. Watch out for rip channels and be very careful if swimming near or west of the groyne, as rips are usually present. Also beware of the river mouth, which has shifting shoals and strong tidal currents.

Surfing: *Noosa* is a prime surfing spot, with only beach breaks at the beach, but some of Australia's best right hand breaks along the north side of the head.

Fishing: The groyne and entrance wall are very popular, as well as rip gutters on the beach north of the groyne.

Summary: A heavily developed, although still very attractive, location, saved by the national park and many beaches.

Noosa National Park

Area:	382 ha
Coastline length:	2 km
Beaches:	1534-1537 (4 beaches)

Noosa National Park is a 3 km² park occupying most of Noosa Head. It extends from immediately east of the town of Noosa Heads, along the northern shoreline to Noosa Head and occupies the older, dune-covered, 100 m high bedrock backing the head and Alexandria Bay beach. The park has a delightful walking track from the town to the head and round via Alexandria Bay and Paradise Caves to Sunshine Beach, as well as four small pocket beaches and the world famous surf at Granite, Tee Tree and National Park.

1533-1536 LITTLE COVE, TEA TREE, GRANITE BAY FAIRY POOLS

No. Beach	Unpatrolled Rating	Type	Length
1533 Little Cove	2	R	100 m
1534 Tea Tree	2	R	100 m
1535 Granite Bay	2	R	80 m
1536 Fairy Pools	2	boulder	50 m

Noosa Head is a 50 m high basalt headland located 2 km east of Noosa Beach. Between the beach and head is a 3 km long rocky shoreline backed by the steep slopes of the Noosa National Park, with a road, car park, then walking track out to the head. In along the base of the north facing slopes are four small pocket beaches.

Little Cove (1533) is a 100 m long pocket of northwest facing sand lying just 100 m east of the main Noosa Beach, around small Johnson Point. It is backed by some beachfront houses between the road and the beach, then wooded slopes. Access is difficult, either around rocks at low tide or from the narrow road that clips the northern end of the beach. The beach is backed by a cobble high tide beach, then a veneer of sand which is awash at high tide, together with steps and a small creek in the centre. Waves average less than 0.5 m and rips are rare.

Tea Tree Bay (1534) is the first small bay inside the national park and is located 500 m east of the car park, with access via a walking track around Boiling Pot Point. The walking track runs right along the wooded slopes behind the 100 m long, north facing beach, providing good access the entire length. The beach has a high tide cobble and boulder beach, with a narrow, sandy high tide beach and sloping low tide sand bar. Waves average over 0.5 m, although rips are usually present.

Granite Bay (1535) lies another 500 m on around another point, and is also accessible via the same walking track. It has a more continuous, 80 m long cobble and boulder high tide beach and a sandy low tide sand bar.

Fairy Pools (1536) lies on the eastern point of Granite Bay, immediately in lee of Noosa Head. Beside the pools on the rock platform, it has a 50 m long cobble and boulder beach, with the boulders continuing underwater. The walking track passes around the slopes behind the beach and onto the head.

Swimming: These four beaches and their backing walking track are very popular with walkers, sunbakers and surfers, who walk out to ride the various breaks along this shore. In summer they are crowded with sunbakers and swimmers, while surfers fly by on the intervening point breaks.

Surfing: This section of coast provides Noosa with its world famous point breaks. There are five recognised breaks on each of the points; beginning out at *Granite* on the eastern side of Granite Bay, and again off *Tea Tree* Bay, *Boiling Pot* point, along the boulder shore paralleling the car park, called *National Park*, and finally off Johnson Point, called *Jonhsons*. A larger east to south swell is required to get all the points working, with length of ride also increasing with swell size.

Fishing: The rocky points also are a favourite spot for usually relatively safe rock fishing, however be careful out on Noosa Head, as the rocks are steeper and more exposed.

Summary: This section of shore, with its four small, pandana-lined beaches, five point surfing breaks and backing national park, is one of Australia's most scenic and most popular. It is a delightful spot, which accounts for the crowds that flock there year round.

1537 ALEXANDRIA BAY

No.	Beach	**Unpatrolled** Rating	Type		Length
			Inner	Outer bar	
1537	Alexandria Bay	6	TBR	RBB	1.2 km
1538	Devils Kitchen	8	REF		60 m

The eastern side of Noosa Head faces into the prevailing southerly swell and winds, and is a far more energetic shoreline. The basalt rocks of the head extend south for 2 km to Paradise Caves, with two beaches in between. The Noosa walking track runs to the head and along the back of both beaches and on to neighbouring Sunshine Beach.

Alexandria Bay is a 1.2 km long, open, east facing bay containing a sandy beach (1537) of the same length. The beach is bordered by Hells Gate to the north and Roaring Cave and the Devils Kitchen to the south, their names confirming the more energetic nature of the waves. The higher waves also maintain a double bar beach system, with the low gradient beach fronted by an inner bar usually cut by three beach rips and permanent rips against each head (Fig. 3.6 a & b), then a deep trough and the outer bar 100 m off the beach. As the rips and waves average 1 to 1.5 m, this is a hazardous beach for swimmers. The beach is backed by densely vegetated sand dunes that rise up over the bedrock to a height of 40 m.

The **Devils Kitchen** is a 60 m long, steep cobble and boulder beach (1538) located in a gap in a the rocks. The rocky beach is, however, fronted by a low tide sand bar,

with a permanent rip against the northern rocks. It is backed by 20 m high bluffs, with access down a steep track. Due to the difficult access, rocks, rips and higher waves, this is a very hazardous little beach.

Swimming: Both beaches are energetic, rip-dominated beaches and care should be used, as there have been numerous rescues from the rips on the secluded, though relatively popular, Alexandria Bay beach, while the Devils Kitchen is not so appealing.

Surfing: Alexandria Bay is where the surfers head when the waves are too low along the Noosa Head breaks. It has several good beach breaks.

Fishing: This exposed rocky section is a popular, though more hazardous, rock fishing location.

Summary: Noosa's more exposed beaches, offering relative seclusion with only foot access, but potentially hazardous swimming.

Sunshine to Coolum beaches

1539   SUNSHINE BEACH (SLSC)
SUNRISE BEACH (Lifeguard)
PEREGIAN BEACH (SLSC)
COOLUM BEACH (SLSC)

Patrols:				
Surf Lifesaving Club:		September to May		
Lifeguard on duty:		all year		

No.	Beach	Rating	Type		Length
			Inner	Outer bar	
1539	**Sunshine - Coolum**				
1539A	**Sunshine**	6	TBR	RBB/LBT	2 km
1539B	**Sunrise**	6	TBR	RBB/LBT	1 km
	weekends + holiday				
1539C	Marcus	6	TBR	RBB/LBT	5 km
1539D	**Peregian**	6	TBR	RBB/LBT	5.5 km
1539E	**Coolum**	6	TBR	RBB/LBT	1.5 km
				Total length	15 km

Sunshine Beach forms the northern end of a 15 km long beach that runs almost due south to Sunrise, Marcus and Peregian Beaches and on to Coolum Beach at the southern end. Surf lifesaving clubs are located at Sunshine (Fig. 4.110), Peregian and Coolum, with lifeguards also patrolling Sunrise Beach. The beach is bordered by Noosa Head and Lion Rock in the north, and runs uninterrupted to Point Perry in the south. It faces essentially due east and receives waves averaging 1.5 m. The waves and fine sand produce an energetic double bar system. The inner bar is usually attached to the beach, but cut by deep rip channels every 200 to 250 m. Beyond this is a deep longshore trough, then further out the outer bar, which is cut by large rips every few hundred metres. On average there are 60 rips operating along the entire length of the beach (Fig. 2.6a), together with strong permanent rips at the north end against Lion Rock and in the south against Point Perry.

There are three main settlements along the beach at Sunshine Beach, Peregian Beach and the larger Coolum Beach. **Sunshine Beach** (1539A) is primarily a holiday and residential area, that extends for 3 km along the beach. It has the newest surf lifesaving club on the beach, established in 1982, reflecting the newness of the development. The club is located in a grassy, bluff-top beach reserve (Fig. 4.111) 1.5 km south of Lion Rock. Steps and a ramp lead down to the patrolled beach.

Figure 4.111 *A view across the wide energetic surf zone to Sunshine Beach and SLSC (centre).*

Peregian Beach (1539D) is 8 km south of Sunshine and parallels the beach for 4 km, down to the small Stumer Creek mouth. The Peregian Surf Life Saving Club is located in a large beachfront reserve opposite the main shopping area (Fig. 4.112). The beach has good parking, access and facilities. The club was established in 1962 when the area was first developed, and patrols a potentially hazardous rip dominated double-bar surf zone (Fig. 4.113).

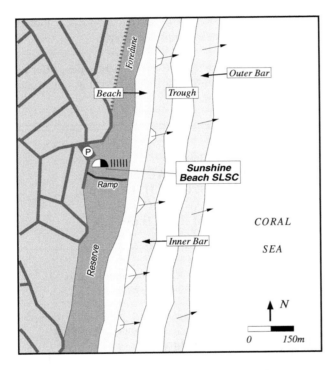

Figure 4.110 *Sunshine Beach SLSC is located in a foreshore reserve overlooking the wide double bar and rip-dominated surf zone.*

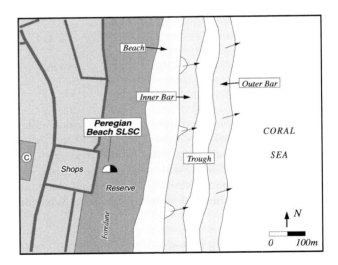

Figure 4.112 *Peregian Beach and SLSC.*

Sunrise Beach (1539B) is located between Sunshine and Peregian Beaches. There is a patrol tower at the base of the track leading from the park and the beach is patrolled by a lifeguard during the Christmas and Easter holidays.

Marcus Beach (1539C) runs for 5 km between Sunrise and Peregian and is backed by a vegetated foreshore reserve, with also some parking and beach access points and sections of beachfront residential areas. It is not patrolled and maintains the same rip-dominated inner and outer bar system.

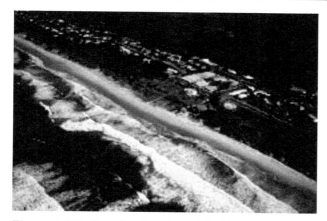

Figure 4.113 *Peregian Beach and SLSC (centre) showing the energetic wave conditions typical of this beach.*

Coolum Beach (1539E) occupies the southern 2 km of the beach, with small Stumer Creek and a few rocks in the surf defining the northern end of this part of the beach. The main Caloundra-Noosa Road passes just behind the beach, which is backed by a large reserve containing a caravan park, as well as the surf lifesaving club at the southern end, close by the rocks of Point Perry (Fig. 4.114). All facilities are available in the adjoining shopping area. This is the traditional surfing beach with the surf lifesaving club formed in 1919. Wave hieight decareses slightly toward the point, however rips still dominated the surf (Fig. 4.115).

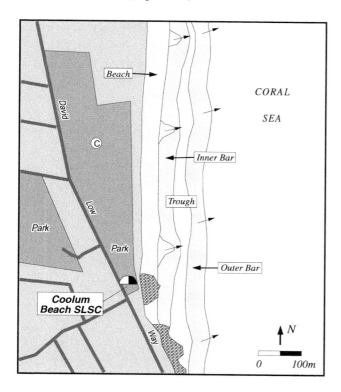

Figure 4.114 *Coolum SLSC patrols a rip-dominated double bar surf zone, with the additional hazard of rocks at the southern end.*

Figure 4.115 *Coolum Beach under energetic wave conditions, showing the surf club and caravan park, and rip dominated surf zone.*

Swimming: Swimming anywhere along the beach is relatively hazardous owing to the persistent strong rips, the deep longshore trough and larger rips further offshore. You should only swim at the four patrolled beaches. Stay between the flags and well clear of the rips.

Surfing: There are numerous beach breaks the length of the beach, with best conditions at high tide on a moderate swell.

Fishing: The rips produce good holes the length of the beach, with the deep longshore trough also accessible with a long cast.

Summary: A long beach offering plenty of surf and a range of settlements and facilities.

1540 POINT PERRY

	Unpatrolled		
No. Beach	Rating	Type	Length
1540 Pt Perry (1)	7	TBR+rocks	100 m
1541 Pt Perry (2)	7	TBR+rocks	100 m
1542 Pt Arkwright	7	TBR+rocks	200 m

Point Perry is a 20 to 30 m high rocky headland that forms the southern boundary of Coolum Beach. The rocks of the headland continue in a south-east direction for 1.5 km to Point Arkwright. Along the base of the bluffs are three small, rock-bound beaches. The main road runs along the top of the bluffs, with small car parks above each. Steep tracks provide access from the car parks to each beach.

Beaches **1540** and **1541** are adjoining sand beaches, both 100 m long and separated by a slight protrusion in the bluffs and a rock platform and reef. The two beaches are both backed by steep, vegetated, 20 m high bluffs.

They receive waves averaging over 1 m, which maintain strong permanent rips that tend to run out along the northern rocks.

Beach **1542** is closer to Point Arkwright and is a 200 m long sand beach, backed by a high tide cobble beach and 30 m high vegetated bluffs. There are numerous rocks in the surf zone and two strong permanent rips at either end.

Swimming: These are three exposed, hazardous beaches, dominated by rocks and strong permanent rips.

Surfing: *Point Perry* has a good right hand break, on the southern point of beach 1541.

Fishing: These beaches are most popular with fishers who fish the rocks, permanent rip holes and rock gutters.

Summary: Three accessible but hazardous beaches, best left to the experienced surfers and rock fishers.

1543	YAROOMBA BEACH (Lifeguard)
	HYATT BEACH RESORT (Lifeguard)
	MARCOOLA (SLSC)
	MUDJIMBA BEACH (SLSC)
1544	TWIN WATERS RESORT BEACH
	(Lifeguard)

SLSC Patrols:					
Surf Lifesaving Club:		September to May			
Lifeguards on duty:		Marcoola,		school holidays	
		Hyatt Coolum Resort		all year	
		Mudjimba		all year	
		Twin Waters Resort		all year	
No.	Beach	Rating	Type		Length
			Inner	Outer bar	
1543A	Yaroomba	6	TBR	RBB/LTT	1 km
1543B	**Hyatt Resort**	6	TBR	RBB/LBT	2 km
1543C	**Marcoola**	6	TBR	RBB/LBT	3 km
1543D	**Mudjimba**	6	TBR	RBB	1.8 km
			Total length		7.8 km
1544A	Ninderry	6	TBR		1.5 km
1544B	**Twin Waters**	6	TBR		1.5 km

From **Point Arkwright** the beach runs almost due south for 8 km to a cuspate foreland, then a further 3 km to the shifting mouth of the Maroochy River. The two beaches have only been developed in the past 30 years, first with the establishment of Marcoola 4 km south of Point Arkwright in the 1960's, including the establishment of the Marcoola Surf Life Saving Club in 1969, and Mudjimba settlement in lee of Mudjimba (or Old Woman) Island. The island forms a wave shadow in lee of which the shoreline protrudes a few hundred metres seaward.

Two more recent resorts have also been developed. Coolum Beach Resort, extending for 2 km south of Point Arkwright, was established in the 1980's, while the southern Twin Waters Resort fronting both the ocean and the river was open in 1994. The two settlements and two resorts offer a wide range of facilities for tourists.

The **Yaroomba-Marcoola** beach (1543) is well exposed to the prevailing southerly swell, which averages 1.5 m for the length of the beach. The waves have combined with the fine to medium sand to build two bars across the 200 m wide surf zone. The inner bar is usually attached to the beach and cut by rip channels every 250 to 300 m, with a deep trough past the first line of breakers, then a deeper outer bar paralleling the beach and cut by more widely spaced rips (Fig. 4.116). On average there are 40 rips along the beach. In addition, there is a strong permanent rip against Point Arkwright and some rocks in the surf at Mudjimba.

Yaroomba beach (1543A) begins at Point Arkwright and extends south for just over 1 km. It is backed by residential development. Beside the beach rips there are two strong permanent rips, one against the point and the second against some rocks in the surf 200 m south of the point. The best access is from the car park and lookout on the 20 m high point and a beachfront reserve 1 km south of the point.

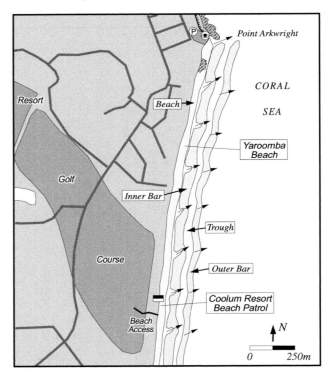

Figure 4.116 *Yaroomba Beach extends south of Point Arkwright, with a lifeguard patrol located at the Coolum Resort beach access point.*

Hyatt Beach Resort and golf course backs the next 2 km of beach (1543B) (Fig. 4.117). The resort provides an access track and a lifeguard patrolling the beach year around. There are rips every 200 to 300 m so ensure you swim in the patrolled area.

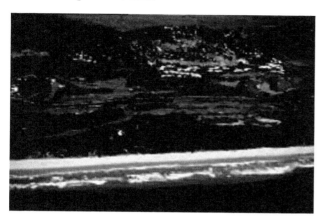

Figure 4.117 *Coolum Resort is fronted by its golf course and public beach, shown here with a trough and small rips crossing the bar.*

Marcoola beach (1543C) is 4 km south of the point, with the Marcoola Surf Life Saving Club located in the centre of the settlement. It is surrounded by a beachfront reserve, with facilities for beachgoers and picnickers. The Esplanade parallels the beachfront reserve, providing parking and access. (Fig. 4.118) At the southern end of Marcoola is the Surfair beachfront resort at Discovery Beach. Besides the usual rips (Fig. 4.119), the beach here has dark coffeerock exposed on either side of the access area.

Mudjimba beach (1543D) is 3 km south of Marcoola and site of Queenslands newest surf club established in 1996. Mudjimba is primarily a residential area. There is a large beachfront reserve that provides good parking and access to the beach (Fig. 4.120). Coffeerock outcrops on the beach in front of the access track. Be careful here as the rock produces additional rips and currents (Fig. 4.121).

The **Ninderry-Twin Waters** beach (1544) runs for 3 km from the cuspate foreland down to the north side of the Maroochy River mouth (Fig. 4.122). Waves are a little lower along this beach, resulting in only one inner bar. However it continues to be dominated by rips all the way to the river mouth, where the shifting tidal shoals and strong tidal currents pose an additional hazard.

Ninderry beach (1544A) extends south of the Mudjimba foreland for 1.5 km. It is backed by a residential area and caravan park, and paralleled by the Esplanade and a foreshore reserve.

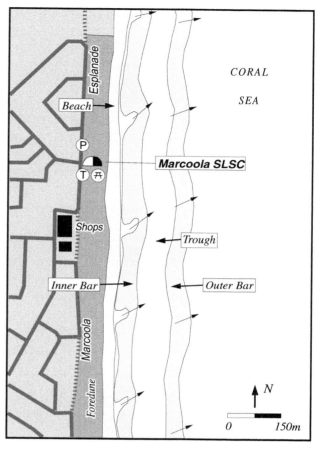

Figure 4.118 *Marcoola SLSC is located in a foreshore reserve just north of the town centre. It patrols a rip-dominated surf zone.*

Figure 4.119 *Marcoola Beach and surf club (centre) usually has a wide, rip-dominated surf zone.*

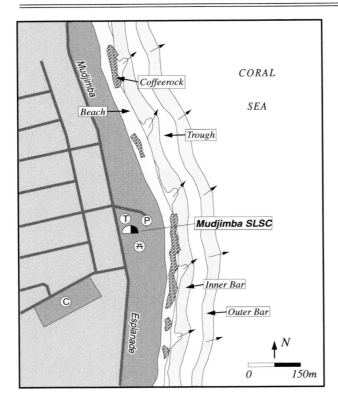

Figure 4.120   *Map of Mudjimba Beach, with main ac-cess in a wide foreshore reserve. Beware of the persis-tent rips and the dark coffeerock.*

Figure 4.121   *Mudjimba Beach lies on a cuspate foreland formed in lee of Mudjimba Island. This results in usually low waves and a narrow surf zone.*

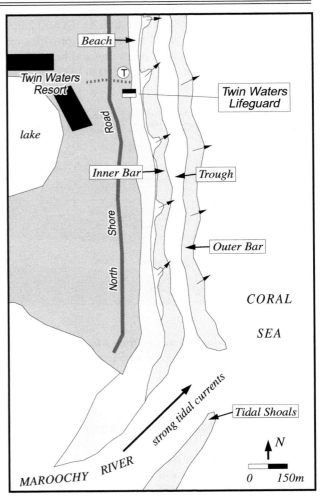

Figure 4.122   *Twin Waters Resort has a lifeguard pa-trolling the adjacent section of ocean beach. South of the patrolled area the tidal currents of the Maroochy River are an additional hazard.*

Twin Waters Resort is built around a large artificial lake, 1 km north of the Maroochy River. It parallels part of the southern 1.5 km of beach (1544B) which runs on down to the river mouth (Fig. 4.123). The Esplanade runs the length of the beach with a car park on the southern spit. The resort has an access track to the beach at the north-ern end of the resort, together with an amenity block. Rips persist right to the river mouth, where there is a deep tidal channel and strong tidal currents.

Swimming: This is a long, potentially hazardous beach with persistent rips and strong currents. You should only swim in the surf at the patrolled Marcoola Beach, or un-der the supervision of the lifeguards at Coolum and Twin Waters resorts. Always check for rips before entering, stay on the attached inner bar and clear of the rip chan-nels and deeper outer trough.

Surfing: There are beach breaks the length of the beach, with better banks usually at Point Arkwright. Best condi-tions are produced by a moderate swell and westerly winds on a high tide. If you have a boat, there is a good left point break on *Mudjimba Island.*

Fishing: There is excellent beach fishing in the numerous rip gutters, together with somewhat more hazardous rock fishing at Point Arkwright and the river mouth south of Twin Waters.

Summary: A long, relatively natural beach offering a range of facilities, including three resorts. Take care if swimming as it is dominated by rips.

Figure 4.123 *View across the mouth of the Maroochy River with the Twin Waters resort and beach located on the northern side.*

1545 MAROOCHYDORE (SLSC)
1546 ALEXANDRA HEADLAND (SLSC)

Patrols:				
Surf Lifesaving Club:		September to May		
Lifeguard on duty:		all year		
No.	Beach	Rating	Type	Length
1545A	**Maroochydore (SLSC)**	6	TBR	2 km
1545B	**Alexandra Headland**	5	TBR-LTT	1 km
1546	Alexandra Head	6	LTT+rocks	500 m

On the south side of the Maroochy River is the town of **Maroochydore**, which merges with the adjoining towns of Alexandra Headlands and Mooloolaba, along a curving 6 km of predominantly sandy coast (Fig. 4.124). The beach usually begins at Pincusion Island, a rocky outcrop that forms the southern entrance to the river. Occasionally the river breaks through on the south side of the island, shortening the beach, as occurred in 1999. When this occurs the deep channel and strong tidal currents pose an addition hazard to bathers wandering north of the patrolled area. The beach, however' is usually attached to the island and 3 km long. It is backed initially by low dunes, then the caravan park, followed by Maroochydore beach and Alexandra Headland beach, which ends at the low, rocky Alexandra Headland.

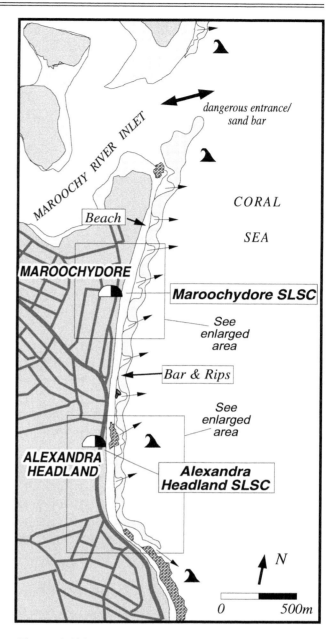

Figure 4.124 *A map of the Maroochydore-Alexandra Headland region showing the location of the two surf clubs.*

Maroochydore beach (1545A) is patrolled by the Maroochydore Surf Life Saving Club; one of the oldest in Queensland, being founded on New Years Day in 1916. The club is located at the end of the main street and there is a large car park adjacent to the club house (Fig. 4.126), with a fenced foredune backing the beach. Rips are common along the beach, so stay in the patrolled area.

The entire beach faces east-north-east and receives waves averaging 1 m at Maroochydore, decreasing slightly toward Alexandra Headland. These produce a single bar along the beach that is cut by rips every 200 m (Fig. 4.125). During and following big seas the bar may detach from the beach, forming a continuous trough that feeds the rips.

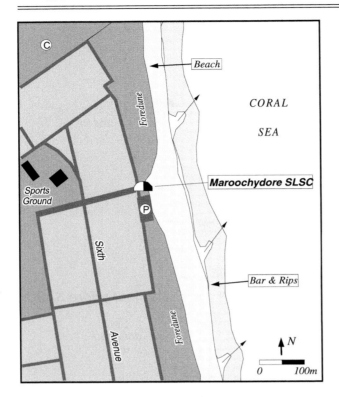

Figure 4.125 *Map of Maroochydore beach showing locaton of surf club and typical beach conditions.*

Figure 4.126 *View of centrally located Maroochydore Surf Club (centre).*

Alexandra Headland beach (1545B) and its surf life-saving club are located 2 km to the south, 300 m north of the headland. The club house and adjoining foreshore reserve are wedged between the main road and the beach (Fig. 4.127). The club was founded in 1924, indicating the long popularity of this beach. At times coffeerock is exposed just north of the surf club, dividing the high tide beach in two (Fig. 4.128).

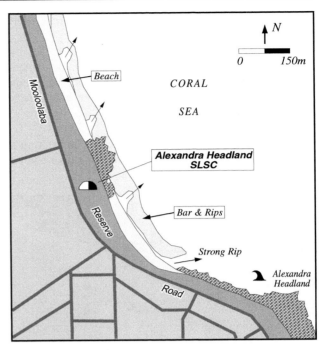

Figure 4.127 *Alexandra Headland Beach and surf club.*

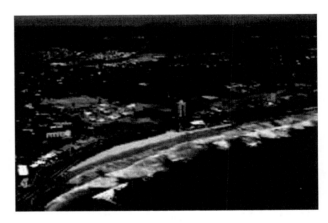

Figure 4.128 *View of Alexandra Headland Beach and surf club (centre), the beach narrows to the north owing to coffeerock outcropping on the beach.*

On the headland is a second beach (**1546**), a 500 m long high tide strip of sand fronted by continuous, 50 m wide low tide rock flats. This beach is backed by a beachfront caravan park and a grassy bluff, with the main road on top of the 10 to 30 m high bluff.

Swimming: While waves are usually 1 m or less, care need be taken on these beaches owing to the common presence of rips and the rocks on the headland beach. The northern end of Maroochydore Beach also contains deep tidal channels and shifting bars and is particularly hazardous. Likewise, there is a strong permanent rip against the rocks at the southern end of Alexandra Headland Beach.

Surfing: The best surfing spot is the break off *Alexandra*

Headland. It works in a moderate to high swell and can provide some good right-handers. There are also beach breaks all the way up to the river mouth.

Fishing: The river mouth and the headland are the most popular locations, together with beach fishing into the rip holes, when present.

Summary: A well established, popular beach offering all facilities for tourists and the added safety of two surf lifesaving clubs.

1547 MOOLOOLABA (SLSC)

Patrols:			
Surf Lifesaving Club:	September to April		
Lifeguard on duty:	all year		

No.	Beach	Rating	Type	Length
1547	**Mooloolaba**	4	LTT/R	2 km

Mooloolaba beach (1547) is a curving, north-east to north facing beach lying in lee of Point Cartwright. The low rocks of the point form the northern boundary, with the Mooloolaba harbour entrance wall at the eastern end (Fig. 4.129). The beach is backed by a continuous foreshore reserve which, toward the northern end, is the site of the Mooloolaba Surf Life Saving Club (founded in 1923), a beachfront caravan park and a large car park. A road runs behind the reserve and provides access points along the beach. On the south side of the road are a number of facilities servicing the boating activities on Mooloolaba Harbour.

The beach receives waves reduced in height by Point Cartwright. They average 0.5 to 1 m at the surf lifesaving club, dropping down to less than 0.5 m by the harbour entrance. The beach changes accordingly, with a shallow, attached bar in front of the club house that is only cut by rips during and following higher seas, while at other times it is continuous with no rips (Fig. 4.130). This bar narrows and finally disappears toward the harbour wall, where usually low waves surge up the beach.

Swimming: This is the safest of the Sunshine Coast beaches, particularly under normal low wave conditions. Rips will form when waves exceed 1 m and there is a permanent rip against the northern rocks.

Surfing: Usually a low beach break that tends to close out with bigger waves. Point Cartwright used to have a great right-hander until the harbour entrance was built.

Fishing: Most anglers head for the harbour side or the entrance walls. Good rip holes only occur following bigger seas.

Summary: The safest beach on the Sunshine Coast, with numerous facilities, including the surf lifesaving club, caravan park and a shady park by the harbour entrance.

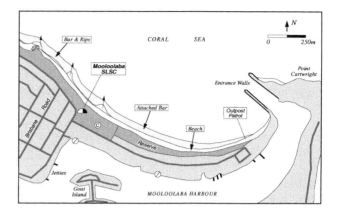

Figure 4.129 *Map of Mooloolaba Beach and harbour. The 2 km long beach is patrolled by the surf club and by an outpost patrol toward the harbour entrance.*

Figure 4.130 *Mooloolaba Beach is protected by Point Cartwright and generally has a low surf, as shown in this view.*

1548 PT CARTWRIGHT

	Unpatrolled		
No. Beach	Rating	Type	Length
1548 Pt Cartwright	5	LTT+rocks	100 m

Following construction of the Mooloolaba harbour training walls in the 1960's, a small beach has accumulated on the point against the eastern wall. The 100 m long beach (**1548**) faces north and consists of a narrow strip of high tide sand, fronted by a continuous intertidal rock flat. It is backed by the grassy reserve on the point, with a walkway leading along the crest of the wall to the beach.

Swimming: This beach is only suitable for swimming at high tide during low wave conditions. At other times, the rocks make it a hazardous location.

Surfing: *Point Cartwright* used to be a classic right-hander prior to construction of the training walls. Today it is a shorter, heavier break ending at the breakwater; for experienced surfers only.

Fishing: A very popular spot for fishing off the break-waters and beach.

Summary: A nice little beach for a walk or picnic in the reserve, but usually unsuitable for swimming.

1549 KAWANA WATERS (SLSC)
POINT CARTWRIGHT BEACH
BUDDINA BEACH
KAWANA BEACH
WARANA BEACH
BOKARINA BEACH
WURTULLA BEACH

Patrols:				
Surf Lifesaving Club:		September to May		
Lifeguard on duty:		Season		

No.	Beach	Rating	Type Inner	Outer bar	Length
1549A	Pt Cartwright	6	LTT/TBR	RBB/LBT	1 km
1549B	Buddina	6	LTT/TBR	RBB/LBT	1 km
1549C	**Kawana**	6	LTT/TBR	RBB/LBT	2 km
1549D	Warana	6	LTT/TBR	RBB/LBT	2 km
1549E	Bokarina	6	LTT/TBR	RBB/LBT	2 km
1549F	Wurtulla	6	LTT/TBR	RBB/LBT	1 km
				Total length	9 km

Kawana Beach is the main recreational section of the 9 km long beach that runs due south from Point Cartwright to Curmmundi Creek mouth. The main Noosa-Caloundra Road runs between 1 and 2 km west of the beach. There are six settlements/sections along the beach. The northern end of the beach, called Point Cartwright Beach (1549A) is backed by high rise apartments. This type of development continues south, with the addition of residential and holiday houses and canal estates developed in the 1970's along Buddina (1549B), Kawana (1549C) and Warana (1549D) Beaches. These settlements all parallel the beach, with a densely vegetated foreshore reserve separating them from the beach. South of Warana, the housing development continues at Bokarina (1549E), with only the southern 1 km bordering Wurtulla Beach (1549F) down to Curmmundi Creek likely to remain natural, as an environmental park.

The beach faces due east and receives waves that average 1.5 m for much of its length, decreasing slightly toward the creek mouth. These waves maintain a 200 m wide double bar system, with the inner bar usually attached to the beach and cut by rips every 250 to 300 m during and following higher waves. A continuous deep trough parallels the inner bar, with the outer bar cut by more widely spaced rips. There is also a strong permanent rip running out against Point Cartwright and tidal shoals and occasional deep channels at Curmmundi Creek mouth, particularly when the creek is open.

The beachfront drive makes a detour at Kawana around the Kawana Waters Surf Life Saving Club, which was established, along with the development, in 1980. The club house is backed by a broad, grassy reserve and a large parking area. The main shopping area is also just behind the beach (Figs. 4.131 & 4.32).

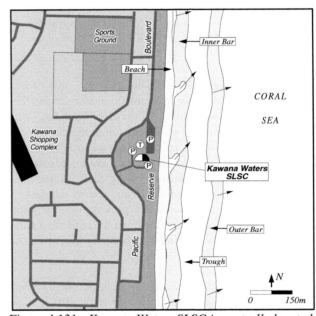

Figure 4.131 *Kawana Waters SLSC is centrally located in a large foreshore reserve adjacent to the Kawana town centre.*

Figure 4.132 *An aerial view of Kawana Surf Club and surrounding reserve.*

Swimming: This is a potentially hazardous beach containing up to 35 rips. The safest place to swim is at Kawana, in the patrolled area. Keep an eye out for the rips and stay on the inner portion of the attached bars. Be careful if swimming near Point Cartwright, owing to the strong rip, or near the creek mouth where tidal currents can be very strong.

Surfing: There are numerous beach breaks along the entire beach. They work best in a moderate swell at high tide.

Fishing: A popular beach for beach fishing either into the rip holes, when present, or casting out into the deep longshore trough. There is also rock fishing off Point Cartwright and at the creek mouth when open.

Summary: A long beach backed by a ribbon of housing development. The beachfront remains in a natural state, with good access points along most of the beach. However the safest swimming is in the area patrolled by the Kawana Waters Surf Life Saving Club.

1550   CURRMUNDI

No.	Beach	Unpatrolled Rating	Type	Length
1550	Currmundi	5	TBR	1.3 km

Currmundi beach (1550) is a 1.3 km long, east facing beach located between the mouth of Currmundi Lagoon and some low rocks that separate it from Dickey Beach. The northern 500 m are backed by the small Currmundi residential area, with a foreshore reserve and beachfront houses. The southern half lies on either side of the small Coondibah Lagoon, that occasionally flows across the beach. The land surrounding the lagoon is a national fitness camp, with a small section of residential houses at the southern end. There is good beach access via the streets and car parks at either end. The beach receives waves averaging just over 1 m, which maintain a series of four to five rips across the bar. These rips occasionally infill during lower waves.

Swimming: A moderately hazardous beach when rips are present and the lagoons are open. Stay clear of rips and on the attached sections of the bar.

Surfing: Usually a low to moderate beach break, with conditions depending on the bars and rips.

Fishing: A popular spot for beach fishing when there are rip holes, and creek fishing when the lagoons are open.

Summary: An accessible, but relatively natural beach with a continuous foredune, two lagoons and a wooded

area in the fitness camp.

1551   NORTH CALOUNDRA (SLSC) DICKEY BEACH

Patrols:			
Surf Lifesaving Club:	September to May		
Lifeguard on duty:	Season		
No. Beach	Rating	Type	Length
1551 **Dickey (North**			
Caloundra SLSC)	**5**	LTT/TBR	800 m

Dickey Beach (1551) is the site of the North Caloundra Surf Life Saving Club. The beach is located 2 km north of the Caloundra town centre and is just off the main Noosa-Caloundra Road. The beach is 800 m long, extending from low rocks at the northern end that separate it from Curmmundi Beach, to a bluff and rock platform at the southern end (Fig. 4.133). The beach faces north-east and receives waves averaging 1 m, which are reduced in size after rounding Caloundra Head and some offshore reefs.

The small Bunbubah Creek crosses the centre of the beach, with a caravan park on its banks, while beachfront houses back either end of the beach. The North Caloundra Surf Life Saving Club, founded in 1950, sits above the north side of the creek, with a watch tower on the beach. There is a large car park just off the main road on the south side of the creek.

The lower waves produce a single 60 m wide bar, that is usually attached to the beach and cut by three to four rips during and following higher waves (Fig. 4.134). Two permanent rips run out against the northern and southern rocks.

Swimming: A moderately safe beach under normal wave conditions and away from the rips at either end. However, high waves produce additional rips along the beach. The patrolled area in the centre of the beach, in front of the surf club, is the safest location.

Surfing: The beach usually has a low beach break. However a moderate to high swell can produce a peaking left and right over the northern rocks, named *Ann Street* after the backing street.

Fishing: A popular spot for people staying in the caravan park, with most fishing off the rocks at each end, or from the beach and in the small creek.

Summary: A very accessible, smaller beach used by locals and people staying at the caravan park. It usually has lower waves than the longer beaches to the north, plus the added protection of the North Caloundra Surf Life Saving Club.

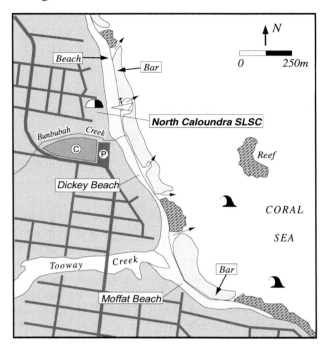

Figure 4.133   *North Caloundra SLSC is located at Dickey Beach, with Moffat Beach around the rocks to the south.*

Figure 4.134   *An oblique view of Dickey Beach, with North Caloundra Surf Club located immediately north of Bunbubah creek mouth.*

1552, 1553   **MOFFAT, SHELLY**

Unpatrolled			
No.	Beach	Rating	Type	Length
1552	Moffat	3	R	800 m
1553	Shelly	4	R+rocks	900 m

Moffat and Shelly Beaches lie north and south of 30 m high Moffat Head. The head, together with Caloundra Head 1.5 km to the south, forms the northern boundary of a protruding bedrock peninsula that is occupied by much of the city of Caloundra.

Moffat beach (1552) is an 800 m long, curving, crenulate, roughly east facing beach that is bordered by a protruding rock platform to the north and the bluffs and rock platforms of the head to the south, with the small Tooway Creek usually dammed in the centre. Two grassy reserves with good access and parking lie to either side, all surrounded by residential housing. Waves average less than 1 m at the beach which, together with the medium to coarse sand, maintain a relatively steep beach with no bar. However, higher waves generate a rip against the northern rocks.

Shelly Beach (1553) runs for 900 m south of Moffat Head. The beach faces east, receiving waves averaging over 1 m. However it is bounded by rock platforms, together with rocks in the centre and rock reefs off the beach (Fig. 2.9b). The end result is a steep reflective beach with no bar, but with strong permanent rips between the central rocks. There is good access to the beach, with a caravan park behind the southern half, a grassy reserve in the centre, a car park to the north and houses on the bluffs behind.

Swimming: Moffat Beach offers the safest swimming under normal low wave conditions, however both beaches have strong rips when waves exceed 1 m.

Surfing: *Moffats* on the southern point can produce a good peeling right during high swell.

Fishing: These are two very popular beaches for rock and beach fishing, as well as the small Tooway lagoon.

Summary: Two local beaches offering good access and parking, and usually quieter conditions.

1554   **KINGS BEACH, CALOUNDRA (SLSC)**

Patrols:	
Surf Lifesaving Club:	September to May
Lifeguard on duty:	all year

No.	Beach	Rating	Type	Length
1554	**Kings** (Caloundra SLSC)	5	LTT/TBR	500 m

Kings Beach (1554) is Caloundra's main surfing beach. It is located adjacent to the centre of town and is ringed by apartments and holiday flats. It is home to both the

Caloundra Surf Life Saving Club, founded in 1924, and the Caloundra Juniors Surf Life Saving Club.

The northern end begins as the rocks of Caloundra Head are replaced by sand. There is an elevated pool on the rock platform, a large car park across the road and a grassy reserve, fronted by a seawall, running from the pool to the surf club, 200 m down the beach (Fig. 4.135). The reserve continues to the centre of the beach, beyond which are beachfront apartments. The beach is 500 m long and faces south-east. However, waves are reduced by Moreton Island and average 0.5 to 1 m, which maintain a continous attached bar usually cut y three rips (Fig. 4.136). The southern end terminates at Deepwater Point; a low, rocky headland and platform, together with a rock groyne. Beyond the point are the shifting sand bars and channels of the Pumicestone Channel mouth, a large tidal creek that runs behind Bribie Island.

Swimming: This is a moderately safe beach under normal conditions. However, permanent rips run out against the rocks at either end and these intensify when waves exceed 1 m, together with an additional two rips forming along the beach. The safest swimming is in the patrolled area.

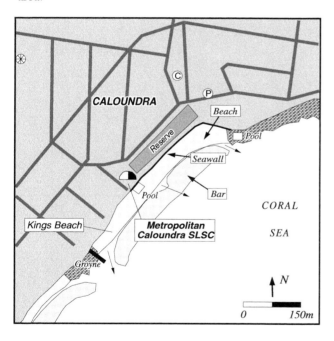

Figure 4.135 *Map of Caloundra's main beach, Kings Beach, the site of the Caloundra SLSC.*

Surfing: Usually reasonable beach breaks, that are best around the rips at each end.

Fishing: A popular beach owing to its central location, together with the prevalence of deeper rip holes off the rocks.

Summary: Caloundra's main and most popular beach,

backed by all facilities, including the surf lifesaving club and a swimming pool.

Figure 4.136   *A view south along Kings Beach, site of the Caloundra Surf Club (centre).*

1555   BULCOCK

No.	Beach	Unpatrolled Rating	Type	Length
1555	**Bulcock**	3	TBR+tidal channel	1 km
Lifeguard: Christmas & Easter				

Bulcock Beach and adjoining Deepwater Point form the highly variable northern bank of the large mouth of Pumicestone Channel, a 30 km long tidal channel that runs the length of Bribie Island to connect with Moreton Bay. Consequently, strong tidal currents run in and out of the entrance (Fig. 3.1c). The point is actually a low, partly vegetated sand spit (1555) that begins on the southern side of the Kings Beach groyne and can grow up to 1 km in length. The Esplanade runs along the base of the spit, providing good access to the ocean and channel sides. The spit faces east and, besides receiving waves averaging 1 m, is daily impacted by the strong tidal currents and shifting channels. The end result is a highly variable beach and surf zone, cut by both wave generated rip currents and tide generated tidal channels and currents.

Swimming: Be very careful if swimming off the point, as it usually has very strong tidal currents, as well as rip currents, together with the deep Pumicestone Channel.

Surfing: The shifting bars and trough do generate some occasional good breaks.

Fishing: A popular spit to fish the deeper tidal channels, as well as deep tidal and rip holes along the beach.

Summary: A dynamic strip of sand and surf best left to the fishers and experienced surfers.

1556-1558   WOORIM BEACH, BRIBIE ISLAND
(SLSC)

Patrols:				
Surf Lifesaving Club:		September to May		
Lifeguard on duty:		November to February		
No. Beach	Rating	Type	Length	
1556 Woorim	**4**	LTT	30.8 km	
(Bribie Is SLSC)				
1557 Skirmish	**2**	R+tidal	4.7 km	
South Pts		channel		
1558 Bongaree	**2**	R+tidal	4 km	
		channel		

Bribie Island is one of the five large south-east Queensland sand islands. However unlike Fraser, Cooloola, Moreton and Stradbroke Islands, which all contain large sand dune systems, Bribie lies protected in lee of Moreton Island and, while extensive, is a low sand island rarely reaching 5 m above sea level. It has been developed by building out of successive shorelines and low foredunes, rather than the massive sand dunes of the other sand islands. The island is 34 km long, extending from the Pumicestone Channel entrance, opposite Caloundra, in a south-easterly direction down to the only town at Woorim on broad Skirmish Point, with South Point forming the southernmost extremity. A bridge across the southern entrance to Pumicestone Channel connects the island to the mainland and the Bruce Highway 20 km to the west. Woorim spreads for 3 km along the beach. It is a thriving town and holiday area offering all facilities for tourists and holidaymakers. Major attractions are the generally safe beaches and abundant opportunities for fishing the beaches, bays or ocean.

Bribie Island is named after the first permanent white settler, an ex-convict named Bribie who lived on the island from 1834.

The Bribie Island Surf Life Saving Club is located at the end of the main entrance to Woorim (Fig. 4.137). The Caloundra Surf Life Saving Club began patrolling the beach in 1924, with the Bribie Island Surf Life Saving Club formed in 1926. The club house is located just off the road in a bushy, but narrow, reserve that overlooks the long beach. It is backed by a large car park and all facilities are within easy walking distance of the beach. The reserve widens south of the club house.

Moreton Island lies 15 km due east of Woorim, with the waters in between forming the wide, shoal dominated entrance to Moreton Bay. Moreton Island protects Bribie Island from most southerly swell, resulting in waves av-

eraging 0.5 to 1 m along **Woorim** beach (1556). These usually produce a narrow high tide beach with a continuous attached bar exposed at low tide (Fig. 2.8a & 4.138). Rip channels are usually absent at Woorim and only form during and following waves greater than 1 m. However southerly winds can produce a strong current running to the north.

One kilometre south of Woorim is **Skirmish Point**, where the shoreline swings south-west for 1 km to Bold Point, then continues on to the west to South Point, the southern tip of the island. This convex, roughly south-south-east facing beach (1557) faces into Moreton Bay. The beach receives low waves and is usually calm inside Bold Point. However strong tidal currents and a deep channel parallel much of the eastern half of the beach.

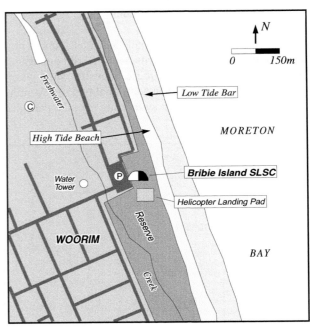

Figure 4.137   *Map of Woorim, the main settlement on Bribie Island and site of the Bribie Island SLSC.*

Figure 4.138   *An aerial view south along Woorim Beach on Bribie Island, with the Bribie Island Surf Club located in front of the tower.*

Bongaree beach (1558) extends north from Skirmish Point for 4 km to the Bribie Island Bridge, with the town of Bongaree backing the northern 3 km of the beach. This is a narrow low energy beach, subject to wave and tidally-driven pulses of sand that move slowly north along the beach. The pulses or sand waves induce a mobile crenulate beach shape, narrowing and eroding the beach in between the waves of sand. As a consequence a sea-wall and some groynes have been build along much of the beach. Between Bongaree town centre and the bridge the beach is backed by a continuous road and foreshore reserve, together with a fishing jetty, boat ramp, and car parking, as well as most facilities for visitors. Calm conditions normally prevail at the shore, however strong tidal currents parallel the beach in the deep Pumicestone Channel.

Matthew Flinders landed on Bongaree Beach in 1799. The beach is named after a Sydney aborigine of that name, who accompanied Flinders and was first to land on the beach to greet the local aborigines.

Swimming: Woorim is a relatively safe beach, particularly under normal low waves. Be careful when waves exceed 1 m as they can produce a heavy shorebreak, rips and a drag along the beach. South of Skirmish Point and along Bongaree Beach the waves are usually low, however tidal currents become an increasing hazard.

Surfing: Usually a low shorebreak, which closes out in bigger surf.

Fishing: A very popular fishing destination for both beach and bay fishing.

Summary: The closest surfing beach north of Brisbane and a popular weekend and holiday destination.

MORETON BAY, MORETON & STRADBROKE ISLANDS
THE GOLD COAST (Fig. 4.139)

Length of Mainland Coast:	299 km
Length of Moreton Island Coast	87km
Mainland Beaches:	42 (1559-1601)
Moreton Island beaches	15
Surf Life Saving Clubs	Point Lookout, Redcliffe Peninsula Southport, Surfers Paradise, Northcliffe, Broadbeach, Kurrawa, Mermaid Beach, Nobbys Beach, Miami Beach, North Burleigh, Burleigh Heads, Tallebudgera, Pacific-North Palm Beach Palm Beach, Currumbin, Tugun, Bilinga, North Kirra, Kirra, Coolangatta, Tweed Heads, Rainbow Bay
Lifeguards:	At all the above surf clubs plus numerous lifeguard towers along Gold Coast beaches

Regional Map 12: Moreton Bay to Rainbow Bay

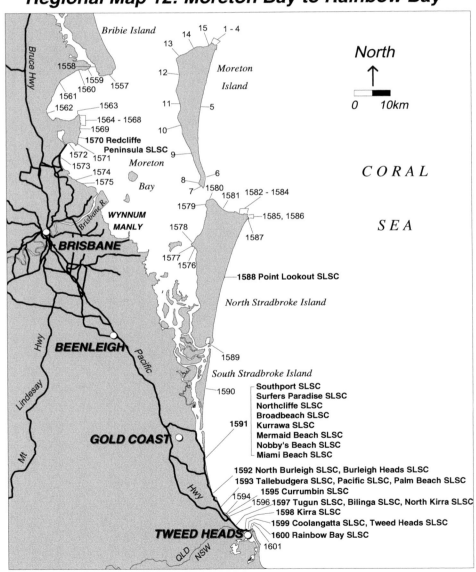

Figure 4.139 *Regional map 12   Moreton Bay, Moreton Island, Stradbroke Island and the Gold Coast. Beaches with Surf Life Saving Clubs and lifeguards indicated in bold.*

MORETON BAY

Moreton Bay is a large, 11 500 km² bay bordered by Moreton and Stradbroke Islands to the east, and the Brisbane River and eastern Brisbane suburbs to the west. The bay is 60 km long and up to 30 km wide. It has three entrances; the small meandering tidal channel in the south that connects with Jumpinpin and Nerang inlets, the 3 km wide South Passage between Moreton and Stradbroke Islands, and the large, 15 km wide North West and North East channels, between Bribie and Moreton Islands. The islands protect the bay from all ocean swell, with only low wind waves dominating the low energy shoreline, while in the bay strong tidal currents work the tidal shoals and channels.

The bay has two halves, the more open northern half and narrower and island-filled southern half. The northern half of the bay is bordered in the north by the southern shores of Bribie Island, and along the western shores by Deception Bay, Redcliffe Peninsula and Bramble Bay which terminates at Brisbane Airport and the mouth of the Brisbane River, while Moreton Island forming the eastern boundary. To the south of the Brisbane River the bay narrows progressively in lee of North and South Stradbroke Island, and becomes increasingly congested with numerous low islands. This section describes the bay beaches between Pebble Beach opposite Bribie Island, down the western shores of the bay to Manly.

1559-1562 PEBBLE BEACH - DECEPTION BAY

No.	Beach	Unpatrolled Rating HT	LT	Type	Length
1559	Pebble Beach	1	1	B+SF	1 km
1560	Godwin	1	1	B+RSF	500 m
1561	Beachmere	1	1	B+RSF	8.5 km
1562	Deception Bay	1	1	B+TSF	2.2 km

Deception Bay is a curving, east facing 10 km wide bay, bordered by Sandstone Point and Bribie Island in the north and Reef Point to the south. It has 24 km of low, low energy shoreline consisting of narrow high tide beaches or mangroves fronted by wide intertidal sand flats, together with several creeks and the Caboolture River at Beachmere. There are four small settlements on the bay shore at the new Pebble Beach subdivision, Godwin Beach, Beachmere and Deception Bay, and the larger Scarborough township on the southeastern point.

Pebble Beach (1559) is the name of a 1990's subdivision located immediately east of Sandstone Point. The 1 km long foreshore consists of a narrow high tide sand and gravel beach and scattered mangroves, fronted by 1 km wide intertidal sandflats. A grassy reserve and bluffs backs the beach, with houses located to the lee of the bluff.

Godwin Beach (1560) is an older 500 m long beachfront settlement located to the west of Sandstone Point. Between the houses and the beach is a road, 30 m wide grassy reserve lined with casuarina trees, then a low sloping seawall and drain and a narrow high tide beach. The beach is fringed by mangroves, and fronted by 1 km wide intertidal ridged sand flats (Fig.139). While there is ample parking and a small boat ramp, there are no facilities at the beach.

Figure 4.139 *Godwin Beach on the northern shore of Deception Bay is typical of the very low energy beach and wide ridged sand flats of Moreton Bay.*

Beachmere is a 3 km long beachfront settlement that occupies a low vegetated spit that terminates at the mouth of the Caboolture River. At the town centre there are a few shops and a large reserve and caravan park. For the most part the beach (1561) is backed by houses, with access via side streets. The beach consists of a narrow high tide strip of sand, fronted by 500 m wide low ridges sand flats. Toward the southern end of the settlement the beach is replaced by a seawall, including a concrete boat ramp.

Deception Bay township is the largest and most rapidly growing settlement on the bay. Most of the settlement occupies gently sloping land that forms 10 m high grassy bluffs near the shore. The main road terminates at a beachfront reserve containing a picnic area and beach facilities, with a high tide boat ramp fronted by channel markers. The beach (1562) consists of a narrow strip of high tide sand and seawall, with intertidal sand flats widening to 1 km in places, and mangroves fringing either end the 2.2 km long beach.

Summary: The entire eastern Deception Bay shoreline consists of wide shallow sand flats, as a consequence of which it is only possible to swim toward high tide. Waves consist of low easterly wind chop with calms common.

1563 SCARBOROUGH

No.	Beach	Unpatrolled Rating		Type	Length
		HT	LT		
1563	Scarborough Jetty	1	1	B+SF	200 m
1564	Scarborough Pt	1	2	R+LT	500 m
1565	Scarborough	1	2	R+LT	500 m

Scarborough is the northern most town on the 7 km long Redcliffe peninsula (Fig. 4.140). The peninsula is for the most part bordered by 10 to 20 m high red lateritic bluffs, which give the peninsula its name. Scarborough is spread over either side of the northern Reef Point. To the west is the Scarborough Boat Harbour and Endeavor Park. Between the northern harbour breakwater and the point is an L-shaped 200 m long low energy, red gravely-sandy high tide beach (**1563**), fronted by 300 m wide rock and sand flats. The beach faces northwest into Deception Bay. The reddish gravel is derived from erosion of the red laterite which also outcrops along the shore as red reefs. The beach is backed by a reserve and a seawall, which continues on around the point.

The Scarborough beaches are located on the east side of the point and face east into Moreton Bay. Between Reef and Scarborough points is a narrow, crenulate 500 m long high tide sand and gravely beach (1564), fronted by a mixture of sand and rock flats. The beach is backed by scarped eroding red lateritic bluffs, with a seawall running along the base of the bluffs and out to Scarborough Point, where it forms the boundary between the northern beach and main southern Scarborough Beach.

The main **Scarborough Beach** (1565) is a 500 m long curving east-facing, low energy high tide sand and gravel beach, fronted by sand and rock (laterite) flats. It is bordered at both ends by seawall-groynes, and crossed toward the south by two storm water pipes, the southern one fronted by rock flats. It is backed by a large grassy foreshore reserve, with toilets, picnic area and playground, a row of Norfolk Island pines, then vegetated bluffs, and the blufftop main road. Shops and a hotel are located on the road behind the centre of the beach.

Summary: The main beach is the only one suitable for swimming, and then only toward high tide, owing to the shallow intertidal rock and sand flats. Waves are usually low wind chop and calms common.

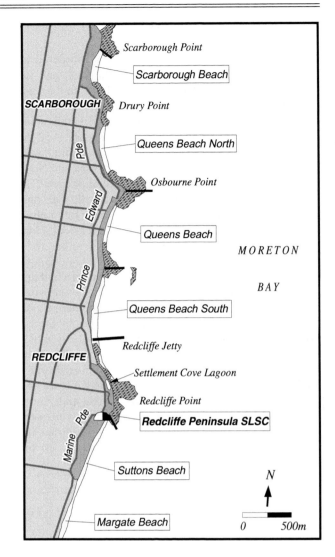

Figure 4.140 *Location map showing the Redcliffe Peninsula beaches between Scarborough and Margate.*

1566-1569 QUEENS BEACH

No.	Beach	Unpatrolled Rating		Type	Length
		HT	LT		
1566	Drury Pt	1	2	R+LT	100 m
1567	Queens North	1	2	R+LT	900 m
1568	Queens	1	2	R+LT	850 m
1569A	Queens South	1	2	R+LT	600 m
1569B	Redcliffe Pier	1	2	R+LT	80 m

Drury Point is a lateritic bluff that separates Scarborough from Queens beach. The eroding bluffs have been protected by a continuous seawall, with groynes located at the southern end of Scarborough Beach and the northern end of Queens Beach North. In between the two groynes and the rock reefs of the point is the 100 m long Drury Point beach (1566).

Queens Beach North (1567) is a curving 900 m long east facing beach which is located between the Drury Point groyne and a more substantial seawall and groyne on the southern Osborne Point. A grassy reserve backs the central-northern section of beach, with a grassy picnic area also on Osborne Point. There is a boat ramp toward the southern end, beyond which the beach narrows and is backed by a seawall. The beach consists of a sandy high tide beach fronted by 200 m wide sand flats, with rocks littering the beach in front of the seawall.

Queens Beach (1568) begins on the south side of the Osborne Point groyne and curves to the south for 850 m to the seawall that surround the lateritic rocks at the mouth of Grants Creek. The sandy beach is backed by a walkway and narrow grassy reserve, then beachfront houses, with vehicle access at either end and via side streets.

Queens Beach South (1569A) is 600 m long sand beach, extending from the Grants Creek seawall to the concrete Redcliffe Pier. The beach consists of a high tide sandy-gravely beach, with lateritic rock along the base of the beach and rock and sand flats off the beach. A shady reserve of Norfolk Island pines and walkway backs the beach, then 20 m high slopes, ontop of which is the main road then the Redcliffe town centre.

The Redcliffe waterfront underwent a major redevelopment during the late 1990's, the focus of which is the rebuilt **Redcliffe Pier** and seaward breakwater, and the Settlement Cove lagoon located between the pier and southern Redcliffe Point. Wedged in on the south side of the pier is an 80 m long pocket of nourished white sand, that now forms a separate small beach (1569B). It is backed by a stepped seawall and walkway. Most people however swim at the lagoon which has excellent shade and all facilities, including pockets of sandy shore.

Summary: The Queens to Redcliffe section contains five low energy high tide beaches, all fronted by shallow sand and in places rock flats, as well as the pier and breakwater and 'lagoon' area.

terminates in the south at Scotts Point. The Redcliffe Peninsula surf club is located at the northern tip of the beach (Fig. 4.141), with a carpark and picnic area on the point. The beach consists of a sandy high tide beach and shallow low tide sand flat. The northern half is backed by a large foreshore reserve called Suttons Beach, with car parking and all facilities for beachgoers. The southern Margate half of the beach is backed by a continuous seawall and walkway, then the main road. Toward the southern end is a boat ramp, then an old dressing pavilion, South of the pavilion a newer seawall has been constructed to protect the backing eroding beach and eroding bluffs of Scotts Point.

Scotts Point is the start of a 1 km long low rocky bluff and reef section that terminates at the southern Woody Point. Along the central northern section is a narrow 200 m long high tide beach (1571), backed by a boulder seawall, walkway and vegetated bluffs, with rocks and reefs littering the intertidal zone.

Summary: Suttons Beach is one of the more open and popular swimming beaches on the peninsula and is patrolled by the Redcliffe Peninsula surf club.

Figure 4.141 *View of Redcliffe Peninsula surf club.*

1570-1571 SUTTONS-MARGATE (REDCLIFFE PENINSULA SLSC)

Patrols **Surf Life Saving Club**: September to May				
No.	Beach	Rating	Type	Length
			HT LT	
1570	Suttons-Margate (Redcliffe Peninsula SLSC)	1	2 R+LTT	2 km
1571	Scotts Pt	1	2 R+LTT	200 m

Redcliffe Point forms the northern boundary of the straight 2 km long Suttons-Margate beach (1570), which

---

1572-1573 CLONTARF

		Unpatrolled		
No.	Beach	Rating HT LT	Type	Length
1572A	Woody Pt	1 1	B+SF	200 m
1572B	Bells Beach	1 1	B+SF	150 m
1572C	Clontarf	1 1	B+SF	300 km

Woody Point jetty was the gateway to the Redcliffe Peninsula before the first bridge was constructed. The new concrete jetty is now used for fishing, and in backed by the Woody Point main street and shopping centre. The low rocky point forms the northern boundary of 10 km wide Bramble Bay, a low energy east facing bay that terminates at Brisbane Airport. Between Woody Point and Hays Inlet is a curving south facing 2 km long shoreline, that originally had a continuous sandy beach, fronted by 200 to 300 m wide sand flats. Today seawalls dominate the shoreline replacing the beach apart from three smaller sandy sections. The first is **Woody Point Beach** (1572A) a crenulate 200 m long southwest facing sand and gravely high tide beach, fronted by the sand and rock flats. It is bordered by the seawall, but backed by a large reserve with a range of facilities for the beach users, together with a large car park and the main shopping area.

One kilometer to the west is a second gap in the seawall that contains 200 m long **Bells Beach** (1572B). This consists of a narrow high tide beach backed by a grassy and casuarina-shaded 30 m wide reserve then the main road, with a caravan park across the road. The beach terminates at the protruding Pelican Park. The park is a major mid 1990's reclamation project that provided landfill for the 500 m long park which contains a range of recreational and picnic facilities, including a boat ramp at the western end. Immediately west of the park is **Clontarf Beach** (1572C) a 400 m long high tide beach, with a backing grassy reserve and main road.

Summary: These are three very low energy beaches only suitable for swimming at high tide, with the best facilities at Woody Point beach and in Pelican Park.

1573-1575 BRIGHTON, SANDGATE, NUDGEE

No.	Beach	Unpatrolled Rating		Type	Length
		HT	LT		
1573	Brighton	1	1	B+SF	80 m
1574	Sandgate Pier	1	1	B+SF	200 m
1575	Nudgee	1	1	B+SF	400 m

Brighton is a bayside suburb that was originally fronted by a 4 km long beach that extended from the mouth of Hays Inlet down to Sandgate. Like much of the bay shore however the beach has been replaced by a continuous seawall. More recently a small artificial beach has been constructed at the main northern entrance to the beach. The present **Brighton Beach** (1573) consists of an 80 m long strip of east facing sand, held in place by rock groynes. It is backed by a small reserve containing shade and picnic and bar-b0que facilities, with 400 m wide sand flats off the beach. Toward the Sandgate end is a swimming pool in lee of the wall.

Sandgate Pier is a 300 m long jetty, to the south of which is a 200 m long beach (1574). The beach is wedged in between the pier and backing seawall, and a southern

200 m long groyne. It is fronted by shallow tidal flats out to the end of the groyne, and backed by a low seawall and shady park with picnic facilities and adjacent car park, then the slopes of Cabbage Tree Head.

Nudgee Beach (1575) is located out on a series of low beach ridges that formed to the north of the Brisbane River mouth. A row of houses occupies the second ridge, with a wide grassy foreshore reserve occupying the first. The beach has largely been replaced by a seawall and high tide boat ramp fronted by 500 m wide sand flats, with only small pockets of beach remaining at high tide, together with some large mangroves. The large reserve is however a popular picnic area and has a small boat ramp.

Summary: Three low energy beaches largely replaced by seawalls, with Sandgate Pier beach offering the best swimming and facilities.

WYNNUM-MANLY

No.	Beach	Unpatrolled Rating		Type	Length
		HT	LT		
-	Wynnum	1	1	B+SF	100 m
-	Manly	1	1	B+SF	200 m

Wynnum is the first bayside suburb south of the Brisbane River mouth. Its foreshore originally consisted of a 2 km long low energy beach and sand flats. A seawall backed by the large Wynnum foreshore reserve however has replaced the beach. The reserve has the full range of facilities including an oval-shaped 100 m long pool-lagoon. Just in front of the pool is a 100 m long artificial beach and foredune formed with nourished sand between two groynes.

The suburb of **Manly**, whose beach has been replaced by the large Manly Boat Harbour, backs the next bay south. The only remnant of the former beach is a 150 m long strip of sand located between the main road and the northern wall of the boat harbour. This is a very low energy beach, often covered with seaweed, and used more by windsufers than swimmers. The beach is backed by a grassy foreshore reserve called Bayside Park.

Summary: Wynnum and Manly beaches are two remnants of once longer beaches, which have largely been replaced by seawalls and the boat harbour.

No.	**Unpatrolled** Beach	Rating	Type	Length
MOR1	Honeymoon Cove	7	TBR+rocks	50 m
MOR2	Tallallebela	7	TBR+rocks	100 m
MOR3	Cape Moreton (W 2)	6	LTT/TBR	50 m
MOR4	Cape Moreton Moreton (W 1)	5	LTT	100 m

The northern tip of Moreton Island consists of a 1.5 km section of high bedrock between 20 m high North Point and 120 m high Cape Moreton. Lighthouses are located on both headlands and the whole area is Commonwealth land. This ridge of bedrock is the only rock on the large and otherwise sandy island. Between the two headlands is an open, north-east facing, predominantly rocky bay, within which are four small pocket beaches. All four are backed by steep cliffs and bluffs rising in places over 60 m. There are steep access tracks down to each of the beaches.

Honeymoon Bay (MOR1) lies immediately south of North Point and is backed by 15 m high bluffs, that permit easy access across the base of the point from backing Yellow Patch Beach. The beach is 50 m long and wedged in between rock platforms, together with some large rock outcrops on the beach. It faces east and receives waves averaging 1 to 1.5 m. These maintain a small high tide beach fronted by a small inner bar, cut by rips against the rocks, then a trough and an outer bar, also cut by the rips 100 m off the beach.

Tallallebela beach (MOR2) lies in the centre of the rocky bay. It consists of a 50 m long high tide beach at the base of 80 m high bluffs, fronted by a low tide beach up to 100 m long, bordered by exposed rock platforms and steep cliffs. Access is via the steep backing bluffs. The beach faces north-east and receives waves averaging over 1 m, which maintain a trough along the beach, usually drained by one rip against the northern rocks.

West of **Cape Moreton** are two small beaches. **MOR3** is a 50 m long, north facing pocket of low tide sand, with waves reaching the base of the backing 60 m high bluffs at high tide. Waves average less than 1 m, but are usually sufficient to maintain a strong rip against the boundary rocks. **MOR4** lies immediately west of the base of the cape and is a slightly more protected, north-west facing, curving strip of low tide sand backed by a mixture of boulders and exposed bedrock of the 50 m high bluffs. The beach is fronted by a wider, relatively shallow bar, which also receives pulses of sand moving around the cape.

Moreton Island

Moreton Island is the fourth largest sand island after Fraser Island, Cooloola 'Island' and Stradbroke Island. It is 38 km long and up to 9 km wide, with an area of 600 km^2. It is composed of a massive, multiple series of nested parabolic dunes, that reach a maximum height of 280 m at Mount Tempest. Like Fraser and Cooloola, it is also anchored in the north-east by a bedrock headland, 60 m high Cape Moreton, which is the only bedrock on the entire island. The island has three shores. The eastern shore faces due east into the southerly waves and winds and is a high energy shore backed by the massive dune systems. The western shore faces into Moreton Bay, is protected from all ocean swell and has low energy sandy beaches, some fronted by extensive sand flats. The main settlement of Tangalooma is located in the centre of the western shore. The 10 km long northern shore faces west-north-west and receives sufficient ocean swell to push occasional sand waves and long sandy spits along the generally low energy shoreline. All the island and island shore is sand, apart from the 2 km of rocky shore between North Point and Cape Cliff, including the central Cape Moreton.

Access to the island is by car ferry or plane and then by 4WD along the network of sandy tracks and beaches. The island has 87 km of shoreline, containing the 31 km long eastern beach, an equally long western beach, that can be divided into 6 beaches, and the north beach.

Mount Tempest National Park covers most of the island, apart from the Cape Moreton area, a strip along the northern half of Braydon to Jason Beaches and small areas around the settlements at Kooringal, Tangalooma, Cowan and Bulwer.

Swimming: These are four difficult to access, hazardous beaches dominated by rips and rocks. The cape beach has the lowest waves, but even here there can be a strong current along the beach, as well as rocks in the surf.

Surfing: The sand waves off the cape beach can produce long right-handers during moderate to higher swell, while Honeymoon Bay has a beach break over the bar.

Fishing: All four beaches offer both deep permanent rip and rock holes, as well as vantage points for rock fishing.

Summary: Four little pockets of rock-bound sand on an otherwise sandy island.

Moreton Island-east coast (ocean beaches)
MOR5 BRAYDON-SOVEREIGN

No.	**Unpatrolled** Beach	Rating	Type Inner Outer Bar	Length
MOR5	**Braydon - Sovereign**			
MOR5A	Braydon	6	TBR RBB/LBT	3.3 km
MOR5B	Spitfire	6	TBR RBB/LBT	2.5 km
MOR5C	Warrajamra	6	TBR RBB/LBT	3.5 km
MOR5D	Eager	6	TBR RBB/LBT	5 km
MOR5E	Jason	6	TBR RBB/LBT	5 km
MOR5F	Gonzales	6	TBR RBB/LBT	4 km
MOR5G	Toompawi	6	TBR RBB/LBT	3 km
MOR5H	Sovereign	6	TBR RBB	5 km
			Total length	31.3 km

The exposed eastern side of Moreton Island runs for 38 km from rocky North Point down to elongate, sandy Reeders Point. In between are the rocks of the cape, then a 31 km long, slightly curving, high energy, east facing beach that extends from the south side of the cape down to the beginning of the large South Passage tidal channel. The beach then continues on for another 6 km inside the passage to Reeders Point, with increasing protection afforded by the offshore tidal shoals.

The long **Braydon-Sovereign** beach (Fig. 1.1a) faces directly into the prevailing south-easterly waves and winds. The entire length consists of a relatively low gradient high tide beach fronted by a continuous inner bar, cut by rips every 200 to 250 m (Fig. 3.2b), then a shore parallel trough and the more continuous outer bar, lying up to 200 m off the beach. The entire beach is backed by a continuous, low, spinifex-covered foredune, then varying degrees of active sand dunes that are widest in the north, where they average 500 m, but decrease in size and extent to discrete blowouts toward the south. Beyond the active dunes are the massive older, vegetated parabolic dunes of the island. In amongst the older dunes are numerous blocked and perched lakes, particularly behind the northern 10 km of beach.

Braydon beach (MOR5A) runs south-south-west of the cape for 3.3 km. It is accessible from the backing Yellow Patch Beach, either over the dunes or via the cape, or up the beach from other tracks. The beach is rip dominated, with a permanent rip against the rocks of the cape.

Spitfire beach (MOR5B) is 2.5 km long and incorporates the small Spitfire Creek and backing lagoon, that drains across the beach. It ends at a small, southern unnamed creek.

Warrajamra beach (MOR5C) is accessible via the main northern 4WD track from Bulwer. The 3.5 km long section is backed by the 40 ha Blue Lagoon and the Blue Lagoon camping area.

Eager beach (MOR5D) is a 5 km long beach section centred on the small Eager Creek, that drains the backing Eager Swamp. This creek also has a popular camping area. The main beach access track from Tangalooma, 5 km to the west, reaches the southern end of this beach section.

Jason beach (MOR5E) extends almost due south from the Tangalooma track. It is backed by increasingly stable dunes, but is still fronted by the rip dominated inner and outer bars.

Gonzales beach (MOR5F) runs due south for 4 km and incorporates the solitary White (or Camel) Rock.

Toompawi beach (MOR5G) is a 3 km long, east facing section that incorporates Rous Battery. It ends at the beginning of the increased dune activity of the Little Sandhills.

Sovereign beach (MOR5H) is the final 5 km long section that runs down to the beginning of the South Passage. The southern few kilometres begin to swing to the south-east in lee of the extensive tidal shoals of the passage, while the active Little Sandhills back the beach.

Swimming: This entire 31 km long beach is dominated by up to 150 rips across the inner bar, together with the deep longshore trough and outer bar. Use extreme care if swimming along the unpatrolled beach. Stay close inshore on the attached portion of the inner bar and clear of the rips.

Surfing: There are literally hundreds of beach breaks along the length of the beach, with conditions depending on the swell, tide and winds.

Fishing: A very popular beach for fishing the rip holes; the reason the majority of campers come over to the island.

Summary: A magnificent, long, natural beach, relatively

safe for driving at low tide, with camping spots at Blue Lagoon and Eager Creek and many kilometres of sand and surf.

Moreton Island –southern end
MOR6 SOUTH PASSAGE

	Unpatrolled			
No.	Beach	Rating	Type	Length
MOR6	South Passage	3	R+tidal channel	6 km

South Passage is one of the two main connections between Moreton Bay and the ocean and each day, massive volumes of water flow through the 3 km wide passage. On the bay side the currents have built a 150 km² flood tidal delta, while on the ocean side the impact of the waves has reduced the ebb tidal delta to 50 km². These still extensive ebb tidal shoals and the 7 km long channel have a tremendous impact on the southern end of the island. First the waves must break over the shoals and fight the currents, substantially reducing their height at the beach; and second, movement in the position of the shoals and channel result in realignment of the beach shoreline, at times building out hundreds of metres and impounding backing lagoons, at other times being cut back to the dunes. Consequently, the 6 km long South Passage Beach (**MOR6**), while relatively low wave energy, is very dynamic over a period of years to decades and is always paralleled by strong tidal currents. The beach ends at Reeders Point, the southern tip of the island. On the western side of the point begin the protected, low energy bay beaches.

Swimming: A low wave energy, but potentially hazardous beach owing to the deep channel and strong tidal currents. Best not to swim off the beach.

Surfing: Usually none.

Fishing: Most popular with boat fishers, but you can also fish the inner channel from the shore.

Summary: The dynamic southern end of the island.

Moreton Island-west coast (bay beaches)
MOR7-12 MAYS HOLE-TANGALOOMA-BULWER

	Unpatrolled			
No.	Beach	Rating	Type	Length
MOR7	Mays Hole	1	B+SF	1.1 km
MOR8	Kooringal	1	B+SF	1.2 km
MOR9	Kounungai	1	B+SF	3.7 km
MOR10	Square Patch	1	B+SF	4.5 km
MOR11	Tangalooma	1	B+SF	8 km
MOR12	Bulwer	1	B+SF	8.5 km

In contrast to the exposed eastern shore of the island, the west facing bay shore is protected from both the ocean waves and dominant easterly winds. Consequently it consists of low energy sand beaches, backed by the vegetated dunes that originated in the east and fronted by sand flats and, in places, tidal channels. The three small island settlements are all located on this side, at Kooringal in the south, the central Tangalooma and northern Bulwer. The ferry 'Moreton Venture' discharges vehicles near Kooringal and at Tangalooma. The 36 km of bay shoreline contain six, near continuous beaches, totalling 27 km in length, with a 9 km section of mangroves between Kooringal and Kounungai beaches.

Mays Hole beach (MOR7) lies immediately west of the southern Reeder Point and runs to the north-west for 1.1 km to low, sandy Campbell Point. The beach receives no ocean swell and is usually calm, with only low wind waves at high tide. The beach is, however, fronted by 100 m wide sand flats, then the deeper water of Mays Hole, then a tidal channel called Days Gutter 300 m off the beach. Shifts in the position of the tidal shoals and gutters result in a shifting shoreline. During the 1990's the beach was eroding, with the backing trees falling down a 10 m high scarp onto the beach.

Kooringal beach (MOR8) fronts the small southern settlement of the same name. The 1 km long beach faces west across the bay. It is fronted by seagrass-covered sand flats, the northern end of Days Gutter, then 4 km of shallow tidal delta, including low, sandy Crab Island. Conditions are usually calm, with only wind waves during westerly conditions at high tide. The ferry lands vehicles at the southern Campbell Point, with 4WD tracks leading to Kooringal and the South Passage Beach.

Kounungai beach (MOR9) runs from the northern end of the bare Little Sandhills for 3.7 km to an inflection in the shore. It faces west-south-west and is a relatively remote part of the island, owing to the absence of backing tracks and the difficulty in driving the narrow beach. The rear of the 80 m high Big Sandhills backs the central portion of the beach.

Square Patch refers to a 60 m high bare patch of otherwise vegetated dune escarpment 1.5 km south of Tangalooma Point. The beach (MOR10) extends for 4.5 km south of the point to the inflection with Kounungai Beach. The beach faces west to south-west and is fronted by a narrow high tide beach and 200 m wide, ridged sand flats, that are only active during strong westerly winds and waves.

Tangalooma beach (MOR11) lies on either side of the main island settlement and entry point. The 8 km long, west facing beach curves gently between the southern Tangalooma Point, past the 1 km long settlement with its two jetties, up to Cowan Cowan Point, site of the main island airstrip. Much of the beach is backed by a 50 to 100 m high, vegetated dune escarpment, with Tangalooma lying at the base of the dunes. The beach consists of a narrow high tide beach, fronted by 100 m wide sand flats, then for much of its length a deep channel called the Tangalooma Road. Two jetties cross the sand flats to reach the channel.

Bulwer beach (MOR12) is a slightly curving, west facing beach that runs for 8.5 km between Cowan Cowan Point and the sandy north-western tip of the island, Comboyuro Point. The small settlement of Cowan Cowan is located just north of the southern point, while the equally small Bulwer settlement is located 2 km south of the northern tip. The houses lie in amongst the dense vegetation. The beach is usually calm, with a low high tide beach fronted by 100 m wide sand flats, then a continuous, deep tidal channel. Two rock groynes on the beach define the Bulwer boat launching area. The beach and two settlements are backed by a continuous swamp up to several hundred metres wide, before the higher dunes of the island are reached.

Swimming: All the bay beaches are usually calm and relatively safe, apart from deep water off some of the beaches and a chance of tidal currents. Seek local advice before venturing too far offshore.

Surfing: None.

Fishing: Excellent fishing for the length of the shore, particularly off the shore in the deeper channels and over the seagrass patches.

Summary: The quieter bay side of the island, home to all the island's small settlements and entry points for those coming by ferry or plane.

Moreton Island-north coast
MOR13-15 **KIANGA-YELLOW PATCH**

	Unpatrolled			
No.	Beach	Rating	Type	Length
MOR13	Kianga Channel	2	R+	2.5 km
MOR14	Heath Island	3	LTT	6.0 km
MOR15	Yellow Patch	4	LTT	2.3 km

The northern shore of Moreton Island faces north-west and extends for 11 km between low, sandy Comboyuro Point and 20 m high, rocky North Point. While this is a lee shore and waves average less than 1 m, decreasing to the west, it is nonetheless a very dynamic shoreline, with the low, sandy Heath Island comprising two crenulate sand spits.

Kianga Channel beach (MOR13) parallels the smaller tidal channel for 2.5 km between Comboyuro Point and the beginning of Heath Island. The beach receives low swell averaging less than 0.5 m and usually has a steep, barless beach, with the deeper water of the channel close inshore. Low, densely vegetated dunes back the beach, while a small creek, with a small backing lagoon, occasionally flows across the centre of the beach.

Heath Island (MOR14) is a convex, crenulate, low sand island that protrudes up to 1 km north of the backing vegetated shore of the main island. In between is an extensive shallow lagoon, drained by a creek across the beach at the northern tip of the protrusion. Sand waves move from east to west along the island shore, causing substantial oscillations in shoreline position and character. Generally, ocean waves average less than 1 m in the east, decreasing to the west. When the swell is up, the waves run westward along the beach, pushing the sand and sand waves to the west.

Yellow Patch (MOR15) is a north-north-west facing beach curving between the eastern end of Heath Island and North Point. Like Heath Island it is a low energy, but very dynamic beach, owing to pulses of sand that move around North Point and are pushed by the waves along the beach. These sand waves at times form elongate spits, backed by temporary lagoons, while in between, the backing vegetated dunes may be eroded and scarped. There is a small collection of shacks and houses in the low vegetated dunes at the eastern end of the beach.

Swimming: Usually relatively safe, apart from the deep water off Kianga. When waves exceed 1 m however, a strong westerly current runs along Yellow Patch and Heath Island Beaches, with occasional rips.

Surfing: During higher swell there can be a good right-hander over the sand waves off *Yellow Patch*.

Fishing: Most fishing is done in boats off the beach, however there is some deep water off the rocks at North Point and at the Heath Island creek mouth.

Summary: A lower energy but still dynamic section of the island shore. Be careful driving these beaches, owing to the variable beach width and creek crossings.

North Stradbroke Island

North Stradbroke Island is the second largest of Queensland's sand islands and its most accessible of the big dune islands. The island forms the south-eastern side of Moreton Bay and its north-east tip lies just 40 km from downtown Brisbane. The island is serviced by car and passenger ferries from Cleveland and Redland. There are three main settlements on the island, at Durwich on the bay side, site of the ferry terminal, at Amity on the north-west tip and at Point Lookout on the north-east tip. Point Lookout is also the site of the main surfing beach and the Point Lookout Surf Life Saving Club (Fig. 2.5a). There are roads linking the three settlements, with sandy 4WD tracks linking the settlements to the long ocean beach.

The island is 35 km in length and up to 12 km wide, with an area of 260 km². It consists of a series of large, densely vegetated sand dunes that average 100 m in height, with sandy Mount Hardgrave reaching 218 m. Individual dunes extend up to several kilometres inland. The island has three distinctive coasts. The east to south-east facing ocean coast receives persistent south-east swell and is dominated by a 33 km long beach, tied at its north end to 40 m high Point Lookout. The north shore receives reduced swell and has lower energy beaches, while the western bay shore is quiescent, with generally low bay waves and beaches fronted by sand flats and tidal channels.

North Stradbroke Island-west coast (bay beaches)
1576-1580 **ADAMS-AMITY PT**

No. Beach	Unpatrolled Rating	Type	Length
1576 Adams	1	B+SF	700 m
1577 Dunwich	1	B+SF	200 m
1578 One Mile	1	B+SF	500 m
1579 Amity	1	B+SF	1.3 km
1580 Amity Pt	1	B+SF	2 km

The western side of North Stradbroke Island faces into the southern half of Moreton Bay. The sandy shore receives no ocean swell, except at the very north-western entrance. Only at the protruding Dunwich and along the north-western Amity shore are the bay waves sufficiently high to maintain low energy sandy beaches fronted by tidal flats and/or tidal channels. As the bay narrows south of Dunwich, the low energy conditions permit mangroves to dominate the shore, while between Dunwich and Amity extensive sand flats and tidal shoals extending 12 km into the bay also maintain calm conditions and a mangrove-fringed shore. In all, there are five bay beaches between Dunwich and Amity Point.

Adams beach (1576) is a 700 m long strip of high tide sand facing south-west into Moreton Bay, located on the southern side of Dunwich. The beach is calm, except during south-westerly wind conditions, and is fronted by seagrass covered sand flats that widen from 100 m in the north to 300 m in the south. The southern end of the beach grades into a mangrove-fringed shore. The northern end of the beach is bordered by the port facilities of Dunwich, including the conveyor belt for the mined sand and jetty for the ferries and shipping.

Dunwich beach (1577) lies in the centre of the Dunwich foreshore and is bordered by the two jetties. It is just 200 m long, faces south-west and is fronted by 200 m wide sand flats, that necessitate the long jetties. The beach is backed by a wide, grassy foreshore reserve, then the town centre.

One Mile beach (1578) forms the northern shore of Dunwich. It is a 500 m long, low energy strip of high tide sand, including a few scattered mangroves, fronted by 1 km wide sand flats. The beach is backed by a foreshore reserve, then houses.

Amity is located on the north-western tip of the island and borders the transition from bay to ocean shore. The main settlement extends along 1 km of the west facing bay shore. Unfortunately, the settlement was located too close to the shore and natural retreat of the beach has caused a 500 m long section to be replaced by a rock seawall. This seawall divides the previously sandy shore into two beaches.

Amity beach (1579) lies south of the seawall and extends for up to 1.3 km, first as a low, crenulate beach backed by the caravan park, then as a low, narrow sand spit that can make up half the beach length. The spit is formed by flooding tidal currents in Rainbow Channel. There is a boat ramp at the northern end of the beach and a jetty across the beach, as well as two rock groynes.

Amity Point beach (1580) extends from the north side of the seawall for 2 km to the low, sandy Amity Point, which forms the northern boundary of the tidal Rainbow Channel. To the east of the point, ocean waves regularly reach the shore. Immediately north of the seawall, the narrow beach is backed by a few houses, with makeshift seawalls, then a broad, wooded reserve. The beach is usually calm, except during very high swell conditions, when low waves move around the point. However the Rainbow Channel carries strong tidal currents just off the beach, which no doubt contribute to the beach erosion.

Swimming: These are five usually calm beaches, with the best swimming at the main Dunwich Beach between the jetties and One Mile, while at Amity the beach in front of the caravan park is relatively safe, so long as you do not swim out into the deep tidal channel with its strong currents. Strong currents also parallel the usually calm Amity Point Beach.

Surfing: None.

Fishing: Most people come over to Dunwich and Amity to fish and relax. They fish from all vantage points into the deep tidal channel, including the jetties, groynes and off the beaches.

Summary: The quieter western side of North Stradbroke and location of two of the main settlements, that are fronted by usually calm beaches.

North Stradbroke Island-north coast
1581-1584 **FLINDERS, CYLINDER, HOME**

No.	Beach	Rating	Type	Length
Unpatrolled				
1581	Flinders	3	R/LTT+sand waves	8.0 km
1582	Rocky Pt	3	LTT+sand waves	1.7 km
1583	**Cylinder**	3	LTT+sand waves	500 m
Lifeguard September to May				
1584	Home (Deadmans)	3	TBR	600 m

The northern shore of North Stradbroke Island faces essentially due north and runs for 13 km from low, sandy Amity Point to 25 m high Point Lookout, the north-east tip of the island. In between are four sandy beaches, the eastern three bordered by rocks and headlands, together with the growing town of Point Lookout, which now backs about 3 km of the north-eastern corner of the island.

Flinders beach (1581) is an 8 km long, north-east to north facing beach, bordered by Amity Point and extensive tidal shoals in the west and the 20 m high Rocky Point in the east. Between the two points, the shoreline has built out 1 to 2 km over the past few thousand years as a series of foredune ridges and spits, backed by an extensive swamp that is drained by two small creeks, one in the centre and one just west of Rocky Point. There is vehicle access to the beach toward the western end along the Flinders Beach Road, which also leads to a camping area and car park, and in the east at Adder Rock, where there is also a beachfront caravan park. The beach has a relatively low gradient and is either barless or fronted by a narrow continuous bar, particularly toward the east. Seaward of the bar, sand moves from east to west in the form of elongate sand waves and bars, which result in a highly variable outer surf zone. During high swell a strong westerly drift runs along and off the beach.

Rocky Point beach (1582) begins on the east side of the point and extends for 1.7 km to a vegetated, 30 m high headland capped by the Point Lookout hotel. The beach faces north-north-east toward Shag Rock, 1.5 km offshore. It receives waves averaging 0.5 m, but these can be considerably higher during big swell. It has a wide, low gradient high tide beach fronted by a continuous low

tide bar, with slowly moving sand waves extending up to 200 m offshore. As the sand waves change over time, so too does the nature of the shoreline, bars and surf. There is access to the beach at Rocky Point, with some access tracks along the beach, and in the east below the hotel.

Cylinder beach (1583) is the most popular swimming beach on the island. The 500 m long, north facing beach is located between the hotel headland and 30 m high Cylinder Head. It is backed by a continuous, shady foreshore reserve, including a picnic and camping area, and then the main Point Lookout Road and settlement. Two small creeks also drain across the beach. The beach receives waves that have been refracted around Point Lookout and average 0.5 to 1 m in height. The lower waves and fine sand maintain a usually wide, low gradient beach fronted by a continuous bar, with the mobile sand waves and bars extending up to 200 m off the beach. As a consequence, like all the north side beaches, beach, bar and surf conditions change considerably over time. The beach is patrolled during the summer and holiday periods.

Deadmans beach (1584) lies on the north-eastern tip of the island, between Cylinder Head and Dune Rocks. It is also known as Home Beach. The 600 m long beach faces due north. It is backed by 20 to 30 m high vegetated bluffs, that are being covered by bluff-top housing. There is a lookout and car park on Cylinder Head and a walking track down to the centre of the beach. The beach is bordered by the points and rocky shore platforms, together with rocks in the centre. Waves averaging 1 m move around Point Lookout and, together with the sand, they move around to maintain a highly variable, 100 m wide surf zone, usually with a deep trough between the beach and outer bar/sand waves.

Swimming: The north shore beaches offer generally low to moderate waves along a dynamic beachfront. The safest swimming is at the patrolled Cylinder Beach. Be careful off all the beaches as there is highly variable bar and trough topography associated with the migrating sand waves, together with often strong westerly drift and rips, when waves exceed 1 m.

Surfing: There are several excellent surfing spots along this section of coast, all of which usually require a larger swell to operate and also a suitable configuration of the sand bars. The main locations are *Cylinders*, off the headland, *Point Lookout* off Deadmans and, during bigger swell, at *Rocky Point*. All can offer low right-handers.

Fishing: The north side is very accessible for beach fishing, usually with a range of holes and gutters associated with the bars and sand waves, as well as a few small creek mouths.

Summary: The northern sun-drenched side of North Stradbroke and site of the largest settlement at Point

Lookout offers north facing beaches of low to moderate energy, with good access to all and accommodation and facilities nearby.

Stradbroke Island-Point Lookout beaches
1585,1586   FRENCHMANS BAY, NORTH GORGE

Unpatrolled			
No. Beach	Rating	Type	Length
1585 Frenchmans Bay	5	TBR	500 m
1586 North Gorge	8	TBR	30 m

South of Dune Rocks are 1.5 km of rock-dominated shore that end at 40 km high Point Lookout. There are three beaches along this section; the longer Frenchmans Bay, backed by 50 to 70 m high bluffs, and two small, rock-bound beaches located in gorges in the cliff face.

Frenchmans Bay beach (1585) faces due east, receiving a little protection from the prevailing south-east waves from Point Lookout, which protrudes 500 m toward the south-east. The beach is 500 m long and is backed by steep, densely vegetated bluffs rising to 70 m, along the crest of which is the main road to the point and main beach. Access to the beach is either around Dune Rocks from Deadmans Beach, or down steep walking tracks from the main road. The beach receives waves averaging between 1 and 1.5 m, which maintain an inner bar usually cut by two rips, including a permanent rip against Dune Rocks. At Dune Rocks, sand blows off the beach and across the point to the back of Deadmans Beach.

North Gorge lies out on Point Lookout and is located in a 50 m wide, 40 m deep, 300 m long gap in the rocks. The small sand beach (**1586**) is wedged deep in the gorge, with access via a walkway from the main road. Rock platforms border each side of the gorge and a sand bar of sorts is usually located toward the entrance. While waves are relatively low by the time they break over the bar and squeeze into the gorge, this is a very hazardous place to swim in the surf. Waves build water up against the beach and then pulse seaward every minute or so, forming a rip that drains the entire gorge.

Swimming: These are two hazardous beaches, suitable for sunbathing, but be very cautious if swimming off the beaches as both are rip-dominated, with the Gorge particularly hazardous.

Surfing: Frenchmans Bay can have good beach breaks over the bars when southerlies are blowing offshore.

Fishing: Both are popular fishing locations owing to the rips and rocks at Frenchmans, and the rather hazardous rock platforms at the Gorge. Rock fishers should use extreme care at both locations as there have been severe drownings.

Summary: Two exposed and very different beaches connected by a scenic walkway around the point.

Stradbroke Island-east coast (ocean beaches)

1587,1588   POINT LOOKOUT (SLSC)
SOUTH HEADLAND
POINT LOOKOUT
BLUE LAKE

Patrols:				
Surf Lifesaving Club:	September to May			
Lifeguard on duty:	School holidays			
No.	Beach	Rating	Type	Length
			Inner Outer Bar	
1587	South Headland	6	TBR RBB	50 m
1588A	**Pt Lookout**	6	TBR RBB/LBT	6 km
1588B	Blue Lake	6	TBR RBB/LBT	27 km
1589	Jumpinpin Inlet	3	R+tidal channel	1 km

Point Lookout is the name of the headland that forms the north-easternmost tip of North Stradbroke Island. It is also the name of the settlement that straddles the north-eastern tip of the island and of the surf lifesaving club, established in 1947, that patrols the beach immediately south of the headland.

The surf club is built on smaller South Headland (Fig. 2.5a), which is separated from Point Lookout by a 50 m long pocket of sand (Fig. 4.142). Both this small beach and the longer beach on the south side of the club are patrolled by the lifesavers. Both beaches are exposed to swell averaging 1.5 m. **South Headland** beach (1586) is a popular beach owing to its protection from most winds and calmer inshore waters. However in the surf and off the beach there is a strong permanent rip, as well as rocks to either side (Fig. 4.143).

The main **Point Lookout** beach (1588A) runs south for 33 km to Jumpinpin Inlet. The waves maintain a 200 m wide double bar system, with the inner bar usually attached to the beach and cut by deep rip channels every 200 to 300 m. A deep trough parallels this bar, with the outer bar on its seaward side cut by more widely spaced rips (Fig. 3.2a). The beach continues like this all the way to Jumpinpin Inlet, only changing its name to **Blue Lake** beach (1588B) 6 km south of Point Lookout. On an average day, up to 150 rips will be operating along the beach (Fig. 2.7b). In addition, there are permanent rips running out of the South Headland pocket beach and the northern end of Point Lookout Beach, below the surf club, together with a rock reef in the rip channel.

At **Jumpinpin Inlet**, the beach (1589) is impacted by the tidal channel and extensive shoals that extend up to 2 km off the inlet. The beach swings around to south-east, then south, facing across the 300 m wide inlet to

South Stradbroke Island. The inlet shoals protect the beach from most ocean waves, and wave height decreases into the inlet. As a consequence, the inlet beach is usually narrow and steep, fronted by deep water of the inlet, with waves breaking offshore on the tidal shoals.

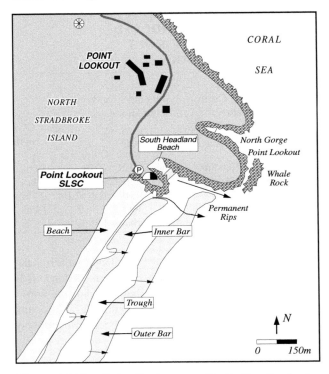

Figure 4.142 *Point Lookout on North Stradbroke Island is the site of the Point Lookout SLSC. It overlooks a 32 km long, energetic beach that runs the length of the island.*

Figure 4.143 *Point Lookout Surf Club and car park are located on a rocky headland overlooking both the small South Headland Beach and the 32 km long beach that extends south of the club house.*

Swimming: These are potentially hazardous beaches and swimmers must be very careful when entering the surf. The small South Headland Beach is relatively safe inshore, however if you move into deeper water you will encounter the strong permanent rip. The main Point

Lookout Beach has the strong northern rip, together with rips down the beach. In all cases only swim in the patrolled area. Down the beach, stay well clear of rips and stay on the inner portion of an attached bar, while at the inlet be careful of the strong tidal currents in the channel.

Surfing: Point Lookout is a popular surfing beach, with good breaks over the outer bar and easy access via the northern rip channel. It works best on a low to moderate swell and north to north-east winds.

Fishing: A very popular beach for both beach fishing into the rip holes and off the rocks of South Headland and Point Lookout. However be careful as the rocks are steep and are washed by large waves. Many beach fishers can be found with their 4WDs down the beach. There is a vehicle beach access track about 500 m south of the point.

Summary: This is North Stradbroke Island's main surfing and fishing beach. It is potentially hazardous for swimming, so take care if entering the surf.

South Stradbroke Island
1590 SOUTH STRADBROKE

No.	Beach	Rating	Type		Length
			Inner	Outer bar	
1590	South Stradbroke	6	TBR	RBB/LBT	21 km
	Lifeguard: Couran Cove Resort Beach				

South Stradbroke Island is a 21 km long, relatively narrow island between Jumpinpin Inlet and the training walls of the Gold Coast Seaway at Nerang. The island is less than 1 km wide for most of its length, widening to 2.5 km along the central section. Unlike its northern neighbour, the entire island is less than 20 m high, consisting of a high frontal foredune complex backed by lower swampy land. There has been a substantial dune stabilisation program on the island that has succeeded in vegetating and stabilising much of the foredunes. There are two settlements on the western side of the island, in the north at Tipplers resort and in the south at the older Currigee settlement, that has a small boat harbour. Access to the island is by boat, with a 4WD required to travel the island's sandy tracks and beach.

The beach (**1590**) faces due east and runs relatively straight between the two inlets. It receives waves averaging 1.5 m, which maintain an energetic, rip dominated double bar system. The inner bar is cut by rips every 200 to 250 m, with up to 100 rips operating along the beach. These rips flow into the longshore trough, then out across the outer bar.

Swimming: This is a relatively isolated, rip dominated

beach, so use caution if swimming anywhere along the beach. The inlets are calmer but contain deep channels and strong tidal currents.

Surfing: The north side of the *Nerang* training wall is a very popular, but highly variable, location, with surfers either paddling across the entrance from The Spit or arriving in small boats. Since the walls were constructed in the 1980's, the southern tip of the island has stabilised and built seaward, as sand from the south side is continuously pumped under the seaway and piped onto this beach,

producing a still growing and mobile shoreline and bars.

Fishing: The inlets and many beach rips provide a wide range of locations for shore based fishing.

Summary: A relatively natural island, right on the doorstep of the great Gold Coast and a nice contrast to its highly developed southern neighbour.

THE GOLD COAST: SOUTHPORT to COOLANGATTA

The **Gold Coast** is indisputably Australia's premier coastal tourist destination. The 37 km of mainly sandy coast contains eight beaches. The Gold Coast Highway parallels much of the coast and intensive tourist development and numerous high rise apartments and hotels back the beaches (Fig. 4.144). The highway and the Coolangatta Airport at the southern end provide the main access for the many hundreds of thousands of tourists who flock to the Gold Coast year round.

The Gold Coast has traditionally been Brisbane's surfing coast, with the main beach known as Surfers Paradise. The combination of popularity with an energetic, if hazardous, surf along the beaches, produced Queensland's first surf lifesaving clubs at Coolangatta in 1911 and Southport in 1912. Since then, an additional 19 surf lifesaving clubs have been established, making the area second only to Sydney in the number of surf clubs to watch over the public as they surf.

Today the Gold Coast not only services the rapidly expanding Brisbane-Logan City population, but a growing number of Australian and overseas tourists. To cope with the high level of public risk associated with the increasing number of swimmers, the Gold Coast City Council also maintains 28 lifeguard stations along the coast, as well as roving vehicle patrols and a helicopter rescue service. All this is necessary for as the name suggests, to be a *paradise for surfers* requires surf and lots of it. Unfortunately the surf is equally a potential hazard for the many inexperienced visitors and non-surfers who venture into the waves.

Figure 4.144 *The Gold Coast.*

1591   **SOUTHPORT - MIAMI**
          THE SPIT
          SOUTHPORT (SLSC)
          SURFERS PARADISE (SLSC)
          MAIN BEACH
          NORTHCLIFFE (SLSC)
          BROADBEACH (SLSC)
          KURRAWA (SLSC)
          MERMAID BEACH (SLSC)
          NOBBYS BEACH (SLSC)
          MAIMI BEACH (SLSC)

Patrols:
Surf Lifesaving Club: September to May
Lifeguard on duty: most all year

No.	Beach	Rating	Type Inner	Outer Bar	Length
1591A	The Spit	6	LTT/TBR	RBB	4 km
1591B	Southport	6	LTT/TBR	RBB	2 km
1591C	Surfers Paradise	6	LTT/TBR	RBB	2 km
1591D	Northcliffe	6	LTT/TBR	RBB/LBT	1 km
1591E	Broadbeach	6	LTT/TBR	RBB	1.5 km
1591F	Kurrawa	6	LTT/TBR	RBB	1.5 km
1591G	Mermaid Beach	6	LTT/TBR	RBB/LBT	2 km
1591H	Nobby Beach	6	LTT/TBR	RBB/LBT	1 km
1591I	Miami Beach	6	LTT/TBR	RBB/LBT	1 km
			Total length		16 km

Lifeguard towers

1591A	Seaway	all year
1591A	Seaworld	school holidays
1591A	Sheraton Tower	all year
1591A	Main Beach	all year
1591A	Breaker St Tower	school holidays
1592B	Narrowneck Tower	all year
1591C	North Surfers	all year
1591C	Elkhorn Tower	all year
1591C	*Surfers Tower	all year
1591C	Clifford St Tower	school holidays
1591D	*Northcliffe Tower	all year
1591D	Wharf Road Tower	school holidays
1591E	*Broadwater Tower	school holidays
1591F	Margaret St.	all year
1591F	*Kurrawa Tower	all year
1591G	*Mermaid Tower	all year
1591H	Hilda St.	Sept - April
1591H	Seashell Av.	Sept - April
1591H	*Nobbys Tower	school holidays
1591I	*Miami Tower	all year
	* Surf Club	

The longest beach on the Gold Coast is the straight stretch of sand that runs for 16 km from the Nerang entrance south to the small South Nobby headland. This beach is the most heavily developed in Australia and is backed for most of its length by beachfront hotels, resorts, apartments and houses. It contains eight surf lifesaving clubs, the most for any beach in Australia, together with 16 lifeguard towers. The beach is very accessible throughout its length and is backed by a tremendous range of facilities.

The beach faces almost due east and receives predominantly southerly swell averaging 1.5 m. This has combined with the generally fine sand to produce a wide, low gradient high tide beach fronted by a 150 to 200 m wide surf zone containing two bars. The inner bar is usually attached to the beach and, depending on wave conditions, may be continuous with a few rips or, during and following higher waves, cut by deep rip channels every 200 to 250 m. A continuous deep trough parallels this bar, beyond which is the deeper outer bar that is cut by more widely spaced rips. In addition, permanent rips lie at the northern end at the Nerang entrance wall and at the sand bypassing jetty just to the south. Deep holes and a permanent rip run out by the wall, while there are strong tidal currents in the inlet. Toward the southern end, both the smaller North Nobby and larger South Nobby headlands also produce permanent rips to either side.

The Spit section of beach (1591A) extends for 4 km from the Nerang training wall to Narrow Neck (Fig. 4.145). The jetty just south of the wall houses a sand bypassing system that pumps sand from the beach across the inlet to South Stradbroke Island. The road from Southport ends at a large car park behind the jetty. The jetty and entrance wall are favourite surfing and fishing spots. The only relatively natural dune section of beach occupies the next 2 km, before the beachfront Mirage Resort is encountered. The Nerang Road links up with Main Beach Parade at Southport and runs right behind the beach. Opposite Seaworld is a beach car park and access, with this section patrolled by the Seaworld lifeguard tower.

Narrow Neck is the location of the Gold Coast *surfing reef*, which will be designed to both widen the precariously narrow beach and dune, as well as provide consistent waves for surfers and surfing contests. It is the most advance surfing reef in the world and should add to the many attractions that already attract millions each year to the Gold Coast.

Immediately south of Narrow Neck is the 2 km long **Southport** (Main Beach) section (1591B). The Southport Surf Life Saving Club, founded in 1919, the Southport Nippers Club and a dressing shed lie just off the beginning of The Esplanade. They are fronted by a large car park, a fenced foredune reserve and the beach (Fig. 4.146). The beach like the entire system has a rip dominated inner bar and outer bar.

Surfers Paradise beach (1591C) is Queensland's best known beach. A lifesaving reel was placed on the beach

in 1921 and the Surfers Paradise Surf Life Saving Club founded in 1924. Today it lies at the heart of the Gold

Coast tourist trade and is frequently visited by hundreds of thousands of tourists from around the world. This 2 km section of beach is backed by a fenced foredune, with numerous access tracks, The Esplanade and the most highly developed section of the Australian coast. The surf club (Fig. 4.147) and three lifeguard towers watch over this rip dominated beach (Fig. 4.148).

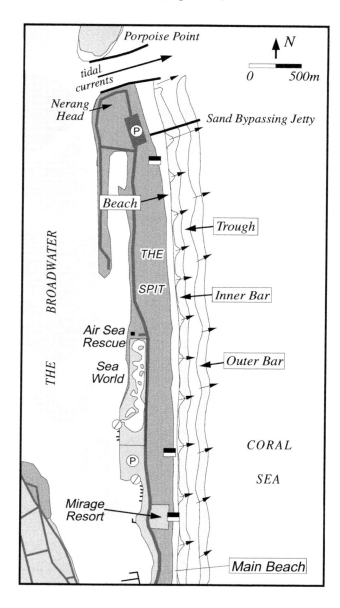

Figure 4.145   *The Spit marks the northern end of the 16 km long Surfers Paradise - Miami Beach. It ends at the Nerang training walls and jetty and is backed by a foreshore reserve.*

Northcliffe Beach (1591D) and its surf lifesaving club are 1 km south of Surfers Paradise. The club, founded in 1947, is surrounded by high rises and development (Fig. 4.146), and parking is limited.

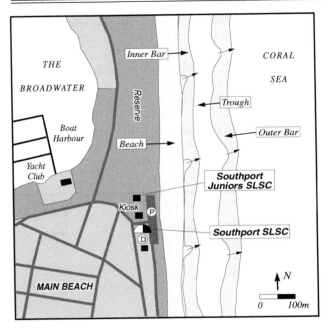

Figure 4.146 *Map showing the location of Southport SLSC and Junior SLSC.*

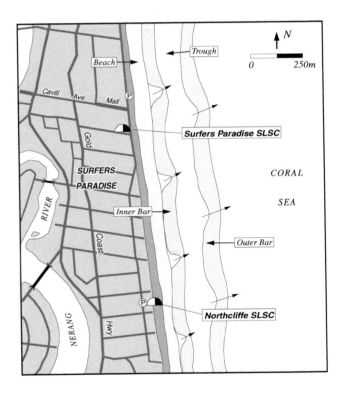

Figure 4.147 *Map showing the location of Surfers Paradise and Northcliffe surf clubs and typical beach conditions.*

Broadbeach Beach (1591E) is a further 1.5 km to the south. Here the beachfront is given over to a reserve, that contains the Broadbeach Surf Life Saving Club, founded in 1935, a large car park and other public amenities (Fig. 4.150 & 4.151).

Kurrawa Beach (1591F) is 1 km further south and has the most extensive beach reserve of all the beach sections. The wide reserve runs for 1.5 km and incorporates the Kurrawa Surf Life Saving Club (Fig. 4.152), the newest on the Gold Coast, being founded in 1958, together with a range of public facilities. This beach, with its wide reserve and facilities, has been chosen as a semi-permanent site for the annual Australian Surf Life Saving Championships. It was the site of the infamous 1996 Championships, held in energetic surf produced by a tropical cyclone, that resulted in the first tragedy in the Championships' 81 year history.

Figure 4.148 *Surfers Paradise Surf Club (far right) looking south toward Northcliffe and Broadbeach. Note the rip dominated inner and outer bars.*

Figure 4.149 *Northcliffe Surf Life Saving Club.*

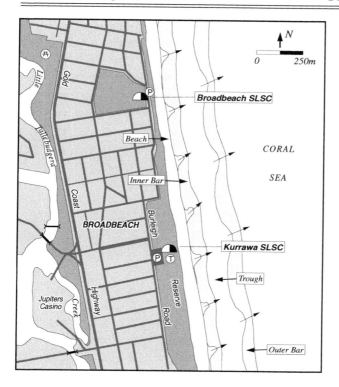

Figure 4.150   *Map showing the location of Broadbeach and Kurrawa Surf Clubs and typical beach conditions.*

Figure 4.151   *Broadbeach Surf Club.*

Figure 4.152   *Kurrawa Surf Club and beach: site of the Australian Surf Life Saving Championships from 1996.*

Mermaid Beach (1591G) runs for 2 km from the end of the Kurrawa reserve. The Mermaid Beach Surf Life Saving Club, founded in 1945, is located on the other side of the road from the beach and is fronted by a large car park and a 200 m long, grassy beach reserve, where the lifesavers maintain a watch tower (Fig. 4.153 & 4.154).

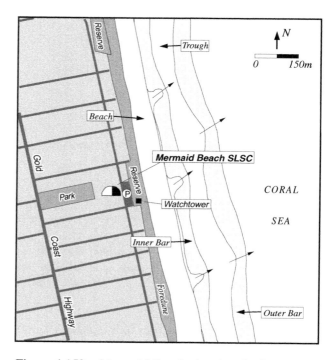

Figure 4.153   *Mermaid Beach showing the location of the surf club and typical beach conditions.*

Figure 4.154   *Mermaid Beach Surf Club.*

Nobbys Beach (1591H) lies at the foot of North Nobby; a 45 m high hillock located just behind the beach and featuring a cable ride to the top. The Nobbys Beach Surf Life Saving Club was established in 1954 and is located on a side street, with only a small car park. (Fig. 4.155)

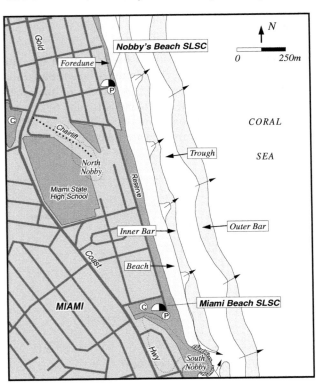

Figure 4.155   *Nobbys Beach and Miami Beach lie at the southern end of the 16 km long beach that begins at the Nerang spit. Surf clubs are located at both beaches.*

Miami Beach (1591I) occupies the southern 1 km of the beach, between North Nobby and South Nobby. Fifty metre high South Nobby protrudes 100 m into the surf, defining the end of the beach. The whole Miami Beach is fronted by a narrow beachfront reserve, with access tracks crossing the reserve from Marine Parade (Fig. 4.156).

The Miami Beach Surf Life Saving Club is located at the southern end. It has a large car park and is backed by a caravan park. The club was founded in 1946, when it was known as the Ipswich Railway Surf Life Saving Club.

Figure 4.156 *Miami Beach Surf Club and backing caravan park, with the slopes of Nobby Head immediately to the south.*

Swimming: The eight surf lifesaving clubs and sixteen lifeguard towers attest to the potential hazards along this long beach. Rips are present whenever waves are breaking and deep rip channels may run out from the shoreline. Swim only in patrolled areas and avoid the rip holes and outer trough. Stay close inshore and on the attached portion of the bar.

Surfing: Beach breaks extend the full length of the beach. Conditions are best on the outer bar, with a moderate swell and offshore winds. The *Nerang* entrance wall can produce some better quality bars and breaks. There are plans for Queensland's first *surfing reef* on the beach at Narrow Neck.

Fishing: Beach fishing is a very popular pastime along the length of the beach, with conditions best when there are rips across the inner bar. The Nerang jetty and entrance wall are very popular spots to fish the channel or surf.

Summary: This is undoubtedly Australia's most heavily developed and utilised beach. It offers a huge range of accommodation, attractions and facilities, together with the surf lifesaving clubs and lifeguard towers, as well as 16 km of beautiful beach and surf. However the surf is hazardous, so be careful as to when and where you swim.

1592   BURLEIGH BEACH
NORTH BURLEIGH (SLSC)
BURLEIGH HEADS/MOWBRAY
PARK (SLSC)

Patrols:
Surf Lifesaving Club: September to May
Lifeguard on duty: all year

No.	Beach	Rating	Type		Leng
			Inner	Outer bar	
1592A	**North Burleigh**	6	LTT	RBB	1 km
	Burleigh		TBR	LBT	
1592B	**Burleigh Heads**	6	TBR	LBT	1 km
	Heads				

Lifeguard Towers
     1592A   *North Burleigh Tower   all year
     1592A   Fourth Av.   all year
     1592B   *Burleigh Tower   all year
     * Surf Club

Burleigh beach is a straight, 2 km long section of sand running between South Nobby headland and the larger Burleigh Heads (Fig. 4.157). The beach is paralleled by The Esplanade, which has apartments and high rise buildings on its western side, with a grassy foreshore reserve located between the road and the beach. The **North Burleigh** Surf Life Saving Club, founded in 1949, is located 300 m south of South Nobby. It is situated in the reserve, with ramps leading down to the beach (1592A).

At the southern end of the beach (1592B) is the **Burleigh Heads**/Mowbray Park Surf Life Saving Club, that dates back to 1919. It is located in the reserve, together with dressing sheds, a kiosk and a car park. The rocks of 90 m high Burleigh Heads begin just south of the club house, with a pool on the rocks. There is also a small rock groyne crossing the beach just to the north of the club house.

The entire beach faces east-north-east and picks up most of the predominantly southerly swell, with waves averaging 1.5 m. These have produced a wide, low high tide beach fronted by a 200 m wide surf zone containing inner and outer bars (Fig. 4.158). The inner bar is usually cut by several deep rip channels that flow into the deeper longshore trough, with the outer bar cut by more widely spaced rips. In addition, there is a permanent rip against South Nobby and a large permanent rip running out against Burleigh Heads (Fig. 4.159).

Swimming: The safest swimming is in the two patrolled areas at North Burleigh and Burleigh. There are usually six to eight rips along the beach, together with the permanent rips at each end, so use care if swimming and stay on the inner portion of the attached bar, clear of the rip channels and outer trough.

Surfing: *Burleigh Heads* is one of Australia's top point breaks, with long, fast right-handers peeling over the sand and boulders that line the head. The beach also has beach breaks, that are best on the outer bar.

Fishing: There are usually good rip holes along the beach, particularly against the rocks at each end.

Summary: A readily accessible beach backed by a continuous reserve, with two surf lifesaving clubs and lifeguard towers to provide added protection.

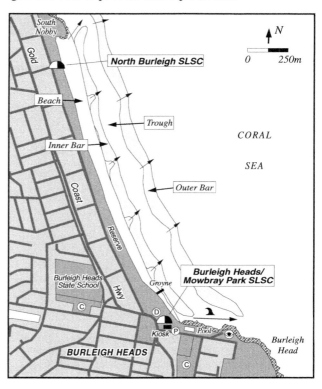

Figure 4.157   *Two kilometre long Burleigh Beach has surf clubs at both ends (North Burleigh and Burleigh Heads/Mowbray Park). The beach typically has inner and outer bars, both cut by rip currents.*

Figure 4.158   *A view south from South Nobby to Burleigh Heads. North Burleigh Surf Club is located immediately south of South Nobby, while Burleigh Heads Surf Club lies at the southern end of the beach.*

Figure. 4.159   *Burleigh Heads beach and surf club, with a deeper rip channel off the rocks,*

1593   PALM BEACH
TALLEBUDGERA (SLSC)
PACIFIC-NORTH PALM BEACH (SLSC)
PALM BEACH (SLSC)

Patrols:
Surf Lifesaving Club: September to May

No.	Beach	Rating	Type		Length
			Inner	Outer bar	
1593A	**Tallebudgera**	6	LTT/TBR	RBB/LBT	1 km
1593B	**Pacific**	6	LTT/TBR	RBB/LBT	1 km
1593C	**Palm Beach**	6	LTT/TBR	LBT	2 km

Lifeguard Towers		
1593A	Tallebudgera Ck	Sept - April, School
1593A	*Tallebudgera Tower	Sept - April, School
1593B	*Pacific Tower	Sept - April, School
1593C	*Palm Beach Tower	Sept - April, School
* Surf Club		

Palm Beach is a 4 km long, north-east facing beach bordered by Burleigh Heads and Tallebudgera Creek to the north, and Currumbin Point and Currumbin Creek in the south (Fig. 4.160). Both creeks have rock groynes/training walls extending out across the surf zone to form the actual beach ends. The Gold Coast Highway parallels the beach 200 m inland and numerous side streets provide good access to the length of the beach. The entire beach is backed by beachfront houses and apartments, including several high rises. A continuous seawall has been built to protect this development from cyclone erosion, together with two small groynes toward the centre of the beach.

The beach is patrolled by three surf lifesaving clubs, at Tallebudgera in the north, Pacific-North Palm Beach in the centre and Palm Beach to the south, together with lifeguards at all three locations. Waves average 1.5 m along the beach, which have produced a 200 m wide double bar system. The inner bar is usually cut by deep

rip channels every 200 to 300 m, in addition to strong permanent rips against the end groynes and smaller rips against the two central groynes. A deep trough parallels the bar, with the outer bar cut by more widely spaced rips. At times, beach erosion cuts the beach back to the seawall.

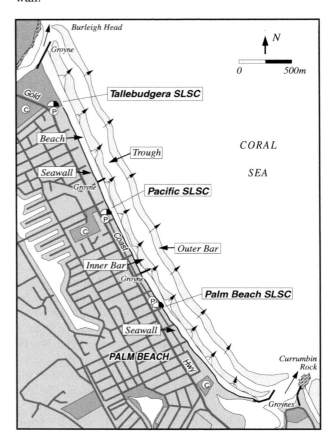

Figure 4.160   *Four kilometre long Palm Beach has three surf clubs (Tallebudgera, Pacific and Palm Beach) and is bordered by groyne-controlled tidal inlets at both ends.*

Tallebudgera Surf Life Saving Club (1593A) is located 400 m south of the Tallebudgera Creek groyne, which extends 300 m out to sea. The club, which was founded in 1946, is perched on top of the foredune with a good view of the beach. It is right off the highway with a large car park next to the club (Fig. 4.161).

Pacific-North Palm Beach Surf Life Saving Club (1593B) is located 1.2 km south of Tallebudgera, toward the centre of the beach. The club was established in 1947 and sits just off the highway on a beachfront building block. It has a small car park, with a seawall fronting the club house (Fig. 4.162).

Figure 4.161 *Tallebudgera Surf Club.*

Figure 4.162   *Pacific Surf Club (centre) fronted by a seawall.*

Palm Beach Surf Life Saving Club (1593C) was established in 1930, when it was known as the City of Brisbane Surf Life Saving Club. The club sits just off the highway and is surrounded by a large car park, with a patch of grassy reserve and a seawall running along the front of the club house.

Swimming: This is a potentially hazardous beach owing to the inner bar rips and is particularly hazardous near any of the rock groynes. Do not swim at the creek mouths, as they have deep tidal channels with strong currents. The safest swimming is in the three patrolled areas, with Tallebudgera providing the best access.

Surfing: There are beach breaks along the length of the beach, both on the inner bar during low waves and on the outer bar during moderate swell.

Fishing: The beach usually provides several rip holes, together with the four groynes and two creeks.

Summary: Heavy development along this beach tends to restrict access and parking, while the seawall, when exposed, also restricts access to the beach and even the size of the beach, particularly at high tide.

1594, 1595   CURRUMBIN SLSC

Patrols:				
Surf Lifesaving Club:		September to May		
Lifeguard on duty:		all year		
No. Beach	Rating	Type		Length
		Inner	Outer bar	
1594 Currumbin	2	R+tidal		300 m
Inlet		inlet		
1595 **Currumbin**	6	LTT/	RBB/	750 m
		TBR	LBT	
Lifeguard Tower:				
1595 Currumbin Alley		Sept - April, School holidays		
1595 *Currumbin Tower		all year		
* Surf Club				

Currumbin beach is a 750 m long, north-east facing beach that extends from 10 m high Currumbin Rock down to 15 m high, humpy Elephant Rock (Fig. 4.163). Currumbin Creek flows out on the north side of the point and, to contain the inlet and shifting tidal shoals, both sides of the creek have been stabilised by the construction of rock groynes. The groyne in front of Currumbin Point extends 300 m from the road in front of the point out to Currumbin Rock. The groyne and point have partly stabilised a 300 m long, low energy beach and low sand spit (1594). This is a popular and very accessible beach, with calm conditions in the inlet and low waves toward the point, however it is fronted by the deep, shifting tidal shoals and channel, together with strong tidal currents.

Pacific Parade runs around the point and parallels the main 750 m long **Currumbin** beach (1595), providing good access to the length of the beach, with large car parks on the point and at Currumbin Surf Life Saving Club, located beside Elephant Rock. The club was founded in 1919, making it one of the oldest on the Gold Coast. A few houses and apartments line the west side of the road, below the 30 m high slopes of the point.

The beach receives waves averaging 1.5 m which produce a 200 m wide double bar system. The inner bar is usually attached to the beach and cut by four rips, including a permanent rip against Currumbin Rock. The deep trough runs out past Elephant Rock all the way to Currumbin Rock, with the outer bar further offshore (Fig. 4.164).

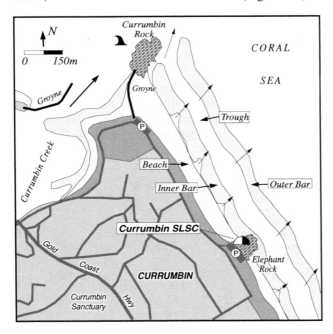

Figure 4.163   *Currumbin Beach extends for 750 m between the low Currumbin Rock and higher Elephant Rock, with the surf club located at the base of Elephant Rock.*

Figure 4.164   *A view south along Currumbin Beach, with the surf club located on the north side of Elephant Rock. Note the strong rip current flowing seaward of the surf zone.*

Swimming: This is a potentially hazardous beach owing to the persistent rips and strong permanent rips against the rocks, particularly at Currumbin Rock. The safest bathing is at the inlet beach and at the southern patrolled end of the surf beach. Stay on the inner bar and well clear of the rocks and rips.

Surfing: There are beach breaks on the inner and outer bars, however Currumbin is best known for the classic right-hand break that wraps around the north side of

Currumbin Point. Most surfers walk out along the northern groyne before paddling out.

Fishing: There is both good beach fishing into the rip holes and rock fishing off Currumbin and Elephant Rocks, as well as fishing into Currumbin Creek.

Summary: This beach is a little off the highway, but a popular spot with swimmers, surfers, windsurfers and fishers.

1596-1598   **TUGUN - KIRRA**
            TUGUN (SLSC)
            BILINGA (SLSC)
            NORTH KIRRA (SLSC)
            KIRRA (SLSC)

Patrols:
Surf Lifesaving Club: September to May
Lifeguard on duty: September to May

No.	Beach	Rating	Type		Length
			Inner	Outer Bar	
1596	Flat Rock	6	LTT/TBR	RBB	600 m
1597A	**Tugun**	6	LTT/TBR	RBB	1.5 km
1597B	**Bilinga**	5	LTT	RBB	1.5 km
1597C	**North Kirra**	5	LTT	RBB	1.5 km
1598	**Kirra**	4	R/LTT	RBB	300 m
			Total length		5.5 km

Lifeguard Towers:
    1597A   *Tugun Tower   Sept - April
    1597A   Biling   Sept - April, School holidays
    1597B   *North Kirra Tower School holidays
    1597C   *Kirra Tower   Sept - April
    * surf club

Elephant Rock forms the northern boundary of a near-continuous, 5.5 km long beach that is broken into three beaches by the small Flat Rock in the north and the constriction of the Kirra groyne in the south. While the beaches are separated by these structures, they share a continuous surf zone and outer bar. The Gold Coast Highway parallels most of the beach, with beachfront development backing the first 2.5 km, then Pacific Parade running behind the beach down to where it joins the highway at North Kirra. The highway then skirts the remainder of North Kirra and Kirra Beach. At times, beach erosion exposes the seawall between North Kirra and Kirra Beaches, separating Kirra into a small pocket beach. Long groynes have been constructed at either end of Kirra Beach. The northern groyne separates it from North Kirra and is designed to keep sand on Kirra Beach. The southern groyne is 200 m long and has been built out on Kirra Point to hold in the sand of adjacent Greenmount Beach.

Flat Rock beach (1596) faces due east and extends for 600 m from Elephant Rock to the lower Flat Rock, with the small Flat Rock Creek draining out across the centre.

The beach runs straight past the rocks, with a continuous surf zone. There is road access to either side of the creek, with a bushy reserve surrounding the creek.

Tugun-Kirra Beach (**1597**) continues on past Flat Rock for 4.5 km to the Kirra groyne. It faces north-east for most of its length, swinging around to face north at Kirra Beach. It receives waves averaging 1.5 m at Tugun, which drop to around 1 m at North Kirra and 0.5 to 1 m at Kirra Beach. The waves produce a 200 m wide double bar system for the entire length, however the inner bar changes as the wave height drops. At Tugun the inner bar is attached to the beach and cut by rips every 200 to 300 m. At North Kirra the bar is attached but rips are less frequent, while at Kirra the bar at times welds to the beach, producing a steep, barless beach. The deep trough runs the length of the entire beach, with the outer bar offshore. This bar can be seen running past Kirra Point, where it helps produce the famous surfing break. It is cut by more widely spaced rips, with a strong permanent rip against the northern Elephant Rock.

Tugun Beach (1597A) is patrolled by Tugun Surf Life Saving Club, located 500 m south of Flat Rock (Fig. 4.165). The club was founded in 1924 and has a large club house located on a small foreshore reserve, with parking areas to either side of the reserve (Fig. 4.166).

Figure 4.166   *Tugun Surf Club.*

Bilinga Surf Life Saving Club (1597B) was established in 1922. The club house occupies a beachfront block, with houses to either side and a seawall in front (Fig. 4.167). When the beach is nourished with sand every few years, the width of the beach increases considerably (Fig. 4.168).

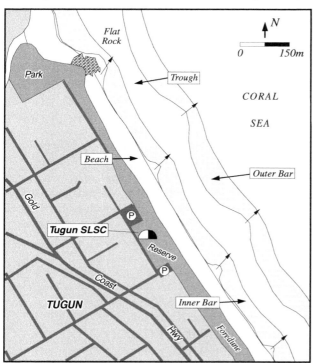

Figure 4.165   *Tugun Beach and surf club are located toward the northern end of a 5.5 km long beach. Rips typically dominate the inner and outer bars.*

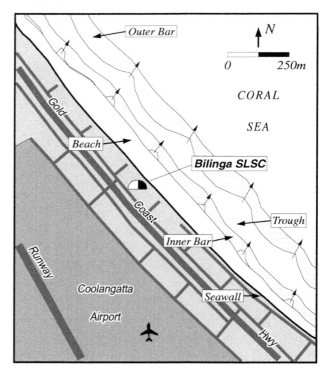

Figure 4.167   *Bilinga SLSC is located just off the Gold Coast Highway and midway along the 5.5 km long Tugun-Kirra Beach.*

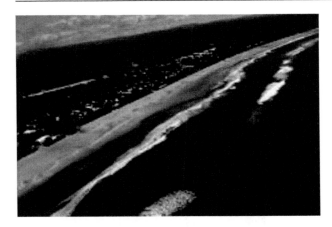

Figure 4.168   *Bilinga Surf Club (far left) fronted by a low tide terrace cut by oblique rips, a deep trough and outer bar.*

North Kirra Surf Life Saving Club (1597C) is the newest on the beach, being established in 1949. It is located on a narrow reserve in front of Pacific Parade. It is also fronted by a continuous seawall which, during erosion periods, becomes exposed a few hundred metres south of the club house (Fig. 4.169).

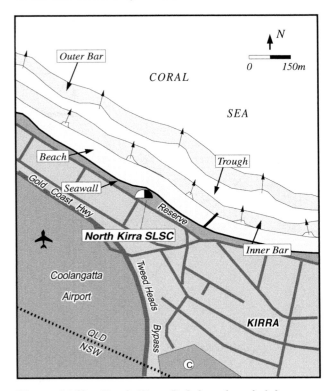

Figure 4.169   *North Kirra Beach and surf club.*

Kirra beach (1598) is tucked in at the southern end of the beach, wedged in between the highway and the two groynes and at times cut off to the north by the exposed sea wall (Fig. 4.170). Kirra Beach is the most protected, north facing part of the beach, with the waves further lowered by the Greenmount groyne. The Kirra Surf Life Saving Club is the oldest on the beach, being established in

1916. Today the 300 m long beach and club house share a patch of land between the road and the beach with a kiosk and car park.

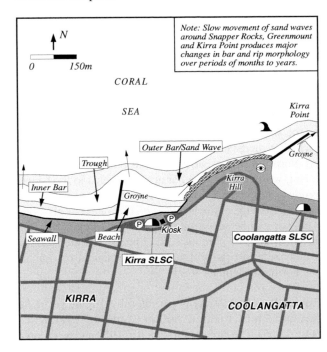

Figure 4.170   *Map showing Kirra Beach and surf club located between Kirra Point and groyne and a northern groyne.*

Swimming: Swimming conditions vary along the beach. Conditions are most hazardous at the Flat Rock and Tugun end owing to the prevalence of rip channels across the inner bar, together with the strong permanent rip against Elephant Rock and smaller rips that can form either side of Flat Rock. The patrolled area at the Tugun Surf Club is the best place to swim. At Bilinga and North Kirra the rips are less frequent, but a deep trough usually parallels the beach. Kirra Beach is the safest, with the lowest waves and usually no rips close to shore. However there is still a deep trough off the beach, with a current often sweeping north past the groyne.

Surfing: There are beach breaks along most of the beach, which to the south are better on the outer bar. The prime surfing spot is *Kirra Point,* famous for its long right-handers breaking over a sand bar produced in part by the Greenmount groyne (Fig. 4.171).

Fishing: There is good beach fishing right along the beach, into the rip holes along the northern end and straight into the longshore trough toward the south. In addition you can fish off the Kirra and Greenmount groynes and, when the beach is eroded, straight off the Kirra seawall.

Summary: This is a long beach that changes its

character from north to south. It has also suffered from beach erosion, resulting in the construction of seawalls and groynes and requiring massive beach nourishment every few years.

Figure 4.171   *Kirra Point, showing the Greenmount groyne (left), the famous Kirra Point break, the Kirra groyne (right) and Kirra Surf Club (centre).*

1599   COOLANGATTA - GREENMOUNT
COOLANGATTA (SLSC)
TWEED HEADS & COOLANGATTA (SLSC)

Patrols:					
Surf Lifesaving Club:		September to May			
Lifeguard on duty:		all year			
No.	Beach	Rating	Type		Length
			Inner	Outer bar	
1599A	**Coolangatta**	5	LTT/TBR	RBB/LBT	400 m
1599B	**Greenmount**	5	LTT/TBR	TBR/RBB	400 m
Lifeguard Towers:					
	1599A	Coolangatta Tower		all year	
	1599B	Greenmount Tower		all year	

Coolangatta-Greenmount Beach is an 800 m long, north facing beach wedged in between the Greenmount groyne, that extends for 200 m off Kirra Point, and the prominent, 30 m high Greenmount Hill (Fig. 4.172). The beach derives its name from the green hill and the wreck of a ship named Coolangatta, which was named after Mount Coolangatta on the Shoalhaven River in New South Wales, where the ship was built. The Gold Coast Highway skirts the western end of the beach and Marine Parade runs the length of the beach, with a seawall protecting the road from episodic erosion.

While the beach faces north and is partially protected by Point Danger from the prevailing southerly swell, it remains a very dynamic beach. This is as a result of large

pulses of sand that move from New South Wales across the Tweed River mouth and around Point Danger. They manifest themselves as large sand waves moving slowly around Greenmount over a period of weeks to months and finally attaching themselves to Greenmount, before moving on eventually around the groyne and on past Kirra to North Kirra Beach. Depending on the size, shape and degree of progress along the beach, bar, surf and rip conditions can vary considerably.

Typically the beach has an attached inner bar, possibly cut by rip channels, with a deeper trough paralleling the beach and an outer bar/sand wave further out. There is often shallow sand around the base of Greenmount and a strong permanent rip against the Greenmount groyne. Waves average 0.5 to 1 m, usually being higher at the western Coolangatta end.

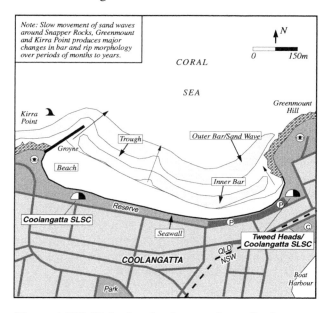

Figure 4.172 *Eight hundred metre long Coolangatta Beach is home to the state's oldest surf club at Tweed Heads/Coolangatta and the newer Coolangatta Surf Club.*

The **Coolangatta** Surf Life Saving Club (1599A) is located at the western end of the beach, 100 m east of the groyne. The club was founded in 1958 and is situated in a foreshore reserve that runs the length of the beach. The club is backed by Marine Parade, high rise apartments and the Coolangatta shopping area (Fig. 4.173).

At **Greenmount Beach**, the Tweed Heads & Coolangatta Surf Life Saving Club (1599B) has the honour of being the oldest in Queensland, being established in 1911. The club house is located at the far eastern end of the beach, close by the slopes of Greenmount. The foreshore reserve gives way to a continuous car park toward the eastern end, providing extensive parking at one of the Gold Coast's safer beaches.

Figure 4.173 *Greenmount Beach and groyne, with Coolangatta Surf Club located just past the groyne and Tweeds Heads & Coolangatta Surf Club at the far end. Note the three prominent rips, including one adjacent to the groyne.* 1599

Swimming: This is a moderately safe beach owing to the usually low waves. However, even under low wave conditions, deep rip channels and troughs occur along the beach and are intensified when waves exceed 1 m. The safest swimming is in the two patrolled areas and clear of the Greenmount groyne.

Surfing: The best surfing is off *Greenmount* Hill. When the bars are right, it produces a long, easy right-hander. There can also be good beach breaks over the inner and outer bars, with conditions depending on the bar shape.

Fishing: There are usually good holes and gutters along the beach, plus the Greenmount groyne and rock fishing off the rock around Greenmount Hill.

Summary: This is a heavily developed and somewhat modified beach since the construction of the groyne. It is continually changing as sand waves move through and past the beach, making it an interesting beach to observe. It has all the usual tourist facilities in backing Coolangatta plus the added safety of usually low waves and two surf lifesaving clubs and lifeguard towers.

1600, 1601 RAINBOW BAY (SLSC) SNAPPER ROCKS

Patrols:
Surf Lifesaving Club: September to May
Lifeguard on duty: all year

No.	Beach	Rating	Type Inner	Outer Bar	Length
1600	**Rainbow Bay**	5	TBR	sand waves	300 m
1601	Snapper Rocks	6	LTT+ rocks	sand waves	100 m

Lifeguard Tower:
      1600 *Rainbow Tower   all year
            *surf club

Rainbow Bay and adjoining Snapper Rocks Beach are the southernmost beaches in Queensland. **Rainbow Bay** (1600) is a relatively small, 300 m long, north facing beach that is bordered in the west by Greenmount Hill. It ends at a low, rocky point, beyond which **Snapper Rocks** beach (1601) continues on for another 100 m to Snapper Rocks; part of the larger 30 m high Point Danger that forms the border with New South Wales (Fig. 4.174).

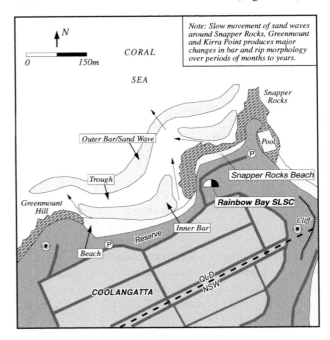

Figure 4.174 *Rainbow Bay and Snapper Rocks Beaches are the southernmost in Queensland and are patrolled by Rainbow Bay SLSC.*

The Rainbow Bay Surf Life Saving Club is the newest on the Gold Coast, being formed in 1963. It is located on the low, rocky point that separates the two beaches. A road runs along the back of both beaches, ending in a parking area behind Snapper Rocks Beach. There is a grassy foreshore reserve behind Rainbow Bay Beach.

Both beaches receive waves that are lowered after moving around Point Danger to an average of about 1 m. The width of both beaches and offshore bar conditions depend on both the prevailing waves and, in particular, the status of sand waves moving around Point Danger. When the bars are present, the beaches are connected by sand, there is a wide surf zone and excellent surfing conditions (Fig. 4.175). When the bars are absent, the beaches are separated by the rocky point, are narrower and there is an attached bar cut by one rip at Snapper Rocks Beach and two at Rainbow Bay. Low rocks dominate the Snapper Rocks surf zone and are a hazard for swimmers.

Figure 4.175 *Snapper Rocks and Rainbow Bay are Queensland's two southernmost beaches. Rainbow Bay Surf Club is located in lee of the rocks separating the two small beaches.*

Swimming: Rainbow Bay in particular is a popular swimming beach, owing to its protected location and usually low waves. However watch for the changing bar conditions and presence of rips, particularly near the rocks and Greenmount Hill. The safest swimming is in the patrolled area on Rainbow Bay Beach. Swimming is not recommended at Snapper Rocks Beach because of the rocks.

Surfing: When the bars are right, the two beaches can combine to produce a classic, long, easy right-hander, that at times has been known to reach adjoining Coolangatta Beach. There are usually reasonable waves that are popular with the surfers during southerlies, when the winds blow offshore.

Fishing: This is a popular fishing area, as the name of Snapper Rocks suggests. There is good beach fishing into the usual rip holes and outer troughs, together with rock fishing off Greenmount and Snapper Rocks.

Summary: Queensland's southernmost beaches are both picturesque, even if somewhat developed. They provide continually changing bar conditions and a nice place to watch the sun set over the Gold Coast high rises.

GLOSSARY

bar (sand bar) - an area of relatively shallow sand upon which waves break. It may be attached to or detached from the beach, and may be parallel (longshore bar) or perpendicular (transverse bar) to the beach.

barrier - a long term (1 000s of years) shore-parallel accumulation of wave, tide and wind deposited sand, that includes the beach and backing sand dunes. It may be 100s to 1000s of metres wide and backed by a lagoon or estuary. The beach is the seaward boundary of all barriers.

beach - a wave deposited accumulation of sediment (sand, cobbles or boulders) lying between modal wave base and the upper limit of wave swash.

beach face - the seaward dipping portion of the beach over which the wave swash and backwash operate.

beach type - refers to the type of beach that occurs under wave dominated (6 types), tide-modified (3 types) and tide-dominated (3 types) beach systems. Each possesses a characteristic combination of hydrodynamic processes and morphological character, as discussed in chapter 2.

beach types
  wave dominated (abbreviations, see Figures 2.3 & 2.4)
    R - reflective
    LTT - low tide terrace
    TBR - transverse bar and rip
    RBB - rhythmic bar and beach
    LBT - longshore bar and trough
    D - dissipative
  tide-modified (abbreviations, see Figure 2.11)
    R+LT – reflective + low tide terrace
    R+BR – R + low tide bar and rip
    UD – ultradissipative
  tide-dominated (abbreviations, see Figure 2.15)
    B+RSF – beach + ridged sand flats
    B+SF – beach + sand flats
    TSF – tidal sand flats

beach hazards - elements of the beach environment that expose the public to danger or harm. Specifically: water depth, breaking waves, variable surf zone topography, and surf zone currents, particularly rip currents. Also include local hazards such as rocks, reefs and inlets.

beach hazard rating - the scaling of a beach according to the hazards associated with its beach type as well as any local hazards.

berm - the nearly horizontal portion of the beach, deposited by wave action, lying immediately landward of the beach face. The rear of the berm marks the limit of spring high tide wave action.

blowout - a section of dune that has been destabilised and is now moving inland. Caused by strong onshore winds breaching the dune.

cusp - a regular undulation in the high tide swash zone (upper beach face), usually occurring in series with spacing of 10 to 40 m. Produced during beach accretion by the interaction of swash and sub-harmonic edge waves.

cuspate foreland - a sandy shoreline protusion formed in lee of an offshore island, reef or structure.

foredune - the first sand dune behind the beach. In Queensland it is usually vegetated by spinifex grass and ipomoea, then casuarina thickets.

groyne - a shore perpendicular structure (rock, timber) built across a beach and/or surf zone.

gutter - a deeper part of the surf zone, close and parallel to shore. It may also be a rip feeder channel.

hole - a localised, deeper part of the surf zone, usually close to shore. It may also be part of a rip channel.

Holocene - the geological time period (or epoch) beginning 10 000 years ago (at the end of the last Glacial or Ice Age period) and extending to the present.

lifeguard - in Australia this refers to a professional person charged with maintaining public safety on the beaches and surf area that they patrol. Also known as *beach inspectors*.

lifesaver - an Australian term referring to an active volunteer member of Surf Life Saving Australia, who patrol the beach to maintain public safety in the surf.

macrotidal - tide range greater 4 m.

megacusp - a longshore undulation in the shoreline and swash zone, with regular spacings between 100 and 500 m, which match the adjacent rips and bars. Produced by wave scouring in lee of rips (megacusp or rip embayment) and shoreline accretion in lee of bars (megacusp horn).

megacusp embayment - see megacusp

megacusp horn - see megacusp

mesotidal - tide range between 2 to 2 m.

microtidal - tide range less than 2 m

parabolic dune - a blowout that has extended beyond the foredune and has a U shape when viewed from above.

Pleistocene - the earlier of the two geological epochs comprising the Quaternary Period. It began 2 million years ago and extends to the beginning of the Holocene Epoch, 10 000 years ago.

rip channel - an elongate area of relatively deep water (1 to 3 m), running seaward, either directly or at an angle, and occupied by a rip current.

rip current - a relatively narrow, concentrated seaward return flow of water. It consists of three parts: the *rip feeder current* flowing inside the breakers, usually close to shore; the *rip neck*, where the feeder currents converge and flow seaward through the breakers in a narrow 'rip'; and the *rip head*, where the current widens and slows as a series of vortices seaward of the breakers.

rip embayment - see megacusp

rip feeder current - a current flowing along and close to shore, which converges with a feeder current arriving from the other direction, to form the basis of a rip current. The two currents converge in the rip (megacusp) embayment, then pulse seaward as a rip current.

rock platform - a relatively horizontal area of rock, lying at the base of sea cliffs, usually lying above mean sea level and often awash at high tide and in storms. The platforms are commonly fronted by deep water (2 to 20 m)

rock pool - a wading or swimming pool constructed on a rock platform and containing sea water.

sea waves - ocean waves actively forming under the influence of wind. Usually relatively short, steep and variable in shape.

set-up - rise in the water level at the beach face resulting from low frequency accumulations of water in the inner surf zone. Seaward return flow results in a *set-down*. Frequency ranges from 30 to 200 seconds.

shore platform - as per rock platform.

swash - the broken part of a wave as it runs up the beach face or swash zone. The return flow is called *backwash*.

swell - ocean waves that have travelled outside the area of wave generation (sea). Compared to sea waves, they are lower, longer and more regular.

trough - an area of deeper water in the surf zone. May be parallel to shore or at an angle.

tidal pool - a naturally occurring hole, depression or channel in a shore platform, that may retain its water during low tide.

wave (ocean) - a regular undulation in the ocean surface produced by wind blowing over the surface. While being formed by the wind it is called a *sea* wave; once it leaves the area of formation or the wind stops blowing it is called a *swell* wave.

wave refraction - the process by which waves moving in shallow water at an angle to the seabed are changed. The part of the wave crest moving in shallower water moves more slowly than other parts moving in deeper water, causing the wave crest to bend toward the shallower seabed.

wave shoaling - the process by which waves moving into shallow water interact with the seabed causing the waves to refract, slow, shorten and increase in height.

wave bore - the turbulent, broken part of a wave that advances shoreward across the surf zone. This is the part between the wave breaking and the wave swash and also that part caught by bodysurfers. Also called *whitewater*.

References

Day, R W, Whitaker, W G, Murray, C G, Wilson, I H and Grimes, K G, 1983, *Queensland Geology*. Geological Survey of Queensland Publication 383, 194pp, plus maps.

Hill, D and Denmead, A K (eds), 1960, *The Geology of Queensland*. Melbourne University Press, Melbourne, 474 pp.

Hopley, D, 1982, *The Geomorphology of the Great Barrier Reef*. Wiley-Interscience, New York, 453pp. The definitive scientific text on the Great Barrier Reef.

Laughlin, G, 1997, *The Users Guide to the Australian Coast*. Reed New Holland, Sydney, 213 pp. An excellent overview of the Australian coastal climate, winds, waves and weather.

Phinn, S R, 1992, Synoptic weather patterns and wave conditions, Surfers Paradise, south-eastern Queensland 1988. Australian Geographical Studies, 30, 142-162.

PWD, 1985, Elevated ocean levels and storms affecting the New South Wales coast 1880 to 1980. NSW Public Works Department, Coastal Section, Report 85041, 287 pp.

Readers Digest, 1983, Guide to the Australian Coast. Readers Digest, Sydney, 479 pp. Excellent aerial photographs and information on the more popular spots along the Queensland coast.

Ross, J (editor), 1995, *Fish Australia*. Viking, Melbourne, 498 pp. An excellent coverage of all Queensland coastal fishing spots.

Short, A D, 1993, *Beaches of the New South Wales Coast*. Australian Beach Safety and Management Program, Sydney, 358 pp. The New South Wales version of this book.

Short, A D, 1996, *Beaches of the Victorian Coast and Port Phillip Bay*. Australian Beach Safety and Management Program, Sydney, 298 pp. The Victorian version of this book.

Short, A D (editor), 1999, Beach and Shoreface Morphodynamics. John Wiley & Sons, Chichester, 379 pp. For those who are interested in the science of the surf.

Surf Life Saving Australia, 1991, Surf Survival; The Complete Guide to Ocean Safety. Surf Life Saving Australia, Sydney, 88 pp. An excellent guide for anyone using the surf zone for swimming or surfing.

Warren, M, 1998, Mark Warren's Atlas of Australian Surfing. Angus & Robertson, Sydney, 232 pp. Covers most Queensland surfing spots.

Williamson, J A, Fenner, P J, Burnett, J W and Rifkin, J F, 1996, *Venomous and Poisonous Marine Animals - a Medical and Biological Handbook*, University of New South Wales Press, Sydney, 504 pp. The definitive book on marine stingers.

Young, N, 1980, Surfing Australia's East Coast. Horwitz, Sydney, 112 pp. Provides maps and a description of most east coast surfing spots.

BEACH INDEX

Note: Patrolled beaches in **BOLD**

See also
GENERAL INDEX p. 349
SURF INDEX   p. 360

GENERAL INDEX

A

aboriginal land
    Yarrabah, 102, 105, 107
airport
    Hamilton Island, 176
airstrip
    Brampton Island, 178
Army Training Area
    Cowley Beach, 113
Australian Beach Safety & Management
    Program, 1
Australian Institute of Marine Science, 145

B

bar, 339
    'collapsing', 40
barrier, 339
barrier systems, 50

bay
    Abbott, 155
    Alexandria, 88, 295
    Alligator, 217
    Alma, 139
    Arthur, 139
    Barbque, 173
    Blue, 187
    Blue Pearl, 173
    Bowling Green, 146, 148
    Bramble, 314
    Bustard, 264
    Catseye, 176
    Cedar, 83
    Cherry Tree, 78
    Chunda, 146
    Cockle, 142
    Coconut, 154
    Cooee, 239
    Corio, 237, 238
    Crab, 176
    Deception, 311
    Double, 168
    Edgecumbe, 161, 163
    Ella, 110
    Etty, 112
    Finch, 79
    Florence, 139
    Frenchmans, 321
    Freshwater, 232
    Funnel, 171
    Gatakers, 281
    Genesta, 179
    Geoffrey, 140
    Greys, 160
    Halifax, 130, 131, 132, 133
    Halliday, 187
    Happy, 176, 177
    Hervey, 274
    Hideaway, 165
    Horseshoe, 138, 143, 161
    Huntingfield, 142
    Ince, 201, 202
    Jonah, 165

    Joyce, 143
    Kennedy, 119
    Keppel, 243, 248
    Kingfisher, 154
    Little Corio, 237
    Little Turtle, 102
    Llewellyn, 199, 200
    Lovers, 143
    Lugger, 118
    Mangrove, 261
    Mission, 102, 117
    Missionary, 121
    Moonlight, 153
    Moreton Bay, 311
    Mulligan, 127
    Murray, 161
    Nelly, 140, 165
    Norris, 143
    Palm, 176
    Paradise, 145, 176
    Pearl, 229
    Picnic, 141
    Pioneer, 168
    Quarantine, 79
    Queens, 160
    Radical, 139
    Ramsay, 124
    Repluse, 180
    Rockingham, 119
    Rodds, 260, 261
    Rollingstone, 142
    Rose, 161
    Sand, 189
    Sand Bank, 213
    Sandringham, 195
    Shark Bay, 153
    Shepherd, 123
    Shoal, 165, 241
    Shoalwater, 221, 222
    Sinclaire, 163
    Slade, 191
    Stanage, 217
    Statue, 240
    Swamp, 171
    The Inlet, 180
    Trinity, 91
    Turtle, 102, 145
    Upstart, 150
    Walker, 80
    Walsh, 83
    Weary, 84
    Wide, 102
    Wilson, 142
    Woodwark, 168
    Young, 142
    Zoe, 126

beach, 1, 9, 339
    activities, 2
    bay, 31
    cusps, 43
    dynamics, 33
    ecology, 29

    face, 339
    hazard rating, 339
    hazards, 57
    hazards, 339
    megacusps, 40
    morphology, 31
    nearshore zone, 32
    number, 1
    ocean, 31
    rips, 40
    sediment, 33
    sediment size, 33
    step, 43
    surf zone, 32
    swash zone, 31
    type, 34, 339
        longshore bar and trough (LBT), 37
        low tide terrace (LTT), 41
        rhythmic bar and beach (RBB), 38
        transverse bar and rip (TBR), 39
    usage, 53
    user, 2
beach hazard rating, 64, 66
beach hazards, 58
    beach-sand flat, 49
    beach-sand ridge, 49
    low tide terrace, 41
    reflective, 43
    reflective-bar and rip, 47
    reflective-low tide terrace, 48
    rhythmic bar and beach, 39
    tidal-sand flats, 50
    transverse bar and rip, 40
    ultradissipative, 46
beach systems, 50
beach type
    beach-sand flat (BSF), 49, 50
    beach-sand ridges (BSR), 48, 49
    dissipative (D), 35
    intermediate, 37
    longshore bar and trough (LBT), 37
    low tide terrace (LTT), 42
    reflective (R), 43
    reflective (R), 41
    reflective-bar and rip (RBR), 46, 47
    reflective-low tide terrace (RLTT), 47
    rhythmic bar and beach (RBB), 38
    tidal sand flats (TSF), 49, 50
    tide-dominated, 48
    tide-modified, 44, 45
    transverse bar and rip (TBR), 39
    ultradissipative (UD), 45, 46
    wave dominated, 44
beachrock, 50, 151, 158, 164, 165, 166, 209,
    216
    Abbott Point, 159
    Maud Bay, 143
    Wreck Rock, 270
berm, 339
bight
    West, 221, 222
blowout, 339
boat harbour

SURF INDEX

Also see surfing locations by beach
name in BEACH INDEX

BEACHES OF THE AUSTRALIAN COAST

Published by the Sydney University Press for the
Australian Beach Safety and Management Program
a joint project of

Coastal Studies Unit, University of Sydney and Surf Life Saving Australia

by

Andrew D Short
Coastal Studies Unit, University of Sydney

BEACHES OF THE NEW SOUTH WALES COAST
Publication: 1993   **ISBN:** 0-646-15055-3
358 pages, 167 original figures, including 18 photographs; glossary, general index, beach index, surf index.

BEACHES OF THE VICTORIAN COAST & PORT PHILLIP BAY
Publication: 1996   **ISBN:** 0-9586504-0-3
298 pages, 132 original figures, including 41 photographs; glossary, general index, beach index, surf index.

BEACHES OF THE QUEENSLAND COAST: COOKTOWN TO COOLANGATTA
Publication: 2000   **ISBN** 0-9586504-1-1
369 pages, 174 original figures, including 137 photographs, glossary, general index, beach index, surf index.

BEACHES OF THE SOUTH AUSTRALIAN COAST & KANGAROO ISLAND
Publication: 2001   **ISBN** 0-9586504-2-X
346 pages, 286 original figures, including 238 photographs, glossary, general index, beach index, surf index.

BEACHES OF THE WESTERN AUSTRALIAN COAST: EUCLA TO ROEBUCK BAY
Publication: 2005   **ISBN** 0-9586504-3-8
433 pages, 517 original figures, including 408 photographs, glossary, general index, beach index, surf index.

Order online from **Sydney University Press** at
http://purl.library.usyd.edu.au/sup/marine

Forthcoming titles:

BEACHES OF THE TASMANIAN COAST AND ISLANDS (publication 2006)   1-920898-12-3

BEACHES OF NORTHERN AUSTRALIA: THE KIMBERLEY, NORTHERN TERRITORY AND CAPE YORK (publication 2007)   1-920898-16-6

BEACHES OF THE NEW SOUTH WALES COAST (2nd edition, 2007)   1-920898-15-8

SYDNEY UNIVERSITY PRESS